thp TRAVIS HOUSE PUBLICATIONS • SONOMA COUNTY • CALIFORNIA

Webb's First Deep Field Image

Photo Credit: NASA, ESA, CSA, and STScI
https://esawebb.org/images/weic2209a/ [retrieved 6-20-25]

Thousands of galaxies flood this near-infrared image of galaxy cluster SMACS 0723. High-resolution imaging from the NASA/ESA/CSA James Webb Space Telescope combined with a natural effect known as gravitational lensing made this finely detailed image possible.

First, focus on the galaxies responsible for the lensing: the bright white elliptical galaxy at the centre of the image and smaller white galaxies throughout the image. Bound together by gravity in a galaxy cluster, they are bending the light from galaxies that appear in the vast distances behind them. The combined mass of the galaxies and dark matter acts as a cosmic telescope, creating magnified, contorted, and sometimes refractively replicated images of individual galaxies.

Clear examples of this replication are found in the prominent orange arcs to the left and right of the brightest cluster galaxy. These are lensed galaxies.

The galaxies in this scene that are farthest away–the tiniest galaxies that are located well behind the cluster–look nothing like the spiral and elliptical galaxies observed in the local universe. They are much clumpier and more irregular. Webb's highly detailed image may help researchers measure the ages and masses of star clusters within these distant galaxies. This might lead to more accurate models of galaxies that existed at cosmic "spring," when galaxies were sprouting tiny "buds" of new growth, actively interacting and merging, and had yet to develop into larger spirals. Ultimately, Webb's upcoming observations will help astronomers better understand how galaxies form and grow in the early Universe.

For a full array of Webb's first images and spectra, including downloadable files, please visit
https://esawebb.org/initiatives/webbs-first-images/

NIRCam was built by a team at the University of Arizona and Lockheed Martin's Advanced Technology Center.

Messier 109 (NGC 3992)

Photo Credit: NASA, ESA
https://www.go-astronomy.com/messier.php?Messier=M109 [retrieved 12-10-25]

M109 is a barred spiral galaxy with a weak inner ring structure around the central bar, approximately 60 million light-years away in the northern constellation Ursa Major. It is the brightest member of the Ursa Major Galaxy Group of roughly eighty galaxies, and is home to a trillion stars.

M109 has three satellite galaxies (UGC 6923, UGC 6940 and UGC 6969) and possibly more. Detailed hydrogen line observations have been obtained from M109 and its satellites.

CIRCUMSTELLAR HABITABLE ZONES

Galaxy Cluster SPT-CL J0615–5746

Photo Credit: ESA/Webb, NASA & CSA, L. Bradley (STSci), A. Adamo (Stockholm University) and the Cosmic Spring collaboration
https://esawebb.org/images/weic2418a/ [retrieved 6/20/25]

An international team of astronomers has used the NASA/ESA/CSA James Webb Space Telescope to discover gravitationally bound star clusters *when the Universe was 460 million years old* [emphasis added]. This is the first discovery of star clusters in an infant galaxy less than 500 million years after the Big Bang. Young galaxies in the early Universe underwent significant burst phases of star formation, generating substantial amounts of ionizing radiation. However, because of their cosmological distances, direct studies of their stellar content have proven challenging. Using Webb, an international team of astronomers has now detected five young massive star clusters in the Cosmic Gems arc (SPT0615-JD1), a strongly lensed galaxy emitting light when the Universe was roughly 460 million years old, looking back across 97% of cosmic time.

The Cosmic Gems arc was initially discovered in NASA/ESA Hubble Space Telescope images obtained by the RELICS (Reionization Lensing Cluster Survey) program of the lensing galaxy cluster SPT-CL J0615–5746.

[Image description: A field of galaxies on the black background of space. In the middle is a collection of dozens of yellowish galaxies that form a foreground galaxy cluster. Among them are distorted linear features, which mostly appear to follow invisible concentric circles curving around the center of the image. The linear features are created when the light of a background galaxy is bent and magnified through gravitational lensing. A variety of brightly colored, red and blue galaxies of various shapes are scattered across the image, making it feel densely populated.]

Circumstellar Habitable Zones

PROCEEDINGS OF THE FIRST INTERNATIONAL CONFERENCE

Thirtieth Anniversary Edition

Edited By
Laurance R. Doyle and J. Ellen Blue

Introduction by Carl Sagan

thp Travis House Publications • Sonoma County • California

Travis House Publications
P. O. Box 99
Bodega Bay, CA 94923
travishousepublications@gmail.com

A copy of the First Edition (1996) may be purchased for $95 plus shipping by contacting Ellen Blue at travishousepublications@gmail.com. It will also soon be available from AbeBooks.com.

Printed in the United States of America

31 30 29 28 27 26 2 3 4 5 6

Thirtieth Anniversary Edition, © 2026

Editing, design and layout by Ellen Blue

Typefaces: Arial Hebrew, Baskerville, Baskerville Old Face, Bodoni 72, Palatino, Times New Roman

Library of Congress Cataloging-in-Publication Data

Circumstellar habitable zones : proceedings of the first international conference / edited by Laurance R. Doyle : introduction by Carl Sagan.

p. cm.

Includes bibliographical references.

ISBN 0-9650896-0-9

1. Life on other planets—Congresses. 2. Exobiology—Congresses.

3. Circumstellar matter—Congresses. I. Doyle, Laurance R.

(Laurance Reeve), 1953– .

QB54.C56 1996

577′.099—dc20 96–12702

CIP

ISBN 0-9650896-0-9

DEDICATION

To all the participants in this habitable zone—Sun and planets, tectonic plates and oceans, mountains and canyons, flora and fauna, swimming, jumping, digging, flying, walking, talking, growing, photosynthesizing—may we traverse each 34 million miles a day through the Universe together in harmony and progress.

L. R. Doyle

Table of Contents

Photos and Images

The Pleiades Star Cluster

Photo Credit: NASA, ESA, and AURA/Caltech
https://science.nasa.gov/asset/hubble/hubble-refines-distance-to-the-pleiades-star-cluster/ .... [retrieved 5-2-26]

The brilliant stars seen in this image are members of the popular open star cluster known as the Pleiades, or Seven Sisters. The Hubble Space Telescope's (HST) Fine Guidance Sensors refined the distance to the Pleiades at about 440 light-years. Hubble's Fine Guidance Sensors measured slight changes in the apparent positions of three stars within the cluster when viewed from different sides of Earth's orbit. Astronomers took their measurements six months apart over a 2 ½ year period. The cluster comprises about 1,000 stars, located in the constellation Taurus.

The color-composite image of the Pleiades star cluster was taken by the Palomar 48-inch Schmidt telescope. The image is from the second Palomar Observatory Sky Survey, and is part of the Digitized Sky Survey.

The Pleiades Cluster includes b-type stars, and contains the reflection nebula NGC 1432, an HII region, and NGC 1435, known as the Merope Nebula. Around 2330 BC the Pleiades marked the vernal point. The cluster is dominated by hot blue luminous stars that have formed within the last 100 million years. Computer simulations have shown that the Pleiades likely formed from a compact configuration that once resembled the Orion Nebula.

Analyzing deep-infrared images obtained by the Spitzer Space Telescope and Gemini North telescope, astronomers discovered that one of the stars in the cluster, HD 23514, which has a mass and luminosity a bit greater than that of the Sun, is surrounded by an extraordinary number of hot dust particles. This could be evidence for planet formation around HD 23514.

Source: Wikipedia

Foreword to the Thirtieth Anniversary Edition

This 2026 reissue, the Thirtieth Anniversary Edition, of the proceedings of the First International Conference on Circumstellar Habitable Zones is a historic volume. It documents the progress in numerous fields over three decades and provides a historical record of a conference that was the basis of the founding of NASA's astrobiology programs. Earth-mass bodies around pulsars were not considered planetary systems, so the conference predates (1994) the first discovery of main-sequence-star extrasolar planets, which now (mostly post–Kepler Mission) number over 6,000.

The conference was divided into seven sections over three days. The first day was planetary aspects: atmospheric considerations in the morning and geological/oceanographic aspects in the afternoon. Day 2 was astrophysical aspects with stellar-defined habitable zones in the morning and planetary systems and extrasolar planetary detection in the afternoon. Then in the evening, four pioneers of (not just circumbinary) habitable zones were honored at the banquet: Charlie Seeger (work with Barney Oliver on SETI target stars), Ray Reynolds (electrical-based ecosystems in Jupiter's moon Europa), Stephen Dole (wrote the first habitable zones book [*Habitable Planets for Man*] in 1964), and Carl Sagan (considerations of giant planet habitability), who also gave the banquet talk: Introduction to Habitable Zones.

Day 3 was about the biological aspects of circumstellar habitable zones. Presented in the morning were aspects of organic material—synthesizing and delivery to Earth. The afternoon was about how biology itself changes the environment for further habitability aspects. There were no organized tours, but many attendants said it was the best conference they had ever been to—perhaps in part because each of the six main sessions was followed by a half-hour open discussion.

The original papers have all been preserved over thirty years, but for this Thirtieth Anniversary Edition, all the photos/images have been brought—more or less—up to date (e.g., James Webb images) to help indicate the scientific progress that has been made in these "short" thirty years (e.g., spectra of post-habitable-zone-conference-discovered planetary atmospheres). A list of the photos or images can be found in the Table of Contents. Three new additional updates will include the following:

- In the year 2000, Frank Drake asked me to teach his Life in the Universe class as he was retiring from the University of California at Santa Cruz. Because no astrobiology textbooks were handy at the time, I used the *Circumstellar Habitable Zones* (CHZ) book as the textbook. The homework questions for the class will be made available as a Teacher's Guide on amazon.com, for both the 1996 edition and this edition.
- The CHZ conference (all but the banquet) was professionally filmed, and a link to the videos of the talks and discussions will be available online (hopefully sometime in the remainder of 2026). Please contact the publisher for information or updates on this project.
- The lead authors or coauthors of the papers presented at the conference were contacted during December 2025 and January 2026, and many are willing to write a short update of the research in their fields. This should make this volume and the update volume (expected in the first half of 2027) particularly valuable for consideration by historians of science.

It is going to be great to compare scientific progress in habitable zone studies over the last thirty years. Thirty years ago just a handful of people were working on the transit of extrasolar planets—Bill Borucki and Dave Koch, Jean Schneider and Michele Chevreton, and me with Jon Jenkins and Hans Deeg—with the later help of Ted Dunham and Zoren Ninkov. Now there must be thousands of transit astronomers including space missions and orbiting observatories doing transit spectroscopy. I should just say that it was a long way from doing photometric observing using the 100-year-old Crossley telescope at Lick Observatory to the discovery of the first circumbinary planet, Kepler-16b, with the Kepler spacecraft. Enjoy the next thirty years of discovery! Let's hope they are every bit as exciting as the last thirty!

Dr. Laurance R. Doyle
Bodega Bay, CA 94923

FOREWORD TO THE FIRST EDITION, 1996

"Some places on Earth are so hot that the inhabitants have to live somewhere else."

[Observation by young habitable zones scientist]

The idea of having a conference on circumstellar habitable zones came from noticing some mistakes in the technical literature that were a simple result of not being aware of the work in other fields. Consequently this conference was held with the express purpose of bringing together planetary scientists, astrophysicists, biologists, geologists, chemists, and those of several other disciplines to educate each other and form collaborations in the appreciation and discovery of what makes a nice place to live. As a result, the conference was extremely productive—discussions continued long after the sessions concluded—and more than half a dozen scientists indicated that it was the best conference they had ever attended. Our hope is that this spirit of discovery—the reason we all went into science to begin with, and so evident at the conference—will be transmitted to the reader and to posterity by these proceedings. Besides delving deeply into one's field of academic endeavor, wonderful discoveries (and more serendipitous ones) can be found on the borders between disciplines. Contrary to the narrow disciplinary trend found in the institutionalization of science and the way by which science is generally funded, most of the real thought breakthroughs have occurred and will in the future occur on these borders. But one must have a bit of a Renaissance outlook to find them—an outlook well represented by the participants in these proceedings.

The field of circumstellar habitable zones (kind of "The-Big-Picture" Ecology) is not new. As Carl Sagan indicates in his introduction, thoughts on the habitability of space (where we all live) have long been considered by great thinkers. As Richard A. Proctor wrote in his 1870 book, *Other Worlds Than Ours* (what might be considered one of the first habitable zones books), "In long past ages Anaximander and Pythagoras studied the subject of other worlds than ours; later, such men as Huygens, Galileo, and Newton have dwelt upon the same interesting theme; . . ." But when it comes to the habitability of other worlds, more than scientists become involved. As Dr. Proctor keenly noticed (*ibid.* p. 3), "The discoveries which are made by astronomers acquire a new interest when they are associated with the subject of life in other worlds. The interest with which the public regard many of those discoveries may, indeed, be said to depend wholly on their bearing on this subject." And this is no doubt true today. When this was written over 100 years ago, the author expressed his incredulity that so eminent a person as William Herschel (discoverer of the planet Uranus) could have believed that there was life on the Moon, before he himself jumped into a discussion of the "Mercurians." In this volume we have chosen to take an admittedly conservative approach to what would make a "comfortable" place to live. Nevertheless, the constraints that such a conservative approach imposes on the habitability of planets around certain stars, given certain geological and atmospheric conditions with biological and biochemical adaptability, are quite encouraging indeed. As far as the results of this conference, virtually none of the *types* of processes that have happened in this Solar System, and in particular on Earth, resulting in inhabitants, is a quirk or highly improbable—from the evolution of our star and the geological processes that recycle our atmosphere, to the abundance of biogenic material and the adaptation of the environment by early oceanic algal biology (what one might call Earth's salad days). No doubt each habitable zone is unique. But the *processes* for habitability appear not to be (that is, there do not seem to be any strong "showstoppers" in the development of biological forms on planets around other stars).

In the final analysis, our intuition leads us to assume that since there are more stars in our Galaxy than one could count in 10,000 years (and hundreds of billions of such galaxies that we can detect), our little Sun may not be unique in having a small liquid-water-covered planet in orbit around it with a multitude of diverse "folks" carrying on the business of being. A look at habitability is thus a look at ourselves, and what a remarkable place we live in. You may begin to draw the same conclusion as you read the diverse and interesting papers that follow.

Laurance R. Doyle
SETI Institute
NASA Ames Research Center
Mountain View, California

ACKNOWLEDGMENTS

Special thanks are due to my fellow organizers of this conference, Dr. John Billingham of SETI Institute and Dr. Donald DeVincenzi of NASA Ames Research Center. This conference would certainly not have been possible without their unwavering support. We would also like to thank Sara Acevedo, Michelle Murray, Vera Buscher, and Thomas Pierson of SETI Institute, and Sandra Owen of NASA Ames Research Center for their continual support. Also, the conference was supported in part by the Exobiology Office at NASA Headquarters and for this we thank Dr. Michael Meyer. Thanks also to the untiring editorial work of the publishers, Travis House Publications, and in particular J. Ellen Blue for this beautiful volume, which she designed. We would also like to thank each one of the Circumstellar Habitable Zones Conference attendees for their wholehearted participation in the conference, which made it such a success, and especially the contributors to this volume for their wonderful articles. Special thanks must also go to Dr. Carl Sagan for his much appreciated scientific participation in the conference and for the enlivening and perspicacious banquet address. I must also thank the session chairs* for their enlivening presence, the overview speakers† for so finely setting the tone for each session, and finally our honorees‡ for pioneering in this field. We hope this volume consequently communicates, to some extent at least, the scientific excitement of discovery about the progress being made in understanding our home in the Universe. To this end we wish to acknowledge you, the reader, and fellow inhabitant of this habitable zone.

LRD

---

* Brian Toon, Carl Sagan, Michael Carr, David Morrison, Icko Iben, Frank Drake, Jill Tarter, William Borucki, Sherwood Chang, Clifford Matthews, Chuck Klein, Jack Farmer.

† James Kasting, Chris McKay, Daniel Whitmire, George Wetherill, Chris Chyba, David Des Marais.

‡ Stephen Dole, Carl Sagan, Ray Reynolds, Charles Seeger.

The Trifid Nebula (M20)

Photo Credit: NASA, ESA, Hubble
https://go-astronomy.com/nebula.php?Nebula=148 [retrieved 12-10-25]

In the heart of the Trifid Nebula, a convergence of three distinctive dark lanes of interstellar dust make the Trifid Nebula one of the most recognizable and striking star birth regions in the night sky. The dust, silhouetted against glowing gas and illuminated by starlight, cradles the bright stars at the heart of the nebula. Opaque dust formations resembling mountains dominate the lower region, while dark dust filaments intricately weave through the nebula. The luminosity of the Trifid is largely attributed to a solitary massive star situated near its center.

JOURNAL ABBREVIATIONS

A&A	*Astronomy and Astrophysics*
A&AS	*Astronomy and Astrophysics Supplement*
A&SS	*Astrophysics and Space Science*
Adv. Appl. Microbiol.	*Advances in Applied Microbiology*
Adv. Phys. Geochem.	*Advances in Physical Geochemistry*
Adv. Space Res.	*Advances in Space Research*
AJ	*The Astronomical Journal*
Amer. J. Sci.	*American Journal of Science*
Amer. Sci.	*American Scientist*
Am. Mineral.	*American Mineralogist*
Anal. Chem.	*Analytical Chemistry*
Angew. Chem.	*Angewandte Chemie*
Ann. NY Acad. Sci.	*Annals of the New York Academy of Sciences*
Ann. Rev. Astron. Astrophys.	*Annual Review of Astronomy and Astrophysics*
Ann. Rev. Earth Planet. Sci.	*Annual Review of Earth and Planetary Science*
Ann. Rev. Microbiol.	*Annual Review of Microbiology*
Ann. Rev. Plant Physiol.	*Annual Review of Plant Physiology*
ApJ	*The Astrophysical Journal*
ApJSupp	*The Astrophysical Journal Supplement*
Appl. Env. Microbiol.	*Applied and Environmental Microbiology*
Appl. Spectr.	*Applied Spectroscopy*
Arch. Biochem. Biophys.	*Archives of Biochemistry and Biophysics*
Biochem. Biophys. Res. Comm.	*Biochemical and Biophysical Research Communications*
Biochem. J.	*Biochemical Journal*
Bull. Amer. Astron. Soc.	*Bulletin of the American Astronomical Society*
Bull. Astron. Inst. Czechosl.	*Bulletin of the Astronomical Institute of Czechoslovakia*
Bull. Math. Biol.	*Bulletin of Mathematical Biology*

Celest. Mech.	*Celestial Mechanics*
Celest. Mech. Dyn. Astron.	*Celestial Mechanics and Dynamical Astronomy*
Chem. Ber.	*Chemische Berichte*
Chem. Geol.	*Chemical Geology*
Chem. Phys. Lett.	*Chemical Physics Letters*
Chem. Soc. Rev.	*Chemical Society Reviews*
Compte. Rend. Acad. Sci.	*Comptes Rendues des Academies des Sciences*
Contrib. Mineral. Petrol.	*Contributions to Mineralogy and Petrology*
CRC Crit. Rev. Biochem.	*CRC Critical Reviews in Biochemistry*
Earth Planet. Sci. Lett.	*Earth and Planetary Science Letters*
Econ. Geol	*Economic Geology*
Eur. J. Biochem.	*European Journal of Biochemistry*
FASEB Journal	*Federation of American Societies for Experimental Biology Journal*
Fed. Eur. Biochem. Soc. Lett.	*Federation of European Biochemical Societies Letters*
First LDEF Symp.	*First Long Duration Exposure Facility Symposium*
Geochim. Cosmochim. Acta	*Geochimica et Cosmochimica Acta*
Geogr. Bull.	*Geographical Bulletin*
Geogr. Rev.	*Geographical Review*
Geol. Soc. Am. Bull.	*Geological Society of America Bulletin*
Geophys. Res. Lett.	*Geophysical Research Letters*
IAU Circ.	*IAU Circular*
Ind. Eng. Chem.	*Industrial and Engineering Chemistry*
Int. Colloq. Venus	International Colloquium on Venus, Aug. 10–12, 1992, Pasadena, CA
J. Am. Chem. Soc.	*Journal of the American Chemical Society*
J. Atmos. Sci.	*Journal of the Atmospheric Sciences*
J. Bacteriol.	*Journal of Bacteriology*
J. Brit. Interplanet. Soc.	*Journal of the British Interplanetary Society*
J. Chem. Eng. Data	*Journal of Chemical and Engineering Data*
J. Chem. Phys.	*Journal of Chemical Physics*
J. Chem. Soc. Chem. Commun.	*Journal of the Chemical Society, Chemical Communications*
J. Chromatog.	*Journal of Chromatography*
J. Exp. Biol.	*The Journal of Experimental Biology*

J. Exp. Theor. Phys.	*Journal of Experimental and Theoretical Physics*
J. Gen. Microbiol.	*Journal of General Microbiology*
J. Geol.	*Journal of Geology*
JGR	*Journal of Geophysical Research*
JGR Planets	*Journal of Geophysical Research—Planets*
J. Mater. Sci.	*Journal of Materials Science*
J. Mol. Biol.	*Journal of Molecular Biology*
J. Mol. Evol.	*Journal of Molecular Evolution*
J. Non-Cryst. Solids	*Journal of Non-Crystalline Solids*
J. Org. Chem.	*Journal of Organic Chemistry*
J. Phys. Chem.	*Journal of Physical Chemistry*
J. Phys. Chem. Solids	*Journal of Physics and Chemistry of Solids*
J. Theor. Biol.	*Journal of Theoretical Biology*
J. Vac. Sci. Technol.	*Journal of Vacuum Science and Technology*
Microbiol. Rev.	*Microbiological Reviews*
Microbiol. Sci.	*Microbiological Sciences*
MNRAS	*Monthly Notices of the Royal Astronomical Society*
Mol. Cell. Biochem.	*Molecular and Cellular Biochemistry*
Mol. Phys.	*Molecular Physics*
Org. Geochem.	*Organic Geochemistry*
Origins of Life	*Origin of Life and Evolution of the Biosphere*
Palaeogeogr., Palaeoclimatol., Palaeoecol.	*Palaeogeography, Palaeoclimatology, Palaeoecology (Global and Planetary Change Section)*
PASP	*Publications of the Astronomical Society of the Pacific*
Phil. Trans. R. Soc. Lond. A	*Philosophical Transactions of the Royal Society of London, Series A*
Photosynth. Res.	*Photosynthesis Research*
Phys. Chem. Minerals	*Physics and Chemistry of Minerals*
Phys. Lett. A	*Physics Letters A*
Phys. Rev. Lett.	*Physical Review, Letters*
Planet. Space Sci.	*Planetary and Space Science*
Plant Physiol.	*Plant Physiology*
Precambrian Res.	*Precambrian Research*

Proc. 14th LPSC	*Proceedings of the Fourteenth Lunar and Planetary Science Conference*
Proc. Natl. Acad. Sci. USA	*Proceedings of the National Academy of Sciences of the United States of America*
Proc. R. Irish Acad.	*Proceedings of the Royal Irish Academy*
Proc. Roy. Inst. Gt. Britain	*Proceedings of the Royal Institution of Great Britain*
Proc. R. Soc. Lond. A	*Proceedings of the Royal Society of London A*
Proc. R. Soc. Lond. B.	*Proceedings of the Royal Society of London B*
Publ. US Naval Obs.	*Publications of the U.S. Naval Observatory*
Rev. Geophys.	*Reviews of Geophysics*
Rev. Geophys. Space Phys.	*Reviews of Geophysics and Space Physics*
Rev. Scientifique	*Revue Scientifique*
Sci. Am.	*Scientific American*
Sci. News	*Science News*
Second LDEF Symp.	*Second Long Duration Exposure Facility Symposium*
Sky & Tel.	*Sky & Telescope*
Solar Phys.	*Solar Physics*
Sov. Astron.	*Soviet Astronomy*
Space Sci. Rev.	*Space Science Reviews*
Stadler Symp.	*Stadler Symposium*
Tectonophys.	*Tectonophysics*
Tetrahedron Lett.	*Tetrahedron Letters*
Trends Biochem. Sci.	*Trends in Biochemical Sciences*

Photo Credits, Special Mentions, and Conference Memories

Covers

Front Cover: Halley's Comet, shot by astronomer Dennis Mammana from the Pinacate Mountains of northern Mexico in March 1986. The "teapot" of Sagittarius and the star clouds of the Milky Way are seen in the upper right. © 1986, Dennis Mammana. Used by permission. This image illustrates the major aspects of circumstellar habitable zones: planetary, astrophysical, and biological.

Back Cover: Earth at Night, 2016; NASA Earth Observatory

Special Mentions

Laurance R. Doyle, the conference organizer, has been an astronomer and Principal Investigator at the SETI Institute for almost forty years. As a selected member of the Kepler Mission Science Team he discovered Kepler 16b, the first known circumbinary extrasolar planet, nicknamed "Tatooine."

J. Ellen Blue is the founder and president of Travis House Publications, the company that published the original conference proceedings. She is a huge fan of the Drake equation and one of the conference's most avid supporters. It was her idea to refresh the First Edition as a Thirtieth Anniversary Edition. Ellen has been in the publishing industry for more than half a century and has worked for a number of major publishers and on a great variety of books.

As the most famous astronomer of the latter part of the 20th century, Carl needs no introduction. It is but right to say that the First International Conference on Circumstellar Habitable Zones could not have taken place without his enthusiastic support. His Introduction in this volume is a unique perspective on historic ideas about planetary habitability. He helped turn this contribution into the precursor of NASA's Astrobiology programs. His vast scientific and educational contributions are well known throughout the world. Thank you Carl!

First International Conference on Circumstellar Habitable Zones

January 19–21, 1994

Dr. Carl Sagan with Dr. David Morrison at the First International Conference on Circumstellar Habitable Zones, NASA Ames Research Center, 1994.
https://images-assets.nasa.gov/image/ARC-1994-A94-0009-1/ARC-1994-A94-0009-1~medium.jpg
[retrieved 4-6-26]

Conference Speakers by Row (opposite page, left to right)

1st Row: Carl Sagan, Don DeVincenzi, Jim Kasting, Bob Haberle, Jon Jenkins, Darren Williams

2nd Row: David Summers, Friedemann Freund, Linda Jahnke

Special Presentation by Dr. Icko Iben

3rd Row: Dan Whitmire, Alan Hale, Laurance Doyle, John Stauffer, George Wetherill, Luke Dones

4th Row: Jack Lissauer, Dave Koch, Bill Borucki, Jean Schneider, Stefan Brandmeier

5th Row: John Billingham, Chris Chyba, John Oró, Jeff Bada, Max Bernstein, Cliff Matthews

6th Row: Ted Bunch, Yvonne Pendleton, Alexandra MacDermott, Dave Des Marais, Maria Gogarten-Boekels, Stanley Miller

Last Row: Chris McKay, David Schwartzman, Martin Heath, Liaofu Luo

Conference Speakers

Introduction and Planetary Aspects

Astrophysical Aspects

Biological Aspects

Rho Ophiuchi Cloud Complex [annotated]

Photo Credit: NASA, ESA, CSA, STScI, K. Pontoppidan (STScI), A. Pagan (STScI)
https://esawebb.org/images/weic2316b/ [retrieved 6-20-25]

This first anniversary image from the NASA/ESA/CSA James Webb Space Telescope shows the Rho Ophiuchi cloud complex. It is a complex of interstellar clouds with different nebulae, particularly a dark nebula centered 1° south of the star ρ Ophiuchi, of the constellation Ophiuchus. At an estimated distance of about 140 parsecs, or 460 light years, it is one of the closest star-forming regions to the Solar System.

Temperatures of the clouds range from 13 to 22 K, and there is a total of about 3,000 times the mass of the Sun in the material. Over half of the mass of the complex is concentrated around the L1688 cloud, and this is the most active star-forming region. There are embedded infrared sources within the complex. A total of 425 infrared sources have been detected near the L1688 cloud. These are presumed to be young stellar objects, including 16 classified as protostars, 123 T Tauri stars with dense circumstellar disks, and 77 weaker T Tauri stars with thinner discs. The last 2 categories of stars have estimated ages ranging from 100,000 to a million years.

[*Image description*: Red dual opposing jets coming from young stars fill the darker top half of the image, while a glowing pale-yellow, cavelike structure is bottom centre, tilted toward two o'clock, with a bright star at its centre. The dust of the cave structure becomes wispy toward eight o'clock. Above the arched top of the dust cave are three groupings of stars with diffraction spikes caused by the telescope structure. A dark cloud sits at the top of the arch of the glowing dust cave, with one streamer curling down the right-hand side. The dark shadow of the cloud appears pinched in the centre, with light emerging in a triangle shape above and below the pinch, revealing the presence of a star inside the dark cloud. The image's largest jets of red material emanate from within this dark cloud, thick and displaying structure like the rough face of a cliff, glowing brighter at the edges. At the top centre of the image, a star displays another, larger pinched dark shadow, this time vertically. To the left of this star is a more wispy, indistinct region.]

Since the field of view of the photo is very small, at 6.4 arc-minutes, it displays just a tiny region of what appears in most other photographs of the Rho Ophiuchi cloud complex.

Sources: NASA, ESA, CSA, STScI; Wikipedia

The Cygnus Loop

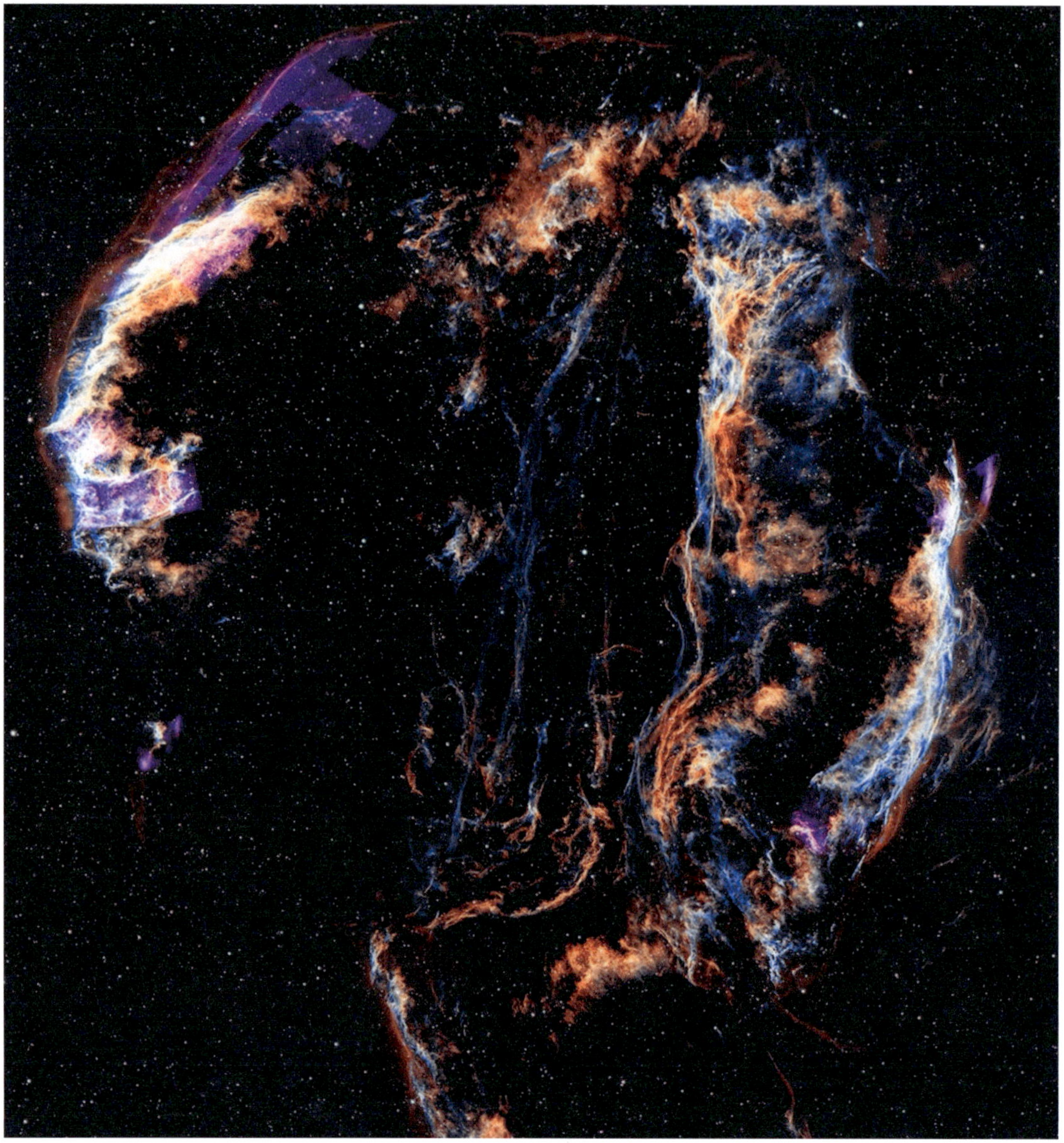

Photo Credit: X-ray: NASA/SAO/CXC; Optical: John Stone (Astrobin)
Image Processing: NASA/SAO/CXC/L. Frattre, N. Wolk
https://www.nasa.gov/image-article/seeing-the-cygnus-loop-in-a-new-way/ [retrieved 12-10-25]

The Cygnus Loop, also known as the Veil Nebula, is the ethereal wreckage of a supernova explosion that occurred some 17,000–25,000 years ago. It is a cloud of dust and gas that is steadily expanding outward from the original blast site. Because it is relatively close to us, it is a popular target for professional astronomers studying the aftereffects of supernovae. Although the supernova that created the Cygnus Loop occurred a long time ago, its shockwave is still traveling through space, pushing clouds of gas in its path quickly forward. As starlight travels through this gas, some of the light is absorbed by the elements that are present. The resulting spectrum has black lines, called 'absorption lines,' indicating the presence of these elements.

Source: McDonald Observatory, The University of Texas at Austin
https://mcdonaldobservatory.org/news/releases/20240719

INTRODUCTION

A Region in Sagittarius

Photo Credit: Dennis Harper

Wikimedia Commons: Sagittarius region, Milky Way on slide film [retrieved 12-26-25]

Circumstellar Habitable Zones: An Introduction

Carl Sagan

Cornell University

Are there other Earths? If we take our own planet as prototypical and consider the 400 billion or so other stars in the Milky Way Galaxy, how many of them have environmental conditions conducive to the sort of life we know? This is a very old question, going back at least to the beginning of the European Enlightenment, when it became clear that there *were* other Suns and that they might have other Earths around them. "What a wonderful and amazing Scheme have we here of the magnificent Vastness of the Universe!" wrote an awed Christianus Huygens in 1670. "So many Suns, so many Earths. . . !" His book was entitled *New Conjectures Concerning the Planetary Worlds, Their Inhabitants and Productions*.

That there are many Suns—perhaps 400 billion of them in our Galaxy—we can readily grant. But are there many planets, many Earths, many biospheres? The question is profound and difficult, touching not only important scientific issues, but also philosophical, religious, and even emotional notes. To answer it clearly we must draw upon our knowledge of stellar evolution, the origin and abundance of planetary systems, geology and geophysics, the early history of the Earth, prebiological organic chemistry, and the origin and evolution of life. The interdisciplinary challenge posed by this subject requires it to be addressed in a sym-

Note: This introduction is based upon Dr. Sagan's banquet talk at the conference, January 20, 1994.

posium such as this, with experts in many fields. Recent progress provides some encouragement. We may really be on the verge of finding out whether there are other Earths.

For me, the most important consensus view of this conference is that many main sequence stars are likely to have one or more terrestrial planets orbiting them at heliocentric distances compatible with surface liquid water. This is certainly not a new conclusion, although it is now better buttressed than ever before.

I'm honored to receive, as a gift from the organizers of the conference, samples of the Murchison and Allende meteorites, and to know that Charles Seeger, Ray Reynolds, and Steve Dole are corecipients. (If he were still alive, Su-shu Huang would certainly also have been a recipient.) The complex organic matter in Murchison, including amino acids and nucleotide bases, aliphatic and aromatic hydrocarbons, is relevant to the origin and cosmic abundance of life. Yvonne Pendleton and colleagues, here at Ames, have recently found that Murchison provides the best fit of any known substance to the organic matter in the interstellar grains. And while the train of argument is still not secure at every step, this finding suggests that the stuff of life is everywhere in the Galaxy, that the same building blocks are accessible to all Earthlike worlds early in their histories, and that if there are abundant Earths, there may be abundant life as well.

Certainly arguments for the first of these propositions—abundant Earths—have been indicated for decades. I want especially to mention Steven Dole's seminal work, *Habitable Planets for Man*, written in 1964. (If it had been written today, it would surely have been called *Habitable Planets for Humans*.) It sets the stage for the considerations on habitable zones at this conference. You can also see in other early work of Dole's his very simple approach to the origin of the planets in our Solar System, and many harbingers of the best of the work on Solar System cosmogony being done today. Steve did this work in time stolen from his position at the Rand Corporation. He pursued these matters not because he was asked to, but because he couldn't help himself.

If I look at *Habitable Planets for Man*, or *Intelligent Life in the Universe*, which I. S. Shklovskii and I wrote in 1966, I'm struck by how much we've learned in the intervening thirty years. But I'm also struck by how, over the decades and the centuries, intellectual fashions change. And finally, in thinking about the conclusions of this meeting, I'm struck by how conservative our opinions largely are—not just about which environments in a given planetary system might be suitable for life, but also about what kind of life we're imagining.

For example, on the issue of the abundance of extrasolar planetary systems, the point of highest-flying optimism was reached around the time of Huygens, when every star was imagined to have a planetary system. The argument was very simple: Why else have the stars been put there, if not to warm habitable planets? The most pessimistic point in recent centuries came in the '20s and '30s of this century, when the preeminent view of how the Solar System began was a grazing collision of two stars—our Sun and one other. The tidal interaction drew long tendrils of star-stuff out, which then condensed into planets. This view, originally suggested by T. C. Chamberlin, became associated with the illustrious names of Sir James Jeans and Sir Harold Jeffreys.

You can immediately see that the number of planetary systems in the Galaxy, if this is the only way they could form, must be tiny. To take some representative numbers, the mean free time, t, between collisions for a given star is $\sim 1/(n\sigma v)$, where n is the number density of stars in the plane of the Galaxy, σ is the stellar collision cross-section, and v is some mean stellar velocity. Taking a mean separation between stars of $\sim$ 1 pc, $\sigma = \pi R^2$ where R is the current solar radius, and $v \sim 20$ km/s, we find $t \sim 10^{19}$ yrs for any given star. For 10^{11} stars this gives $\sim 10^8$ yrs per collision of *some* pair of stars and thus $\sim$ 100 pairs that have collided in the history of the Galaxy.

Given some other choices of parameters, especially a lower value of n and a younger age for the Galaxy, as was popular in the '20s and '30s, you might even come out with only one such collision since the formation of the Milky Way. The only planets would be around the Sun and around its long-lost coparenting collision partner, now perhaps on the other side of the Galaxy. In that case, we were obviously very lucky, and there are few if any other Earths.

In the early '40s, Kaj Strand of Northwestern University announced that he had found a low-mass companion to the star 61 Cygni. But 61 Cygni is quite close—about 11 light years away. If it had a planetary system, then there had to be something wrong with the Chamberlin/Jeans/Jeffreys hypothesis. This was a sufficiently discordant fact that the aged Sir James Jeans in one of his last published papers, in the June 20, 1942 number of *Nature*, revised his hypothesis. He argued that stars begin their lives as greatly distended objects, contracting out of the interstellar medium. Maybe *that* was the condition under which stars collided and planets formed. If stars were originally the size of the orbit of Uranus, then R is some 4,000 times more, σ is 2×10^7 times more, and the number of collision pairs on the tidal hypothesis is $\sim 2 \times 10^9$, or $\sim 1\%$ of what we think today is the number of stars in the Galaxy. (Jeans' calculation gave him an order of magnitude still more; and he satisfied himself that the collisions would occur before the Kelvin-Helmholtz contraction significantly reduced the radius of the protostar.) He concluded, "A far larger proportion of the stars than we have hitherto imagined must be accompanied by planets; life may be incomparably more abundant in the universe than we have thought."

In fact, by this time the Chamberlin/Jeans/Jeffreys hypothesis had been definitively refuted—both because it provided no way to explain the current distribution of angular momentum in the Solar System, and because the temperatures in the filamentary material drawn out from the stars was so high that the material would not have condensed into planets, but would have dissipated into interstellar space. Lyman Spitzer of Princeton University played a leading role in refuting the hypothesis. But Jeans' work does remind us that collisions of protostars, especially when formed in clusters, might be an important process.

I've spent a little time on this historical curiosity because on the possibility of other planetary systems and other life it held sway for about a generation. It was taken very seriously. Its proponents were extremely distinguished. But in the absence of good data, wholly erroneous ideas can become established, infiltrate textbooks, and discourage competent refutation. The key lack in this whole subject has been data. (Incidentally, the announcement of a large Jupiter around 61 Cygni was also erroneous; Jeans revised his erroneous theory in response to a spurious datum.)

One of the key findings relevant to extrasolar planets in the last fifteen years has been the repeated discovery of circumstellar disks of gas and dust, especially around young stars on their way to becoming suns of solar spectral type. These disks are precisely what Immanuel Kant and Pierre Simon, the Marquis de Laplace, had postulated must be the structures out of which the planets formed. In retrospect, it's an astonishing achievement that these two giants of philosophy and science were able—merely from the near-coplanarity of the orbits of the planets and the equator of the Sun, and a few other facts—to deduce that there was, long ago, a flattened solar nebula out of which the planets formed. (Kant thought that spiral nebulae were the hypothesized preplanetary disks. No one knew that they were distant galaxies.)

We now find that something like half of the nearby young stars of solar spectral type in the IRAS survey have circumstellar disks. This is truly exciting. And then there's the wonderful picture obtained by Brad Smith and Rich Terrile, in which we can actually *see* the disk around the star Beta Pictoris. There have been some hints of the detection of feeding zones in which forming planets are accreting nebular material. Time variability in these distant nebulae has been attributed by some to planetesimal-sized condensates colliding with the local star; but there has not yet been any compelling direct demonstration that planets are in fact forming in these circumstellar nebulae. Perhaps there will be soon. Nevertheless, these discoveries are very supportive of the notion of many Earths. Suddenly the subject has moved from vague philosophizing and a small number of excellent papers in astrophysics to a concrete reality.

Many attempts are now under way to detect fully formed planets around nearby stars—using astrometric methods, radial velocity techniques, and photometry. None of them has succeeded yet, but we are clearly on the verge of being able to detect at least Jupiter-sized planets around the nearest stars—if there are any to be found. A most important and serendipitous recent discovery is of a bona fide planetary system around a distant star (620 pc away) found by a most unlikely technique—from the timing residuals of the radio emissions of the 6.2-millisecond pulsar 1257+12, planets were tentatively announced two years ago by Alexander Wolszczan, now at Pennsylvania State University. Recently, the predicted mutual gravitational interactions of these planets have been confirmed from a study of long time baseline observations; the evidence that this is truly a new planetary system and not starquakes on the neutron star surface (or something) is now overwhelming. Unlike all the other techniques, this method makes close-in terrestrial planets comparatively easy and more distant Jovian planets comparatively difficult to detect.

A 2.8 Earth-mass planet orbits the pulsar every 98 Earth-days at roughly half an AU; and another with 3.4 Earth mass, or maybe somewhat more, has a 67 Earth-day year at about one-third of an AU. A small world, still closer to the star, of 0.015 Earth mass, is at about 0.2 AU. Crudely speaking, one of these worlds is approximately at the distance of Mercury from our Sun; another is midway between the distances of Mercury and Venus. Interior to them is a world of roughly the mass of the Moon at about half Mercury's distance from our Sun. Whether these planets are the remnants of an earlier planetary system that somehow survived the supernova explosion that produced the pulsar, or whether they formed from the resulting circumstellar accretion disk subsequently to the supernova explosion, we do not know. But in either case, we have now learned that there are other Earths.

Although we are ignorant of the exosphere temperatures and radii of these worlds, it may be possible for them to retain any endogenous or exogenous atmospheres. But the charged particle luminosity of 1257+12 is about 4.7 times that of the Sun. So even a planet at one AU, with the Earth's albedo and no greenhouse effect, would have a surface temperature around 550 K. The planet at 0.5 AU will have a mean surface temperature without a greenhouse effect greater than that of Venus with its massive greenhouse effect.

These planets do not seem hospitable for life, but there may be others, farther from 1257+12, that are. Of course, we don't even know that such worlds would have atmospheres; perhaps any atmospheres were stripped away in the supernova explosion, if they date back that far.

Fashions change on the biological side as well. Until about the seventeenth century, there was no problem about the origin of life. Everybody knew that life spontaneously arose all around us—mice from the mud of the Nile, lice from sweat, maggots from putrefying meat, and fireflies from conflagrations. Of course, it was different for "higher" animals like us, but the "lower" animals arose all by themselves. When microbes were discovered they were thought to be spontaneously forming as well. These attitudes were not fully reversed until Pasteur performed his famous experiment with a swan-necked flask, which soon led to the doctrine that life came only from life.

Now there was a major problem about the origin of life—in fact, an insoluble problem: If life can come only from life, and the Universe has a finite age, how can life arise? There were serious scientists who proposed that life was coextensive with matter, that when the Universe was made life was made, and that's the end of it, or, alternatively, that life and the Universe were both infinitely old.

First-rate scientists such as Helmholtz and Arrhenius suggested that life began on the Earth at a particular moment in the early history of our planet, but from interstellar microorganisms. We are accustomed to say that the panspermia hypothesis does not solve the problem of the origin of life, but merely contrives a formulation that releases us from the obligation of finding an answer. (In our time, a few scientists as illustrious as Francis Crick have at least flirted with the panspermia hypothesis, and Fred Hoyle has passionately embraced it.) The change in fashion on this issue is traceable to the 1920s—to the work of A. I. Oparin and J. B. S. Haldane; but the experimental demonstration that significant quantities of biologically relevant organic matter might have been produced on the early Earth was not performed until 1953 (by Stanley Miller under Harold Urey's direction).

Today we can debate whether the origin of life depended on imported organic matter or home manufactures. (If the Earth's early atmosphere was even mildly reducing, the production rate of organics was, as Chris Chyba and I have shown, perhaps three orders of magnitude greater than what was available from interplanetary organics that survived entry.) But especially if we imagine likely concentration mechanisms on Earth, where pools and even lakes are quickly forming and evaporating, even concentrations of 10^{-6} grams per gram of organic matter might be enough. Comparable numbers might well apply to other Earthlike planets.

In a series of papers in the '60s and '70s, I pointed out that the time available for the origin of life on Earth is remarkably constricted—by the paleontological record on the one side, and by the inhospitability of the late-bombardment early Earth (as revealed by the dating of returned lunar samples) on the other. There is an obvious danger in extrapolating from a single example. But the fact that the origin of life happened a few hundred million years or less from the time that conditions were at all favorable for life does suggest that the processes leading to life are highly probable in environments like that of the early Earth. If so, then there's at least a fair chance that life began on early, warm, wet Mars and on innumerable other Earthlike planets in the Galaxy.

But if we were somehow to know that the time available was too short for the origin of life—or that the earliest organisms, as deduced from attempts to root the phylogenetic trees, were too omnicompetent to have arisen so quickly—then we might be tempted by the panspermia hypothesis. But for panspermia to work, there must be enough interstellar seeds of life so that they fall on the Earth very early in its history. If so, there should be enough of them to fall on all those other

Earths early in their histories. The least likely circumstance, it seems to me, is that the rapid origin of life on Earth is a statistical fluke, and that we are making a mistake in extrapolating from a single example. But of course I might be wrong.

Or consider the changing fashions about life on other planets. In the eighteenth century you had charming little works like that of Bernard de Fontenelle, in which he assures the high-born lady whom he is instructing in astronomy that every planet is inhabited. You could tell something about the quality of the inhabitants. It was very easy: People who lived on Mercury were mercurial. People who lived on Venus were amorous. People who lived on Mars were martial. Those who lived on Jupiter were jovial. People who lived on Saturn were saturnine. Easy: just words.

But even a figure as prominent and important in the history of science as Christian Huygens—the inventor of the wave theory of light and the first pendulum clock, the person who first understood the rings of Saturn, the discoverer of Titan, a true scientific giant—could be caught up in arguments that we can only shake our heads at today. Let me give you his argument for marijuana on Jupiter. It went like this: Jupiter has four moons. And then he asks the question that nobody today would ask: *Why* does Jupiter have four moons? They have to be there for a purpose. What is the purpose? Well, what's the purpose of *our* Moon? The purpose of our Moon is to serve as a navigational guide to sailors. The four Galilean satellites of Jupiter can only mean that there's an awful lot of sailing on Jupiter. If there's sailing, there's boats; if there's boats, there's sails; if there's sails, there's ropes; and if there's ropes, there's hemp.

I think it's safe to conclude that comparably fallacious arguments are being made today, even by the best of us. It's possible that some of the things we've been saying at this meeting are wrong. And it's *even* possible that 300 years from now people will be laughing at us as we laugh at Huygens for this lapse. I sincerely hope so, because that will be a hallmark of the progress of our science.

* * * * * * *

In this meeting, we've intentionally focused on circumstellar zones compatible with "life as we know it," liquid water–based life. Since we ourselves are composed of 75% or so water, this is clearly a chauvinism. But it is an understandable and excusable chauvinism. We are sure that there is such a thing as life dependent upon liquid water, and we are not sure that other kinds of life, based say on liquid NH_3 or HF, exist—although they might. If we reach optimistic conclusions with a narrow focus, then whatever we have ignored could only serve to broaden the biological arena.

What's more, there are some reasons to think that liquid water *is* the ideal medium for life. Lawrence J. Henderson was a physiologist who, in a famous book called *The Fitness of the Environment* (1913), argued from first principles that many of the characteristics of life on Earth are universally optimum. However much the arguments at first sight seem to be in the spirit of Voltaire's *Candide*, they must be taken seriously. The argument for water is: You want a liquid, so that solute molecules can interact; you want a high dielectric constant, so many molecules can be dissolved in it; you want a medium with a high heat capacity, so it's difficult to change its temperature; you want something with a large difference between the melting and the boiling points; and so on. You look through the entire list of materials known, and water comes out on top. (A more modern version of this argument can be found in Harold F. Blum's now-forgotten book, *Time's Arrow and Evolution* [1951].) Plus, there's something that Henderson didn't know about: Water has a very high cosmic abundance. And if we thought that some other liquid—hydrogen fluoride, say—was an excellent medium for life, a cogent objection would be that we're not likely to find many worlds with pools, much less oceans, of HF.

There is certainly an argument here. But still, how much are we hobbled by our ignorance? When we talk about "life as we know it," who is the "we"? It depends very much on our experience. Suppose in the narrowest sense "life as we know it" meant Presbyterians. Would we insist that if organisms on other planets were not Presbyterians, they are not alive? Are we forced to be narrow just because the circle of our acquaintances is narrow? After all, all life on Earth is the same—with the same biochemical ground plan, the same proteins for catalysis, the same nucleic acids for replication (you can argue about DNA or RNA, but the chemical difference is minor), the same genetic code (with a few minor exceptions) for translating nucleic acid information into protein information. Is that the way it has to be everywhere?

Is our genetic code, as Francis Crick argues, merely a "frozen accident," the first thing that worked, so life has stuck with it ever since? Or is there an underlying chemical reason that certain nucleotide triplets code for certain amino acids and not others? Must all biologies in the Galaxy have a code similar to ours? Do all use nucleic acids and proteins? Do all use organic molecules? We have hardly a clue, and we may never know unless we find life elsewhere. Necessarily, our current perspective is very limited.

Even if we reconcile ourselves to liquid water and organic matter chauvinism, there are some possibilities we've neglected at this meeting. Because of the Jupiter/Io/Europa tidal lock, a warm, underground ocean on Europa is a possibility, a very friendly environment for life. We may need an additional energy source, but Voyager data hint that there are cracks where the sunlight gets in. If not, there's a thermal gradient. Maybe there's a kind of water life there. If so, what does that say about the definition of a habitable zone? Jim Kasting, in his very stimulating work, extends its outer limit maybe as far as Mars. With Europa, it's out to 5 AU.

Or, let's consider the Jovian planets. All of them must have abundant water, for cosmic abundance reasons and because of the impossibility of significant atmospheric escape. Standard models of Jupiter have massive oceans of liquid water or NH_4OH floating in the sky—dense clouds at several or more bars pressure. The interaction volume is orders of magnitude more than in the oceans of the Earth. Are we sure there's no life there? In 1976, Ed Salpeter and I published a paper in *The Astrophysical Journal Supplement* about ecological niches on the Jovian planets. We tried to show that there were physically and chemically reasonable ecostyles on such a planet. We were not insisting they exist, but arguing that it would be a mistake to prematurely exclude such organisms. If that's true, the habitable zone might extend out to 30 AU. Why are we excluding worlds other than those with thin atmospheres and solid surfaces warmed by sunlight?

Or consider Titan, at a mean surface temperature of 95 K, covered with what looks like enormous quantities of very interesting complex organic molecules. Important evidence on the early stages of prebiological organic chemistry might be there, but you might well argue that at 95 K and with no liquid water, the origin of life cannot happen. (Even if that's all, it's a very important place to go to and examine.) But Reid Thompson and I propose that it's much more interesting than that, because we calculate from very simple, one-dimensional conduction physics that impacts melt surface ice, and the melts last, not for the age of the Solar System, but for something like a thousand years (for the average impact site). How much chemistry can you get in a thousand years, given an enormous buildup of prebiological organics before that? Maybe you can get quite far. Should the habitable zone extend to worlds like Titan?

There are still more exotic possibilities. For example, there is Tommy Gold's suggestion, in a 1993 number of the *Proceedings of the National Academy of Sciences*, that the microorganisms brought up from deep drilling rigs in continental crust show the existence of a rich population of extreme thermophiles. Maybe they extend even to the upper mantle. Those organisms are even more decoupled, it appears, from the photosynthetically based surface life than the organisms that live in the deep-sea vents. Gold goes so far as to propose that this is an independent ecosystem that can arise on every other solid body in the Solar System. If you're decoupled from the Sun, then the interiors of Pluto and large comets may be within the habitable zone.

Or spend a moment thinking about brown dwarfs, hypothetical stars, considerably more massive than Jupiter, but considerably less massive than the Sun, that you might imagine as the tail of the stellar luminosity function. There's a wide range of stellar masses, many more little ones than big ones. Nobody knows if brown dwarfs exist. We've heard one expert opinion today that they don't because the observations do not conform to a particular theory. On the other hand, we also have two independent teams in the last few months who, using gravitational lensing, claim to have found evidence of such objects in the line of sight. From the tiny fraction of 4 π sterradians that have so far been observed by gravitational lensing, an enormous number of brown dwarfs is deduced—according to some estimates, even enough to explain the missing mass in the Galaxy or (who knows?) the Universe. If this is right, there are more brown dwarfs than there are planets. In the 1950s, it was suggested by no less a figure than Harlow Shapley that brown dwarfs—he called them "Lilliputian stars"—were inhabited. He pictured them in the 2-, 3-, 400-K surface temperature range, with lots of volume, nice and warm.

* * * * * * *

Such proposals amount to speculation piled on speculation. It would be foolish to spend much time debating them. But they remind us to keep our eyes and minds open: Our present formulation of the notion of habitable zones may be fundamentally short-sighted. The main answer might lie not in a certain range of heliocentric distance, but in some other parameter space altogether. It follows that we are very lucky, with so conservative and chauvinistic a definition of life and of habitable zones, to find anything at all to talk about. The arena of discourse might be much broader than even optimists imagine.

On the other hand, there are factors that go in the other direction. For example, active galactic nuclei are common—centers of galaxies explode. There is evidence that the center of *our* Galaxy has exploded: A vast expanding cloud of hydrogen has already passed us, moving out from the center of the Milky Way. A massive black hole, perhaps a miniquasar at 10^6 solar masses, seems to lie at the center of the Galaxy. Every planetary system close to the center of the Galaxy may be vulnerable in 10^7 to 10^9 years. It may be necessary to exclude from our considerations, at least of long-lived biospheres, the central portions of the Galaxy.

David Morrison and Clark Chapman have played an important and salutary role in calling our attention to the danger of collisions from asteroids and comets. The prototypical example is the Cretaceous-Tertiary (K-T) impact of 65 million years ago, which not only wiped out the dinosaurs but 75% of the species of life on Earth from a 10-kilometer-diameter impactor. The further back in time you go, the more frequent should be the impacts and the larger their mass. In our case, we are the beneficiaries of the K-T impact. Our ancestors were timid, nocturnal, mouselike insectivores on a planet ruled by dinosaurs. The extinction of the dinosaurs permitted the adaptive radiation that led to us. So from our point of view, a certain level of interplanetary collision is a good thing. But too much clearly is risky.

Now, what about some other planetary system? Is the risk of asteroidal and cometary collision more or less? That depends on whether large populations of asteroids are generated and retained, or accreted into larger worlds, or gravitationally ejected into interstellar space. It depends on whether there is an Oort cloud of comets, which in turn depends on whether there is a Uranus or a Neptune to scatter planetesimals out into the cloud, so that they subsequently can be perturbed back in by passing stars and interstellar clouds. It depends, as George

Wetherill has stressed, on whether there is a Jupiter to scatter comets away from the Earth: If you don't have a Jupiter—and some theorists are having trouble understanding how Jupiters are made—then the cometary flux is a few orders of magnitude greater. Some planetary systems that might seem, by the criteria discussed at this meeting, perfectly habitable, might be barren because of the impact hazard.

* * * * * * *

I apologize for making remarks that are entirely obvious, which merely come down to the point that we are limited and fallible, that our ideas will be revised when new data are available, that the likelihood of other Earths might be greater than or smaller than we now estimate. Of course. But still, it might have been that the bulk of the current evidence spoke compellingly against abundant other worlds and other life. Instead, current data, limited as they surely are, speak to the most exciting research opportunities in this emerging field.

I conclude by paying tribute to a particular habitable zone, the only one we're really sure about—the surface of our own world. It is beautifully appropriate for us. We don't want to make the mistake of confusing cause and effect. The reason it's so appropriate is that we've evolved on it and are delicately attuned to it. But nevertheless, you might have to go a great many light years to find one as lovely as ours.

The penultimate chapter of Steven Dole's book, written back in the polluting '60s, was called "An Appreciation of the Earth," another example of the prescience of our honoree. Surely it is our responsibility to keep this zone habitable.

A Composite Image of the Western Hemisphere—2002

Photo Credit: NASA's Earth Observatory
https://www.nasa.gov/missions/terra/earth-from-space-15-amazing-things-in-15-years

[retrieved 10-30-25]

Planetary Aspects of Habitable Zones

Last Gasp of the Sun–Santa Cruz Mountains

Photo Credit: © Matt Tilghman
https://www.matttilghman.com/photos/suns-last-gasp-sunset-santa-cruz-mountains [retrieved 11-4-25]

A G2 dwarf star, one of 30 billion in the Milky Way Galaxy, sets in the nitrogen-oxygen atmosphere of a terrestrial planet. Photosynthesizing vegetation on the California hills is shown.

HABITABLE ZONES AROUND STARS: AN UPDATE

James F. Kasting

Department of Geosciences
Pennsylvania State University
University Park, PA 16802

Abstract

The habitable zone (HZ) around a star is defined as the region in which a planet can maintain liquid water on its surface. The continuously habitable zone (CHZ) is the region where habitability can be maintained over some finite time period. Constraints on the width of the HZ and CHZ around our Sun and around other main sequence stars have been estimated from calculations performed with a one-dimensional climate model. The instantaneous HZ around our Sun is at least 0.42 AU wide, and the 4.6-Gyr CHZ is at least 0.2 AU wide. Stars ranging from F0 to M0 have HZs that are of the same apparent width, if distances are expressed on a logarithmic scale. If extrasolar planets exist and if they are spaced logarithmically, as seems likely, the chances of finding habitable planets around other stars are reasonably good.

The topic of habitable zones is likely to become increasingly important in the next five to ten years, as that is about how long many astronomers think will be required before extrasolar planets are detected around other main sequence stars. The subject has been discussed by a number of authors over the past thirty years, including Huang (1959, 1960), Dole (1964), Shklovskii and Sagan (1966), Hart (1978,

1979), Whitmire et al. (1991), and Kasting et al. (1993). What follows is a brief description of the results of the most recent of these studies, along with some comments on new developments over the past year and a half.

Assumptions Made in Defining the Habitable Zone

The habitable zone (HZ) around a star can be defined in different ways, which are useful for different purposes. From the standpoint of harboring life, the simplest definition is that the HZ is the region in space where an Earthlike planet could support liquid water at its surface. This definition presupposes that extraterrestrial life will itself be Earthlike and will require liquid water.

Habitable zones tend to migrate outward with time because main sequence stars become brighter as they age and convert hydrogen into helium and heavier elements. Thus, a second useful concept is the continuously habitable zone, or CHZ, which is the region in space where a planet can remain habitable for some specified time period (Hart 1978). A common choice for this time is 4.6 Gyr, the age of the Earth, because this was the time required for intelligent life to evolve on our planet. Other choices are possible, however. Microbial life apparently required 1 Gyr or less to appear on Earth, so this shorter time interval might be appropriate for studies dealing with the possibility of finding simple forms of life elsewhere in the Galaxy.

Some authors (e.g., Dole 1964) are more concerned with the prospects for finding planets that could be inhabited by humans. In this case, the allowed temperature limits are smaller and the corresponding HZs will be somewhat narrower. Some discussion of this issue is provided in the paper by Williams et al. included in this volume.

The problem of defining the liquid water regime for planets has been tackled most recently by Kasting et al. (1993). The theory developed there is based on four essential ideas.

(1) Extrasolar planets, if they exist, are likely to resemble those in our own Solar System. Planets that form in the inner parts of such systems will be rocky bodies endowed with appreciable inventories of volatiles, including carbon dioxide and water. Planets that are Earth-sized or larger will also have long-lived internal heat sources that drive plate tectonics on their surfaces. Plate tectonics, or some equivalent form of planetary resurfacing mechanism, plays an important role in planetary habitability because it controls the long-term cycle of carbon at the surface.

(2) A planet that is too close to its primary will lose its water by either a "moist" or a "runaway" greenhouse. In the first case, liquid water is present on the planet's surface as it is being lost; in the second, all water is present in the atmo-

sphere as steam. In both situations, excessive buildup of water in the lower atmosphere results in a wet stratosphere. This, in turn, permits rapid photodissociation of water vapor by ultraviolet radiation, followed by escape of hydrogen to space. Getting water vapor into the stratosphere is essential to this loss process because this allows hydrogen to escape rapidly despite limitations imposed by diffusion (Hunten 1973, Kasting and Pollack 1983).

(3) Planetary surface temperatures are buffered by a negative feedback mechanism involving atmospheric CO_2 and rates of silicate weathering (Walker et al. 1981). Silicate weathering, followed by deposition of carbonate sediments on the seafloor, is the long-term sink for atmospheric CO_2. Vulcanism at midocean spreading ridges and at plate boundaries is the source. The cycling time for carbon in this system is a relatively short 400,000 years (not the 400 million years that was mistakenly quoted in Kasting et al. 1993). Silicate weathering rates increase with increasing temperature, largely as a consequence of increased evaporation and precipitation. This draws down atmospheric CO_2, thereby decreasing the greenhouse effect and decreasing planetary surface temperature. Conversely, a cold planet should have a slow silicate weathering rate and will therefore tend to accumulate CO_2 in its atmosphere. This cycle is capable of stabilizing climate over a fairly broad range of solar forcings.

(4) Beyond some critical distance from the star, the stabilization mechanism breaks down because CO_2 begins to condense in the atmosphere and/or at the poles (Kasting 1991). CO_2 clouds further cool the planet by enhancing the planetary albedo and by decreasing the lapse rate in the region where they form. The latter effect is relatively simple to parameterize in climate models and can be used to locate the outer boundary of the HZ.

Constraints Derived from Climate Models

Estimates for the positions of the inner and outer boundaries of the HZ for various stars were derived by Kasting et al. (1993) using a one-dimensional, radiative-convective climate model. The results for three representative stellar types are shown graphically in Figure 1. Numerical values for our own Sun are given in Table 1. Here, S_{eff} is the solar flux relative to the present value at Earth's orbit (1360 W m^{-2}), and *d* is the mean distance from the Sun in astronomical units. Parameters for the three stars are given in Table 2.

Three different limits are shown for both the inner and outer edges. The two inner sets (dashed lines in Fig. 1) are values derived directly from the one-dimensional climate model. On the inner edge, the "runaway greenhouse" limit is the point at which the oceans vaporize completely; the "water loss" limit is the point at which the stratosphere becomes sufficiently wet (an H_2O mixing ratio > 10^{-3})

that rapid water loss can occur. On the outer edge, the "maximum greenhouse" limit is the point beyond which a CO_2/H_2O atmosphere is incapable of raising the mean planetary surface temperature to 0°C; the "1st CO_2 condensation" limit is the point at which CO_2 clouds first appear. The "recent Venus" limit is based on the observation that Venus does not appear to have had liquid water on its surface for the last one billion years (Solomon and Head 1991) and has therefore been outside the HZ for at least that time. Solar luminosity was ~ 8% lower 1 Gyr ago (Gough 1981), so the flux at Venus' orbit was 1.76 (= 0.92 × 1.91) times that at Earth. The "early Mars" limit comes from the observation that Mars may have been within the HZ some 3.8 Gyr ago, based on the valley networks preserved on its surface (Pollack et al. 1987). Solar luminosity at that time was ~ 25% lower than today.

Two questions usually arise when these different sets of limits are presented—Which limits should one use, and, How do clouds affect the results? These questions are related: uncertainties regarding clouds are the primary reason that it is difficult to define the HZ boundaries more reliably. Although one cannot give a definitive answer to either question, the following remarks may help to illuminate them.

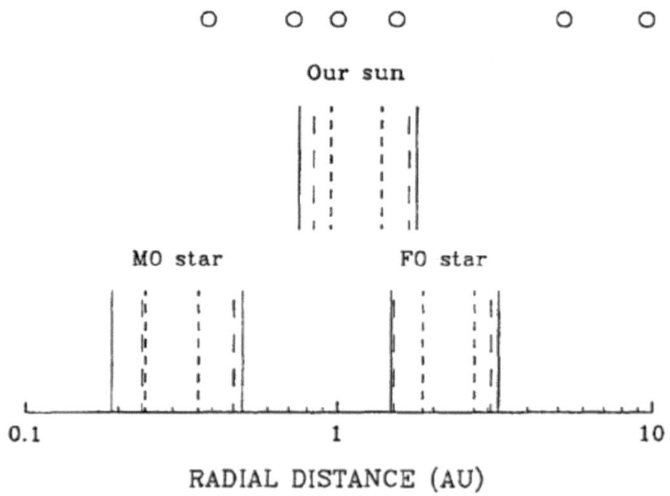

FIG. 1. *Habitable zone limits for the three stars listed in Table 2. The solid lines represent the "recent Venus" and "early Mars" limits; the long-dashed lines represent the "runaway greenhouse" and "maximum greenhouse" limits; and the short-dashed lines represent the "water loss" and "1st CO_2 condensation" limits. "Recent Venus" and "early Mars" limits for the M0 and F0 stars were scaled to our own Sun using the S_{eff} values for "water loss" and "maximum greenhouse," respectively. The circles at the top represent the six innermost planets in our Solar System.*

TABLE 1. ***Habitable Zone Limits for Our Sun***

	S_{eff}	d(AU)
Inner Edge Limit		
Recent Venus*	1.76	0.75
Runaway Greenhouse	1.41	0.84
Water Loss	1.10	0.95
Outer Edge Limit		
1st CO_2 Condensation	0.53	1.37
Maximum Greenhouse	0.36	1.67
Early Mars†	0.32	1.77

* *Solar flux at Venus' orbit 1 Gyr ago.*

† *Solar flux at Mars' orbit 3.8 Gyr ago. Venus and Mars orbit the Sun at 0.72 AU and 1.52 AU, respectively.*

TABLE 2. ***Stellar Parameters***

Stellar Type	$T_{eff}(K)$	$M/M_\odot$	$L/L_\odot$
M0	3700	0.5	0.06
G2	5700	1.0	1.0
F0	7200	1.3	4.3

Luminosities correspond to estimated values midway through the star's main sequence lifetime.

On the inner edge of the HZ, the "water loss" limit is the one of primary physical concern. A planet that orbits its primary closer than this limit would lose its water within a few hundred million years (Kasting and Pollack 1983), which is faster than the probable time scale for originating life. The actual solar flux required to fuel rapid water loss, however, is probably higher than the value shown here because cloud feedback was neglected in the climate model. Water clouds in this model were parameterized by adopting a high surface albedo (0.22) and holding it fixed throughout the simulations; thus, the cloud layer is effectively at the surface. Physical intuition suggests that cloud cover and cloud optical depth would increase at higher surface temperatures because more water vapor would be available to form clouds and because the depth of the convective region would also increase. The dominant effect of these clouds should be to increase the planetary albedo, thereby cooling the surface. Increased cloudiness would also increase the absorption of outgoing infrared radiation, warming the surface; however, this effect is likely to be small because water vapor itself is already a good absorber of infrared radiation. Thus, the inner edge of the HZ probably lies considerably closer to the star than the "water loss" limit would imply. The "recent Venus" limit is a fairly firm upper bound on the critical solar flux, although an optimist who believed that Venus was completely resurfaced 0.5 Gyr ago (Schaber et al. 1992) might want to raise this value by ~ 4% to account for the increase in solar luminosity by that time.

Near the outer edge of the HZ, it is CO_2 clouds rather than H_2O clouds that give rise to the greatest uncertainty. (Once CO_2 begins to condense, its much larger abundance compared with water vapor ought to produce thicker and more spatially extensive clouds.) The increase in planetary albedo caused by these clouds, which has been neglected in the climate model, ought to dominate their greenhouse effect because CO_2 ice is a poor absorber of infrared radiation, except near 15 μm. Thus, the "maximum greenhouse" limit derived from the climate model is in one sense too optimistic, because the middle part of such an atmosphere would be filled with highly reflective CO_2 clouds. Conversely, the "1st CO_2 condensation" limit is too pessimistic because cloud albedo is only beginning to have an effect at this point. So, by this reasoning, the outer edge of the HZ should lie somewhere between these two limits. On the other hand, an optimist would point out that early Mars appears to have been within the HZ despite having a solar flux that is below either of these two values. One possible explanation (Kasting 1991, Sagan and Chyba, abstract, this volume) is that early Mars was warmed by reduced greenhouse gases in addition to CO_2 and H_2O. If all planets had reducing atmospheres early in their histories, as assumed in Hart's models (Hart 1978, 1979), then the HZ might be substantially wider than calculated here. It is difficult to construct a robust model of this type, however, because the planet's climate evolution depends critically on the timing of the switch from a reducing to an oxidizing atmosphere. The HZ outer limits shown here are robust in the sense that they depend primarily on the existence of a stabilizing feedback between surface temperature and atmospheric CO_2, an idea which is widely accepted.

Time-Dependent Considerations

Going from the concept of the HZ to the CHZ requires that one integrate the climatic results discussed above with theoretical models of stellar evolution. Once again, a preliminary attempt to do so has been made by Kasting et al. (1993). The stellar evolution models used were based on calculations by Iben (1967a,b). Stars of masses 0.5 $M_\odot$, 0.75 $M_\odot$, 1.0 $M_\odot$, 1.25 $M_\odot$, and 1.5 $M_\odot$ were modeled explicitly. A note of caution should have been inserted in the Kasting et al. paper: Iben's calculated stellar fluxes do not agree exactly with those shown in Table 2, which were from a different source. Any investigator who wishes to apply the climate model results to a particular type of star should use the calculated limits on S_{eff}, rather than on d. Appropriate values of S_{eff} for stars of spectral types between M0 and F0 (3700 K < T_{eff} < 7200 K) can be derived by fitting quadratic expressions in effective radiating temperature, T_{eff}, to the values listed in Kasting et al. (1993), Table III. Distances can then be easily worked out using the actual luminosity of the star in combination with the inverse square law.

It is illuminating to use our own Sun as an example of how one goes from the HZ to the CHZ. Following Hart (1978), let us define the CHZ as the region around the Sun where the Earth could remain habitable for the last 4.6 Gyr. Then the inner edge of the CHZ is exactly the same as the inner edge of the present day HZ. This follows immediately from the postulate that our Sun is brighter now than it was at any time in the past. The outer edge of the CHZ must, however, be closer to the Sun than the outer edge of the present day HZ because the young Sun was not as bright. If the Sun was ~ 30% dimmer 4.6 Gyr ago (Gough 1981), then the outer edge fluxes must be raised from those shown in Table 1 by a factor of 1/0.7, or 1.43. The "1st CO_2 condensation" limit, for example, moves to $S_{eff} = 0.53/0.7 = 0.76$, or $d = 1.15$ AU. Thus, the most conservative (i.e., narrowest) estimate for the width of the 4.6 Gyr CHZ around our Sun is 0.95 AU to 1.15 AU, or 0.2 AU. The actual width of the CHZ is probably somewhat wider, for reasons discussed earlier.

A complication that was pointed out by Kasting et al. (1993) concerns planetary "cold starts." The question here is: Would a planet that was initially located just outside the HZ become habitable as its primary brightened? The significance of this issue was first recognized by Caldeira and Kasting (1992). Once CO_2 clouds begin to form in a terrestrial planet atmosphere, the planetary albedo may become so high that the surface will remain cold even at relatively high solar heating rates. Thus, an initially frozen planet may remain that way even after it begins receiving sufficient starlight to be within the HZ limits shown here. Large bolide impacts have been suggested as a way out of this climatic dead end (G. Bada and S. Miller, pers. comm.); nonetheless, this type of situation is clearly more complicated than cases in which the planet begins its existence within the HZ.

Planetary Spacing and the Frequency of Habitable Planets

Whether habitable planets exist around other stars depends on whether other planets exist, where they form, how big they are, and how they are spaced. All of these questions are best addressed by observation. However, as the data are not yet available, some theoretical speculation on these questions may be warranted.

Figure 1 gives some idea as to why the last question is important. The planets in our own system are spaced according to Bode's Law, one version of which is given in Kasting et al. (1993). Most astronomers do not believe that Bode's Law has any real significance; it is simply an empirical relation without any theoretical basis. A watered-down version of Bode's Law that may have theoretical significance, however, is this—planetary semimajor axes are spaced more or less evenly if distances are plotted on a logarithmic scale. The gap between Mars and Jupiter is the obvious exception to this rule, but this gap is widely accepted to be a consequence of resonances set up by Jupiter's large gravitational field.

Why the planets in our Solar System are spaced logarithmically has yet to be determined. It should be noted that Jupiter's moons are also distributed in this manner (Kasting et al. 1993, Fig. 16), as are the moons in the Saturnian system. Most likely, the reason has to do with maintaining orbital stability over long time scales. Planetary (or satellite) masses should also affect the spacing, which is why Bode's Law is unlikely to be a general relation. The article by J. Lissauer in this volume may provide some insight into this question.

Where planets form around different stars and how big they are are also important. Preliminary work (Wetherill 1994) indicates that the terrestrial planet formation region is fairly independent of stellar mass for stars between 0.5 $M_\odot$ and 1.5 $M_\odot$. The size of the planets formed in Wetherill's model depends strongly on the assumed surface density of the circumstellar nebula. The latter conclusion seems intuitive but not very useful, as we do not know the mass distribution of circumstellar nebulae. One suspects that the first conclusion may depend strongly on the assumed angular momentum distribution of the nebula, which is also poorly constrained. For the present, we should put both of these down as unsolved questions.

If planets form within the HZ of a star, if they are big enough to experience plate tectonics, and if they are logarithmically spaced, the following observations apply.

(1) All stars between F0 and M0 have instantaneous HZs of roughly the same width, if distances are expressed on a logarithmic scale (as in Fig. 1). Thus, from this standpoint the chances of finding a habitable planet around an M star are about the same as for an F, G, or K star.

(2) Stars earlier than F0 have short-lived CHZs because they evolve very rapidly. The main sequence lifetime of an F0 star is about 2 Gyr. Early-type stars also emit large amounts of UV radiation, which could pose an additional problem for life.

(3) Stars somewhat smaller than our own Sun (a G2 star) have wider and longer-lived CHZs because they evolve more slowly. Early K stars may be even better candidates than G stars for fostering intelligent life because their planets could remain habitable for longer periods of time.

(4) Stars later than ~ K5 are poor candidates for harboring habitable planets because their HZs lie within the classical tidal locking radius; thus, otherwise Earthlike planets may develop synchronous rotation. This conclusion is not ironclad—in our own Solar System, Mercury lies within the tidal locking radius, yet is caught in a spin-orbit resonance that keeps it rotating slowly. Synchronously

rotating planets may themselves be habitable if interhemispheric heat transport is sufficient to prevent volatiles from freezing out entirely on their dark sides.

(5) Multiple star systems may also harbor habitable planets, but more work needs to be done to study orbital stability in these systems.

Bearing in mind the uncertainties mentioned above and those inherent in the climate modeling, the author is personally optimistic that habitable planets exist around many other late-F- to mid-K-type stars. Whether any of these planets are actually inhabited is, of course, a separate question. Again, I am optimistic, although here the grounds are more philosophical than scientific. While these questions may appear highly speculative at the moment, anticipated advances in interferometer power may make them testable within the next few years to decades. In particular, a large, cooled, space-based or lunar-based interferometer that could operate in the infrared could potentially image Earth-sized planets around nearby stars and perform spectroscopy on their atmospheres (TOPS 1992). Following a suggestion made originally by J. Lovelock for Mars (see, e.g., Lovelock 1979), the presence of life could be inferred from the simultaneous detection of methane and molecular oxygen in a planet's atmosphere. (These gases are far out of chemical equilibrium and are unlikely to coexist in the absence of a biological source.) Even easier to detect would be the 9.6 μm band of ozone, O_3. O_3 is a good indicator of O_2 and, hence, a nearly unambiguous indicator of life. (The one exception is a runaway greenhouse-type planet, like Venus, that may build up a transient O_2-rich atmosphere abiotically.) This is an exciting prospect with far better chances of success than the search for intelligent life using radio communication.

Further Considerations Regarding Planetary Habitability

Within two months after the Kasting et al. (1993) paper appeared, three other papers were published that raised additional questions about planetary habitability. Laskar and Robutel (1993) performed direct integration of the equations of motion for the Solar System and showed that the obliquities of three of the terrestrial planets varied chaotically over large ranges. Earth is an exception, but only because it has a large moon that accelerates its rate of spin axis precession (Laskar et al. 1993). The calculations for Mars' chaotic obliquity were confirmed independently by Touma and Wisdom (1993).

The paper that bears most directly on the question of planetary habitability is that of Laskar et al. (1993). If their numerical calculations are correct, Earth's obliquity would vary chaotically from 0° to 85° on a time scale of tens of millions of years were it not for the presence of the Moon. This calculation presumes a 24-hour spin period for the Earth, which might not be the case were the Moon absent. But their point is well taken: Earth's climatic stability is dependent to a large extent on the existence of the Moon. The Moon is now generally believed to have formed as a consequence of a glancing collision with a Mars-sized body during the latter stages of the Earth's formation (Hartmann et al. 1986). If such moon-forming collisions are rare, argued Laskar et al., habitable planets might be equally rare.

Some preliminary calculations of the effects of large obliquity variations on planetary habitability have now been performed and are discussed in this volume (Williams et al. 1996). Obliquity variations would probably have little effect on the existence or nonexistence of liquid water. The reason is that liquid water has a high heat capacity; thus, the existence of oceans on an Earthlike planet should tend to damp out seasonal temperature fluctuations. More interesting is the question of whether such planets could be inhabited by humans. This requires tighter climatic constraints (Dole 1964) and might appear to be more problematical. Williams et al. (1996) argue, however, that most Earthlike planets would still be habitable by Dole's more rigorous criteria.

Clearly, future work on the topic of planetary habitability should investigate the consequences of chaotic obliquity variations and the circumstances under which they occur. Additional work also needs to be performed on planetary formation and the question of planetary spacing. Both of these areas are becoming ripe for theoretical investigation as computational power continues to increase. Ultimately, multidimensional climate models could be used to improve on the climatic constraints on HZs as well, but such work may be premature until a real need for it arises. Once extrasolar planets are actually observed, the question of planetary habitability should move quickly into the realm of mainstream scientific inquiry.

References

Caldeira, K., and J.F. Kasting (1992), "Susceptibility of the Early Earth to Irreversible Glaciation Caused by Carbon Dioxide Clouds," *Nature* **359:** 226–228.

Dole, S.H. (1964), *Habitable Planets for Man* (New York: Blaisdell Publishing Company).

Gough, D.O. (1981), "Solar Interior Structure and Luminosity Variations," *Solar Phys.* **74:** 21–34.

Hart, M.H. (1978), "The Evolution of the Atmosphere of the Earth," *Icarus* **33:** 23–39.

Hart, M.H. (1979), "Habitable Zones Around Main Sequence Stars," *Icarus* **37:** 351–357.

Hartmann, W.K., R.J. Phillips, and G.J. Taylor (1986), *Origin of the Moon* (Houston: Lunar and Planetary Science Institute).

Huang, S.-S. (1959), "Occurrence of Life in the Universe," *Amer. Sci.* **47:** 397–402.

Huang, S.-S. (1960), "Life Outside the Solar System," *Sci. Am.* **202 (4):** 55–63.

Hunten, D.M. (1973), "The Escape of Light Gases from Planetary Atmospheres," *J. Atmos. Sci.* **30:** 1481–1494.

IBEN, I. (1967a), "Stellar Evolution Within and Off the Main Sequence," *Ann. Rev. Astron. Astrophys.* **5:** 571–626.

IBEN, I. (1967b), "Stellar Evolution. VI. Evolution from the Main Sequence to the Red Giant Branch for Stars of Mass 1 $M_\odot$, 1.25 $M_\odot$, and 1.5 $M_\odot$," *ApJ* **147:** 624–649.

KASTING, J.F. (1991), "CO_2 Condensation and the Climate of Early Mars," *Icarus* **94:** 1–13.

KASTING, J.F., AND J.B. POLLACK (1983), "Loss of Water from Venus. I. Hydrodynamic Escape of Hydrogen," *Icarus* **53:** 479–508.

KASTING, J.F., D.P. Whitmire, AND R.T. REYNOLDS (1993), "Habitable Zones Around Main Sequence Stars," *Icarus* **101:** 108–128.

LASKAR, J., AND P. ROBUTEL (1993), "The Chaotic Obliquity of the Planets," *Nature* **361:** 608–614.

LASKAR, J., F. JOUTEL, AND P. ROBUTEL (1993), "Stabilization of the Earth's Obliquity by the Moon," *Nature* **361:** 615–617.

LISSAUER, J.J. (1996), "Spacing of Planetary Orbits About Stars of Differing Masses," this volume.

LOVELOCK, J.E. (1979), *Gaia: A New Look at Life on Earth* (Oxford: Oxford University Press).

POLLACK, J.B., J.F. KASTING, S.M. RICHARDSON, AND K. POLIAKOFF (1987), "The Case for a Wet, Warm Climate on Early Mars," *Icarus* **71:** 203–224.

SAGAN, C., AND C. CHYBA (1996), "Organic Shielding of UV-Labile Greenhouse Gases on Early Earth and Mars," abstract, this volume.

SCHABER, G.G., R.G. STROM, H.J. MOORE, L.A. SODERBLOM, R.L. KIRK, D.J. CHADWICK, D.D. DAWSON, L.R. GADDIS, J.M. BOYCE, AND J. RUSSELL (1992), "Geology and Distribution of Impact Craters on Venus: What Are They Telling Us?," *JGR* **97:** 13257–13301.

SHKLOVSKII, I.S., AND C. SAGAN (1966), *Intelligent Life in the Universe* (San Francisco: Holden-Day, Inc.).

SOLOMON, S.C., AND J.W. HEAD (1991), "Fundamental Issues in the Geology and Geophysics of Venus," *Science* **252:** 252–260.

TOPS (1992), *Toward Other Planetary Systems*, NASA.

TOUMA, J., AND J. WISDOM (1993), "The Chaotic Obliquity of Mars," *Science* **259:** 1294–1297.

WALKER, J.C.G., P.B. HAYS, AND J.F. KASTING (1981), "A Negative Feedback Mechanism for the Long-Term Stabilization of Earth's Surface Temperature," *JGR* **86:** 9776–9782.

WETHERILL, G.W. (1994), "The Plurality of Habitable Worlds," in *Fifth Exobiology Symposium and Mars Workshop*, NASA Ames Research Center, April.

WHITMIRE, D.P., R.T. REYNOLDS, AND J.F. KASTING (1991), "Habitable Zones for Earth-like Planets Around Main Sequence Stars," in *Bioastronomy: The Search for Extraterrestrial Life*, J. Heidmann and M.J. Klein, eds. (Berlin: Springer-Verlag), pp. 173–178.

WILLIAMS, D.M., J.F. KASTING, AND K. CALDEIRA (1996), "Chaotic Obliquity Variations and Planetary Habitability," this volume.

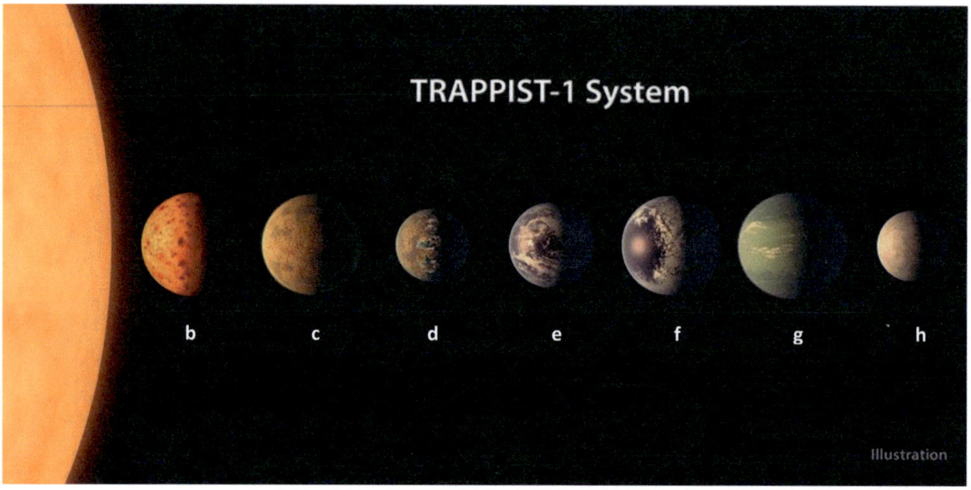

TRAPPIST-1e	TRAPPIST-1f	TRAPPIST-1g
Mass: 0.692 Earth	Mass: 1.039 Earths	Mass: 1.321 Earths
Orbit: 6.1 days	Orbit: 9.2 days	Orbit: 12.4 days

Image Credit: NASA/JPL-Caltech
https://science.nasa.gov/exoplanets/trappist1/#hds-sidebar-nav-4 [retrieved 4-14-26]

TRAPPIST-1 is a red dwarf star. The TRAPPIST-1 system contains a total of seven planets, all around the size of Earth. Three of them–TRAPPIST-1e, f, and g–are in their star's habitable zone, where temperatures are just right for liquid water to pool on an Earth-like planet's surface.

While TRAPPIST-1b, c, and d are too close to be in the system's likely habitable zone, and TRAPPIST-1h is too far away, the planets' discoverers say that more optimistic scenarios could allow any or all of the planets to have liquid water. The strikingly small orbits of these worlds make it likely that most, if not all, of them perpetually show the same face to their star. If none of these planets had an atmosphere, this could result in an extreme range of temperatures from the day to the night sides, allowing for situations not factored into the traditional definition of a habitable zone.

The system was revealed through observations from NASA's Spitzer Space Telescope and the ground-based TRAPPIST (TRAnsiting Planets and Planetesimals Small Telescope) telescope, as well as other ground-based observatories.

Source: NASA Science

The following article by Haberle et al. was the first detailed atmospheric modeling of synchronously rotating planets.

Can Synchronously Rotating Planets Support an Atmosphere?

Robert M. Haberle

Space Science Division
NASA Ames Research Center
Moffett Field, CA 94035

Christopher P. McKay

NASA Ames Research Center

Daniel Tyler

San José State University
NASA Ames Research Center

Ray T. Reynolds

NASA Ames Research Center

Approximately 75% of the main sequence stellar population consists of M stars (Allen 1976). These stars are the smallest and least luminous of main sequence stars so if they have planetary systems, their habitable zone must be in close. However, if these zones are too close, tidal torques acting on any planets within them could ultimately force them to synchronously rotate (Dole 1964, Kasting et al. 1993). Thus, these planets would have a side that is always in sunlight, and a side that is always in darkness. It would appear then, that synchronously rotating planets could not sustain life since heat loss on the dark side would eventually freeze out the atmosphere. In this case, M stars would not be suitable candidates in the search for extraterrestrial life.

However, when atmospheres are differentially heated they generate a circulation that transports heat from regions of net gain to regions of net loss. Thus, a synchronously rotating planet would generate a circulation that would transport heat from the light side to the dark side. If the transport is sufficiently vigorous, temperatures on the dark side could be kept high enough to prevent the atmosphere from freezing out. The amount of transport required depends on the mass and composition of the atmosphere, and the kind of circulation system it develops. Mass and composition are important because they determine the heat capacity of the atmosphere and the temperature at which freezing begins. The circulation is important because it determines how rapidly heat is carried from the light side to the dark side. Thus, a synchronously rotating planet with enough atmospheric mass and a vigorous enough circulation system should be able to maintain its atmosphere against collapse. The question then is, How much of an atmosphere is required?

The principles just outlined can be illustrated with a simple model (Fig. 1). On the light side, sunlight is absorbed at the surface and reradiated in the infrared. The atmosphere absorbs some of this infrared radiation and then emits it both to space and back to the surface (the greenhouse effect). The situation is the same on the dark side except for the lack of sunlight. These two regions are coupled by heat transport which depends on a parameter, $A = p_s c_p / gt$, where p_s is the surface pressure, c_p is the specific heat capacity, g is gravity, and t is a transport time scale. It is this parameter that embodies the mass (p_s/g), composition (c_p), and heat transport (t) properties mentioned in the preceding paragraph.

Solutions to this simple model for an absorbed solar flux comparable to that for Venus are shown in Figure 2. For low values of A (low mass, specific heat, and/or large transport time scales) the dark side is nearly decoupled from the light side and atmospheric temperatures are very low—low enough to freeze out an atmosphere of almost any composition. For large values of A (high mass, specific heat, and/or small transport time scales) the two regions are strongly coupled and freeze-out is less likely. In the limit as A exceeds 100, the temperature difference between the light side and dark side vanishes and the planet becomes isothermal—in spite of the fact that one side is perpetually in darkness. This is, of course, the situation for Venus today (though in the strict sense, Venus is not a synchronous rotator). If we take $A = 10^3$, $p_s = 100$ bars, $c_p = 10^3$ J/kg/K, and $g = 10$ m/s^2, then our simple model suggests that the Venusian atmosphere has a transport time scale of roughly ten days, a value that is consistent with its famous cloud-top four-day rotation period.

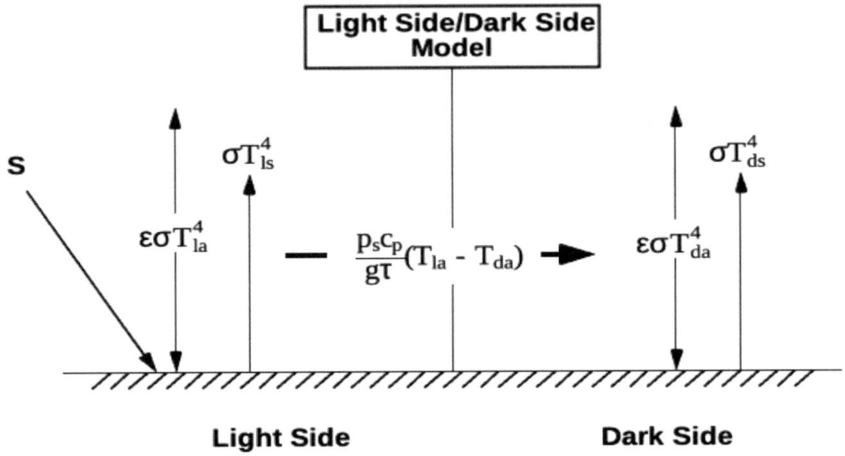

FIG. 1. *A simple energy balance model for a synchronously rotating planet. Writing the balance for the surface and atmosphere on both the light side and dark side yields four equations in four unknowns:* T_{la}, T_{ls}, T_{da}, *and* T_{ds}, *which represent the atmosphere and surface temperatures for the light side and dark side, respectively.*

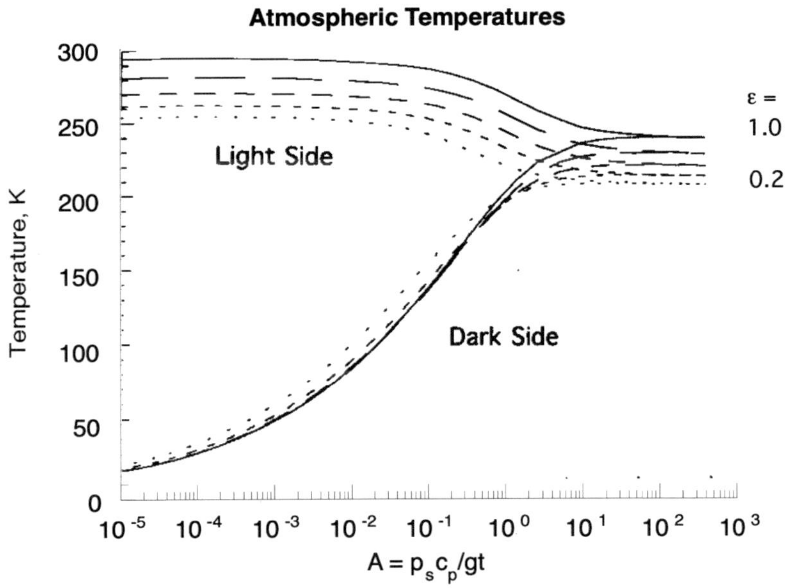

FIG. 2. *Atmospheric temperatures for a planet absorbing 425 W* m^2 *from its sun. Upper curves are for the light side, lower curves are for the dark side. The different curves correspond to different emissivities ranging from* $\varepsilon = 0.2$ *(dotted line) to 1 (solid line).*

However, to actually determine the surface pressure required to prevent the atmosphere from freezing out, a more specific model is required. The simple model discussed above illustrates the potential for maintaining an atmosphere, but it cannot distinguish between the effects of heat capacity (mass and composition) and transport since these have been lumped together into a single parameter (A). To estimate the required surface pressure, we therefore use the model of Haberle et al. (1994), which was designed to study the evolution of the Martian atmosphere.

Their model, which has a similar energy balance to the one above, is specific to a CO_2-H_2O atmosphere in which heat transport is accomplished by a class of motions known as baroclinic waves. Such waves have been detected in the Martian atmosphere (Barnes 1980), and are responsible for the midlatitude storm systems familiar to us here on Earth. To make their model appropriate for our purpose we have set the obliquity to zero and have increased the fractional area of the polar region to 50% of the total area. With these changes, the poles in their model become analogous to the dark side of a synchronously rotating planet—that is, they receive no insolation and occupy half the planet's surface.

The required surface pressure for a range of available insolations is shown in Figure 3. For surface pressures less than the required value, polar caps form and the atmosphere freezes out. Note that the required value increases with increasing distance from the sun (decreasing insolation). However, the increase is modest at first and rapid later.

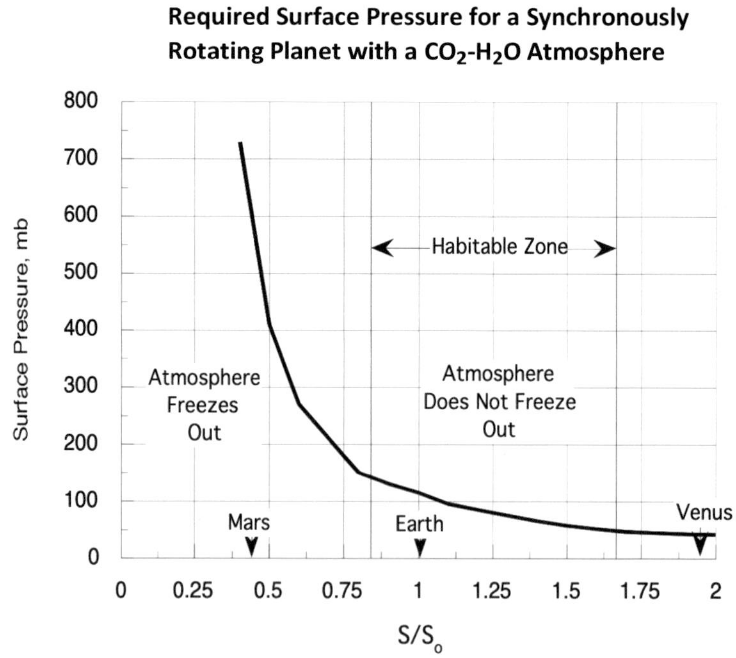

FIG. 3. *Surface pressure required to prevent the atmosphere from freezing out on a synchronously rotating planet as a function of the available insolation expressed as a fraction of the contemporary solar constant. The indicated Habitable Zone is based on the most optimistic estimates of Kasting et al. (1993) for their Standard Model.*

At distances roughly equivalent to where Mars is located today the upturn is very steep. This suggests that there is a well-defined maximum distance at which a synchronously rotating planet can still support an atmosphere. Beyond that distance, the required amount of CO_2 becomes implausibly large. According to Figure 3, that distance cannot yield less than ~ 30% of the insolation available at Earth today. For M-like stars, this insolation would correspond to a distance somewhere between 0.04 and 0.46 AU, depending on its spectral type. Note also from Figure 3 that once inside that limit, the amount of CO_2 required to prevent freeze-out is not that large. A synchronously rotating planet, receiving about the same insolation as Earth does today, would require only about 100 mbars of CO_2. This is well within estimates of the CO_2 inventory for terrestrial-like planets.

There are, of course, many uncertainties in these calculations, the most notable being the treatment of atmospheric heat transport. Clearly, more sophisticated models will be needed to better address this issue. However, these preliminary results are encouraging and suggest that M stars should not be ruled out when searching for extraterrestrial life.

References

ALLEN, C.W. (1976), *Astrophysical Quantities*, 3d ed. (London: The Athlone Press), p. 249 (Table 1).

BARNES, J.R. (1980), "Time Spectral Analysis of Midlatitude Disturbances in the Martian Atmosphere," *J. Atmos. Sci.* **37:** 2002–2015.

DOLE, S.H. (1964), *Habitable Planets for Man* (New York: Blaisdell Publishing Company).

HABERLE, R.M., D. TYLER, C.P. MCKAY, AND W.L. DAVIS (1994), "A Model for the Evolution of CO_2 on Mars," *Icarus* **109:** 102–120.

KASTING, J.F., D.P. WHITMIRE, AND R.T. REYNOLDS (1993), "Habitable Zones Around Main Sequence Stars," *Icarus* **101:** 108–128.

Radar Observations of Venus

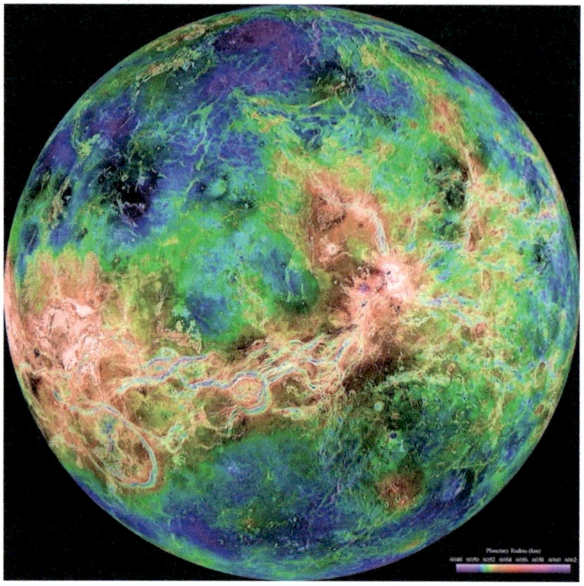

Image Credit: NASA/JPL-Caltech/USGS
https://science.nasa.gov/image-detail/amf-ed32cd82-638b-4693-867b-d0574d769e54/ [retrieved 7-23-25]

This hemispheric view of Venus was created using radar observations, including images from NASA's Magellan spacecraft. Magellan imaged more than 98% of Venus. Gaps in the Magellan coverage were filled with images from the Earth-based Arecibo radar. The composite image was processed to improve contrast and to emphasize small features, and was color-coded to represent elevation. Magellan launched on May 4, 1989, and was deployed from the cargo bay of Space Shuttle Atlantis on May 5, 1989.

Our 'sister' planet Venus could serve as a model for many exoplanets. Venus's climate may have at one time been far more Earthlike than its current climate, which is inhospitable and more like that inside a pressure cooker, with surface temperatures of 450°C and pressures equivalent to 1000m under the sea. How did Venus evolve from a past 'habitable' state to its present one, and does that provide indications of our own possible future?

Venus

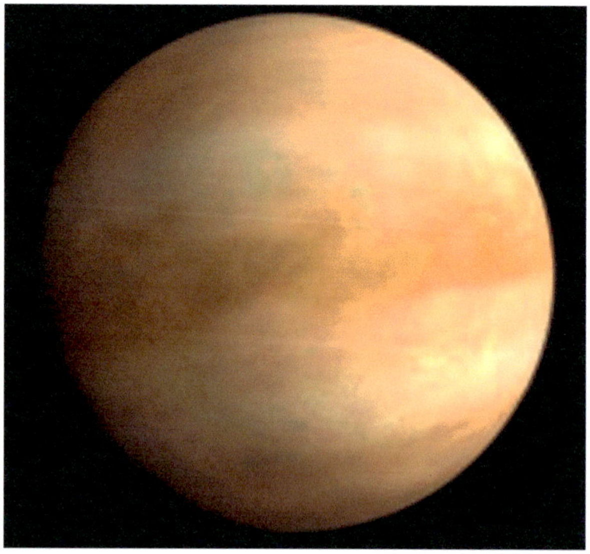

Photo Credit: NASA's Goddard Space Flight Center, Scientific Visualization Studio
https://svs.gsfc.nasa.gov/13640/ [retrieved 1-12-26]

VENUS: A FAILED EARTH?

Jon M. Jenkins

SETI Institute

NASA Ames Research Center

Moffett Field, CA 94035

Abstract

While Earth and its "twin," Venus, are of comparable size and mass, and were each formed by the same process of planetary accretion, Earth supports intelligent life, while Venus is not capable of sustaining life in any known form. The closer position of Venus to the Sun constrains water to appear only in its gas phase in the atmosphere of Venus. How much water was present in the early atmosphere of Venus? Is the Venus water vapor abundance in steady state, or is it a signature of an early ocean? The explanations for these phenomena, while elusive, have implications for the evolution of atmospheres on terrestrial planets, and their capacity to support life. The enhancement of deuterium to hydrogen in the Venus atmosphere compared with Earth's atmosphere has been interpreted by some as evidence of a lost primordial ocean. New measurements and theoretical advances have not settled the issue but have succeeded only in muddying the waters.

Once romanticized as Earth's twin, Venus has revealed itself to be a barren, arid desert. Why is Venus so different from Earth? One key question is whether Venus once had a substantial abundance of water, and if so, how it lost this primordial ocean. If Venus did not originally possess an abundance of water similar to that of Earth, there are serious implications for the size and nature of liquid water–defined habitable zones. This topic has been hotly debated within the planetary science community for decades. This paper seeks to review the competing theories and weigh them against recent theoretical advances and new observations.

A brief comparison of Earth and Venus is shown in Table 1. Venus and Earth have nearly the same mass and size and, hence, density, and similar inventories of carbon and nitrogen.

TABLE 1. ***Comparison of Venus and Earth***

	Earth	*Venus*
*Mean Distance from Sun (AU)	1	0.72
*Mass (kg)	5.976(24)	4.871(24)
*Equatorial Radius (km)	6370	6052
†N (Earth = 1)	1	1.02–1.7
†C (Earth = 1)	1	0.6–1.8
$T_{surface}$ (K)	300	740
†CO_2 (Atmospheric)	330 ppm	96%
H_2O (Earth = 1)	1	0.00042%
†N_2 (Atmospheric)	78%	4%
†^{36}Ar (Earth = 1)	1	72 ± 10
†^{40}Ar (Earth = 1)	1	0.25 ± 0.04
†D/H (Earth = 1)	1	120

* *Adapted from Colin 1983.*

† *Adapted from Donahue and Pollack 1983.*

The bulk of Earth's carbon inventory is locked within carbonate rocks ($CaCO_3$), while most of Venus' carbon resides in the atmosphere in the form of carbon dioxide (CO_2). This massive, dense CO_2 atmosphere is responsible for the current greenhouse on Venus that produces a mean surface temperature of 740 K (compared with Earth's hospitable 330 K mean surface temperature). Earth's atmosphere is relatively depleted in nonradiogenic argon, ^{36}Ar, with respect to Venus, but is highly enriched in radiogenic argon, ^{40}Ar (a decay product of radioactive potassium, ^{40}K). This is due to inefficient outgassing from the interior of Venus over the span of geologic history compared with the Earth. Of greatest interest to liquid water–defined habitable zones is the minute quantity of water in the Venus atmosphere, compared with the great abundance of water on Earth. The highly enriched Venus deuterium-to-hydrogen (D/H) ratio has been attributed to fractionation following the loss of a primordial ocean similar in size to the present Earth ocean (Donahue et al. 1982, Donahue and Pollack 1983, Donahue and Hartle 1992). By what process did Earth and Venus acquire their original water reservoirs, and how has Venus lost its inventory?

The terrestrial planets were formed by the process of planetary accretion (Ahrens et al. 1989). According to this theory, grains in the solar nebula merged and developed by unclear processes into ~ 10 km diameter planetesimals. Collisions and mergers, enhanced by gravitational focusing, allowed the largest to grow rapidly, sweeping up the smaller bodies in their feeding zones. The great amounts of energy released during accretion caused the loss of most of the volatiles in the developing planets. Thus, the terrestrial planets acquired their inventories of volatiles after the initial formation of the actual planets. Several theories attempt to explain the origin of the atmospheres of the terrestrial planets (Donahue and Pollack 1983, Ahrens et al. 1989). Some theories maintain that Venus was a "wet" planet at birth and lost its water early on; others posit instead that Venus has always been a dry planet. None of these theories, however, is able to accommodate all of the data.

On the basis of calculations of the equilibrium condensation state of material formed in different regions of the solar nebula, Lewis concluded that condensates formed in the present day orbits of Venus and Mercury could not contain water-bearing minerals, while water-bearing minerals could form at and beyond approximately one AU (Lewis 1972, 1973; Barshay and Lewis 1975, 1976, 1978; Fegley and Lewis 1979, 1980). Thus, Lewis and his colleagues argued that Venus has always been dry.

The grain-accretion models of Wetherill (1985), however, demonstrated that all terrestrial planets shared essentially the same feeding ground during accretion. This makes it unlikely that Venus did not receive an abundance of water comparable to the Earth's, unless some exotic mechanism prevented water-bearing materials from penetrating to 0.7 AU. Many researchers believe that Venus, Earth and Mars formed with similar inventories of H_2O (Ingersoll 1969, Walker et al. 1970, Rasool and de Bergh 1970, Smith and Gross 1972, Walker 1975). In addition, Walker et al. (1970) proposed that the formation of Venus' CO_2 atmosphere from the oxidation of hydrocarbons would produce an amount of water equal to one-sixth of the present terrestrial ocean.

Since it is likely that Venus did possess a large quantity of water at one time in the distant past, how did it disappear? Ingersoll (1969) proposed a runaway greenhouse process in which the stratospheric cold-trap was washed out when the fraction of H_2O in the atmosphere exceeded 0.2. This process would have allowed the photodissociation of water in the stratosphere followed by rapid hydrodynamic escape of H (Watson et al. 1981, Kasting and Pollack 1983). Some of the resulting oxygen could have been dragged along and could have escaped with the hydrogen, or it could have reacted with the crust (Walker 1975). These models, however, neglect the effects of convection and fail to account for the loss of the 20 bar of water remaining after this process shuts down due to reestablishment of the stratospheric cold-trap. The moist greenhouse process proposed by Kasting et al. (1984) and further developed in Kasting (1988) is more successful at eliminating the last 20 bar of H_2O. This model is based on an initial atmosphere of one bar N_2 and

O_2 near 100° in which the CO_2 (g) is initially suppressed in carbonate rocks. In this model, liquid water is present in large quantities on the Venus surface when hydrodynamic escape initiates. Once the hydrodynamic escape tapers off, it is possible for hydrogen to escape faster than deuterium, which would result in an enhanced D/H ratio.

Estimates of the size of the original Venus H_2O reservoir based on the observed D/H ratio are dependent on the fractionation factor, *f*, and on assumptions regarding the source of water (*f* is a measure of the efficiency of the fractional deuterium escape rate relative to the fractional hydrogen escape rate). Recently, Gurwell and Yung (1993) revised the efficiency of collision with hot oxygen atoms for deuterium escape upward by a factor of ten. This changes the picture drastically. Donahue and Hartle (1992) postulate two hypotheses to explain the current D/H ratio in light of these new escape efficiencies: (1) assuming no exogenous sources of water, at least 4.2–14 m of early low-D/H water existed on Venus that subsequently experienced Rayleigh fractionation (if the escape flux was sufficiently large in earlier times); (2) water vapor is near or approaching steady state with the escape flux balanced by endogenous or exogenous sources with a high D/H ratio. Further, Donahue and Hartle concluded that this second hypothesis precluded comets and outgassing from the mantle as possible sources, since cometary D/H is too low.

The surface of Venus provides additional information for us to consider. The Magellan spacecraft used radar to map over 99% of the planet's surface. Schaber et al. (1992) examined the cratering record revealed by these images to obtain the following results: (1) the distribution of craters is highly uniform; (2) the size-density distribution implies a young surface age of 500 Myr; (3) 62% of the craters are pristine, 4% are embayed or partially embayed by lava, and 34% are affected by tectonism; and (4) only 10% of the surface has been resurfaced since the last cratering event. Monte Carlo simulations support the 500 Myr surface age (Bullock et al. 1992), which has led some researchers to argue for cyclic plate tectonism on Venus (Parmentier and Hess 1992, Turcotte 1992). This style of tectonics would provide opportunities for water in the magma to escape to the atmosphere at periodic intervals.

In light of the revised escape flux for deuterium, the emerging consensus on the abundance of water vapor (30 ppm), and the evidence for a geologically young surface from Magellan images, Grinspoon (1993) proposed two explanations: (1) water vapor is near steady state, with the loss balanced by volcanic outgassing from a highly fractionated mantle (caused by massive early hydrogen escape), or (2) water vapor is in or near steady state, with loss balanced by Rayleigh fractionation after catastrophic resurfacing in the past 0.5–1 Gyr.

It seems likely that Venus evolved with a substantial abundance of water, much like the Earth, and that it lost this ocean through the moist greenhouse process early in its history. Recent findings revealed by Magellan, coupled with new estimates for the escape of deuterium, make it unlikely that the D/H ratio is the signature of a primordial ocean. As Grinspoon (1993) states, the signature of the lost ocean may be absent in the atmosphere, but may well be present in the mantle.

REFERENCES

AHRENS, T.J., J.D. O'KEEFE, AND M.A. LANGE (1989), "Formation of Atmospheres During Accretion of the Terrestrial Planets," in *Origin and Evolution of Planetary and Satellite Atmospheres*, S.K. Atreya, J.B. Pollack, and M.S. Mathews, eds. (Tucson: University of Arizona Press), pp. 328–385.

BARSHAY, S.S., AND J.S. LEWIS (1975), "Chemistry of Solar Material," in *The Dusty Universe*, G.B. Field and A.G.W. Cameron, eds. (New York: Neale Watson Academic Publ.), pp. 33–40.

BARSHAY, S.S., AND J.S. LEWIS (1976), "Chemistry of Primitive Solar Material," *Ann. Rev. Astron. Astrophys.* **14:** 81–90.

BARSHAY, S.S., AND J.S. LEWIS (1978), "Chemical Structure of the Deep Atmosphere of Jupiter," *Icarus* **33:** 593–611.

BULLOCK, M.A., D.H. GRINSPOON, AND J.W. HEAD (1992), "Monte Carlo Modeling of the Resurfacing of Venus," paper presented at the International Colloquium on Venus, August 10–12, Pasadena, California.

DONAHUE, T.M., AND J.B. POLLACK (1983), "Origin and Evolution of the Atmosphere of Venus," in *Venus*, D.M. Hunten, L. Colin, T.M. Donahue, and V.I. Moroz, eds. (Tucson: University of Arizona Press), pp. 1003–1036.

DONAHUE, T.M., AND R.E. HARTLE (1992), "Solar Cycle Variations in H^+ and D^+ Densities in the Venus Ionosphere: Implications for Escape," *Geophys. Res. Lett.* **19:** 2449–2452.

DONAHUE, T.M., J.H. HOFFMAN, R.R. HODGES, AND A.J. WATSON (1982), "Venus Was Wet: A Measurement of the Ratio of Deuterium to Hydrogen," *Science* **216:** 630–633.

FEGLEY, M.B., JR., AND J.S. LEWIS (1979), "Thermodynamics of Selected Trace Elements in the Jovian Atmosphere," *Icarus* **38:** 166–179.

FEGLEY, M.B., JR., AND J.S. LEWIS (1980), "Volatile Element Chemistry in the Solar Nebula: Na, K, F, Cl, Br, and P," *Icarus* **41:** 439–455.

GRINSPOON, D.H. (1993), "Implications of the High D/H Ratio for the Sources of Water in Venus' Atmosphere," *Nature* **363:** 428–431.

GURWELL, M.A., AND Y.L. YUNG (1993), "Fractionation of Hydrogen and Deuterium on Venus Due to Collisional Ejection," *Planet. Space Sci.* **41:** 91–104.

HUNTEN, D.M., T.M. DONAHUE, J.C.G. WALKER, AND J.F. KASTING (1989), "Escape of Atmospheres and Loss of Water," in *Origin and Evolution of Planetary and Satellite Atmospheres*, S.K. Atreya, J.B. Pollack, and M.S. Mathews, eds. (Tucson: University of Arizona Press), pp. 386–422.

INGERSOLL, A.P. (1969), "The Runaway Greenhouse on Venus: A History of Water on Venus," *J. Atmos. Sci.* **26:** 1191–1198.

KASTING, J.F. (1988), "Runaway and Moist Greenhouse Atmospheres and the Evolution of Earth and Venus," *Icarus* **74:** 472–494.

KASTING, J.F., AND J.B. POLLACK (1983), "Loss of Water from Venus. 1. Hydrodynamic Escape of Hydrogen," *Icarus* **53:** 479–508.

KASTING, J.F., J.B. POLLACK, AND T.P. ACKERMAN (1984), "Response of Earth's Surface Temperature to Increases in Solar Flux and Implications for Loss of Water from Venus," *Icarus* **57:** 335–355.

LEWIS, J.S. (1972), "Low-Temperature Condensation from the Solar Nebula," *Icarus* **16:** 241–252.

LEWIS, J.S. (1973), "Origin and Composition of the Terrestrial Planets and Satellites of the Outer Planets," in *The Origin of the Solar System*, H. Reeves, ed. (Paris: CNRS), pp. 202–205.

PARMENTIER, E.M., AND P.C. HESS (1992), "Chemical Differentiation of a Convecting Planetary Interior: Consequences for a One-Plate Planet Such as Venus," paper presented at the International Colloquium on Venus, August 10–12, Pasadena, California.

RASOOL, S.I., AND C. DE BERGH (1970), "The Runaway Greenhouse and Accumulation of CO_2 in the Venus Atmosphere," *Nature* **226:** 1037–1039.

SCHABER, G.G., R.G. STROM, H.J. MOORE, L.A. SODERBLOM, R.L. KIRK, D.J. CHADWICK, D.D. DAWSON, L.R. GADDIS, J.M. BOYCE, AND J. RUSSEL (1992), "Geology and Distribution of Impact Craters on Venus: What Are They Telling Us?," *JGR* **97:** 13257–13301.

SMITH, L.L., AND S.H. GROSS (1972), "The Evolution of Water Vapor in the Atmosphere of Venus," *J. Atmos. Sci.* **29:** 173–178.

TURCOTTE, D. (1992), "Episodic Plate Tectonics on Venus," paper presented at the International Colloquium on Venus, August 10–12, Pasadena, California.

WALKER, J.C.G. (1975), "Evolution of the Atmosphere of Venus," *J. Atmos. Sci.* **32:** 1248–1256.

WALKER, J.C.G., K.K. TUREKIAN, AND D.M. HUNTEN (1970), "An Estimate of the Present-Day Deep-Mantle Degassing Rate from Data on the Atmosphere of Venus," *JGR* **75:** 3558–3561.

WATSON, A.J., T.M. DONAHUE, AND J.C.G. WALKER (1981), "The Dynamics of a Rapidly Escaping Atmosphere: Applications to the Evolution of Earth and Venus," *Icarus* **48:** 150–166.

WETHERILL, G.W. (1985), "Occurrence of Giant Impacts During the Growth of the Terrestrial Planets," *Science* **228:** 877–879.

CHAOTIC OBLIQUITY VARIATIONS AND PLANETARY HABITABILITY

Darren M. Williams

Department of Astronomy & Astrophysics
Pennsylvania State University
University Park, PA 16802

James F. Kasting

Department of Geosciences
Pennsylvania State University

Ken Caldeira

Global Climate Research Division
Lawrence Livermore National Laboratory
Livermore, CA 94550

Abstract

A one-dimensional energy balance climate model is used to place habitability constraints on Earthlike worlds with large obliquities. We consider planets with Earthlike land/sea ratios for three geographic cases: (a) Earthlike continents, (b) a supercontinent at the equator, and (c) a supercontinent at one pole. Case (a) and (b) planets with predominantly low-latitude continents are found to be habitable on most land areas at obliquities less than 90°. The reason is that seasonal temperature fluctuations, which would otherwise be large at high obliquities, are damped by the large heat capacity of the oceans in the upper latitudes. Planets with polar continents are only marginally habitable at low obliquities and become almost completely uninhabitable for obliquities greater than 55°.

INTRODUCTION

Earth's obliquity is thought to vary regularly between 22.1° and 24.5° with a dominant period of 41,000 years (Berger 1988). Obliquity variations affect climate by reducing average solar elevation and hence, insolation for some latitudes while increasing elevation and insolation for others. Times of low obliquity are times of reduced insolation for higher latitudes, which can lead to glaciations in these areas. Geologic evidence suggests that obliquity variations were the dominant cause of glacial/interglacial oscillation during the Pleistocene (Ruddiman et al. 1989). (More recently, the largest fluctuations have been on the 100,000-year time scale, which corresponds to an eccentricity cycle.) Larger obliquity variations could augment the severity of climate extremes, possibly to the extent of affecting planetary habitability.

Laskar et al. (1993) have shown that large, chaotic obliquity variations should occur for three of the four terrestrial planets in times less than 10 million years. The varying obliquity of Mars has long been studied by Ward (1973, 1974, 1979, 1992), who has shown that Mars' obliquity (currently 25°) could reach 45° during times when its spin precession is in resonance with its orbit precession. Touma and Wisdom (1993) and Laskar et al. (1993) later extended the range of possible Martian obliquities to 50° and 60°, respectively. Laskar et al. went on to demonstrate that substantially larger obliquity variations (approaching 90°) are possible for Mercury and Venus. Mercury and Venus are thought to have fixed obliquities today because their rotation rates are in resonance with their sidereal periods, but their obliquities should have varied chaotically before this state was reached. Earth's obliquity does not follow this pattern: it varies regularly and by only small amounts. The reason, as demonstrated in a follow-up article by Laskar and Robutel (1993), is that it is stabilized by the Moon. If the Moon were not present, and if Earth's spin period remained unchanged, Earth's obliquity is predicted to vary chaotically between 0° and 85°.

This poses an interesting question concerning the frequency of habitable planets around other stars. (We shall assume here that extrasolar planets exist, although this has not yet been demonstrated for main sequence stars.) Earth's Moon is thought to have formed as the result of a glancing collision with a Mars-sized body during the process of planetary accretion (Hartmann et al. 1986). Such moon-forming impacts may be statistically unlikely; of the four inner planets in our Solar System, only Earth has a large moon. Otherwise Earthlike planets orbiting other stars may be subject to large, chaotic obliquity variations if the celestial mechanics of such planetary systems resembles our own. This thought led Laskar and Robutel (1993) to suggest that habitable planets might be statistically rare.

Here, we investigate this question quantitatively using a simple energy balance climate model that is described in the next section. The calculations are preliminary: we do not claim to have examined all possibilities. For simplicity, we have assumed that our hypothetical planets have Earthlike atmospheres containing 300 ppm of CO_2 and that they have Earthlike land/sea ratios. (About 30% of the Earth is covered by land.) We have, however, taken plate tectonics into account by allowing the continents to rearrange themselves into different positions. As we shall demonstrate later on, the location of the continents is a prime determinant of whether a planet's climate will be stable at high planetary obliquities.

To study planetary habitability, we need a definition of what is meant by the term *habitable*. Often, this term is used simply to mean that a planet is capable of supporting liquid water on its surface (e.g., Hart 1978, Kasting et al. 1993). This is not a very interesting definition for our problem: the presence of liquid water is not likely to be greatly affected by a planet's obliquity. What Laskar and Robutel had in mind, we are sure, was whether other planets might exist that would be habitable by humans.

Rather than dithering about exactly what climatic conditions are required for human habitability, we have chosen simply to adopt the criteria suggested by Dole (1964): A habitable planet is one over which 10% of the surface exhibits mean annual temperatures between 0°C and 30°C, with temperature extremes not exceeding −10°C and 40°C. We realize that large, populated areas of the Earth (our home town of State College, Pennsylvania, for example) would not be considered habitable by Dole's criteria. As will become clear later on, however, it is differences in climate, rather than absolute limits on climate, that are most relevant to estimating whether a particular planet might be habitable. We can use Dole's criteria to help frame our question succinctly, as long as we remember that our results will retain an element of ambiguity.

THE ENERGY BALANCE CLIMATE MODEL

We used a one-dimensional energy balance climate model to investigate the effects of large obliquity on otherwise habitable planets. Apart from geographic redistribution of continental surfaces, all of Earth's parameters were maintained. Those familiar with energy balance climate modeling may jump forward to the Results section at this point without losing an understanding of what has been done.

Earth is divided into eighteen latitudinal belts, each 10° wide, centered at latitudes of 5°, 15°, 25°, . . . and so on. Latitude is denoted by ϕ and parameterized by $x = \sin \phi$. Energy flow into and out of each latitude belt was balanced using the time-dependent energy balance equation given by

$$C(x,T)\frac{\delta T(x,t)}{\delta t} - \frac{\delta}{\delta x}D(1-x^2)\frac{\delta T(x,t)}{\delta x} + I(x) = S(x,t)(1-A[x,t]) \qquad (1)$$

where S is the bolometric solar energy flux incident upon Earth's upper atmosphere and A is known as the top-of-atmosphere albedo. The product $S(1 - A)$ is a measure of the energy absorbed in a given latitude belt. Incident solar energy is balanced by infrared energy, I, lost to space, poleward diffusion of heat by atmospheric and oceanic currents, and energy stored by each latitude belt. Heat diffusion was parameterized using an adjustable diffusion coefficient D, whose effects will be discussed later. Similarly, heat storage is governed by a latitudinally averaged heat capacity C, which is proportional to the amount of water contained in each belt. Individual terms in the energy balance equation were treated as follows.

Insolation

Insolation depends on both the solar energy flux at Earth and the angle at which sunlight strikes the surface. $S(x,t)$ is equal to the product $Q\mu$ where Q is the solar constant at Earth, taken to be 1360 W m^{-2}, and μ is cos θ where θ is the angle between local zenith and the Sun. Local zenith angle depends on latitude ϕ, and solar declination δ, which in turn depends on obliquity δ_o. The relationship between solar declination and obliquity is

$$\sin \delta(\mathrm{t}) = -\sin \delta_o \cos \omega t \tag{2}$$

where ω is the orbital velocity of the Earth. We ignored the diurnal insolation cycle, and at the same time, resolved seasonal temperature cycles by choosing modeling time steps of one day. Average daily insolation was found by multiplying the solar constant Q by a diurnally averaged solar zenith angle to obtain

$$S(x,t) = \frac{Q}{\pi}\left(\sin\phi \sin\delta + \cos\phi \cos\delta \frac{\sin H}{H}\right) \tag{3}$$

where H is the radian half-day length at latitude ϕ given by

$$\cos H = -\tan\phi \tan\delta \qquad \text{for } 0 < H < \pi \tag{4}$$

H becomes equal to π for a twelve-hour half-day, allowing for times and locations when the Sun never sets. This occurs poleward of 66.5° on Earth for present obliquity; however, the long days do not appreciably warm the climate in these polar areas because the solar zenith angle is never less than 66.5°. Planets with larger obliquities would have lower zenith angles at high latitudes, raising insolation in these regions. Ward (1974) recognized this after calculating and plotting annual average insolation versus latitude for obliquities between 0° and 90°. We include Ward's plot as Figure 1 and observe that for obliquities greater than about 54°, annual average insolation is greatest for the high latitudes. This is easily understood by considering a planet like Uranus with an obliquity near 90°. The poles of Uranus spend half the Uranian year in sunlight and the other half in shadow, exactly as would a point on the equator of a planet at 0° obliquity! Both regions receive an annual average insolation of ~ $0.32Q$ (Q/π), which is 27% higher than the planetary average of $0.25Q$.

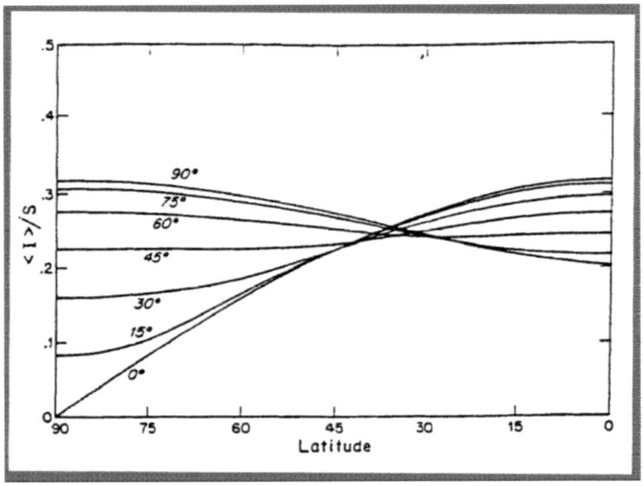

FIG. 1. *Average annual insolation as a function of latitude for different obliquities (from Ward 1974).*

Albedo

Top-of-atmosphere albedo depends appreciably on surface albedo, which, for watery surfaces, increases with increasing zenith angles. We adopt the albedo model of Thompson and Barron (1981), which includes the necessary zenith angle dependence. We assume that sea ice has a surface albedo of 0.55 and begins to form at 273.15 K. The ice/ocean fraction, f_i, was presumed to depend on temperature, T, in the following way: $f_i = 0.1(273.15 - T)$, with the ocean becoming 100% frozen at 263.15 K. The land albedo parameterizations (Eqns. A9 and A10 in Thompson and Barron 1981) for direct and diffuse radiation were adjusted by replacing the term "$-0.017(T-283.15)$" with "$-0.07(T-273.15)$." This causes land albedo to increase for temperatures less than 273.15 K, as would be expected if the land was covered with snow. The variation of surface albedo with temperature, $\delta a_s/\delta T$, was decreased from −0.017 to −0.07 to yield surface albedos similar to those of Thompson and Barron (1981), but with our new frozen surface assumption. Base land albedos for unfrozen surfaces are taken from Thompson and Barron (1981;

Table 1). Base land albedo is assigned a value of 0.16 for all land areas on planets with geography different from Earth's.

Clouds scatter solar radiation back to space, thereby raising the planctary albedo. Without a more rigorous model that can parameterize the formation, structure, and scattering properties of clouds, we can only make empirical assumptions about cloud fraction and cloud albedo. Cloud fraction tends to increase with latitude and with decreasing temperature. Hence, we assume a cloud fraction of 0.5 for unfrozen latitudes and 0.65 for latitudes with temperatures less than 273.15 K. We follow Thompson and Barron (1981), who assumed a cloud albedo of 0.5 for the diffuse insolation component and parameterized cloud albedo for the direct component by

$$\alpha_c = 0.323 + 0.535(Z - 0.79), \tag{5}$$

where Z is the solar zenith angle. Scattering of the direct component is greatest when sunlight traverses clouds at large solar zenith angles because the paths through the clouds are longest at glancing incidence. We tuned Thompson and Barron's albedo parameterization to our model by adjusting the coefficients of Equation 5 to

$$\alpha_c = 0.355 + 0.620\,(Z - 0.79) \tag{6}$$

This form allows our model to reproduce Earth's present, mean-planetary albedo of ~ 0.31.

Outgoing Infrared Radiation

The outgoing infrared flux is typically approximated by $A + BT$, as in other energy balance models. Modelers parameterize outgoing infrared this way to facilitate finding analytic solutions to the energy balance equation (Eqn. 1). We adopt the A and B coefficients of Caldeira and Kasting (1992). These were derived from a one-dimensional, radiative-convective model (e.g., Kasting and Ackerman 1986), which includes atmospheric absorption of infrared radiation by CO_2 and H_2O vapor. The infrared coefficients A and B reported by Caldeira and Kasting are -326.4 W m^{-2} and 1.953 W m^{-2}K^{-1}, respectively, for 300 ppm of atmospheric CO_2. The coefficients are given with an rms error under 5 W m^{-2} in the temperature range 194 K $< T <$ 303 K. Their value of A has had 15.56 W m^{-2} subtracted from it to simulate absorption of outgoing IR by clouds. We augmented A by 1.5 W m^{-2} to lower the global average surface temperature computed by our model to the observed value of ~ 288 K.

Diffusion

Latitudinal heat transport is parameterized by the term (North et al. 1981)

$$-\frac{\delta}{\delta x}D(1-x^2)\frac{\delta T(x,t)}{\delta x} \tag{7}$$

Heat flow between two adjacent latitude belts is proportional to the area, and hence the circumference, of each belt and to the temperature gradient between them. *D* is the proportionality constant or diffusion coefficient. Figure 2 illustrates the model dependence on *D*. Setting *D* to zero eliminates poleward diffusion of heat, causing the tropics to warm and the poles to cool. Increasing *D* raises diffusion efficiency causing the Earth to become isothermal. Observations suggest that *D* should be somewhere in between. We take *D* (Caldeira and Kasting 1992) to be 0.6334 W $m^{-2}K^{-1}$, which places the present day ice line at *x* 0.95, or 72° latitude. Everything poleward of this is assumed frozen, with a sea ice fraction of 100%.

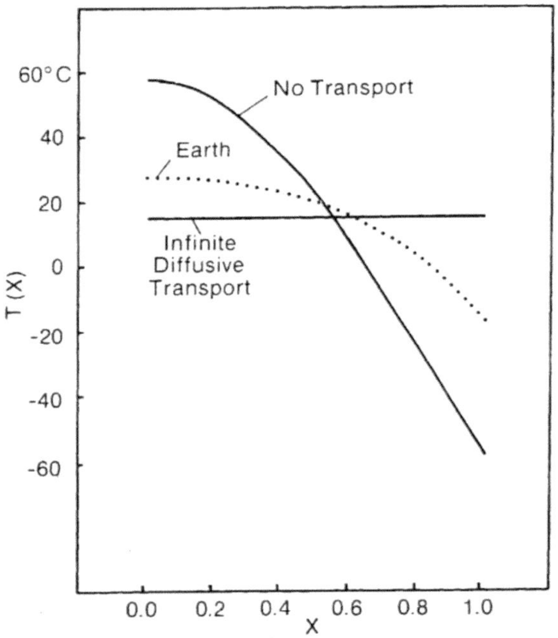

FIG. 2. *Energy balance temperature (°C) for the cases of no diffusive transport, infinite diffusive transport, and present case (schematic) (from North et al. 1981). Courtesy of American Geophysical Union.*

Heat Capacity

The storage of energy in a given latitude belt is governed by the average surface heat capacity *C*(*x*, *T*). We assume, following North et al. (1983), that the relative heat capacities of land, sea ice, and ocean are 1:6.5:60, respectively. Oceanic heat capacity for an ocean layer 70 m deep is taken to be 3.031×10^8 W m^{-2}°C^{-1} (North and Coakley 1979). Average heat capacities for different latitude belts are found by weighting the heat capacities of land and ocean, if unfrozen, by their respective areas. Ocean percentages within each latitude belt for present Earth were taken from Sellers (1965). Below 263.15 K, oceanic heat capacity is replaced by that of sea ice.

Results

Solutions to the energy balance equation (1) were obtained in less than five minutes of CPU time using a SPARC Station 10 (Model 41) computer. The numerical integration was performed using a two-step iteration procedure. First, the diffusion term (Eqn. 7) was spatially differenced to second order using an arbitrary initial temperature profile. The energy balance equation (1) was then explicitly differenced to first order in time yielding a new temperature profile, which could be used in a second spatial differencing. Over thirty modeling years were needed to converge on a final temperature profile.

We required our model to reasonably reproduce Earth's present climate before considering additional cases. Figure 3 gives global seasonal cycles for Earth near present obliquity. Given the simplicity of our model, we believe it fits Earth's current seasonal temperature profile fairly well. It gives an approximate annual average polar ice line of 75° in the northern hemisphere, which nearly matches the observed value of 72°. Seasonal cycle amplitudes grow, as expected, with increasing latitude because of larger insolation swings in these areas. Apart from the polar latitudes, the difference between model annual average belt temperatures and those reported by Sellers (1965) is everywhere less than 5 K. Seasonal amplitudes are smaller than expected. According to Figure 3, residents of the latitude belt centered on 35° can expect summer temperatures to reach only 72°F and winter temperatures to dip no lower than 59°F. Those quick to object, perhaps citing personal experience with winter temperatures below 0°F and summer temperatures above 100°F near this latitude, are reminded that our one-dimensional model calculates average temperatures over an entire latitude belt. Oceans sharing the same belt damp the seasonal extremes because of their large heat capacity.

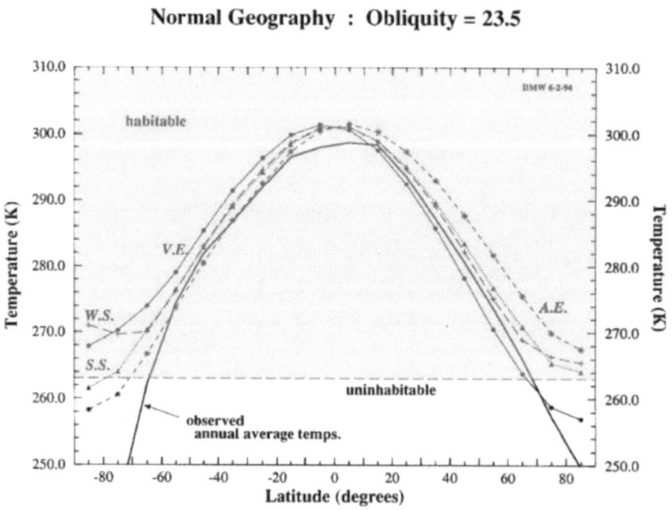

FIG. 3. *Latitudinal seasonal cycles for Earth, at 23.5° obliquity, with present geography. The lower habitability limit at 263 K is marked with a dashed line. Observed annual average temperatures, taken from Sellers (1965), are indicated by a thick solid line. Each curve is labeled by season with abbreviations having the following meanings: V.E. = vernal equinox, S.S. = summer solstice, A.E. = autumnal equinox, W.S. = winter solstice.*

The thermal inertia of oceanic surfaces is also responsible for a phase lag between a belt's insolation extremum and temperature extremum. This effect is most obvious at the South Pole, which has the lowest heat capacity of any latitude belt. In Figure 3, Antarctica is shown to be warmest near the peak of Antarctic summer, whereas all other latitude belts reach maximum temperatures near the beginning of local autumn.

Seasonal temperature cycles for −85° and 45° are plotted in Figure 4 against insolation cycles to better illustrate temperature phase lags. In Figure 4A, Antarctic temperatures respond within thirty days to insolation changes, whereas the lag between insolation and temperature maximum seen in Figure 4B is nearly 2.5 times this long. Normalized planetary average insolation is 0.25. Figure 4A shows that when Earth's obliquity is nearly 90°, the poles receive a maximum daily insolation of one solar constant at the solstices.

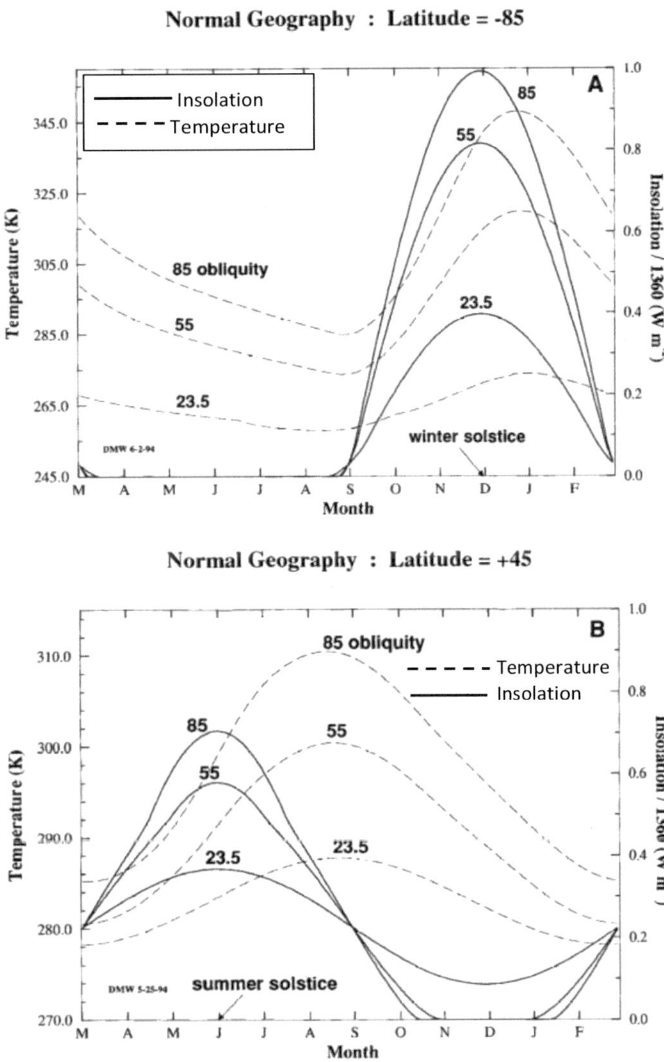

FIG. 4. *Seasonal comparison of insolation and surface temperature cycles for 23.5°, 55°, and 85° obliquity and present Earth geography at (A) −85° and (B) 45° latitude. Insolation is normalized with a solar irradiance at one AU of 1360 W m^{-2}. Planetary average insolation is 0.25.*

Obliquity Variations

We first ran our model for Earth at 55° and 85° obliquity to examine the obliquity space and confirm Ward's prediction that planets near 55° obliquity receive similar average insolations at all latitudes. Figure 5A supports this prediction by showing that seasonal cycles at all latitudes are approximately centered on an average temperature of 290 K. Despite equatorial temperatures that are ~ 10 K cooler than today, global average temperature has increased from 288 K at 23.5° to 289.5 K at 55° obliquity, largely because the poles have become ice free.

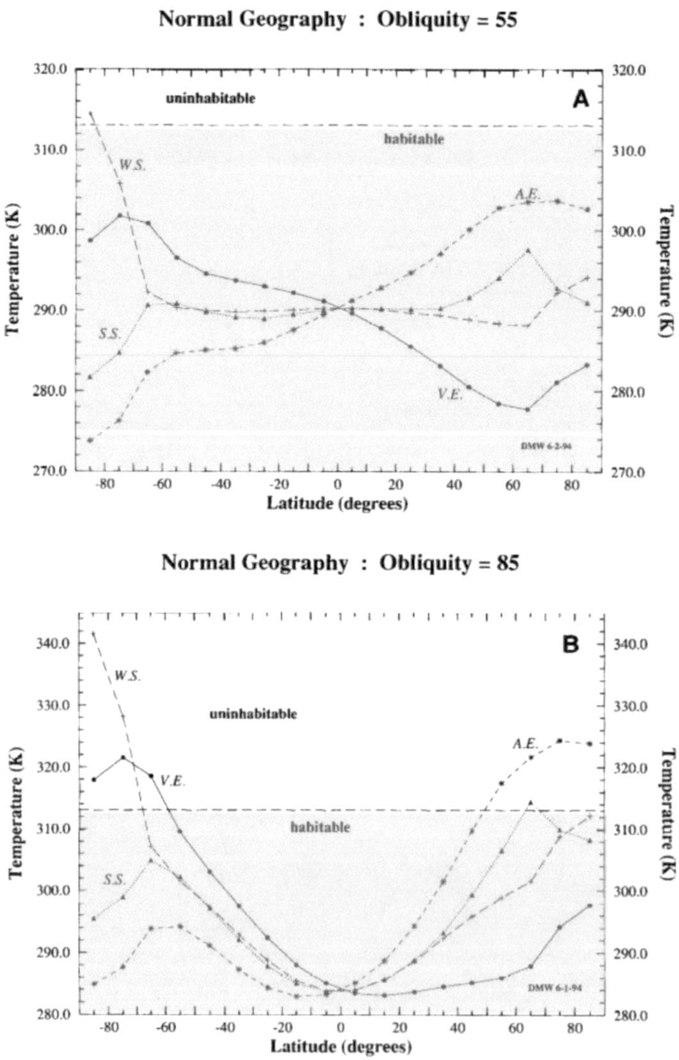

FIG. 5. *Latitudinal seasonal cycles for Earth, with present geography, and at (A) 55° and (B) 85° obliquity. The upper limit of habitability, denoted by a dashed line, is 313 K.*

Ward also showed that at high obliquities the poles should receive more insolation than the lower latitude belts. This behavior is seen in the results for 85°

obliquity, which are shown in Figure 5B. Average planetary albedo has fallen below 0.29 because of decreasing solar zenith angles at the unfrozen poles. This causes the average global temperature to increase to 292 K. Under Dole's habitability criteria, most land surfaces remain habitable even at 85° obliquity. A caveat should be inserted at this point. The infrared coefficients, A and B, taken from Caldeira and Kasting (1992), are valid only for temperatures below 303 K. Thus, summer temperatures in the polar regions of this high-obliquity "Earth" may be unrealistic. Future work is planned to extend the range of our parameterizations.

Geographic Variation

We investigated in subsequent modeling runs whether planetary habitability depends on continental location. Modeling experiments were performed by placing one supercontinent covering 30% of Earth's surface at the equator and then at one pole. We set the obliquity equal to 23.5°, 55°, and 85°, as before. Temperature profiles for Earth with an equatorial supercontinent were similar to those of present Earth (Figs. 3 and 5), with one exception. The continental temperature cycle is nearly in phase with the seasonal insolation cycle because of the low heat capacity at these latitudes. While not important for small obliquities (the Sun remains near the zenith year-round and low-latitude insolation is nearly constant), large solar excursions from equatorial zenith for a planet at high obliquity will appreciably reduce low-latitude insolation every six months. For planets with low-latitude continents, temperatures will vary with the insolation yielding a semiannual seasonal cycle (see Figs. 6 and 7). Despite the frequent temperature swings seen on the supercontinent, it was found to be 100% habitable at all obliquities, using Dole's temperature criterion.

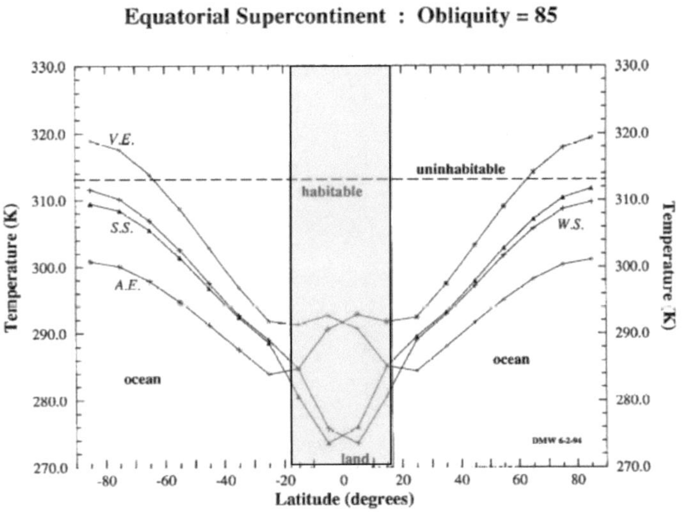

FIG. 6. *Latitudinal seasonal cycles for an Earthlike planet at 85° obliquity with a supercontinent covering 30% of its surface centered at the equator.*

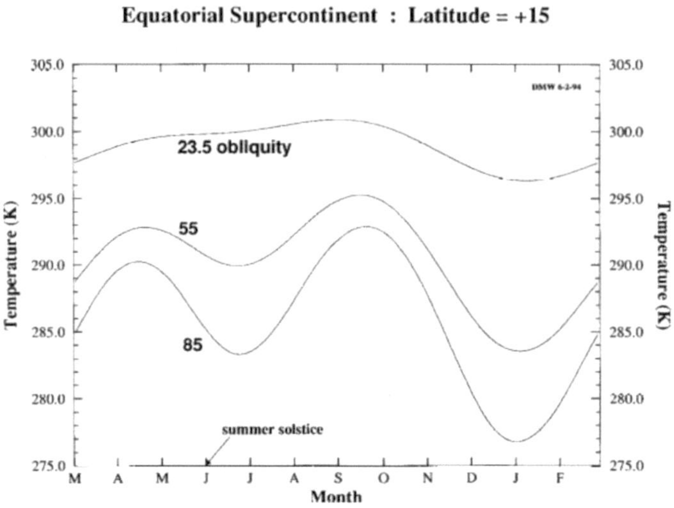

FIG. 7. *Seasonal cycles at +15° latitude for a planet with an equatorial supercontinent and obliquity of 23.5°, 55°, and 85°.*

Earthlike planets with polar supercontinents are less likely to be habitable. The reason is twofold. First, latitude belts with no water have a much lower heat capacity than those with water and, therefore, display large-amplitude seasonal cycles. Figure 8A shows that, even for present obliquity, the seasonal extremes on the polar supercontinent are much greater than those found for present Antarctica with less than one-tenth the supercontinent's area.

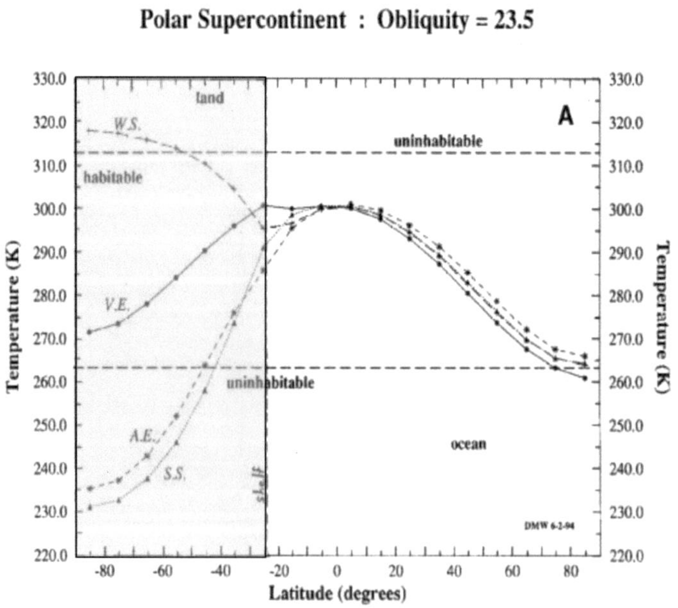

FIG. 8A and B (opposite). *Latitudinal seasonal cycles for an Earthlike planet at (A) 23.5°, and (B) 55° obliquity with a supercontinent covering 30% of its surface centered on the South Pole. The continent lies south of –23° latitude.*

Second, the polar seasonal cycle is amplified at higher obliquities. Insolation at high obliquity in the polar latitudes is plotted in both Figures 1 and 4A. These figures demonstrate that insolation is eliminated for approximately half the year and becomes nearly four times planetary average six months later at the solstice. Figure 8A also shows that, for present obliquity, on the rapidly cooling polar supercontinent, temperatures should reach 230 K during times of low insolation. Polar summer temperatures are predicted to reach 315 K. Seasonal extremes on the supercontinent are even larger at 55° obliquity (Fig. 8B) when less than 15% of the polar supercontinent remains habitable. Although polar supercontinent seasonal cycles were obtained for obliquities up to 60°, temperatures were well beyond the limits of our parameterizations and largely outside the range of habitability.

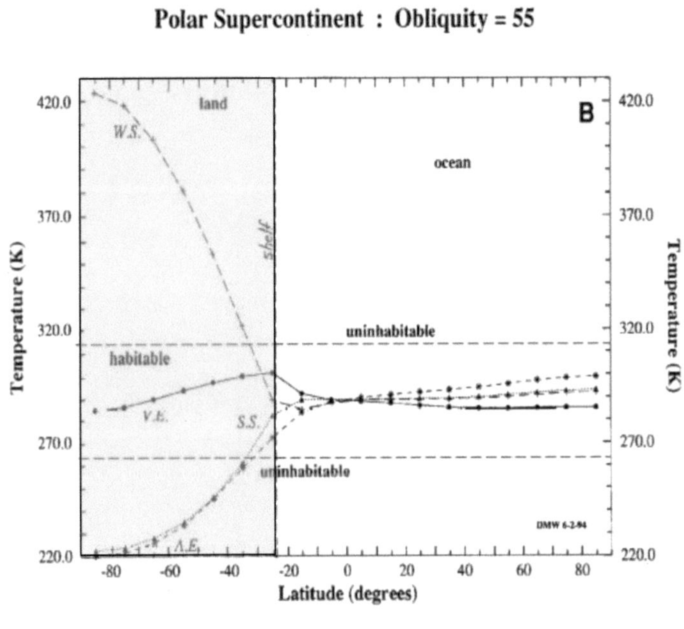

FIG. 8B.

A graphic summary of our results is given in Figure 9A, B, and C.

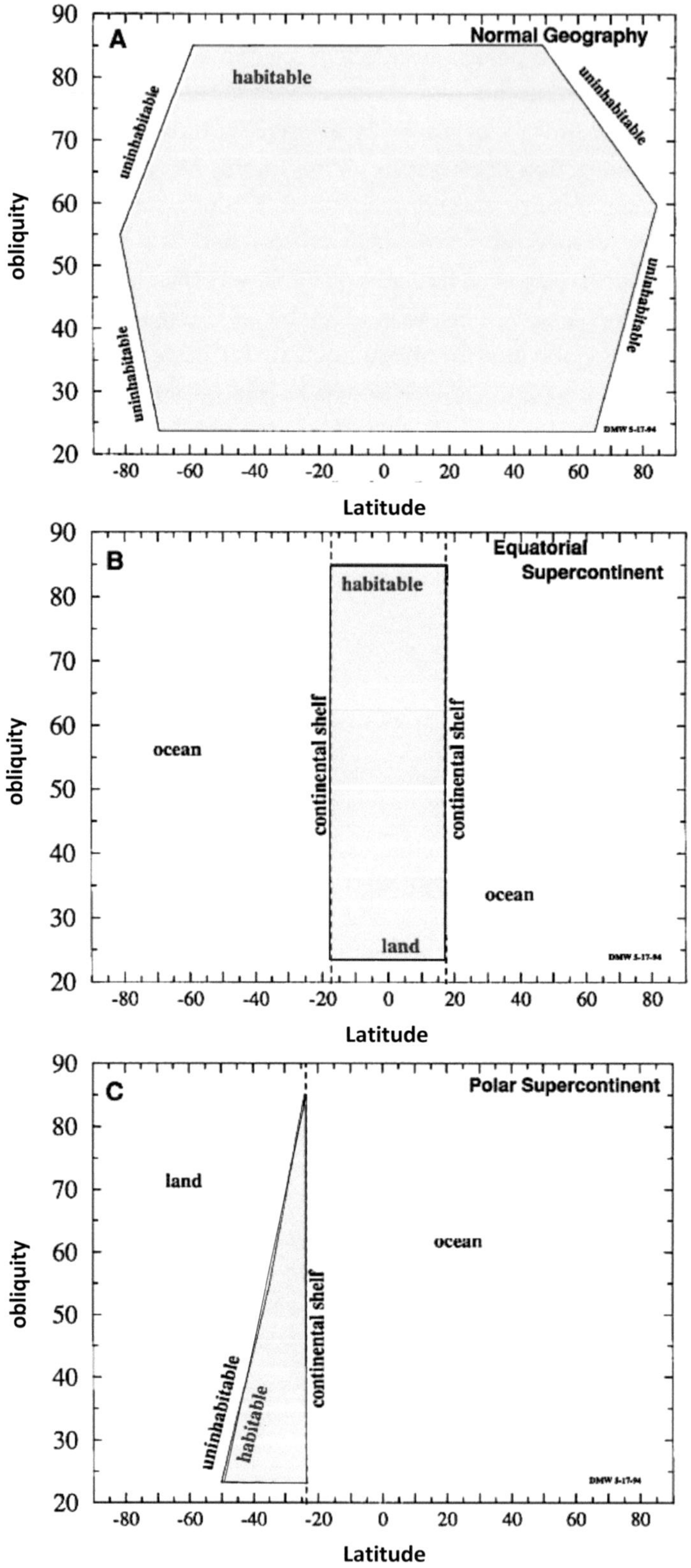
A
Normal Geography
habitable
uninhabitable
uninhabitable
uninhabitable
uninhabitable
obliquity
Latitude
B
Equatorial Supercontinent
habitable
land
ocean
ocean
continental shelf
continental shelf
obliquity
Latitude
C
Polar Supercontinent
land
ocean
uninhabitable
habitable
continental shelf
obliquity
Latitude

FIG. 9A, B, and C (opposite). *Schematic illustrating habitable latitudes for 23.5°, 55°, and 85° obliquity and three geographies: (A) normal geography for present Earth, and supercontinents covering 30% of Earth's surface centered on (B) the equator and (C) the South Pole. Latitudes with land and temperature extremes in the range 263 K < T < 313 K are considered habitable and are shaded gray for obliquities between 23.5° and 85°. Habitable region boundaries were extrapolated using results shown in Figures 3, 5, 6, and 8.*

CONCLUSIONS AND FUTURE WORK

We have investigated the effects of large obliquity variations on planetary habitability. A one-dimensional energy balance climate model has been used to demonstrate that Earthlike planets could remain habitable at all obliquities. The major stabilizing influence is the large heat capacity of the oceans. While planets with equatorial continents are equally habitable, those with high-latitude continents may be uninhabitable at modest obliquities because of large seasonal fluctuations in insolation and surface temperatures.

These calculations were performed for constant atmospheric CO_2 levels and thus do not include the possible negative feedback provided by the carbonate-silicate cycle. High continental temperatures should increase the rate at which carbon dioxide is removed from the atmosphere by the weathering of silicate minerals. Enhanced weathering should lower atmospheric CO_2 and reduce greenhouse warming. Conversely, weathering should slow on continents as they cool. If all land surfaces are covered with ice and snow, weathering should cease entirely until enough volcanic CO_2 accumulates to melt some of the ice. Temperature extremes on planets with large obliquities may be damped as a result, as the extremes occur precisely on the continents where weathering occurs. Future work is planned to integrate the carbonate-silicate cycle into our climate model and to extend the infrared parameterizations to a larger temperature range.

Our model may then be used to make new estimates of the outer edges of circumstellar habitable zones. Earthlike planets near the outer edge of habitability will have a slower weathering rate and, as a result, should develop dense CO_2 atmospheres. Gierasch and Toon (1973) suggested that planets with dense CO_2 atmospheres should exhibit small seasonal cycles and little or no equator-to-pole temperature gradient, because latitudinal heat transport becomes very efficient in an atmosphere that is optically thick at infrared wavelengths. Thus, a planet like Earth, but with a denser CO_2 atmosphere, might exhibit relatively mild seasonal temperature fluctuations, even at very high obliquities.

Finally, we will use our model to investigate valley formation on early Mars. Viking and Mariner data show that most of the Martian valley networks are found on heavily cratered terrain estimated to be over 3.8 billion years old. If these valleys were formed by liquid water, as most investigators believe, it is difficult to explain them using climate models that rely on CO_2-H_2O greenhouses (e.g., Kasting 1991). Such models predict that Mars should not have been warm enough at that time to have liquid water flow over its surface.

Mars could conceivably attain seasonally warm polar temperatures at high obliquity. The polar caps should melt or sublime under these conditions, possibly causing seasonal flooding of the polar regions. At high obliquities, the Martian equator may develop an equatorial ice cap that could migrate back and forth between pole and equator with the chaotically varying obliquity. Such migrations may have played a role in producing the fluvial features and highly eroded crater morphologies that are observed on some parts of the planet.

REFERENCES

BERGER, A. (1988), "Milankovitch Theory and Climate," *Rev. Geophys.* **26:** 624–657.

CALDEIRA, K., AND J.F. KASTING (1992), "Susceptibility of Early Earth to Irreversible Glaciation Caused by Carbon Dioxide Clouds," *Nature* **359:** 226–228.

DOLE, S.H. (1964), *Habitable Planets for Man* (New York: Blaisdell Publishing Company).

GIERASCH, P.J., AND O.B. TOON (1973), "Atmospheric Pressure Variation and the Climate of Mars," *J. Atmos. Sci.* **30:** 1502–1508.

HART, M.H. (1978), "The Evolution of the Atmosphere of the Earth," *Icarus* **33:** 23–39.

HARTMANN, W.K., R.J. PHILLIPS, AND G.J. TAYLOR (1986), *Origin of the Moon* (Houston: Lunar and Planetary Institute).

KASTING, J.F. (1991), "CO_2 Condensation and the Climate of Early Mars," *Icarus* **94:** 1–13.

KASTING, J.F., AND T.P. ACKERMAN (1986), "Climatic Consequences of Very High Carbon Dioxide Levels in the Earth's Early Atmosphere," *Science* **234:** 1383–1385.

KASTING, J.F., D.P. WHITMIRE, AND R.T. REYNOLDS (1993), "Habitable Zones Around Main Sequence Stars," *Icarus* **101:** 108–128.

LASKAR, J., AND P. ROBUTEL (1993), "The Chaotic Obliquity of the Planets," *Nature* **361:** 608–614.

LASKAR, J., F. JOUTEL, AND P. ROBUTEL (1993), "Stabilization of the Earth's Obliquity by the Moon," *Nature* **361:** 615–617.

NORTH, G.R., AND J.A. COAKLEY (1979), "Differences Between Seasonal and Mean Annual Energy Balance Model Calculations of Climate and Climate Sensitivity," *J. Atmos. Sci.* **36:** 1189–1203.

NORTH, G.R., R.F. CALAHAN, AND J.A. COAKLEY (1981), "Energy Balance Climate Models," *Rev. Geophys. Space Phys.* **19:** 91–121.

NORTH, G.R., J.G. MENGEL, AND D.A. SHORT (1983), "Simple Energy Balance Model Resolving the Seasons and the Continents: Application to the Astronomical Theory of the Ice Ages," *JGR* **88:** 6576–6586.

RUDDIMAN, W.F., M.E. RAYMO, D.G. MARTINSON, B.M. CLEMENT, AND J. BACKMAN (1989), "Pleistocene Evolution of Northern Hemisphere Climate," *Paleoceanography* **4:** 353–412.

SELLERS, W.D. (1965), *Physical Climatology* (London: University of Chicago Press).

THOMPSON, L.S., AND E.J. BARRON (1981), "Comparison of Cretaceous and Present Earth Albedos: Implications for the Causes of Paleoclimates," *J. Geol.* **89:** 143–165.

TOUMA, J., AND J. WISDOM (1993), "The Chaotic Obliquity of Mars," *Science* **259:** 1294–1297.

WARD, W.R. (1973), "Large Scale Variations in the Obliquity of Mars," *Science* **181:** 260–262.

WARD, W.R. (1974), "Climatic Variations on Mars. I. Astronomical Theory of Insolation," *JGR* **79:** 3375–3386.

WARD, W.R. (1979), "Present Obliquity Oscillations of Mars: Fourth-Order Accuracy in Orbital e and I," *JGR* **84:** 237–241.

WARD, W.R. (1992), "Long-Term Orbital and Spin Dynamics of Mars," in *Mars*, H. Kieffer, B. Jakosky, and C. Snyder, eds. (Tucson: University of Arizona Press), pp. 298–320.

The Small Sagittarius Star Cloud (M24)

Photo Credit: © Patrick A. Cosgrove
https://cosgrovescosmos.com/projects/messier-24

The Small Sagittarius Star Cloud (M24) is located in the constellation Sagittarius. It appears as a large and dense population of stars with a few areas of dark dust. Looking toward Sagittarius, we are looking toward the highly populated core of our own galaxy. Unfortunately, we don't see the core, as many molecular gases and dark dust clouds obscure our view.

But in the case of M24, what we are seeing is actually a hole in the obscuring gas and dust, which provides a glimpse of the dense stars of the Sagittarius-Carina arms of the Milky Way. The visible starfield is quite large–roughly 1.5 degrees across and located about 10,000 light-years away.

The volume of space filled by M24 extends a further 10,000–16,000 light-years. It is also quite dense–in binoculars alone, about 1,000 stars can be seen. In this image, you can see an incredible density of stars of all types. The dark rifts and globules are areas of high dust concentration that have their own designation in the Barnard catalog.

Source: Cosgrove's Cosmos.com

EXPERIMENTAL RESULTS ON THE PREBIOTIC FIXATION OF NITROGEN UNDER A NEUTRAL ATMOSPHERE BY IRON(II)

David P. Summers

SETI Institute
Mountain View, CA 94043

Sherwood Chang

NASA Ames Research Center
Moffett Field, CA 94035-1000

Abstract

What conditions are necessary for the origin of life? One problem encountered in theories of the evolution of life under a neutral atmosphere is the lack of routes for the prebiotic fixation of nitrogen. One possible route is the reduction of nitrite by aqueous Fe(II) in the ocean. Both aqueous Fe^{+2} and FeS will reduce nitrite to ammonia. The reduction by Fe^{+2} would require ocean pHs to be greater than 7.3. The reaction is also strongly favored by warmer ocean temperatures and is not significantly inhibited by solutes such as are present in Earth's oceans. More generally, this process requires the presence of liquid water on the planet in contact with a source of Fe(II) (implying volcanic/hydrothermal activity?). Shock heating of the atmosphere is also necessary to produce nitrite. This implies either electric discharges or meteors passing through the atmosphere.

INTRODUCTION

Can life form under a neutral atmosphere? What conditions are needed? The origin and evolution of life depends on the availability of reduced or "fixed" nitrogen. In a strongly reducing atmosphere this is not a problem. Compounds essential to the chemical evolution of life can be formed by reactions between HCN, NH_3, and carbonyl compounds produced in spark discharges, as well as by other Miller-Urey processes (Schopf 1983, Stribling and Miller 1987). However, NO is produced from the shock heating of an N_2/CO_2 atmosphere (Yung and McElroy 1979, Chameides and Walker 1981, Fegley et al. 1986, Kasting 1990). Current geochemical evidence seems to point to such a nonreducing atmosphere on the Earth (Schopf 1983, Walker 1985, Mattioli and Wood 1986, Kasting 1987, Gregor et al. 1988). So one question is, Can reduced nitrogen be formed under a neutral atmosphere and, if so, under what conditions?

The NO produced by shock heating would be converted to nitric and nitrous acids and rained into the oceans (Mancinelli and McKay 1988). Thus, one approach to the question of prebiotic nitrogen fixation/reduction with a neutral atmosphere may be the conversion of nitrites and nitrates to ammonia in the ocean. One source of reducing equivalents is Fe(II) (Eqn. 1).

$$6\ \mathrm{Fe(II)} + 7\ \mathrm{H_2O} + \mathrm{NO_2}^- \text{ Æ } 6\ \mathrm{Fe(III)} + 11\ \mathrm{H}^+ + \mathrm{NH_3} \quad (1)$$

In prebiotic oceans, much higher concentrations of dissolved Fe(II) would have existed than today, and it has been postulated that the Archean banded iron formations result from the oxidation of aqueous Fe(II) (Holland 1973, Holland 1984, Walker and Brimblecombe 1985, Derry and Jacobsen 1990). Increased hydrothermal activity would also have resulted in increased FeS (Windley 1976, Turcotte 1980).

RESULTS AND DISCUSSION

As the data in Table 1 indicate, iron (II) will reduce nitrite to ammonia. All ammonia concentrations reported are relative to samples taken from the reaction at time zero. No ammonia is detected in the absence of nitrite or Fe(II). The amount of ammonia produced, when the reaction was complete, was used to calculate the percentage of nitrite converted to ammonia (the product yield). In addition to our results on nitrite reduction, some reduction of nitrate to ammonia was detected. However, ammonia was not reproducibly formed from nitrate in every experiment, although in experiments where ammonia was produced, product yields were

~95%. In all cases, ammonia formation from nitrate was slower, by at least a factor of 8, than from nitrite.

Data in Table 1 show the dependence of the rate of reaction on pH (entries 1–8). At pH $\leq$ 7.3, the reaction does not proceed at any detectable rate. The maximum rate is found at a pH of 7.6.

TABLE 1. ***Rate of Reduction of Nitrite to Ammonia by Aqueous*** Fe^{+2}*

Entry	*T (°C)*	*pH*	*Additions*†	*Rate (μM min^{-1})*	*Product Yield*‡ *(%)*
1)	22	7.1	None	< 0.004§	–
2)	22	7.3	None	< 0.008§	–
3)	22	7.5	None	10	35
4)	22	7.6	None	15	37
5)	22	7.9	None	8.8	28
6)	22	8.0	None	8.5	33
7)	22	8.0	NaCl	44	64
8)	22	8.4	NaCl	6.6	25
9)	22	8.0	Cations	67	85
10)	22	8.0	Anions	36	52
11)	0	7.6	None	~ 0.1	3.8
12)	11	7.6	None	0.45	24
13)	23	7.6	None	12	44
14)	30	7.6	None	61	74
15)	40	7.6	None	630	60
16)	22	7.5	CO_2	0	0
17)	80	7.5	CO_2	0.006	75

* *A volume of a nitrogen-purged $NaNO_2$ solution (23 mM) sufficient to make up a 0.32 mM solution was added by syringe to a 12.3 mM $FeCl_2$ (Aldrich) solution (350 ml) stirred under nitrogen purging. In occasional experiments a volume of $FeCl_2$ solution sufficient to make up 12.3 mM was added to a 0.32 mM solution of $NaNO_2$ with no change in results. After a t = 0 blank was withdrawn and purging stopped (to avoid sweeping out ammonia), the reaction was allowed to proceed under nitrogen. During the reaction, aliquots were withdrawn by syringe and analyzed for ammonia by colormetric methods (Doyle and Mahapatro 1984).*

† *NaCl = 0.53 M NaCl. Anions = 5 mM NaBr, 28 mM Na_2SO_4, and 0.53 M NaCl. Cations = 51 mM $MgCl_2$, 12 mM $CaCl_2$, 13 mM KCl, and 0.53 M NaCl. CO_2 = saturated with respect to $FeCO_3$, one atm CO_2, and 0.53 M NaCl.*

‡ *The percentage of reacted nitrite that is converted to ammonia.*

§ *These values are within experimental noise and represent upper limits. All other numbers and changes from experiment to experiment are significant.*

As the pH increases past 7.6 the rate falls off, making reactivity above ~ pH 9 low. An Arrhenius plot yields a slope of 22,000 K^{-1}. Data in Figure 1 and Table 2 show that the rate varies proportionally to the concentration of nitrite but not to the concentration of iron. A least-squares fit of the equation $y = kx^n$ to the data on rate vs. $[Fe^{+2}]$ yields a value of 1.8 for n, thus giving an apparent rate law (at pH 7.9) of the form rate = k[nitrite] $[iron]^{1.8}$ where $k = 35 \times 10^{-6}\ min^{-1}\ M^{-1.8}$.

At higher temperature the solubility of siderite (Freier 1976) also increases. If the concentration of Fe(II) is fixed by the solubility of $FeCO_3$, then we can define a pseudo first-order rate law: rate = $k'[NO_2^-]$ where $k' = k[Fe(II)]^{1.8}$. A plot of k' vs. temperature (Fig. 2) shows that, between 10°C and 80°C, the rate of nitrite reduction in the ocean increases by eleven orders of magnitude. The reduction is strongly favored by warm ocean environments.

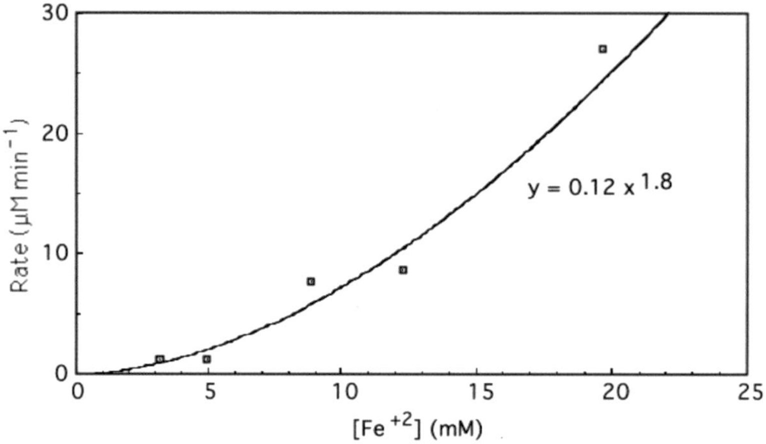

FIG. 1. *Plot of the rate of nitrite reduction as a function of Fe^{+2} concentration; 0.32 mM $NaNO_2$, room temperature, pH 7.9. The curve and equation represent the best fit* of y = x^n *to the data.*

TABLE 2. ***Dependence of the Rate of Nitrite Reduction by Aqueous Fe^{+2} on Nitrite and Iron(II) Concentration****

$[NO_2^-]$ (mM)	*$[Fe^+]$ (mM)*	*Rate (μM/min)*	*pH*	*Relative Concentration*	*Relative Rate*	*Product Yield*† *(%)*
			Nitrite Concentration Dependence			
0.0327	12.3	0.9	7.9	1	1	–
0.131	12.3	5.0	7.9	4	5.6	34
0.327	12.3	8.8	7.9	10	9.8	28
0.654	12.3	18	7.9	20	20.0	30
0.327	12.3	4.2	8.0	1	1	28
0.654	12.3	7.8	8.0	2	1.9	26
			Iron Concentration Dependence			
0.327	3.2	1.3	7.9	1	1	–
0.327	4.9	1.3	7.9	1.5	1	27
0.327	8.9	7.7	7.9	2.8	6.1	30
0.327	12.3	8.8	7.9	4	6.9	28
0.327	19.7	27	7.9	6.2	21	42

TABLE 2. (cont.)

* *At room temperature. Variable volumes of nitrogen-purged 23 mM $NaNO_2$ (Baker) were added by syringe to a 12.3 mM $FeCl_2$ (Aldrich) solution (350 ml) stirred under nitrogen purging. After a t = 0 blank was withdrawn and purging stopped (to avoid sweeping out ammonia) the reaction was allowed to proceed under nitrogen. During the reaction, aliquots were withdrawn by syringe and analyzed for ammonia by colormetric methods (Doyle and Mahapatro 1984).*

† *The percentage of reacted nitrite that is converted to ammonia.*

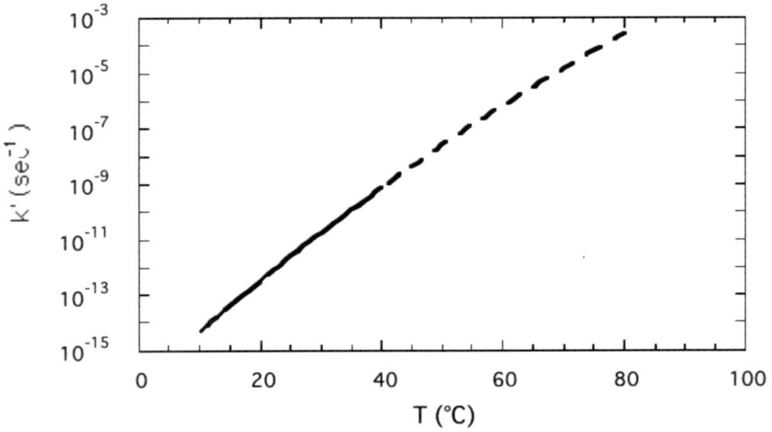

FIG. 2. *Plot of the pseudo first-order rate constant* (k' = *k[Fe(II)]$^{1.8}$) as a function of ocean temperature (pH 7.6, 0.2 atm CO_2, saturated with respect to siderite). The rate constant,* k, *was measured experimentally between 0 and 40°C (solid line) and extrapolated to 80°C (dashed line); see text.*

To evaluate the importance of this reaction in an ocean environment, one must consider the effect of likely dissolved species in such an ocean. For example, anions may interfere with the reaction, particularly by binding to the Fe(II) centers and deactivating them. Cations may form nitrite or anionic iron complexes. The predominate salt in Earth's ocean is NaCl. When 0.5 M NaCl is added to the reaction, a fivefold increase in the rate is observed (entries 6 and 7, Table 1). When other cations present in present day seawater are added, an additional rate increase is observed (entries 7 and 9). The presence of sulfate and bromide ions at levels equal to those found in present day environments reduced the rate of reduction slightly (entries 7 and 10). Under a predominately carbon dioxide/nitrogen atmosphere, the concentration of carbonate/bicarbonate would have been significant. The experiment with bicarbonate/carbon dioxide had to be run at 80°C because of the low solubility of $FeCO_3$. However, extrapolation to the conditions of the other experiments indicates that there is, within experimental error, no inhibition by bicarbonate/carbonate.

In addition to experiments with dissolved Fe^{+2}, experiments were also conducted with suspensions of FeS particles. FeS showed activity toward the reduction of nitrite to ammonia. Preliminary results indicate that the reaction proceeds at pH = 7.2 in the dark. The rate is 0.2 μM liter/min g with an ammonia yield of 78%. The reaction proceeds under carbon dioxide but carbon dioxide is not reduced.

SUMMARY

What conditions are necessary for the formation of reduced nitrogen as outlined above? There are two main requirements—production of nitrite and/or nitrate and delivery of the nitrite/nitrate to an aqueous solution/suspension of Fe(II).

To produce nitrites and nitrates we need the formation of NO by shock heating of the atmosphere. This can be caused by meteors passing through the atmosphere and by electric discharges. We also need atmospheric water to photochemically generate H· (necessary for the formation of nitrite and nitrate from NO [Mancinelli and McKay 1988]).

The overriding requirement for the formation of an aqueous solution/suspension of Fe^{+2}/FeS is the presence of liquid water in the first place. Additionally, that water must have been mixed with nitrite/nitrate being rained out of the atmosphere. It must also be in contact with a source of Fe(II). On the Earth, this implies hydrothermal activity.

The formation of ammonia by aqueous Fe^{+2} solutions requires slightly basic (< 7.3) conditions (though formation by FeS appears to permit more acidic conditions). Formation is favored by higher ocean temperatures and the presence of ionic species similar to those present in the modern ocean.

REFERENCES

CHAMEIDES, W.L., AND J.C.G. WALKER (1981), "Rates of Fixation by Lightning of Carbon and Nitrogen in Possible Primitive Atmospheres," *Origins of Life* **11:** 291–302.

DERRY, L.A., AND S.B. JACOBSEN (1990), "The Chemical Evolution of Precambrian Seawater: Evidence from REEs in Banded Iron Formations," *Geochim. Cosmochim. Acta* **54:** 2965–2975.

DOYLE, M.P., AND S.N. MAHAPATRO (1984), "Nitric Oxide Dissociation from Trioxodinitrate (II) in Aqueous Solution," *J. Am. Chem. Soc.* **106:** 3678–3679.

FEGLEY, B., JR., R.G. PRINN, H. HARTMAN, AND G.H. WATKINS (1986), "Chemical Effects of Large Impacts on the Earth's Primitive Atmosphere," *Nature* **319:** 305–308.

FREIER, R.K., ED. (1976), *Aqueous Solutions*, vol. 1 (Berlin: Walter de Gruyter).

GREGOR, C.B., R.M. GARRELS, F.T. MACKENZIE, AND J.B. MAYNARD, EDS. (1988), *Chemical Cycles in the Evolution of the Earth* (New York: John Wiley & Sons).

HOLLAND, H.D. (1973), "The Oceans: A Possible Source of Iron in Iron Formations," *Econ. Geol.* **68:** 1169–1172.

HOLLAND, H.D. (1984), *The Chemical Evolution of the Atmosphere and Oceans* (Princeton, NJ: Princeton University Press), ch. 4.

KASTING, J.F. (1987), "Theoretical Constraints on Oxygen and Carbon Dioxide Concentrations in the Precambrian Atmosphere," *Precambrian Res.* **34:** 205–229.

KASTING, J.F. (1990), "Bolide Impacts and the Oxidation State of Carbon in the Earth's Early Atmosphere," *Origins of Life* **20:** 199–231.

MANCINELLI, R.L., AND C.P. MCKAY (1988), "The Evolution of Nitrogen Cycling," *Origins of Life* **18:** 311–325.

MATTIOLI, G.S., AND B.J. WOOD (1986), "Upper Mantle Oxygen Fugacity Recorded by Spinel Iherzolites," *Nature* **322:** 626–628.

SCHOPF, J.W., ED. (1983), *The Earth's Earliest Biosphere: Its Origin and Evolution* (Princeton, NJ: Princeton University Press), pp. 53–92, and references therein.

STRIBLING, R.E., AND S.L. MILLER (1987), "Energy Yields for Hydrogen Cyanide and Formaldehyde Syntheses: The HCN and Amino Acid Concentrations in the Primitive Ocean," *Origins of Life* **17:** 261–273.

TURCOTTE, D.L. (1980), "On the Thermal Evolution of the Earth," *Earth Planet. Sci. Lett.* **48:** 53–58.

WALKER, J.C.G. (1985), "Carbon Dioxide on the Early Earth," *Origins of Life* **16:** 117–127.

WALKER, J.C.G., AND P. BRIMBLECOMBE (1985), "Iron and Sulfur in the Pre-Biologic Ocean," *Precambrian Res.* **28:** 205–222.

WINDLEY, B.F., ED. (1976), *The Early History of the Earth* (New York: John Wiley & Sons).

YUNG, Y.L., AND M.B. MCELROY (1979), "Fixation of Nitrogen in the Prebiotic Atmosphere," *Science* **203:** 1002–1004.

The Rosette Nebula

Photo Credit: NASA, ESA, STScI, DSS
Image Processing: Joseph DePasquale (STScI)
https://science.nasa.gov/asset/hubble/rosette-nebula-context-image/
[retrieved 12-10-25]

The Rosette Nebula is a vast star-forming region, 100 light-years across, that lies at one end of a giant molecular cloud in the constellation Monoceros. The nebula is estimated to contain around 10,000 solar masses. It is located about 5,000 light-years away from Earth. Intense radiation from the young stars inside a cluster in the nebula causes the gasses to glow. The background image is from the Digitized Sky Survey, while the inset is a small portion of the nebula as photographed by the Hubble Space Telescope. Dark clouds of hydrogen gas laced with dust are silhouetted across the image. The colors come from the presence of hydrogen, oxygen, and nitrogen. The nebula's multiwavelength appearance can be explored on Viewspace.org.

Organic Matter Supplied to a Planet by Tectonic and Volcanic Activity

Friedemann Freund

Department of Physics
San José State University
San José, CA 95192-0106
and
NASA Ames Research Center
Moffett Field, CA 94035

Richard Ho

Department of Physics
San José State University
San José, CA 95192-0106
and
NASA Ames Research Center
Moffett Field, CA 94035

Note: Dr. Freund is also with the SETI Institute, 2035 Landings Drive [currently located at 339 N. Bernardo Ave., Suite 200], Mountain View, CA 94043.

Abstract

For Life to emerge, the early Earth must have acquired a large stock of abiogenically synthesized organic molecules, either from terrestrial sources or from space. The work reported here considers processes of a fundamental nature by which organic molecules may be formed in the solid state. Recent studies of common rock-forming minerals have shown that, during crystallization from fluid-laden magmas, H_2O and CO_2 enter into the mineral matrix, forming a solid solution. When rocks are uplifted by volcanic or tectonic forces, thus cooled and decompressed, the solid solutions become supersaturated. At the same time, however, a redox reaction takes place by which the solute H_2O is split into H_2 plus O_2^{2-}, and the solute CO_2 is split into CO_2^{2-} plus O_2^{2-}, where the peroxy represents a dissolved O atom. Spatial separation between the reduced and the oxidized parts occurs within the mineral matrix, because the diffusively mobile H_2 and C segregate to structural defects such as subgrain boundaries, where they begin forming protomolecular entities $H_xC_yO_z^{n-}$, while the peroxy remain lattice-bound. Upon weathering, these protomolecular entities are expected to hydrolyze, releasing H–C–O molecules into the environment. We report on organic molecules released (1) during impact fracture of upper-mantle-derived olivine single crystals, using a fast-scanning mass spectrometer, (2) during heating of freshly crushed olivine crystals, using a field ionization mass spectrometer, and (3) on gas chromatography of the gases collected during fracture of olivine crystals and of the basalt by which they were brought to the surface during a volcanic eruption.

SOURCES OF ABIOGENIC ORGANIC MATTER

For Life to emerge some 3.7 billion years ago the early Earth must have acquired a stock of abiogenically synthesized organic molecules either from an indigenous terrestrial source or from space.

When Miller published his pioneering study in 1953 (see References) on the synthesis of amino acids through electric discharges in a simulated early atmosphere, the early atmosphere was thought to have contained free H_2, CH_4 and NH_3 besides CO_2, N_2 and H_2O vapor. In such a gas environment, reduced compounds dominate the suite of reaction products. However, it is now widely believed that the early atmosphere was not reducing (e.g., Walker 1977, Chang et al. 1983, Walker 1985, Chang 1988), but attempts to produce organic molecules in a neutral atmosphere by energetic processes like lightning, shock waves, glow discharges or UV radiation have not met with great success (Pinto et al. 1980, Chittenden and Schwartz 1981, Stribling and Miller 1987, Henet et al. 1992). The reason is that, though reduced and oxidized intermediates may form in abundance, they do not become spatially separated fast enough to allow for the formation of stable compounds. Many other synthesis reaction pathways have since been considered—photochemical, catalytic or template mediated—where the conditions for spatial separation may be more favorable (for a review see Chang 1993). The possibility, already suggested by Oró (1961) that organic matter may have been supplied from space, in particular during early heavy bombardment, has gained new prominence lately. Maybe carbonaceous chondritic meteorites, comets and interplanetary dust delivered organics in sufficiently large quantity and complexity to give Life a chance (e.g., Mukhin et al. 1989, Anders 1989, Chyba 1990, Chyba et al. 1990, Chyba and Sagan 1992, Lawler and Brownlee 1992).

In this report we direct the attention to one area of scientific endeavor that, in our opinion, has not yet been well explored: the solid Earth. We argue that, on a planet like Earth, which has been volcanically and probably also tectonically active since its earliest time (Ingmanson and Dowler 1977, Drever et al. 1988, Kröner and Layer 1992), rocks from great depth are continuously cycled through uplift, weathering, sedimentation and subduction. However, having crystallized in the presence of H_2O and CO_2, the minerals in these rocks may contain H–C–O precipitates that turn into organic matter during weathering.

We address, first, thermodynamic questions about crystallization in H_2O/CO_2-rich environments. Second, we report on an interesting in situ redox reaction affecting solute H_2O and CO_2 during cooling and decompression that accompanies the uplift. Third, we present experimental data obtained with olivine crystals from the upper mantle that are consistent with our assertion that minerals represent a matrix in which the abiogenic synthesis of organic molecules can take place.

Thermodynamics of Mineral-Gas-Fluid Interactions

Figure 1 shows an isobaric section through a two-component phase diagram where AO is MeO and BO is H_2O and/or CO_2. Under equilibrium conditions, at any given pressure, the melt contains δ H_2O and/or CO_2. T_m is the melting temperature of the pure solid, and T_{cryst} is the crystallization temperature of a mineral crystallizing in an H_2O/CO_2-laden melt. This mineral is a solid solution (*ss*) containing δ H_2O and/or CO_2 components in its matrix.

Initially, close to T_{cryst}, the solute concentration δ is large, but upon cooling and decompression δ decreases along the bold line. If it were possible to maintain thermodynamic equilibrium, δ would reach zero at 0 K, with K being the absolute temperature. However, equilibrium can never be fully maintained, even at very slow cooling rates. Therefore, a critical temperature T_{crit} exists at which the solid solution (*ss*) becomes supersaturated (*sss*). In the *sss* state the system tends toward exsolution. Microscopically, exsolution is driven by strains that the solute entities, atoms or ions, exert on the surrounding mineral matrix, thus raising its total energy. For the system to regain energy, the solute atoms or ions segregate into those regions in the crystal structure where strains can be relaxed. Dislocations, subgrain boundaries, grain boundaries and the crystal surface are such regions. They act as sinks for excess solutes. The inset in Figure 1 shows a solute concentration profile across a small volume of an *sss* crystal bordered by subgrain boundaries toward which segregation takes place.

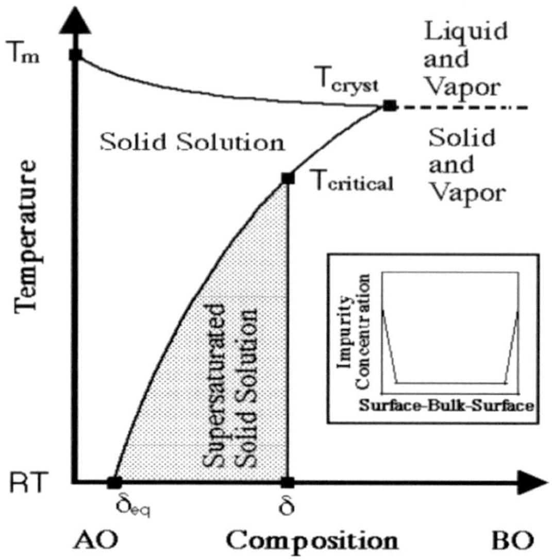

FIG. 1. *Isobaric section through an AO-BO phase diagram where* AO = MeO *and* BO = H_2O *and/or* CO_2. *δ is the amount of gas/fluid phase component in solid solution. The lightly shaded area designates the solid solution, the stippled area the supersaturated solid solution, which is thermodynamically unstable with respect to exsolution. Inset: Schematic representation of a segregation profile within a grain or subgrain.*

If the solute is a gas/fluid component such as H_2O and/or CO_2, exsolution may simply mean degassing. However, in real crystals, the process can be more complicated. In the following section we take a closer look at solid solutions and chemical reactions that may occur in the solid state.

Redox Reactions in the Solid State

We shall briefly discuss the dissolution of H_2O and CO_2. Inside dense mineral structures solute H_2O molecules are torn apart, transferring one proton onto an O^{2-} in the matrix, thus creating two hydroxyl anions:

$$H_2O + O^{2-} \Leftrightarrow 2\,OH^- \qquad (1)$$

The dissolution of an H_2O molecule in a mineral *MeO*, where *Me* represents a divalent metal, can be described by replacement of one Me^{2+} cation by two H^+. Thus, with 2 δ OH^- chargewise balancing δ Me^{2+} vacancies, we have:

$$MeO + \delta\, H_2O \Leftrightarrow Me_{1-\delta}\, O_{1-2\delta}\, (OH)_{2\delta} + \delta\, MeO \qquad (2)$$

Most geologists regard Equation 2 as the final stage of the H_2O dissolution process and assume the OH^- concentration as determined by infrared (IR) spectroscopy to be equal to the total H_2O content (Stolper 1982, Aines and Rossman 1984, Miller et al. 1987, Bell and Rossman 1992). This is wrong because not all solute OH^- remain as OH^-.

Local charge compensation for the Me^{2+} vacancy is best achieved by two OH^- directly adjacent to the vacancy [$OH^-\Diamond^-HO$], where $\Diamond$ symbolizes a cation vacancy. The question arises whether this defect is stable. The short answer is no because the strong electric field near the vacancy exerts a force on the OH^- causing them to electronically rearrange. Two O^{2-} donate one electron each to H^+ oxidizing to O^- and combining to a peroxy anion, O_2^{2-}, while two H^+ accept an electron each and turn into an H_2 molecule:

$$[OH^-\Diamond^-HO] \Leftrightarrow [O^-(H_2)^-O] \qquad (3a)$$

$$[O^-(H_2)^-O] \Leftrightarrow O_2^{2-}(H_2) \qquad (3b)$$

Equations 3a and 3b describe a "water splitting" reaction in the solid state, $H_2O = H_2 + \frac{1}{2}\, O_2$, which may be of great interest to questions relating to the origin of Life. It creates a reduced and an oxidized solute, H_2 and O_2^{2-}, both differing with respect to their diffusive mobility. While O_2^{2-} is lattice-bound and immobile, H_2 may diffuse away:

$$[O^-(H_2)^-O] \Leftrightarrow O_2^{2-}\ \Diamond\ + H_2 \qquad (3c)$$

The sequence described by Equations 3a–3c has been documented in a series of papers describing experiments by independent techniques, using MgO as a predominantly ionic, SiO_2 as a predominantly covalent, and olivine, $(Mg,Fe)_2SiO_4$, as an intermediate model compound (Freund and Wengeler 1982, Freund 1985, Freund and Oberheuser 1986, Freund et al. 1991). One of the unexpected features of this OH^- redox conversion seems to be that it does not require the participation of transition metal cations. It proceeds even when reduced transition metal cations are present in the same mineral matrix, without affecting their oxidation states.

Data on the dissolution of CO_2 in solid matrices are still spotty. According to some authors CO_2 does not enter into solid solution at all (Mathez et al. 1987). This view, however, violates basic thermodynamic principles. It is not a question whether CO_2 dissolves in a given mineral that crystallizes from the CO_2-containing melt but how much CO_2 enters into solid solution at given temperatures and pressures. One way to incorporate CO_2 into a mineral structure would be to electronically couple the CO_2 molecule to one lattice O^{2-}, forming a carbonate anion:

$$CO_2 + O^{2-} \Leftrightarrow CO_3^{2-} \quad (4a)$$

This reaction is considered by many to be true simply because carbonates such as calcite, $CaCO_3$; dolomite, $(MgCa)CO_3$; and siderite, $FeCO_3$, are known rock-forming minerals (Mathez et al. 1987). However, we are not concerned here with bulk minerals but with CO_2 in dilute solid solution (Freund et al. 1980, Freund 1986). As a solute, carboxy anions other than CO_3^{2-} may become energetically favorable. For instance, if CO_2 couples to two O^{2-} during dissolution, two oxyanions may be generated:

$$CO_2 + 2\,O^{2-} \Leftrightarrow CO_2^{2-} + O_2^{2-} \quad (4b)$$

Equation 4b also describes a redox reaction. The reduced part, CO_2^{2-}, is a bent complex with an OCO angle of about 130°, equivalent to a dissolved CO molecule, $CO + O^{2-} \Leftrightarrow CO_2^{2-}$. The oxidized part is the same O_2^{2-} as in Equations 3b and c, equivalent to a dissolved O atom: $O + O^{2-} \Leftrightarrow O_2^{2-}$. The CO_2^{2-} anion may further split and become even more reduced according to

$$CO_2^{2-} \Leftrightarrow CO^- + O^- \quad (4c)$$

The CO^- is probably important for understanding the diffusion of C in dense mineral matrices. The positive C atom makes transient, but short, C–O– bonds, lowering the activation energy for the diffusion of interstitial C (Freund 1987).

In summary, in this section we presented evidence that H_2O and CO_2, when dissolved in a mineral matrix, can undergo an in situ redox reaction, formally splitting them into H_2 + O and CO + O. Only the reduced parts, H_2 molecules and interstitial C atoms, are expected to be diffusively mobile.

Exsolution of the Most Mobile Species

Supersaturation is a state of higher energy due to strains associated with excess solutes. To return to equilibrium, all excess solutes have to segregate. If this is kinetically not possible, partial energy can be gained by segregation of those solutes that are diffusively mobile. H_2 molecules are mobile, even in a dense mineral structure. The interstitial diffusion of C atoms, though not known yet in detail, may be facilitated by forming transient C–O^- bonds as indicated above. By contrast, O_2^{2-} represent tightly coupled, diamagnetic O^- pairs that are diffusively immobile.*

If the reduced solutes H_2 and C (diffusing as CO^-) segregate while the oxidized solutes remain, we are faced with an exciting possibility for organic chemistry. This is because segregation concentrates the exsolved solutes wherever local strains are relaxed.

Figure 2a depicts schematically two subgrain boundaries meeting in a dense cubic oxide structure. For the sake of simplicity cations are omitted and only O^{2-} anions are shown. Some O^{2-} sites close to the subgrain boundary are shaded indicating admixed O^-.† Also shown are two CO_2^{2-}, one CO^- and four H_2, segregating toward the subgrain boundaries which provide strain relaxation.

As the C and H_2 concentrations along the subgrain boundaries increase, C–C and C–H bond formation can be expected to occur, competing with and exchanging for the existing C–O^- bonds. Eventually, as illustrated in Figure 2b, the subgrain boundary may become decorated with protomolecular entities of the general formula $(H_xC_yO_z)^{n-}$. These entities may consist of short chains or may contain aromatic rings and side chains that are either H-terminated or linked to the surrounding lattice by C–O– bonds.

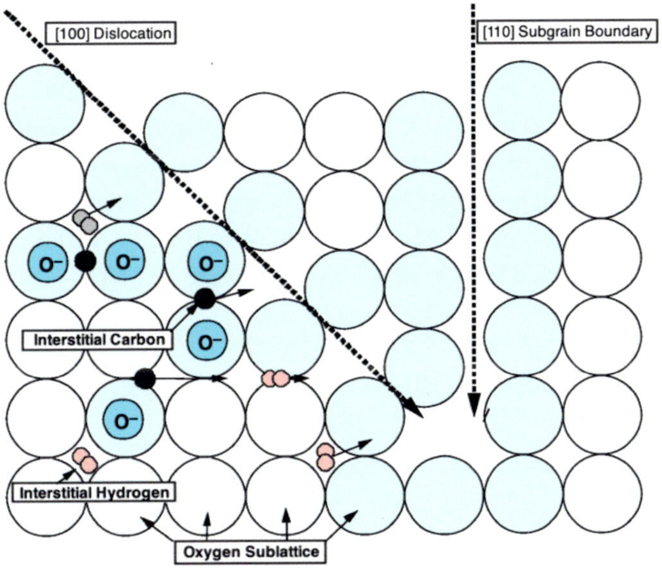

FIG. 2a. *Projection of two subgrain boundaries (dashed lines) in a dense cubic oxide structure (only O^{2-} anions shown). Some O^{2-} are shaded indicating admixed O^-. Lower left: two CO_2^{2-}, one CO^- and four H_2, segregating toward the subgrain boundaries.*

* The O^- are immobile only as long as they remain coupled in pairs forming a peroxy bond. When the peroxy bond breaks, typically at 450–500°C, the O^- diffuse. Their diffusion, however, is not that of an ion but that of an electronic charge, a defect electron also called a positive hole, which delocalizes over the O^{2-} sublattice (Freund et al. 1993).

† The delocalized, wavelike nature of O^- allows for the partial admixing of O^- density to compensate for the local nonstoichiometry that is encountered at defects such as dislocations or subgrain boundaries.

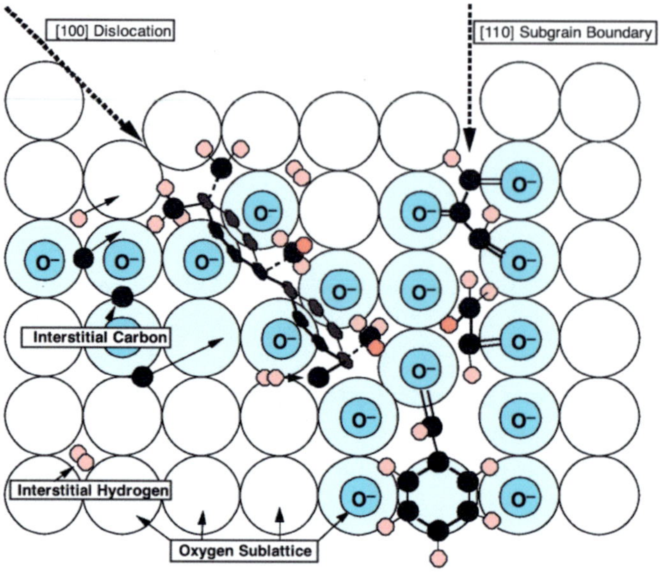

FIG. 2b. *Possible result of C and H segregation to the subgrain boundaries and of the formation of C–C and C–H bonds in exchange for C–O⁻ bonds, leading to $(H_xC_yO_z)^{n-}$ protomolecular entities.*

For the sake of illustration and with no claim to accuracy, a C_2 and a C_3 unit are shown in the upper right. Aromatic C_6 rings with a C–C bond length of ≈1.40 Å fit like a crown around O^{2-}, which has typically a diameter of 2.8 Å. This is illustrated on the center right by a protomolecular entity, $(H_4C_2O)^-$, that resembles an acetaldehyde molecule, H_3C–CHO, and in the lower right by a protomolecular entity, $(H_6C_7O)^-$, that resembles a benzaldehyde molecule, H_5C_6CHO, except that in both cases the terminal oxygens are O^-. Since the geometry of the segregation site is a two-dimensional plane, C–C bond formation may proceed further, giving rise to polyaromatic systems with and without side chains as illustrated schematically in the upper left.

A Synthesis Pathway for Complex Organic Molecules

We have reason to believe that the concept presented here for the generation of protomolecular entities in the solid state is well founded. The thermodynamics involving solid solution formation during crystallization, supersaturation, and solute segregation during cooling are universal. The in situ redox reactions are experimentally well documented, at least for H_2O. Equally well founded is the prediction that the reduced solutes, H_2 and CO_2^{2-} or CO^-, should be diffusively more mobile and will therefore preferentially segregate, while peroxy oxygen remains lattice-bound.

The question then arises how the postulated protomolecules may be converted into full-fledged organic molecules. For this to happen, C–O^- bonds have to be broken. Two processes come to mind—hydrolysis, for instance during acid dissolution, $C–O^- + H_3O^+ \Rightarrow C–H + H_2O + OH^-$, or additional electron transfer involving O^-. To illustrate the latter we take the protomolecular unit depicted in the lower right of Figure 2b and assume that the ring is already fully hydrogenated. We can write an electron exchange reaction for the terminal $C–O^-$ with a nearby O^- leading to benzaldehyde:

$$(C_6H_5)–\overset{H}{C}=O^- + O^- \Rightarrow (C_6H_5)–\overset{H}{C}=O + O^{2-} \quad (5)$$

The next question is: How can such an electron exchange reaction be initiated in the crystal matrix? One way is to break the $O^-–O^-$ bonds of O_2^{2-}, either thermally or by other means, to produce delocalized O^-. It has been shown (Dickinson et al. 1986, Dickinson et al. 1987) that the mechanical shock generated during crystal fracture is enough to generate O^-, which diffuse to the fracture surface within milliseconds and lead to the emission of O atoms and Mg atoms. Possible reaction steps for these processes are:

$$O_2^{2-} \Leftrightarrow O^- + O^- \Rightarrow O^{2-} + O \quad (6a)$$

$$O^- + O^- + Mg^{2+} \Rightarrow Mg + 2\,O \quad (6b)$$

In a similar manner, protomolecular entities that happen to lie on the fracture surface will start to interact with the delocalized O^- generated by the fracture process and react as hypothesized by Equation 5.

Experimental Approach

Infrared Spectroscopy

We report here on studies of olivine crystals from the San Carlos locality in Arizona. These olivine crystals come from upper mantle peridotite nodules that were brought up by a volcanic eruption. The nodules are imbedded in cavity-rich basalt, once the vesiculating magma that filled the volcanic pipe. The olivine crystals have an approximate composition $(Mg_{0.9},Fe^{2+}_{0.1})_2SiO_4$ and range in size from a few mm to 10–20 mm. Coming from a depth of at least 80 km they have been exposed to > 25 KBar hydrostatic pressures and to substantial partial pressures of those gases that drive volcanic eruptions: H_2O and CO_2 with additions of H_2S, HCl and other constituents. Having been brought up rapidly in the rising magma the olivine crystals were decompressed and cooled moderately quickly. Optically, they appear slightly turbid, revealing in thin sections a dense mosaic of subgrains, of micron size or smaller. On fracture surfaces the subgrains form steps and terraces of the same size range (Brown 1980, Green and Guegen 1983, Mackwell et al. 1990, Jaoul 1990, Thompson 1992).

Figure 3 shows an infrared (IR) spectrum of a 9.6 mm thick olivine single crystal in the 2000–4000 cm^{-1} region where the characteristic O–H, C–H and C–O stretching bands are expected to appear. No O–H bands are visible though a weak signal for an H–H stretching band, indicative of molecular H_2, has been observed around 4150 cm^{-} (Freund and Oberheuser 1986). A broad, weak feature in the C–H stretching region can be interpreted as a sign for some C–H bonds. The double band in the C–O stretching region indicates the presence of C–O bonds but does not allow for an assignment as to whether it belongs to CO^-, CO_2^{2-} or CO_3^{2-} units or to C–O bonds in other local environments.

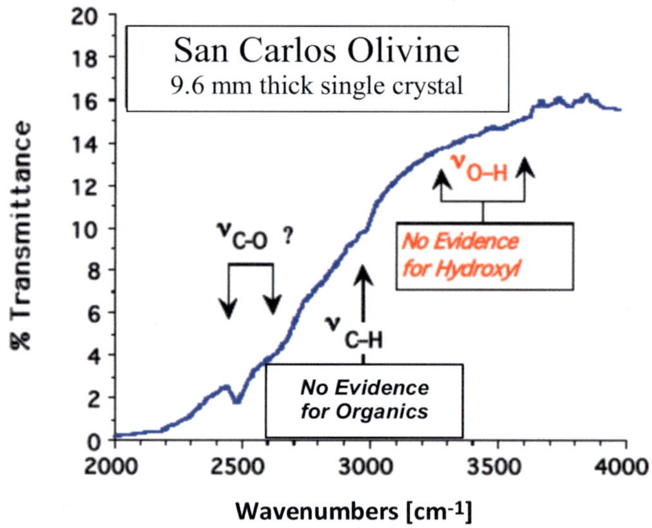

FIG. 3. *Infrared spectrum of a 9.6 mm thick olivine single crystal in the 2000–4000 cm^{-1} region where O–H, C–H and C–O stretching bands are expected to appear. (The "noisy" background between 3700 and 3900 cm^{-1} is due to residual water vapor in the spectrometer.)*

The indistinct nature of the IR spectrum does not rule out the presence of protomolecular entities which are expected to be strongly coupled to the mineral matrix, thus leading to extreme IR band broadening. However, it is quite clear that the olivine crystal does not contain the fully developed organic molecules that Tingle et al. (1990, 1991) postulated under the assumption that volcanic gases had penetrated existing microfissures, filling them with catalytic reaction products. Such fillings would give rise to much sharper IR bands, even if the concentration of organics, averaged over the thickness of the crystal, were in the parts-per-billion range.

Impact Fracture Experiments

Another way to see whether protomolecular entities may be present in the olivine crystals under study is to impact-fracture in vacuum and measure the organic molecules released from the fracture surface. In collaboration with J. T. Dickinson of Washington State University, single crystal fracture experiments were conducted in an ultra-high-vacuum system at a base pressure in the 10^{-9} torr range, using a fast-scanning quadrupole mass spectrometer (QMS), which recorded the 1–110 amu mass range every 80 msec (Dickinson et al. 1985, Dickinson et al. 1986). The data from the last scan before fracture are stored in a transient recorder together with twenty subsequent scans, spanning 1.6 sec total data collection time. Rectangular pieces, approximately 6 × 10 × 1 mm, cut from large olivine single crystals and cleaned in chlorinated hydrocarbons, were mounted between two supporting stainless steel plates, spaced by 2 mm, about 10 mm from the 70 V ionization source of the QMS. After thorough evacuation at ambient temperature the crystals were fractured with a stainless steel wedge, triggering the data acquisition system. It is important to note that, following the impact fracture, the gases desorb directly from the fracture surface into the QMS ionization source without incurring wall collisions. No chlorinated hydrocarbons were detected after fracture, indicating that the cleaning procedure did not penetrate fissures or that fissures into which the chlorinated hydrocarbons could have penetrated were absent.

Figure 4 shows a typical QMS sequence. Spectrum 1 shows the residual gas composition before fracture. The intensities of the H_2O, N_2/CO, and CO_2 signals at 18, 28, and 44 amu, respectively, correspond to partial pressures in the 10–10 torr range. The fracture lasted less than 5 μsec and occurred nearly halfway through Spectrum 2. Immediately after fracture a strong emission of higher-molecular-weight gas compounds is observed. Spectrum 3 shows increased intensity over the entire mass region. The intensity continues to grow in part during Spectrum 4 and Spectrum 5. The total time after fracture covered by Figure 4 is 275 msecs.

Figure 5 (p. 79) shows digitized mass data on a logarithmic scale for four scans during another room temperature fracture run. The residual gas immediately before fracture is shown in (a). The time of fracture is marked by the dashed line in (b). Thereafter the spectra are rich in low- to medium-mass signals, suggesting a complex gas mixture including organic molecules. Clusters of peaks of masses separated by 13 or 14 amu fill the entire spectrum, indicating increments of –CH– and –CH_2– units. The upward arrows in (c) and (d) indicate off-scale signals. The most prominent heavy mass is at 105 amu, most likely $C_7H_5O^+$, the main mass spectrometer fragment of benzaldehyde, C_7H_6O (106 amu).

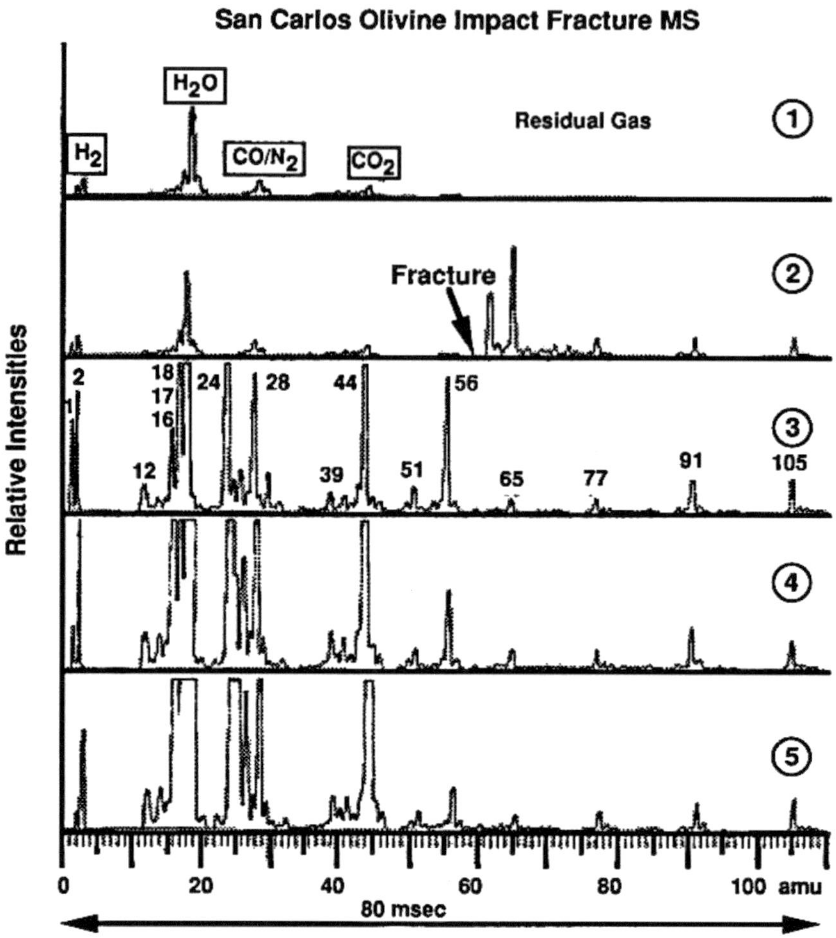

FIG. 4 (above). *Gas evolution just prior to and after impact fracture of an upper mantle–derived olivine single crystal measured with a fast-scanning quadrupole mass spectrometer over the mass range 1–110 amu. Intensities are linear in scale. Each of the five consecutive scans shown took 80 msec. Spectrum 1 shows the residual gas composition before fracture. The fracture that lasted < 5 μsec is indicated in Spectrum 2.*

FIG. 5 (opposite). *Four selected 1–110 amu scans just prior to, during and after impact fracture of an upper mantle–derived olivine single crystal. Intensities are logarithmic in scale. The time intervals over which the data were acquired are indicated with the time of fracture set at 0. Spectrum 1 shows the residual gas composition just before fracture.*

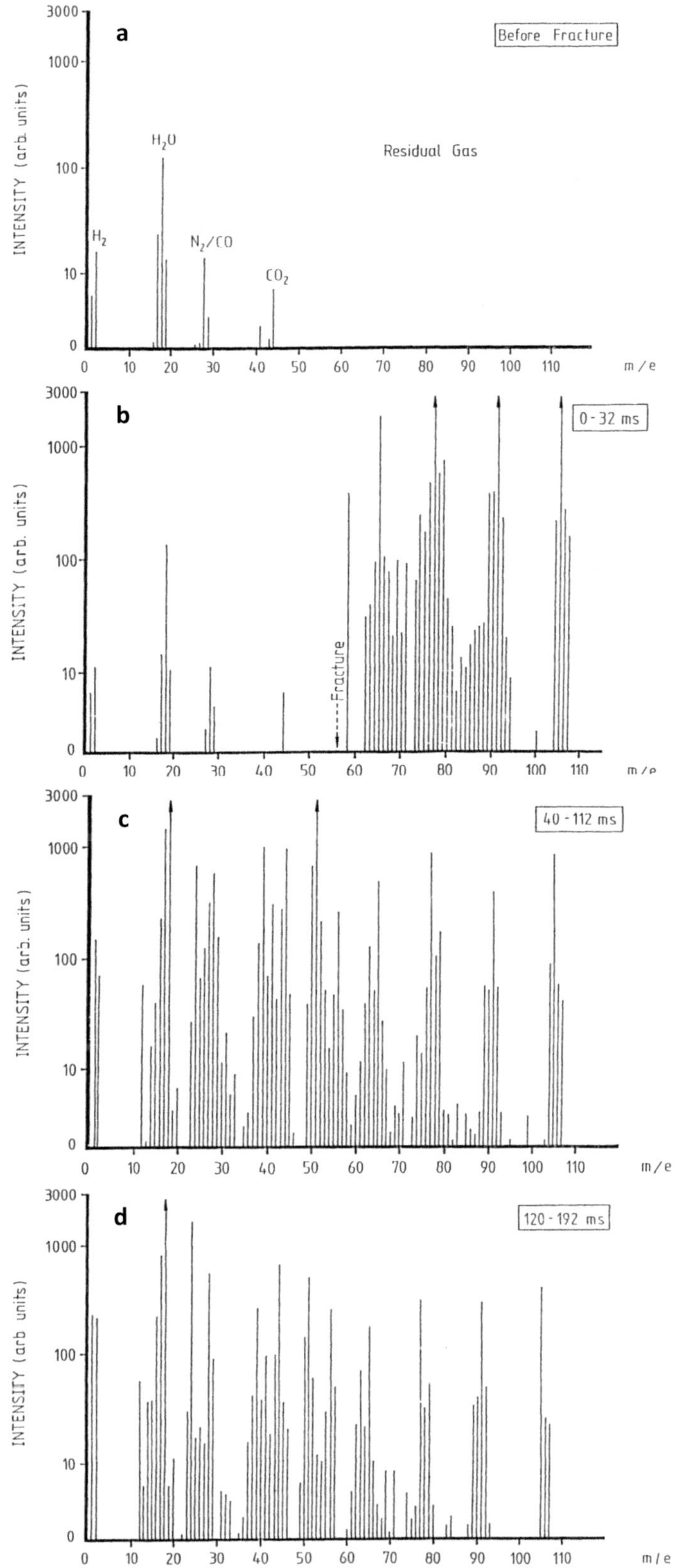
a
Before Fracture
Residual Gas
H_2
H_2O
N_2/CO
CO_2
b
0 - 32 ms
Fracture
c
40 - 112 ms
d
120 - 192 ms
INTENSITY (arb. units)
m / e

The fresh fracture surface acts as a sink for any mobile solute other than O^- that may exist in the bulk. For instance, residual interstitial H_2 molecules will diffuse toward the surface after the fracture. As shown in Figures 4 and 5 the intensity of the 2 amu signal, due to molecular H_2, rises after fracture, indicating release of molecular H_2 from the fracture surface. However, diffusing toward and passing through the surface is not all that the H_2 molecules can do. They also react at the surface, for instance with protomolecular entities, hydrogenating $C{-}O^-$ bonds: $H_2 + 2\ C{-}O^- \Rightarrow 2\ C{-}H + 2\ OH^-$. Such a reaction may result in the formation of C_1 molecules such as formaldehyde. Such or similar reactions are supported by observations during flash desorption experiments where the evolving organic molecules had different degrees of hydrogenation—first saturated (mostly aliphatic), then moderately saturated (olefinic) and then highly unsaturated (possibly with $-C{\equiv}C$ bonds) (Knobel 1983).

Alternatively, H_2 molecules that try to pass through the fracture surface can be scavenged by O^-, leading to the formation of H_2O molecules:

$$H_2 + 2\,O^- \Rightarrow H_2O + O^{2-} \quad (7)$$

That this reaction indeed seems to occur is illustrated by Figure 6 where the signal intensities of 1, 2, and 17 amu, due to H^+, H_2^+, and OH^+, respectively, are plotted versus time. H^+ can be mass spectrometer fragments of H_2O and any hydrocarbon molecules, but H_2^+ is characteristically due to molecular H_2. As shown in Figure 6, immediately after fracture H_2 molecules pass through the surface but at the same time H_2O molecules begin to evolve. Since we know from the IR spectra that H_2O molecules are definitely absent from the olivine crystal prior to fracture, their appearance in the mass spectrum signals some chemical reaction at the fracture surface. The H_2O evolution increases at first, reaches a maximum after about 300 msec (off scale) and then decreases. The H_2 evolution shows a different time dependence. After the initial rise the 2 amu intensity decreases, passes through a minimum near 300 msec and increases again thereafter before falling off slowly after a longer time.

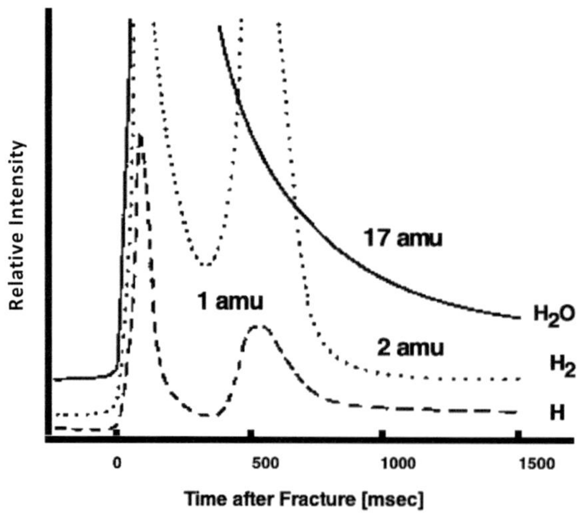

FIG. 6. *Intensity variations of the mass spectrometer species 1, 2, and 17 amu, representing H^+, H_2^+, and OH^+, the latter deriving from H_2O, as a function of time over 1.6 sec with $t = 0$ being the time of fracture of the olivine crystal.*

This time-dependent sequence is consistent with Equation 7 and suggests that an O^- pulse is generated in the bulk through the fracture event and reaches the surface within a very short time after fracture. The O^- immediately begin to oxidize H_2 to H_2O. This diminishes the number of H_2 molecules that make it through the surface. However, as the O^- pulse subsides, H_2 molecules can again pass through the fracture surface and reappear in the mass spectrum. This lasts until the supply of H_2 is exhausted. Similar correlations in the time-dependent evolution of CO/CO_2 and different hydrocarbon molecules have been noted but more work is needed, especially with better time resolution.

In summary, the QMS experiments have revealed a complex chemistry that appears to take place at the olivine surface within fractions of a second after fracture. Much of this chemistry seems to be driven by the diffusion of H_2 molecules and O^- states to the fracture surface and by the reactions among H_2 and O^-. Similar reactions appear to control the transformation of protomolecular entities into desorbable molecules. Though the number of organic molecules coming off the surface is difficult to measure, the best estimate is that they are on the order of at least one monolayer equivalent.

Field Ionization Mass Spectrometry

Next we present data on the release of complex organic molecules from freshly crushed olivine single crystals during heating, using a field ionization mass spectrometer (FIMS). This work was done in collaboration with R. Malhotra, SRI International, Menlo Park, California. Field ionization does not ionize H_2O, N_2, CO, CO_2 and some of the simplest hydrocarbons but it is well suited for the study of more complex organics. FIMS produces almost exclusively parent molecule ions, thus facilitating the assignment of the observed amu values to specific molecular species.

Olivine crystals from the same batch used for the fracture experiments were crushed to a coarse powder in a clean alumina mortar under a laminar flow hood. The crushed powder was loaded into a fused silica capillary, about 0.5 mm diameter and 10 mm long, and inserted into the source chamber of the magnetic sector FIMS instrument. After evacuation at 50°C source temperature, the temperature was raised at 4°C/min up to 500°C, while the mass spectra were continuously recorded.

Figure 7 shows the sum of all gases released over the temperature interval 50–500°C, plotted as intensity versus molecular weight. The peak for benzaldehyde at 106 amu is specifically marked. Other strong peaks occur at lower and higher masses, in particular at 126, 254, and 302 amu. If we plot the intensities as a function of temperature we obtain a graph like that shown in Figure 8 for selected compounds with molecular masses ranging from 78 amu (benzene) and 106 amu (benzaldehyde) to 302 amu.

Closer examination of the FIMS data of the higher-molecular-weight compounds suggests aromatics, some probably polyaromatics, others with side chains. Above 400°C benzene, $C_6H_6^+$ at 78 amu, becomes prominent, probably as a pyrolysis product of more complex aromatics. Some masses suggest oxygenated hydrocarbons or molecules containing nitrogen. The analysis of most mass peaks has not yet been completed because, despite nearly identical sample preparation and heating rates, the reactions at the freshly fractured olivine surface are complex. More work will be needed to unravel which compound or set of compounds forms in any given temperature interval. The situation is complicated by the fact that, before reaching the FIMS ionization source, the gases released from the olivine surface reside in the pore space of the powder bed packed into the fused silica capillary where they undergo collisions with gases and grain surfaces, leading to reactions and eventually pyrolysis.

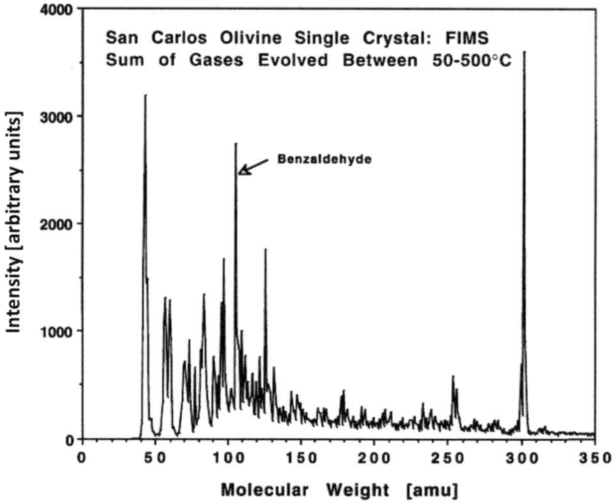

FIG. 7 (left). *Sum of all gases released from a freshly crushed San Carlos olivine single crystal as measured by FIMS over the temperature interval 50–500°C. Besides several prominent mass peaks at lower molecular weights there are several strong peaks suggesting higher-molecular-weight compounds such as at 106 amu, probably benzaldehyde.*

FIG. 8 (right). *Selected molecular species released from a freshly crushed San Carlos olivine single crystal as measured by FIMS, plotted as a function of temperature. In some cases species with one amu higher mass are included (dotted line), due to the natural* ^{13}C *abundance. Benzene at 78 amu is probably a pyrolysis product of polyaromatic compounds decomposed during heating.*

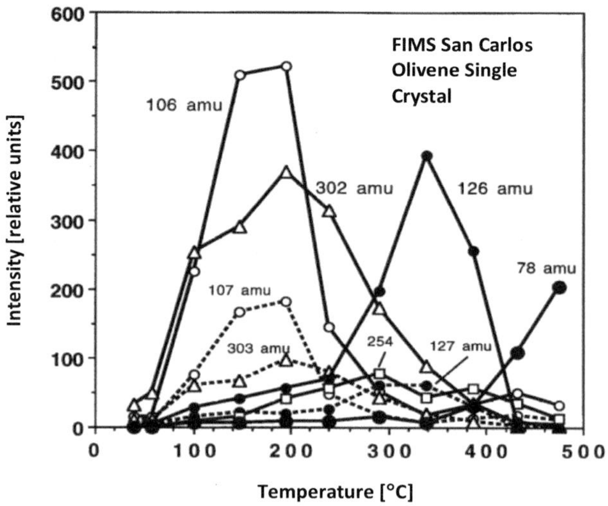

Gas Chromatography–Mass Spectrometry

To collect larger gas samples than available through the experiments described so far we built an ultra-high-vacuum (UHV) jaw crusher with which we can crush typically 50–100 g of olivine single crystals or other mineral and rock samples. The crusher consists of a 25 × 50 × 2 mm tungsten carbide plate mounted on a stationary steel piston and another tungsten carbide plate of the same dimensions, mounted at a 10° angle on a steel piston that is activated from the outside via bellows. The UHV chamber is attached to a turbomolecular pump and a residual gas analyzer. A horizontal glass tube above the crusher allows one to magnetically feed in the sample, either single crystals or chunks of rocks. The crushed material falls into a 25 mm diameter glass tube that can be either cooled to liquid nitrogen temperature or heated up to 500°C. Several 6 mm tubes are attached to collect gases released during fracture or heating. The system is also equipped with a gas manifold allowing for fracture under reactive gases such as H_2 or H_2S or under isotopically labeled gases such as D_2, ^{13}CO and $^{13}CO_2$.

We have used the jaw crusher to fracture (a) upper mantle–derived olivine single crystals and (b) the vesiculated basalt that brought them up during a volcanic eruption. In both cases, approximately 100 g single crystals or pieces of basalt measuring 5–15 mm were thoroughly cleaned in CH_3Cl and evacuated for several hours until achieving a base pressure of $< 10^{-6}$ torr. The pieces were then fed one by one into the jaw crusher by means of a magnetically operated pusher and crushed to a coarse powder to a grain size of < 0.5 mm. The gases collected were analyzed in collaboration with Peter Palmer, NASA Ames Research Center, using a VARIAN gas chromatograph (GC), equipped with a TEKMAR 6000 gas injection system and linked to a Finnigan–MAT ITS 40 mass spectrometer (MS).

Figure 9 shows four GC traces (a) of the gases released during crushing of the basalt and (b) to (d) of the gases released in three temperature intervals from the crushed olivine single crystals. The GC trace of the basalt in (a) is dominated by residual solvent, used to clean the chunks of basalt prior to crushing, but also shows a small peak identified by MS as due to benzene. The GC trace (b) of the gases collected from the crushed olivine between room temperature and 230°C also contains some solvent, but in addition benzene and other as yet not fully identified organic compounds. The amount of benzene increases with increasing temperature in (c) and (d) but so do the other as yet not fully characterized organic compounds. The analysis of the concurrent MS data is still in progress, but we already see that the gases collected during basalt fracture are distinct from those collected during fracture of the olivine crystals imbedded in the same basalt.

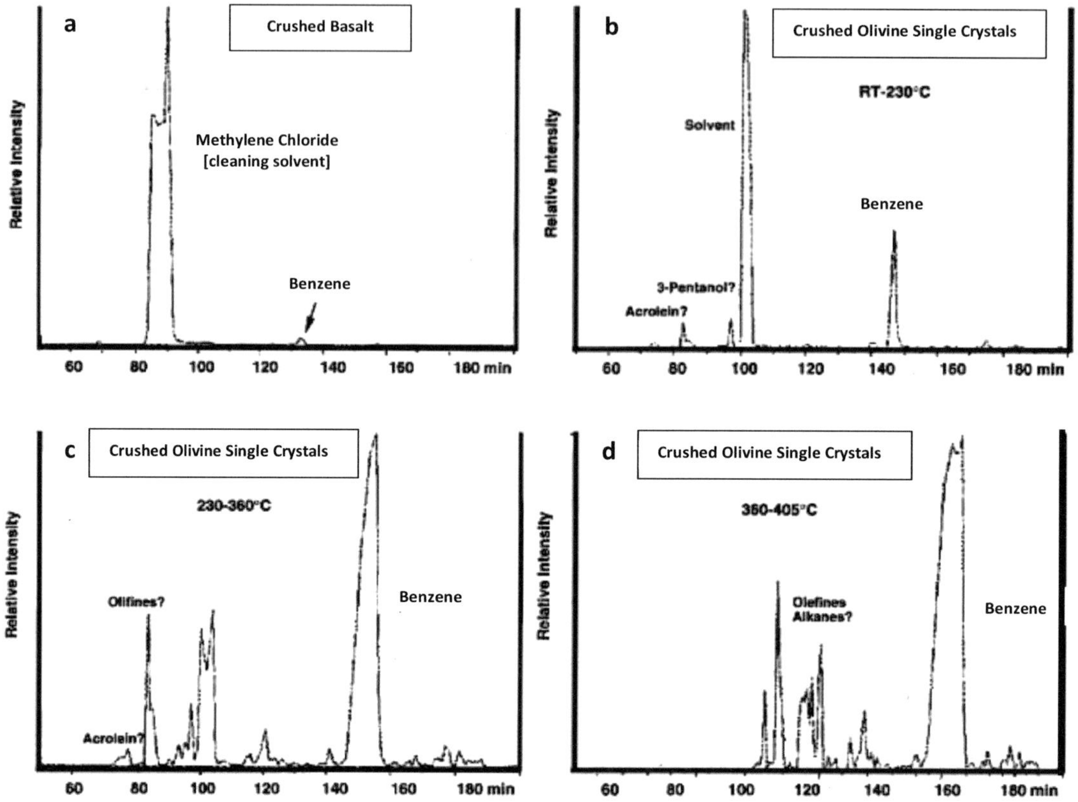

FIG. 9. *Gas chromatography (GC) traces (a) of the gases released during crushing of vesiculated basalt that transported the olivine from the upper mantle, (b) to (d) of the gases released from crushed olivine crystals at increasing temperatures.*

DISCUSSION

The fact that complex organic molecules evolve from the fracture surfaces of olivine single crystals is consistent with the thermodynamic model outlined above comprising the sequential steps—solid solution between mineral and gases, supersaturation during cooling/decompression, redox conversion of the solute entities, and selective exsolution of the reduced entities to subgrain boundaries. Though difficult to estimate, the amount of organic molecules evolved during fracture is on the order of a monolayer equivalent or $\approx 10^{15}$ cm^{-2}. For that many molecules to evolve within milliseconds of a fracture, C and H must already have been present at the internal plane that became a fracture surface.

If C and H were distributed at random in the olivine structure, monolayer desorption is impossible, unless the C and H concentrations were very high. In silicate melts at high pressures, solute H_2O and CO_2 concentrations are known to reach several weight-percent (Holloway 1981, Stolper 1982). However, the solubility in the solid state is believed to be low—less than 100 ppm H_2O, maybe less than 10 ppm (Aines and Rossman 1984, Miller et al. 1987)—and the solubility of the CO_2 component is believed to be even closer to zero (Mathez et al. 1987, Tingle et al. 1990, Tingle et al. 1991). Even at solute H and C concentrations on the order of 100 ppm, the number of H and C on a random fracture surface would be $\approx 10^{11}$ cm^{-2}, significantly less than observed during our fracture experiments. Hence, a mechanism must be in operation leading to an enhancement of C and H concentration where the fracture forms.

When olivine crystals are brought up by molten magma at 1000–1200°C, organic molecules or precursors thereof cannot exist in the olivine matrix. The molecules that are eventually released during fracture must therefore have been introduced or formed at a lower temperature, subsequent to the emplacement and cooling of the olivine crystals in the volcanic pipe. There are two possibilities—microfissures into which organic molecules were introduced from an extraneous contamination source, or subgrain boundaries, mechanically weakened by the segregation of C and H.

The first possibility, that of extraneous contamination, has been proposed by Tingle and coworkers (Tingle et al. 1990, Tingle et al. 1991). Referring to Mathez et al. (1987), who interpreted C-lined internal planes in olivine crystals as evidence for microfissures into which volcanic gases had penetrated, Tingle et al. (1990, 1991) assumed that these gases catalytically reacted with the walls and produced organics. Small amounts of Na^+, K^+, Al^{3+} and other cations normally not found in the olivine structure, were taken as evidence for the volcanic gas intrusion. However, the absence of a distinct C–H infrared band around 3000 cm^{-1}, even in as thick a section of an olivine crystal as shown in Figure 2, speaks against preformed organic molecules having been deposited into microfissures. In addition, the complexity of the molecular species released upon fracture requires that they were formed significantly below 500°C. Such a low temperature would not allow for microfissures to close and seal off so efficiently that even boiling in chlorinated solvents could not reach the organic molecules allegedly deposited in them.

We therefore favor the second possibility, namely that the organic molecules derive from protomolecular entities, $(H_xC_yO_z)^{n-}$, produced by the dissolution-exsolution-segregation of solute H and C to subgrain boundaries as described above. Such a process applies to any "impurities" in the olivine structure such as Na^+, K^+, Al^{3+} and so forth, that have a finite solubility at high temperatures and pressures but become less soluble during cooling/decompression. If subgrain boundaries represent a sink, as argued above, such "impurities" will also become enriched where the fracture eventually cuts through the olivine structure.

In this context it is noteworthy that olivine typically cleaves in a conchoidal and smooth fracture (Brown 1980). However, the San Carlos olivine crystals under study here give terraced fracture surfaces, suggesting subgrain boundaries that are weakened by exsolution of "impurities." In thin sections in the optical microscope under crossed polarizers these subgrain boundaries show up as pervasive μm-sized mosaics—a feature that cannot be formed by adventitious opening and closing of microfractures.

Further evidence for the formation of organic molecules by exsolution is given by comparing the gases released during basalt fracture and olivine fracture. The basalt is strongly vesiculated, indicating active degassing during ascent in the volcanic pipe. If the volcanic gases were capable of catalyzing complex organic molecules on mineral surfaces, the basalt should be a fertile source. This is not the case as shown by the paucity of the GC trace (a) in Figure 9 (p. 84). Further work will be needed to identify the organics released from basalt and to separate them from adventitious contamination.

One other series of tests currently under way is to acid-dissolve minerals and recover the organics that may be generated upon hydrolysis. If our assumption is correct that protomolecular entities $(H_xC_yO_z)^{n-}$ exist in the mineral matrix, hydrolysis is likely to produce different organic molecules from those released upon fracture or during heating (Freund 1984, Knobel et al. 1984).

At the end of this preliminary report we may speculate about some implications of the abiogenic synthesis pathway outlined here. On a volcanically and tectonically active planet like the Earth, rocks are continuously uplifted to build mountains. Thus they become exposed to surface and groundwater that dissolves the minerals. If those minerals contain protomolecular entities formed through exsolution, their protomolecular entities will hydrolyze and become molecules. Considering that every year several cubic kilometers of rock are weathered, the amounts of organics potentially released in this way could be quite large. Assuming ≈ 10 ppm of organics released, a weathering rate of $\approx 3\ km^3/yr$ and a rock density of $\approx 3\ g/cm^3$, the amount would be 10^{11} g/yr. If accumulated over 10^8 years, the total mass of organics released becomes 10^{19} g. Though still speculative, the process warrants attention.

Acknowledgments

This work is being supported by the NASA Ames Research Center Director's Discretionary Fund and through the NASA–SETI Institute Cooperative Agreement NCC 2-572.

REFERENCES

AINES, R.D., AND G.R. ROSSMAN (1984), "Water in Minerals—A Peak in the Infrared?," *JGR* **89:** 4059–4071.

ANDERS, A. (1989), "Pre-biotic Organic Matter from Comets and Asteroids," *Nature* **342:** 255–257.

BELL, D.R., AND G.R. ROSSMAN (1992), "Water in Earth's Mantle: The Role of Nominally Anhydrous Minerals," *Science* **255:** 1391–1397.

BROWN, G.E. (1980), *Olivine and Silicate Spinels, Reviews in Mineralogy*, R.H. Ribbe, ed. (Washington, DC: American Mineralogical Society), pp. 275–381.

CHANG, S. (1988), "Planetary Environments and the Conditions of Life," *Phil. Trans. R. Soc. Lond. A* **325:** 601–610.

CHANG, S. (1993), "Prebiotic Synthesis in Planetary Environments," in *The Chemistry of Life's Origins*, J.M. Greenberg, ed. (Amsterdam: Kluwer Academic Publishers), pp. 259–299.

CHANG, S., D. DES MARAIS, R. MACK, S.L. MILLER, AND G.E. STRATHEARN (1983), "Prebiotic Organic Synthesis and the Origin of Life," in *Earth's Earliest Biosphere*, J.W. Schopf, ed. (Princeton, NJ: Princeton University Press), pp. 53–92.

CHITTENDEN, G.J.F., AND A.W. SCHWARTZ (1981), "Prebiotic Photosynthetic Reactions," *BioSystems* **14:** 15–32.

CHYBA, C.F. (1990), "Impact Delivery and Erosion of Planetary Oceans in the Early Inner Solar System," *Nature* **343:** 129–133.

CHYBA, C., AND C. SAGAN (1992), "Endogenous Production, Exogenous Delivery and Impact-Shock Synthesis of Organic Molecules: An Inventory for the Origins of Life," *Nature* **355:** 125–132.

CHYBA, C.F., P.J. THOMAS, L. BROOKSHAW, AND C. SAGAN (1990), "Cometary Delivery of Organic Molecules to the Early Earth," *Science* **249:** 366–373.

DICKINSON, J.T., M. DRESSER, AND L.C. JENSEN (1985), "Time Correlation of Ion and Electron Emission from Surfaces Following Fracture," Springer Series Surface Science, *Desorption Induced Electron Transitions (DIET)* (Heidelberg: Springer-Verlag), pp. 281–289.

DICKINSON, J.T., L.C. JENSEN, M.R. MCKAY, AND F. FREUND (1986), "The Emission of Atoms and Molecules Accompanying Fracture of Single Crystal Magnesium Oxide," *J. Vac. Sci. Technol. A* **1986.4(3, Pt. 2):** 1648–1652.

DICKINSON, J.T., L.C. JENSEN, AND M.R. MCKAY (1987), "Neutral Molecule Emission from the Fracture of Crystalline Magnesium Oxide," *J. Vac. Sci. Technol. A* **1987.5(4, Pt. 2):** 1162–1168.

DREVER, J.I., Y.H. LI, AND J.B. MAYNARD (1988), "Geochemical Cycles: The Continental Crust and the Oceans," in *Chemical Cycles in the Evolution of the Earth*, C.B. Gregor, R.M. Garrels, F.T. Mackenzie, and J.B. Maynard, eds. (New York: John Wiley & Sons), pp. 17–51.

FREUND, F. (1984), "Abiotic Synthesis of Organic Molecules from Minerals Containing Traces of Dissolved H_2O, CO_2 and N_2. Part I: Basic Principles," in *Proc. 7th Int. Conf. Origins of Life*, K. Dose, A.W. Schwarz, and W.H.P. Tieman, eds. (Dordrecht: D. Reidel Publishing Company), pp. 185–196.

FREUND, F. (1985), "Conversion of Dissolved 'Water' into Molecular Hydrogen and Peroxy Linkages," *J. Non-Cryst. Solids* **71:** 195–202.

FREUND, F. (1986), "Solute Carbon and Carbon Segregation in Magnesium Oxide Single Crystals—Secondary Ion Mass Spectrometer Study," *Phys. Chem. Minerals* **13:** 262–276.

FREUND, F. (1987), "Hydrogen and Carbon in Solid Solution in Oxides and Silicates," *Phys. Chem. Minerals* **15:** 1–18.

FREUND, F., AND H. WENGELER (1982), "The Infrared Spectrum of OH-Compensated Defect Sites in C-Doped MgO and CaO Single Crystals," *J. Phys. Chem. Solids* **43:** 129–145.

FREUND, F., AND G. OBERHEUSER (1986), "Water Dissolved in Olivine: A Single Crystal Infrared Study," *JGR* **91:** 745–761.

FREUND, F., H. KATHREIN, H. WENGELER, R. KNOBEL, AND H.J. HEINEN (1980), "Carbon in Solid Solution in Forsterite—A Key to the Untractable Nature of Reduced Carbon in Terrestrial and Cosmogenic Rocks," *Geochim. Cosmochim. Acta* **44:** 1319–1333.

FREUND, F., M.M. MASUDA, AND M.M. FREUND (1991), "Highly Mobile Oxygen Hole-type Charge Carriers in Fused Silica," *J. Mater. Sci.* **8:** 1619–1622.

FREUND, F., M.M. FREUND, AND F. BATLLO (1993), "Critical Review of Electrical Conductivity Measurements and Charge Distribution Analysis of Magnesium Oxide," *JGR* **98:** 22,209–22,229.

GREEN, H.W., AND Y. GUEGEN (1983), "Deformation of Peridotite in the Mantle and Extraction by Kimberlite: A Case History Documented by Fluid and Solid Precipitates in Olivine," *Tectonophys.* **92:** 71–92.

HENNET, J.C., N.G. HOLM, AND M.H. ENGEL (1992), "Abiotic Synthesis of Amino Acids Under Hydrothermal Conditions and the Origin of Life: A Perpetual Phenomenon?," *Naturwiss* **79:** 361–365.

HOLLOWAY, J.R. (1981), "Volatile Interactions in Magmas," *Adv. Phys. Geochem.* **1:** 273–293.

INGMANSON, D.E., AND M.J. DOWLER (1977), "Chemical Evolution and the Evolution of the Earth's Crust," *Origins of Life* **8:** 221–224.

JAOUL, O. (1990), "Multicomponent Diffusion and Creep in Olivine," *JGR* **95:** 17,631–17,642.

KNOBEL, R. (1983), *Gaseous H–C–N–O Compounds Released from Magnesium Oxide and Silicate Surfaces*, Ph.D. dissertation, Cologne (in German).

KNOBEL, R., H. BREUER, AND F. FREUND (1984), "Abiotic Synthesis of Organic Molecules from Minerals Containing Traces of Dissolved H_2O, CO_2 and N_2. Part II: Experimental Data," in *Proc. 7th Int. Conf. Origins of Life*, K. Dose, A.W. Schwarz, and W.H.P. Tieman, eds. (Dordrecht: D. Reidel Publishing Company), pp. 197–202.

KRÖNER, A., AND P.W. LAYER (1992), "Crust Formation and Plate Motion in the Early Archean," *Science* **256:** 1405–1411.

LAWLER, M.E., AND D.E. BROWNLEE (1992), "CHON as a Component of Dust from Comet Halley," *Nature* **359:** 810–812.

MACKWELL, S.J., Q. BAI, AND D.L. KOHLSTEDT (1990), "Rheology of Olivine and the Strength of the Lithosphere," *Geophys. Res. Lett.* **17:** 9–12.

MATHEZ, E.A., J.D. BLACIC, J. BERRY, C. MAGGIORE, AND M. HOLLANDER (1987), "Carbon in Olivine: Results from Nuclear Reaction Analysis," *JGR* **92:** 3500–3506.

MILLER, G.H., G.R. ROSSMAN, AND G.E. HARLOW (1987), "The Natural Occurrence of Hydroxide in Olivine," *Phys. Chem. Minerals* **14:** 461–472.

MILLER, S.L. (1953), "A Production of Amino Acids Under Possible Primitive Earth Conditions," *Science* **117:** 527–528.

MUKHIN, L.M., M.V. GERASIMOV, AND E.N. SAFONOVA (1989), "Origin of Precursors of Organic Molecules During Evaporation of Meteorites and Mafic Terrestrial Rocks," *Nature* **340:** 46–48.

ORÓ, J. (1961), "Comets and the Formation of Biochemical Compounds on the Primitive Earth," *Nature* **190:** 389–390.

PINTO, J.P., G.R. GLADSTONE, AND Y.L. YUNG (1980), "Photochemical Reduction of Carbon Dioxide to Formaldehyde in the Earth's Primitive Atmosphere," *Science* **210:** 183–185.

STOLPER, E.M. (1982), "Water in Silicate Glasses: An Infrared Spectroscopic Study," *Contrib. Mineral. Petrol.* **81:** 451–462.

STRIBLING, R., AND S.L. MILLER (1987), "Energy Yields for Hydrogen Cyanide and Formaldehyde Syntheses: The HCN and Amino Acid Concentrations in the Primitive Ocean," *Origins of Life* **17:** 261–273.

THOMPSON, A.B. (1992), "Water in the Earth's Upper Mantle," *Nature* **358:** 295–302.

TINGLE, T.N., M.F. HOCHELLA, C.H. BECKER, AND R. MALHOTRA (1990), "Organic Compounds on Crack Surfaces in Olivine from San Carlos, Arizona, and Hualalai Volcano, Hawaii," *Geochim. Cosmochim. Acta* **54:** 477–485.

TINGLE, T.N., E.A. MATHEZ, AND M.F. HOCHELLA (1991), "Carbonaceous Matter in Peridotites and Basalts Studied by XPS, SALI, and LEED," *Geochim. Cosmochim. Acta* **55:** 1345–1352.

WALKER, J.C.G. (1977), *Evolution of the Atmosphere* (London: Macmillan Publishing Company), p. 205.

WALKER, J.C.G. (1985), "Carbon Dioxide on the Early Earth," *Origins of Life* **16:** 117–127.

The Whirlpool Galaxy M51 and Its Companion

Photo Credit: NASA, ESA, S. Beckwith (STScI), and the Hubble Heritage Team (STScI/AURA)
https://science.nasa.gov/image-detail/40370191930-5e438cf3ca-o/ [retrieved 12-10-25]

M51's two curving arms, a hallmark of so-called grand-design spiral galaxies, are home to young stars, while its yellow core is where older stars reside. These arms are star-formation factories, compressing hydrogen gas and creating clusters of new stars. Astronomers used to believe that the arms of the Whirlpool Galaxy were so prominent because of the effects of a close encounter with NGC 5195, the smaller, yellowish galaxy at the outermost tip of one of the Whirlpool's arms [right side of the photo]. At first glance, the compact galaxy appears to be tugging on the arm. However, Hubble's clearer view shows that NGC 595 is passing behind the Whirlpool Galaxy.

For more information, see hubblesite.org/image/1677.

See also https://science.nasa.gov/asset/hubble/out-of-this-whirl-the-whirlpool-galaxy-m51-and-companion-galaxy/

THE SYNTHESIS OF LIPID BIOMARKERS BY METHANE-OXIDIZING BACTERIA AND INTERPRETATION OF PALEO-ENVIRONMENTS ON EARLY EARTH

Linda L. Jahnke

NASA Ames Research Center
Moffett Field, CA 94035

Harold P. Klein

Santa Clara University
Santa Clara, CA 95053
and
SETI Institute
Mountain View, CA 94043

Abstract

Over the course of Earth's history, the most important biological influence has been from the evolution of oxygenic photosynthesis and the development of an oxygen-rich biosphere. Although the availability of free oxygen had profound effects on subsequent biological and planetary evolution, clear paleobiological evidence of the timing of this transition is lacking. Recent technical advances in the microanalysis of organic matter have made it possible to detect residual molecules (biomarkers) in proterozoic sedimentary rock characteristic of specific groups of microorganisms. When coupled with the carbon isotopic fractionations characteristic of biological systems and the new field of compound specific isotope analysis, biomarkers could prove to be a powerful tool for decoding ancient biochemistry from the geochemical record. We have been studying the synthesis of organic biomarker molecules in several types of methane-oxidizing bacteria that should have been key to carbon cycling in paleoenvironments after the evolution of oxygenic photosynthesis.

INTRODUCTION

Key to our understanding of environmental and microbial evolution is information relating the geological and biological records. One such source of information is contained in the complex organic material that has escaped microbial and diagenetic degradation and is now trapped in terrestrial sediments. These biological remnants, or biological fossils, can provide information about the organisms that were probably present at the time of deposition of the parent compounds and also may be useful in elucidating the nature of the environmental conditions under which these molecules were formed.

Recent advances in analytic technology have allowed the identification of a variety of terpenoid hydrocarbons, including hopanoids and steranes, in ancient sediments (Summons and Walter 1990). These substances appear to be particularly resistant to thermal degradation, thus allowing for their detection in samples from precambrian and proterozoic sediments (Summons and Walter 1990, Summons et al. 1988). The magnitude of the contribution of these terpenoids to the inventory of organic matter that has accumulated on the Earth over geologic time is illustrated by the assertion that "The fossil hopanoids alone contain at least as much organic carbon as all the living animals, plants and microorganisms (about 10^{12} t of reduced carbon)" (Ourisson et al. 1987).

While many eubacterial types undoubtedly contributed to the incorporation of such terpenoids into ancient sediments by virtue of their ability to synthesize hopanoids, the production of sterols appears to have been restricted to relatively few species of bacteria (Rohmer et al. 1984, Ourisson et al. 1987). Among the few prokaryotes that could have contributed both of these chemical fossils to early sediments are methanotrophic bacteria, which are known to contain both hopanoids and sterols. In addition to establishing the presence of these organisms in terrestrial sediments, analysis of the carbon isotopic signatures of these bacteria may also provide significant insights into the geological history of this planet (Hayes 1983, 1994).

METHANOTROPHS AND THE GEOLOGIC RECORD

Biomarker Synthesis in the Methanotrophic Bacteria

The methane-oxidizing bacteria (methanotrophs) are of particular importance to the history of oxygen on Earth since they provided a means for more complete cycling of carbon in paleoenvironments. These bacteria are obligate aerobes and require oxygen both for the oxidation of methane via the enzyme methane mono-oxygenase (MMO) and for the generation of an ATP-dependent proton motive force (Anthony 1982). Methane is oxidized through the sequential enzymatic steps shown in Equation 1 to carbon dioxide.

$$CH_4 + O_2 \xrightarrow{\text{MMO}} CH_3OH \rightarrow HCHO \rightarrow HCOOH \rightarrow CO_2 \quad (1)$$

Methane is assimilated into biomass at the formaldehyde level by one of two different mechanisms, either the ribulose monophosphate pathway (RuMP) or the serine pathway (Anthony 1982). With RuMP-type methanotrophs, formaldehyde undergoes an aldol condensation with ribulose 5-phosphate to form a hexulose phosphate. There follows a series of cleavage and rearrangement steps in many respects similar to the Calvin cycle of CO_2 fixation. All of the C-biomass derives ultimately from methane. Methanotrophs having the serine pathway are quite different in that the intermediates are all carboxylic and amino acids, and carbon is derived from two sources—an aldol condensation of formaldehyde and glycine (serine transhydroxymethylase) and a carboxylation of phosphoenol pyruvate (PEP carboxylase). The C3 metabolic intermediate (phosphoglycerate) is generated from two molecules of formaldehyde and one molecule of CO_2. Thus C-biomass may derive from either methane or CO_2 in organisms utilizing this pathway.

As noted above, methanotrophs synthesize both hopanoid and sterol biomarkers. Large amounts of a complex amphiphilic hopanoid, bacteriohopane-polyol, are found together with methyl sterols in the outer membrane of these bacteria (Jahnke et al. 1992). Sterols found in methanotrophs are methylated in the 4-position unlike C27 sterols, such as cholesterol, that are common in eukaryotes (Bird et al. 1971, Bouvier et al. 1976).

In these organisms, all of the enzymatic steps leading to hopanoids are anaerobic. Squalene is the precursor molecule for both hopanoid and sterol synthesis (Fig. 1). Squalene can be directly converted to the pentacyclic hopane skeleton by a proton mediated cyclization. For the synthesis of sterols, the squalene can also be converted via an oxygen-requiring enzyme to squalene epoxide and then cyclized to a 3-hydroxy tetracyclic skeleton (lanosterol). Demethylation of lanosterol involves further oxygen-requiring enzymes and results in the 4,4-dimethyl and 4-methyl sterols found in methanotrophs such as *Methylococcus capsulatus*.

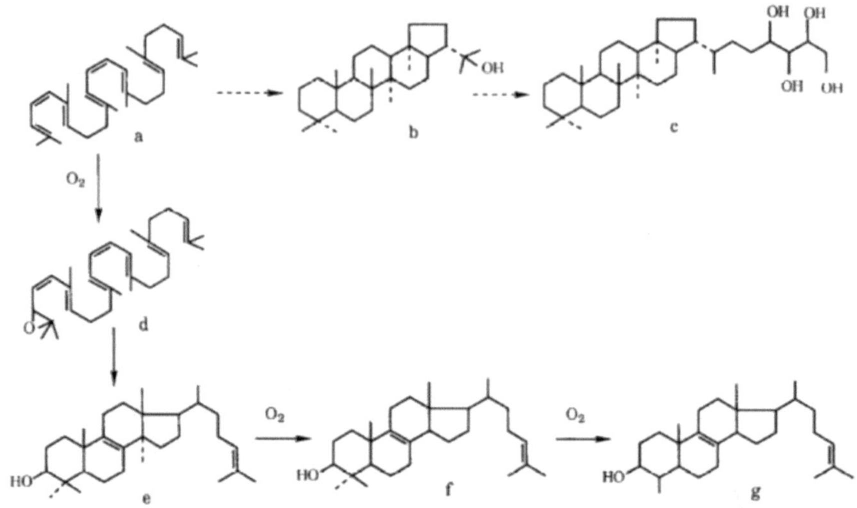

FIG. 1. *General biosynthetic pathway for cyclic triterpenoids in methanotrophic bacteria. (a) squalene, (b) diplopterol, (c) bacteriohopanetetrol, (d) squalene epoxide, (e) lanosterol, (f) 4,4-dimethylsterol, (g) 4 α-monomethylsterol.*

Role of Oxygen

In terms of understanding the environments under which these methane oxidizers may have functioned in the past, it is important to consider the levels of oxygen required in their metabolism. Since these are known to be "aerobic" organisms, requiring oxygen for the formation of some of their lipids, it has been suggested that such aerobic processes must have arisen after the Earth's atmosphere became oxygenic (cf. Bloch 1983). It should be noted, however, that "aerobic" processes do not necessarily imply oxygen concentrations anywhere near present atmospheric levels (PAL) (Jahnke and Klein 1979). While methanotrophs are normally grown in the laboratory under relatively high oxygen levels (10%), they are found in nature in zones approaching 10^{-2} PAL (Hanson 1980). Dalton (1992) has reported K*m* values for oxygen of 16 μM and 0.14 μM for MMO in these organisms. These values correspond to 10^{-1} to 10^{-3} PAL.

Low levels of oxygen may also be adequate for sterol synthesis, even in eukaryotes. For example, the K*m* for oxygen for the squalene epoxidase of *Saccharomyces cerevisiae* was found to be only 0.3% O_2 (Jahnke and Klein 1983). We have also found that sterol synthesis in *M. capsulatus* occurs to the same extent whether cells are grown with high or low levels of oxygen (Jahnke and Nichols 1986). The only difference noted was that above 0.5% O_2, 4-methyl sterol is the predominant end-product and below 0.1% O_2, it is the 4,4-dimethyl sterol. The lack of stereospecificity of the squalene/squalene epoxide cyclase(s) in *M. capsulatus* has led Rohmer et al. (1980) to suggest that the sterol synthetic pathway originated within this bacterial group. And indeed, the broad substrate specificity of the MMO should result in epoxide formation with almost any alkene (Colby et al. 1977), suggesting a potential mechanism for the synthesis of squalene epoxide. The origins of such synthetic mechanisms at oxygen concentrations well below that of the present atmosphere are clearly implied by these studies.

The Stable Carbon Isotopic Record

Throughout much of Earth's history, before the evolution of eukaryotic algae and land plants, the oxygen and carbon cycles were driven by purely bacterial processes. Then as now, the primary biological production of reduced carbon compounds was driven by the fixation of carbon dioxide by organisms containing ribulose bisphosphate carboxylase. Initially, in the absence of atmospheric oxygen, fermentative bacteria could degrade the resultant reduced carbon compounds to carbon dioxide and simpler organic compounds, and the obligately anaerobic methanogens could reduce many of the products of fermentation to methane, using CO_2 as the terminal oxidant. Under these anaerobic environments, residual organics could be deposited into local sediments and methane escaped to the atmosphere. After the evolution of the cyanobacteria, when oxygen began to accumulate, aerobic metabolic processes were added to this simple carbon cycle (Fig. 2). Now, microbial methane oxidation provided a mechanism for the conversion of methane to complex molecules like the triterpenes discussed above. The survival of these methanotrophic lipids (or their molecular fossils) in sediments provides insights not only into the probable origins of methanotrophy but also offers clues about the ambient conditions at the time of their production (Hayes 1983, 1994). These insights derive not only from the identification of methanotrophic lipids, but also from the fact that the products of these organisms are severely depleted in ^{13}C.

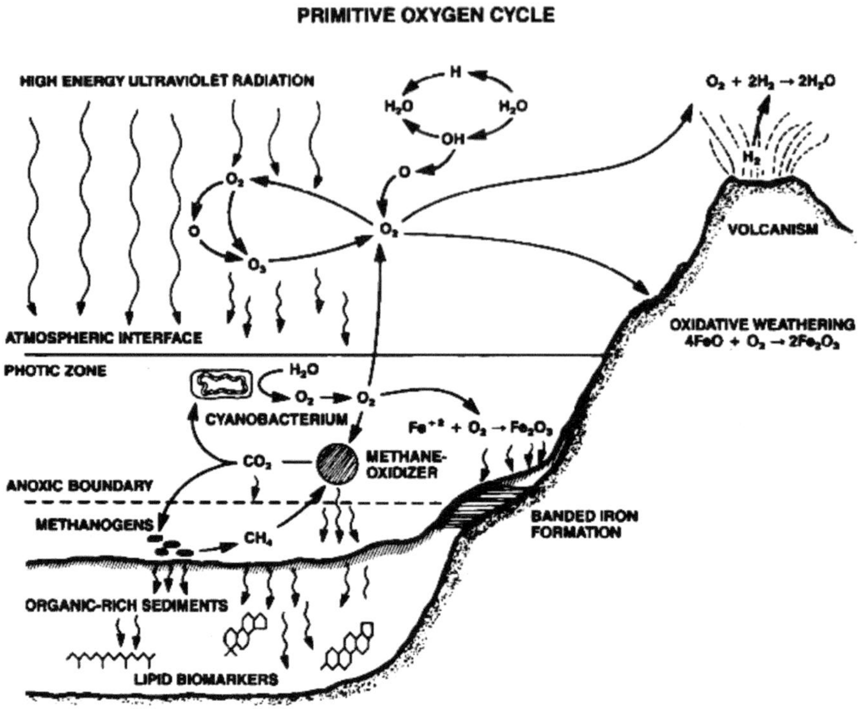

FIG. 2. *Cycling of carbon and the deposition of lipid biomarker molecules in microbially mediated paleoenvironments.*

During the process of carbon dioxide fixation by ribulose bisphosphate carboxylase a kinetic isotope effect results in depletion of carbon-13, so that the $\delta\ ^{13}C$ of the photosynthetic biomass as defined by Equation 2 generally ranges from a negative 20‰ to 30‰ (Schidlowski and Aharon 1992).

$$\delta\ ^{13}C = \left[\frac{^{13}C/^{12}C \text{ sample}}{^{13}C/^{12}C \text{ standard}} - 1 \right] \times 1000\ (‰, \text{PDB}) \qquad (2)$$

Microbial degradation of this depleted biomass by the anaerobic methanogens results in another isotopic effect, and the methane that is produced by these bacteria is now between −60‰ and −110‰ (Whiticar et al. 1986).

We have found that incorporation of such depleted methane into the biomass of methane-oxidizing bacteria results in a further depletion of carbon-13 by an additional 15–30‰ (Summons et al. 1994). This results in extremely ^{13}C-depleted organic molecules (Table 1). The degree of depletion depends on a number of factors relating to environmental conditions and the metabolic nature of the methane-oxidizer involved (serine type vs. RuMP type). In the methanotrophs, the initial oxidation of methane results in a kinetic isotope effect that appears to be primarily associated with the MMO enzyme and whether the cells express a soluble form of this enzyme (sMMO) or a particulate, membrane-bound one (pMMO) (Summons et al. 1994). In RuMP-type methanotrophs such as *Methylomonas methanica* or *M. capsulatus*, the greatest fractionation results when cells grow on methane using a pMMO. In these cases, hopanoid and methyl sterol biomarkers are depleted in ^{13}C by as much as 37‰. While fractionation is not as great in cells using the sMMO or growing with low levels of oxygen, considerable depletion is also apparent.

In the case of methanotrophs using the serine pathway, interpretation of any carbon isotopic fractionation is not as straightforward. Biomass carbon may originate from methane via a pMMO or sMMO as for RuMP-types, or from CO_2 after fixation by PEP carboxylase. This enzyme is reported to have only a minimal kinetic isotope effect with a ^{13}C discrimination of about 2% (Reibach and Benedict 1977). In a growth experiment with a serine type, *Methylosinus trichosporium*, when cells were grown with methane having a $\delta\ ^{13}C$ of −34.2% and CO_2 having a $\delta\ ^{13}C$ of −40.5%, the difference between the resulting biomass carbon (−43.7%), and total lipid carbon (−53.1%) suggested the presence of two metabolic pools of carbon (Table 1). Given a methane fractionation similar to RuMP types, these data imply that the lipid carbon and hopanoid biomarkers were derived almost exclusively from formaldehyde, while overall biomass was derived from both CO_2 and HCHO.

TABLE 1. ***Carbon Isotopic Fractionation of Methanotrophic Biomass and Lipids****

Methanotroph†	Carbon Fixation Pathway	MMO Type	Oxygen (%, v/v)	Carbon Isotopic Fractionation ($\delta\ ^{13}C$, ‰) Biomass	Lipid	Hopanol	Sterol
Methylomonas methanica	RuMP	Particulate	10%	−61.9	−66.8	−72.2	–‡
Methylococcus capsulatus	RuMP	Particulate	10%	−60.9	nd	−68.7	−71.1
M. capsulatus	RuMP	Soluble	10%	−49.5	−51.7	−53.7	−54.8
M. capsulatus	RuMP	nd	0.5%	−48.8	−51.3	−52.1	−52.8
Methylosinus trichosporium	Serine	nd	10%	−43.9	−53.1	−54.8	–‡

* *Data adapted from Summons et al. 1994. MMO, methane monooxygenase; RuMP, ribulose monophosphate pathway; nd, not done.*

† *Cells were grown in an open system with constant sparging of a gas mixture (v/v) of 50% methane ($\delta^{13}C$ −34.2‰), 0.2% carbon dioxide ($\delta^{13}C$ −40.5‰), oxygen (as indicated) and nitrogen as diluent.*

‡ *Sterol in these organisms either is not present or amount is insufficient for isotopic analysis.*

Paleoenvironmental Significance of Methanotrophic Cyclic Triterpenes

Methane-oxidizing bacteria would have played an important role in cycling carbon in early Earth environments (Fig. 2). Although there is evidence for a sulfate-mediated oxidation of methane in anaerobic environments today, this process does not result in assimilation of methane carbon into cell material (Panganiban et al. 1979). Therefore, before the appearance of MMO, no residual evidence of biogenically incorporated methane would occur in the geochemical record. Cyclic triterpene molecules such as sterols and hopanoids preferentially survive microbial and diagenetic degradation and tend to make up a greater portion of buried organic matter (Boon 1984). As pointed out above, geohopanoids are one of the most prevalent compounds extracted from ancient sedimentary rocks and oils. Recognizable hopanoid molecules have been extracted from 1.7 Gya rocks of the Barney Creek formation in Australia (Jackson et al. 1986), attesting to the survival of these organic molecules. The new technique of compound-specific isotopic analysis is providing direct evidence for specific metabolic carbon input to ancient sediments (Hayes et al. 1990, Freeman et al. 1990).

The fact that the metabolism of the methanotrophs results in the production of lipids severely depleted in ^{13}C, coupled with carbon isotope measurements of geological materials from the Archean and Proterozoic eras, has been used by Hayes (1983, 1994) to suggest that atmospheric oxygen would have been available to methanotrophs as early as 2.75 Gya.

References

ANTHONY, C. (1982), *The Biochemistry of Methylotrophs* (London: Academic Press).

BIRD, C.W., J.M. LYNCH, F.J. PIRT, W.W. REID, C.J.W. BROOKS, AND B.S. MIDDLEDITCH (1971), "Steroids and Squalene in *Methylococcus capsulatus* Grown on Methane," *Nature* **230:** 473–474.

BLOCH, K. (1983), "Sterol Structure and Membrane Function," *CRC Crit. Rev. Biochem.* **14:** 47–92.

BOON, J.J. (1984), "Tracing the Origin of Chemical Fossils in Microbial Mats: Biogeochemical Investigations of Solar Lake Cyanobacterial Mats Using Analytical Pyrolysis Methods," in *MBL Lectures in Biology: Microbial Mats: Stromatolites*, vol. 3, Y. Cohen, R.W. Castenholz, and H.O. Halvorson, eds. (New York: Alan R. Liss, Inc.), pp. 313–342.

BOUVIER, P., M. ROHMER, P. BENVENISTE, AND G. OURISSON (1976), "$\Delta^{8(14)}$-Steroids in the Bacterium *Methylococcus capsulatus*," *Biochem. J.* **159:** 267–271.

COLBY, J., D.I. STIRLING, AND H. DALTON (1977), "The Soluble Methane Monooxygenase of *Methylococcus capsulatus* (Bath): Its Ability to Oxygenate *n*-Alkanes, *n*-Alkenes, Ether, and Alicyclic, Aromatic and Heterocyclic Compounds," *Biochem. J.* **165:** 395–402.

DALTON, H. (1992), "Methane Oxidation by Methanotrophs," in *Biotechnology Handbooks: Methane and Methanol Utilizers*, J.C. Murrell and H. Dalton, eds. (New York: Plenum Press), pp. 85–114.

FREEMAN, K.H., J.M. HAYES, J.-M. TRENDEL, AND P. ALBRECHT (1990), "Evidence from Carbon Isotope Measurements for Diverse Origins of Sedimentary Hydrocarbons," *Nature* **343:** 254–256.

HANSON, R.S. (1980), "Ecology and Diversity of Methylotrophic Organisms," *Adv. Appl. Microbiol.* **26:** 3–39.

HAYES, J.M. (1983), "Geochemical Evidence Bearing on the Origin of Aerobiosis, a Speculative Hypothesis," in *Earth's Earliest Biosphere: Its Origin and Evolution*, J.W. Schopf, ed. (Princeton, NJ: Princeton University Press), pp. 291–301.

HAYES, J.M. (1994), "Global Methanotrophy at the Archean-Proterozoic Transition," in *Early Life on Earth*, Proceedings of Nobel Symposium 84, S. Bengtson, ed. (New York: Columbia University Press).

HAYES, J.M., K.H. FREEMAN, B.N. POPP, AND C.H. HOHAM (1990), "Compound-Specific Isotopic Analyses: A Novel Tool for Reconstruction of Ancient Biogeochemical Processes," *Org. Geochem.* **16:** 1115–1128.

JACKSON, M.J., T.G. POWELL, R.E. SUMMONS, AND I.P. SWEET (1986), "Hydrocarbon Show and Petroleum Source Rocks in Sediments as Old as 1.7×10^9 Years," *Nature* **322:** 727–729.

JAHNKE, L.L., AND H.P. KLEIN (1979), "Oxygen as a Factor in Eukaryote Evolution: Some Effects of Low Levels of Oxygen on *Saccharomyces cerevisiae*," *Origins of Life* **9:** 329–334.

JAHNKE, L.L., AND H.P. KLEIN (1983), "Oxygen Requirements for Formation and Activity of the Squalene Epoxidase in *Saccharomyces cerevisiae*," *J. Bacteriol.* **155:** 488–492.

JAHNKE, L.L., AND P.D. NICHOLS (1986), "Methyl Sterol and Cyclopropane Fatty Acid Composition of *Methylococcus capsulatus* Grown at Low Oxygen Tensions," *J. Bacteriol.* **167:** 238–242.

JAHNKE, L.L., H. STAN-LOTTER, K. KATO, AND L.I. HOCHSTEIN (1992), "Presence of Methyl Sterol and Bacteriohopanepolyol in an Outer-Membrane Preparation from *Methylococcus capsulatus* (Bath)," *J. Gen. Microbiol.* **138:** 1759–1766.

OURISSON, G., M. ROHMER, AND K. PORALLA (1987), "Procaryotic Hopanoids and Other Polyterpenoid Sterol Surrogates," *Ann. Rev. Microbiol.* **41:** 301–333.

PANGANIBAN, A.T., JR., T.E. PATT, W. HART, AND R.S. HANSON (1979), "Oxidation of Methane in the Absence of Oxygen in Lake Water Samples," *Appl. Env. Microbiol.* **37:** 303–309.

REIBACH, P.H., AND C.R. BENEDICT (1977), "Fractionation of Stable Carbon Isotopes by Phosphoenolpyruvate Carboxylase from C_4 Plants," *Plant Physiol.* **59:** 564–568.

ROHMER, M., P. BOUVIER, AND G. OURISSON (1980), "Non-Specific Lanosterol and Hopanoid Biosynthesis by a Cell-Free System from the Bacterium *Methylococcus capsulatus*," *Eur. J. Biochem.* **112:** 557–560.

ROHMER, M., P. BOUVIER-NAVE, AND G. OURISSON (1984), "Distribution of Hopanoid Triterpenes in Prokaryotes," *J. Gen. Microbiol.* **130:** 1137–1150.

SCHIDLOWSKI, M., AND P. AHARON (1992), "Carbon Cycle and Carbon Isotope Record: Geochemical Impact of Life over 3.8 Ga of Earth History," in *Early Organic Evolution: Implications for Mineral and Energy Resources*, M. Schidlowski, S. Golubic, M.M. Kimberley, D.M. McKirdy, and P.A. Trudinger, eds. (Berlin: Springer-Verlag), pp. 147–175.

SUMMONS, R.E., AND M.R. WALTER (1990), "Molecular Fossils and Microfossils of Prokaryotes and Protists from Proterozoic Sediments," *Amer. J. Sci.* **290A:** 212–244.

SUMMONS, R.E., T.G. POWELL, AND C.J. BOREHAM (1988), "Petroleum Geology and Geochemistry of the Middle Proterozoic McArthur Basin, Northern Australia. III. Composition of Extractable Hydrocarbons," *Geochim. Cosmochim. Acta* **52:** 1747–1763.

SUMMONS, R.E., L.L. JAHNKE, AND Z. ROKSANDIC (1994), "Carbon Isotopic Fractionation in Lipids from Methanotrophic Bacteria: Relevance for Interpretation of the Geochemical Record of Biomarkers," *Geochim. Cosmochim. Acta* **58:** 2853–2863.

WHITICAR, M.J., F. FARBER, AND M. SCHOELL (1986), "Biogenic Methane Formation in Marine and Freshwater Environments: CO_2 Reduction vs. Acetate Fermentation—Isotope Evidence," *Geochim. Cosmochim. Acta* **50:** 693–709.

WAS THE PRIMITIVE SEA GENTLE FOR LIFE? ITS EXCESS SALT IS TOXIC TO THE CELL

Hakobu Nakamura

Biological Institute, Faculty of Science
Konan University
Kobe 658, Japan

It has been believed that the first life system was born in the primitive sea as a result of chemical evolution. Further, it has also been said in the same meaning that the sea was a mother that cultivated the life system. Surely, the primitive sea could supply inorganic and organic materials needed for the living of the protocells that developed after the origin of life. Here, we have to reconsider whether all the minerals of seawater could positively support the life of the protocells, although it is a fact that the extant cells require in abundance the mineral elements present in seawater. The present seawater contains sodium chloride at a concentration of 3.5% on the average in the world's oceans.

The plasma membranes of living cells have several types of efflux systems that pump out excess sodium ions contained in the cells. However, as shown in this presentation, excess sodium ions are fundamentally toxic to the life mechanism compared with potassium ions. Therefore, all the cells possess so-called sodium pumps and some of them exchange sodium ions for potassium ions, protons, or amino acids.

Generally, sodium ion is believed to be an essential element for the extant living cells. On the other hand, it has been said also that, although sodium ion is contained in all plants, the element is not essential for general land plants with stems and leaves, namely *Cormophyta*. Halophilus plants, namely *Halophyta*, which grow on the seashore, are halophilic and thus some amounts of sodium ions are essential for active growth.

The sodium and potassium balances in the living cells of various organisms demonstrate that sodium concentrations in the cells are significantly lower than those in the medium. On the other hand, potassium concentrations are reversed in the ionic balance between the inner and outer sides of the cells. This is a rule not only in marine organisms but also in terrestrial organisms. These facts mean that normal cells always pump out sodium ion to the outer medium, while the sodium gradient between the inner and outer sides of the cells is osmotically balanced by uptake of potassium, and further by accumulation of low-molecular substances such as free amino acids, particularly glutamate and proline.

There are many kinds of organisms in nature that are halophilic rather than halotolerant. The halophilism is dependent on the species of organism and it is currently classified into three categories—slight halophiles requiring 1–4% NaCl, moderate halophiles requiring 5–10% NaCl, and extreme halophiles requiring 12–30% NaCl. Contrariwise, all organisms except these are called nonhalophiles. Bacterium *Escherichia coli* has been classified as a typical nonhalophile. However, even this bacterium shows better growth in the presence of very low concentrations of sodium chloride, for example 0.3%. Therefore, this bacterium does not show growth in the marine level (3.5%) of salt.

We have isolated an *E. coli acrA* mutant that is supersensitive to sodium chloride. Gene *acrA* is located at min 10.5 on the *E. coli* chromosome. The *acrA* mutant is quite sensitive to sodium ion and cannot grow in a rich (natural) medium containing 5% sodium chloride, as compared with a wild type $acrA^+$ strain. When various kinds of amino acids are added to the minimal media of 20μg/ml, addition of glutamate or proline makes them effectively tolerant to the excess salt. In amino acid metabolism, proline is synthesized from glutamate through only two reaction steps. Salt sensitivities of recombinants have various combinations of alleles of *acrA* and *proA* in the 3% sodium chloride–containing minimal media. Here, the *proA* is a mutant that requires proline for growth. It is remarkable that a single mutant of *acrA* or *proA* becomes more sensitive to excess salt than the wild type of strain. However, the double mutant of the *acrA proA* becomes extremely sensitive to excess salt, even in the presence of 20μg/ml of proline in the medium.

Similar results were obtained with the *acrA* and *proB* mutant. The *proB* is another mutant requiring proline. That is, a single mutation of *acrA* or *proB* leads to sensitivities to excess salt, and the double mutant becomes much more sensitive. On the other hand, data on sodium incorporation and accumulation of allelic recombinants for genes *acrA* and *proA* showed the interesting fact that (1) the wild type (*acrA*$^+$ or *proA*$^+$) strain accumulates sodium ions at the lowest level, (2) the single *proA* mutant accumulates more sodium ion, (3) the single *acrA* mutant accumulates much more sodium ion, and (4) the *acrA proA* double mutant accumulates sodium ion at the highest level. In other words, the more sensitive, the more the cells accumulate sodium ions. Similar results were obtained with the *acrA* and *proB* mutants. That is, the wild-type cells accumulate a small amount of sodium ion. However, mutation of *acrA* or *proB* gene leads to more accumulation of sodium ions. Further, the *acrA proB* double mutant accumulates sodium ion at the highest level. Also in the recombinants of alleles of *acrA* and *proB*, the more sensitive, the more sodium the cells accumulate.

Figure 1 shows the kinetics of the sodium ion's efflux of the *acrA* mutant and the wild-type cells which have already accumulated sodium ions. The data indicate clearly that the wild-type cells actively pump out more sodium ions than the mutant cells. It is apparent that the *acrA* and *pro* (*A* and *B*) genes are responsible for the sodium efflux. Dr. Midgley (1987) also has demonstrated that the *acrA* gene participates effectively for the efflux of positively charged chemicals, such as acriflavine.

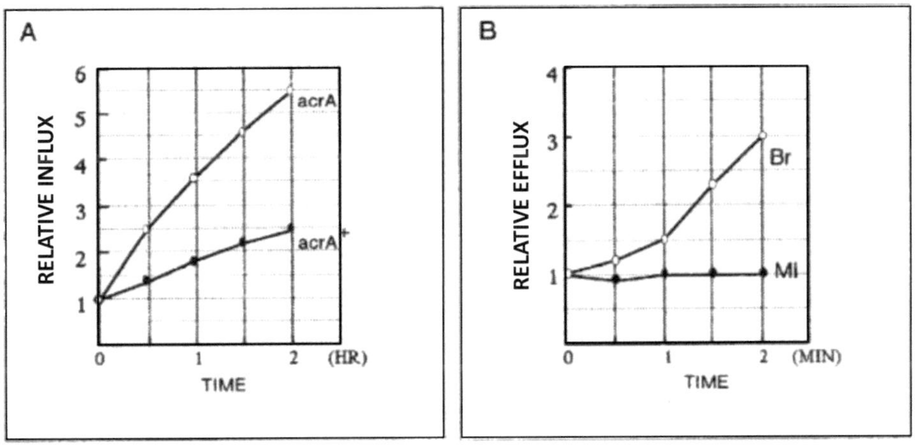

FIG. 1. *Influx and efflux of Na*$^+$ *ions in the* acrA$^+$ *(wild type) and* acrA *(mutant) cells.* A: *Relative influx of Na*$^+$ *ions in the* acrA$^+$ *(○) and* acrA *(•) cells.* B: *Relative efflux of Na*$^+$ *ions in a complete (Br) and a minimal (Mi) media. The data are shown as the ratio of* acrA$^+$ *to* acrA *cells. When incubated in a nutritionally rich medium (Br), the* acrA$^+$ *cells are much more active in the efflux activity than the* acrA *cells. However, under nutritionally poor conditions (Mi), the efflux is not very active.*

Although research is now in progress, it could be reasonably concluded that the efflux of sodium ions from the cells depends on the sodium pumps located in the plasma membrane, for example Na^+-K^+ ATPase, which is the most popular enzyme for the efflux of sodium ions and the influx of potassium ions in the biological world. The existence of the enzyme has been demonstrated from bacteria to higher organisms. Na^+-K^+ ATPase must play an important role in cellular homeostasis under salt-hypertonic conditions. The *acrA* mutant cells of *E. coli* are defective in the homeostatic system and thus must become supersensitive to excess salt. The efflux pump of sodium ions is useful to decrease the amount of the sodium ion accumulated in the cells, namely, useful for detoxification of excess sodium ions. Further, the influx pump of potassium ions is useful to compensate the resulting osmotic imbalance within the cells. We have demonstrated that the *acrA* gene product is a plasma membrane protein possessing a molecular weight of 60,000. Furthermore, we can observe electron-microscopically the disappearance of the larger-class protein particles from the mutant plasma membrane freeze-fractured. In a true revertant from the *acrA* mutant, larger particles recovered to the wild-type level of the plasma membrane. Therefore, we can conclude that the *acrA* mutation led to the loss of a larger class of the particles.

About 4 billion years ago, protocells were generated in the primitive sea. It has generally been said that the seawater was very gentle for the newly born life. Conversely speaking, could life originate and develop only under such circumstances?

We showed that the *acrA* mutant cells were supersensitive to excess sodium ions and they could grow most actively in the presence of about 0.3% sodium chloride. The salt concentration of 3.5% in present seawater must be absolutely toxic for the protocells. Therefore, it is apparent that if the primitive seawater contained such excess salt, life would not have been generated. The salt concentration of the primitive seawater would be much lower than that of the present seawater. Perhaps the primitive seawater had to contain around 0.3% sodium chloride—namely one-tenth as much as the present seawater.

The reasonable supports for this conclusion may be as follows.

(1) The protocells would not have any efflux system for toxic sodium ions, as they contained poor mechanisms only for energy production and reproduction.

(2) The sodium efflux system in the extant cells can work only by expending very much energy. For example, the Na^+-K^+ ATPase requires one mole ATP for 3 sodium ion efflux. Therefore, the protocells would not have such an extra amount of ATP.

(3) The present data show clearly that the *acrA proA* or *acrA proB* double mutant cells are supersensitive to 3.5% sodium chloride, which is the concentration of the present seawater. It seems to be due mainly to a specific toxicity of sodium ion but not hypertoxicity, because sodium toxicity can be cured by the addition of a large amount of potassium chloride, calcium chloride, or a certain amino acid such as proline.

(4) Na^+-K^+ ATPase is a multifunctional enzyme. The amino acid sequence of the enzyme can be divided into at least three functional regions—sodium efflux channel, potassium influx channel, and ATPase. It has been currently demon-

strated that the multifunctional enzymes were formed by genetic rearrangement, namely reconnection of the base sequence of monofunctional primitive genes. In other words, it is possible to say that Na^+-K^+ ATPase is a highly evolved enzyme. It must be a product of the later age of cellular evolution. We can say again that the protocells did not have such an evolved enzyme in the plasma membrane. In other words, surely the protocells did not need the sodium pump, as the salt concentration of the primitive seawater was much lower than at present.

After all, we can conclude that the primitive seawater was very gentle for the protocells and that the salt in seawater gradually accumulated with time through washing of the rock salt by rainfall. Therefore, the sodium efflux system in the plasma membrane was an adaptation to an increasing accumulation of salt in the sea.

References

MIDGLEY, M. (1987), "An Efflux System for Cationic Dyes and Related Compounds in *Escherichia coli*," *Microbiol. Sci.* **4:** 125–127.

The Orion Nebula

Photo Credit: NASA, ESA, Hubble Space Telescope Orion Treasury Project Team, Massimo Robberto (STScI, ESA)

https://science.nasa.gov/asset/hubble/hubbles-sharpest-view-of-the-orion-nebula/

[retrieved 12-11-25]

The Orion Nebula is 1,500 light-years away, and is the nearest star-forming region to Earth. It is a picture book of star formation, from the massive, young stars that are shaping the nebula to the pillars of dense gas that may be the homes of budding stars. The bright central region is the home of the four heftiest stars in the nebula, which are called the Trapezium. Ultraviolet light unleashed by these stars is carving a cavity in the nebula and disrupting the growth of hundreds of smaller stars. Located near the Trapezium stars are stars still young enough to have disks of material encircling them. These disks, called 'protoplanetary disks' [and too small to see in this image], are the building blocks of solar systems.

The faint red stars near the bottom of the image are the myriad brown dwarfs that Hubble spied for the first time in the nebula in visible light. Sometimes called 'failed stars,' brown dwarfs are cool objects that are too small to be ordinary stars because they cannot sustain nuclear fusion in their cores the way our Sun does. The dark red column, below and left, shows an illuminated edge of the cavity wall.

Astrophysical Aspects of Habitable Zones

Eagle Nebula M16 and Comet C/2025 R2 SWAN

Photo Credit: Chuck Ayoub
https://commons.wikimedia.org/wiki/File:Eagle_and_Comet.png [retrieved 12-11-25]

C/2025 R2 (SWAN) was discovered by Ukrainian astronomer Vladimir Bezgly on September 10, 2025, in data collected by the Solar and Heliospheric Wind Anisotropies (SWAN) camera on NASA's Solar and Heliospheric Observer (SOHO) spacecraft. After surviving its close approach to our Sun on September 12, the comet headed out toward the farthest reaches of the solar system, and won't return for approximately 1,400 years.

CIRCUMSTELLAR HABITABLE ZONES: ASTRONOMICAL CONSIDERATIONS

Daniel P. Whitmire

University of Southwestern Louisiana
Lafayette, LA 70504-4210

Ray T. Reynolds

Theoretical Studies
NASA Ames Research Center
Moffett Field, CA 94035

Abstract

From an astronomical perspective, stellar luminosity and its evolution in time are the dominant factors in determining the location, size and evolution of circumstellar habitable zones (HZ). Other stellar properties that can affect HZs are stellar temperature, mass, mass loss, spot coverage and multiplicity. Whether planets are likely to form within a star's HZ and under what conditions their orbits would be dynamically stable for significant periods of time are separate important questions. In this paper we review these topics and suggest areas for future work in light of the recent proposal (Goldin 1994) that NASA adopt as a new unifying focus the search for a habitable planet around a nearby star.

INTRODUCTION

Following Kasting, Whitmire and Reynolds (1993; hereafter KWR) and others, we define the habitable zone as the radial shell around a star within which an Earthlike planet could support surface liquid water. For our purposes an Earthlike planet is one similar in mass and composition to Earth and having comparable surface inventories of CO_2, H_2O, and N_2. The upper mass limit is not well constrained but the lower mass must be sufficient to maintain adequate geological activity in order for the carbonate-silicate cycle to stabilize the climate. If we require stability for $\geq$ 1 Gyr then a mass greater than $\sim 0.1\ M_\oplus$ is necessary based on the geological history of Mars (see below). We restrict our discussion to Earth-type habitats and terrestrial-type life, ignoring the possibility that life may have evolved in such marginal habitats as small planets (McKay and Davis 1991) or tidally heated environments (Reynolds et al. 1983, 1987).

KWR investigated climatic constraints on the inner and outer boundaries of the HZ. The inner boundary was assumed to be defined by the loss of water from the planet by either the moist or runaway greenhouse and the outer boundary by the freezing of all water on the planet's surface. Exactly when these circumstances occur was shown to be sensitive to the effects of H_2O and CO_2 clouds on the planet's radiation budget. Three different sets of flux limits—two theoretical and one observational—were identified for each boundary. The inner and outer boundaries of the HZ around mature main sequence stars depend primarily on stellar luminosity and to a lesser extent on temperature, both of which vary in time as a star evolves. Because of this evolution a planet will remain within the HZ only for a finite amount of time, τ, which is generally less than the main sequence lifetime of the star.

In the most elaborate previous modeling study of the extent and duration of HZs around stars, Hart (1978, 1979) defined τ to be 4.5 Gyr, and called the radial range within which an Earthlike planet could remain habitable for this length of time the continuously habitable zone (CHZ). We continue to use this terminology with the stipulation that τ is a parameter and that it need not necessarily begin at the zero age main sequence (ZAMS).

Hart (1978) concluded that the 4.5 Gyr CHZ for the Sun extended from 0.95 AU to 1.01 AU. This estimate was subsequently revised to 0.96 AU to 1.00 AU (Hart 1979). According to Hart, if the Earth had been closer than 0.96 AU to the Sun it would have experienced a runaway greenhouse within 4.5 Gyr, while if it had been farther from the Sun it would have undergone runaway glaciation. In his model, the Earth's habitable climate survived by sheer coincidence. The appearance of free oxygen and the resulting loss of an assumed reducing primordial atmosphere occurred before a runaway greenhouse developed, but after the danger of runaway glaciation had subsided. As discussed in KWR, the most important distinction between our model and Hart's is our assumption that the climate is stabilized by the carbonate-silicate cycle (Walker et al. 1981, Kasting 1988). This cycle provides a negative feedback by which the ambient amount of atmospheric CO_2 naturally regulates itself to maintain an average surface temperature above 273 K. In our model the Earth's stable, long-term climate is not an accident of precise planetary placement, or a result of the timing of atmospheric events, but is rather a consequence of stable equilibrium. We do not rule out the possible presence of some reduced greenhouse gases early in Earth's history, but we do not require them in order to explain Earth's long-term habitability. On Earth, today, life also contributes to the regulation of atmospheric CO_2 (Lovelock and Watson 1982, Schwartzman and Volk 1989). However, the geophysical carbonate-silicate cycle can keep a planet habitable even in the absence of life.

Hart (1979) also calculated the 4.5 Gyr CHZ around stars with masses different from the Sun and found that it was generally narrower for stars less massive than the Sun and disappeared altogether around stars less massive than 0.85 $M_\odot$, corresponding to spectral types later than K0, which includes the majority of solar-type stars in the Galaxy. The reason for this result was the assumption that the original greenhouse gases were lost at exactly the same time as on Earth, about 2 Gyr after the planet formed. At this stage in their evolution, the lower-mass stars have not yet increased much in luminosity, so their planets would undergo runaway glaciation. Hart's results were subsequently used to support arguments that the probability of complex life in the Galaxy is negligible (e.g., Pollard 1979, Rood and Trefil 1981, Carter 1983, Barrow and Tipler 1986). As discussed below, our climate model is not subject to this type of instability and, consequently, our conclusions are quite different.

Since the early work of Hart, there have been many improvements in climate modeling. Based on the operation of the carbonate-silicate cycle and the geological evidence of water on early Mars, Kasting et al. (1988) and Fogg (1992) suggested that the outer boundary of the Sun's 4.5 Gyr CHZ extended to at least the orbit of Mars. However, these studies neglected the effect of CO_2 condensation, which places a serious constraint on the amount of greenhouse warming available at low solar fluxes (Kasting 1991). Whitmire et al. (1991) incorporated this effect in preliminary HZ calculations. KWR (see also Kasting 1996) used a more complete model to directly examine the climatic constraints on the inner and outer HZ boundaries. Here we use those results, along with stellar evolutionary models, to calculate the evolution of the HZ for a range of stellar masses.

The next two sections describe our model calculations for the Sun and other single main sequence stars. We then consider where planets might form relative to HZs, synchronously rotating planets around M stars, and binary star HZs. In the final section we give our main conclusions and point out several areas for future work in the search for habitable planets around nearby stars.

EVOLUTION OF THE SUN'S HABITABLE ZONE

In the standard solar model, the Sun began its life on the zero age main sequence (ZAMS) with a luminosity of 0.71 $L_\odot$, and a photospheric temperature a few hundred degrees less than the present value. Compared with today, it is likely that the early Sun rotated much faster, had a stronger and more active magnetic field and solar wind, had a greater emission of X-ray and UV radiation, and had a larger fraction of its surface covered with spots.

By the end of core hydrogen burning in about 6.5 Gyr, the Sun's luminosity will have increased to 2.2 $L_\odot$. It will then soar to ~ 2300 $L_\odot$ over the next ~ 1 Gyr as the Sun ascends the first red giant branch (Sackmann et al. 1993). During this phase the Sun will lose a significant amount of mass (~ 0.275 $M_\odot$), most of it just prior to the helium flash that terminates the red giant phase. At maximum luminosity the solar radius will extend to ~ 0.78 AU, depending on the rate and amount of mass loss. The next phase (~ 0.1 Gyr duration) is the horizontal branch in which the Sun's luminosity is ~ 40 $L_\odot$. This is followed by the second red giant or AGB (asymptotic giant branch) phase, in which the luminosity peaks at ~ 5200 $L_\odot$, depending on mass loss. After the thermally pulsing AGB phase and the planetary nebula phase, the ultimate fate of the Sun is to become a cooling white dwarf of mass ~ 0.54 $M_\odot$. Mercury and Venus will have been engulfed by the Sun during the red giant phases. The Earth's fate is less certain. Although the surface of the Sun will extend to one AU during the second red giant phase, prior mass loss will have caused the planet orbits to expand, ultimately by a factor of two. Sackmann et al. (1993) find that the Earth will marginally survive; however, they do not discuss the tidal decay of Earth's orbit, which would tend to counter the expansion due to mass loss.

These phases in the Sun's luminosity evolution were illustrated in KWR and are reproduced here in Figure 1; they are based on the models of Iben (1967a, 1967b, 1974) and Iben and Renzini (1983). These models do not agree in detail with the more recent model of Sackmann et al. (1993), which includes post–main sequence mass loss, especially in the peak luminosity of the first red giant phase. However, our primary interest here is in the main sequence phase, for which there is reasonable agreement.

Given the luminosity and temperature as a function of time, the evolution of the HZ can be calculated for a specific choice of critical fluxes defining the inner and outer boundaries from

$$r_1 = \sqrt{\frac{L(t)}{S_1(T)}}$$

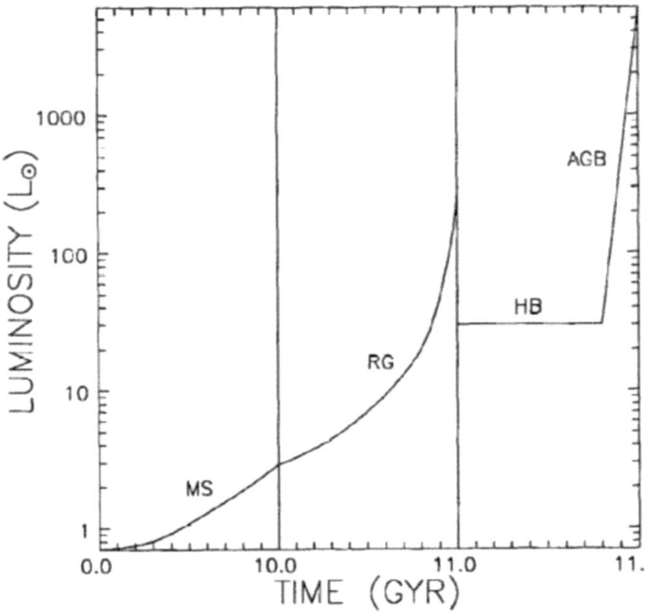

FIG. 1. *Luminosity evolution of a 1 $M_\odot$ star of solar composition. Note that the scale is broken at 10 and 11 Gyr. The discontinuity at 11 Gyr is real and is due to the helium flash. The phases are: MS = main sequence, RG = red giant, HB = horizontal branch, and AGB = asymptotic giant branch. The final white dwarf phase is not illustrated. Details of the evolution during the HB and AGB phases are omitted.*

$$r_2 = \sqrt{\frac{L(t)}{S_2(T)}}$$

where $r_1(r_2)$ is the inner (outer) HZ boundary radius, $L(t)$ is the solar luminosity as a function of time obtained from the stellar evolutionary model, and $S_1(S_2)$ is the appropriate critical solar flux, which in general is a function of stellar temperature. The HZ radii will be given in units of AU when the luminosity is in solar units and the flux is in units normalized to the present solar constant at Earth's orbit, which is taken to be 1360 W m^{-2}.

KWR identified three critical fluxes for the inner radius of the HZ. In flux units normalized to the present solar constant these limits are 1.1, 1.4, and 1.76. The first two result from the loss of planetary water by the moist greenhouse and runaway greenhouse, respectively (Kasting 1988). These are theoretical values obtained from a one-dimensional radiative-convective climate model in which the radiative effect of clouds is parameterized by use of a high surface albedo. This assumption is expected to result in conservative limits: a real planet can probably maintain liquid water at a higher solar flux than calculated. The third limit is empirical and corresponds to the flux at Venus 1 Gyr ago. Spacecraft observations show no evidence that liquid water has flowed on the surface of Venus for at least the last 1 Gyr (Solomon and Head 1991). This flux limit is an upper bound since Venus could not have had significant water since then. The inner HZ radius is thus reasonably well bracketed. The evolution of these inner HZ boundary radii for the Sun is shown in the bottom three curves of Figures 2 and 3.

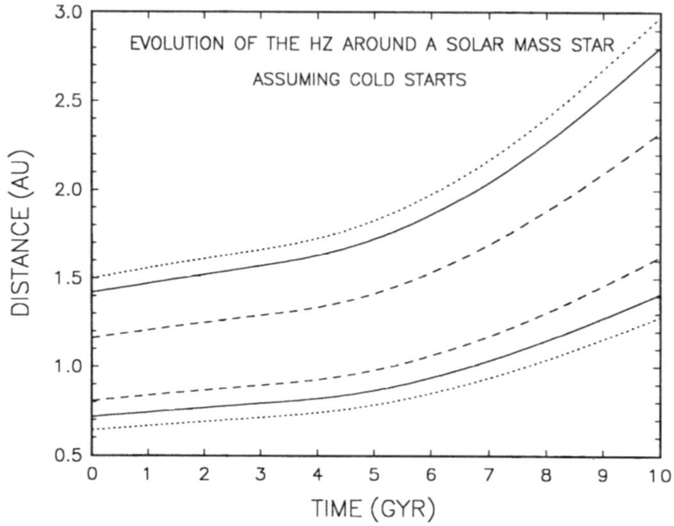

FIG. 2. *Evolution of the HZ around a 1 $M_\odot$ star assuming that an ice covered planet that was initially beyond the outer HZ boundary can be cold started. The three cases shown are discussed in the text and correspond to the most conservative Case 1 (long dashes), the intermediate Case 2 (solid curves), and the least conservative Case 3 (short dashes).*

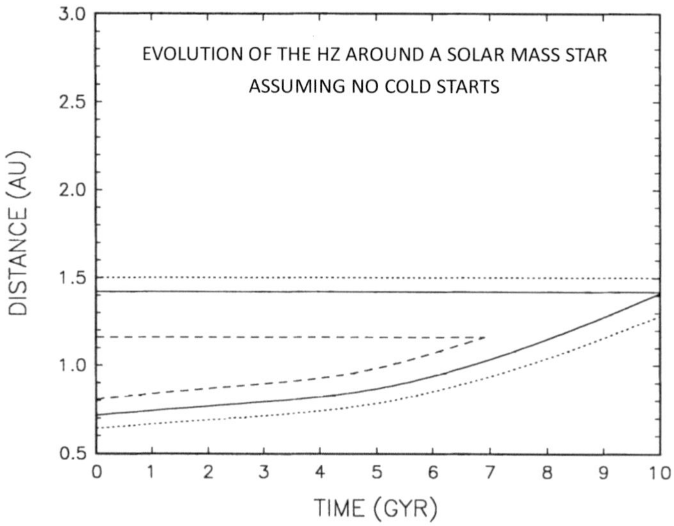

FIG. 3. *Evolution of the HZ around a 1 $M_\odot$ star assuming that an ice covered planet that was initially beyond the outer HZ boundary cannot be cold started until the stellar flux is greater than the critical greenhouse value. The three cases shown are discussed in the text and correspond*

to the most conservative Case 1 (long dashes), the intermediate Case 2 (solid curves), and the least conservative Case 3 (short dashes).

Three critical fluxes corresponding to the outer radius of the HZ were also identified by KWR. In normalized units these flux limits are 0.32, 0.36, and 0.53. The smallest and least conservative limit corresponds to the flux at early Mars 3.8 Gyr ago in the standard solar model. This limit is based on observations and arguments (e.g., Pollack et al. 1987) that early Mars had a warm and wet climate and therefore must have been in the Sun's HZ. (Due to its low mass of 0.1 $M_{\oplus}$ Mars is not an Earthlike planet as we define it. An Earth-mass planet at the distance of Mars from the Sun might have maintained a stable climate for a much longer period, as did the Earth.) The intermediate theoretical flux limit of 0.36 corresponds to the maximum possible CO_2 greenhouse heating, which occurs for a CO_2 partial pressure of ~ 8 bars (KWR, Kasting 1991). At higher CO_2 partial pressures, the increase in planetary albedo outweighs the increase in greenhouse heating. Since the actual solar flux at Mars 3.8 Gyr ago was 0.32 in the standard model, there is a theoretical discrepancy between these two limits. This discrepancy is at present unresolved, but might be explained by the presence of other greenhouse gases in Mars' early atmosphere (Kasting 1991, Sagan and Chyba, abstract, this volume) or by a nonstandard solar model in which the Sun's ZAMS mass was somewhat greater than the present value (Graedel et al. 1991; Whitmire et al. 1995a,b; Doyle et al. 1995, 1996). In addition to this model, other possible astronomical explanations for an increase in the solar flux at early Mars (such as sunspot focusing) have been discussed by Whitmire et al. (1995a,b).

The most conservative flux limit for the outer boundary is the "first condensation" limit of 0.53. This is the flux at which CO_2 first begins to condense and to increase the planetary albedo. This limit is probably too conservative, but is difficult to improve on without a climate model that can handle CO_2 clouds. These three estimates of the critical minimum flux necessary to sustain a mean surface temperature of 273 K were used, along with the 1 $M_{\odot}$ main sequence evolution model of Iben (1967a), to generate the evolution of the outer HZ boundary radii (three upper curves) in Figures 2 and 3.

These figures show the evolution of the Sun's HZ—the inner and outer radii versus time—for sets of the paired critical fluxes. The three sets are paired in the following way. Case number 1 corresponds to the two most conservative flux limits, the moist greenhouse (1.1) and the first condensation point (0.53). Case number 2 is the intermediate case and corresponds to limits determined by the runaway greenhouse (1.4) and the maximum greenhouse (0.36). Case number 3 is our least conservative case and corresponds to limits inferred from the lack of water on Venus as of 1 Gyr ago (1.76) and the evidence for the presence of liquid water on early Mars (0.32).

The time evolution of the three HZ cases shown in Figure 2 is based on the assumption that a planet that is initially beyond the outer boundary of the HZ and therefore frozen, will melt (deglaciate) once the solar flux reaches a value equal to our calculated climatic limits. As discussed by Caldeira and Kasting (1992), this assumption may not be valid. A planet that formed beyond the outer boundary of the HZ would develop a reflective blanket of ice. This ice layer might increase the planetary albedo, reducing the absorption of solar radiation. Such a planet would also be likely to develop thick, highly reflective, CO_2 clouds (Kasting 1991), further reducing the solar flux at the surface. A planet that undergoes global glaciation soon after accretion might remain in that state even when the solar flux increases beyond the appropriate (lower) critical value discussed above. Thus, it may be impossible to "cold start" an initially frozen planet before the solar flux increases by a factor of ~ 2 and is then greater than the critical greenhouse value (Caldeira and Kasting 1992). If cold starts are possible it can be seen from Figure 2 that the HZ moves outward and becomes broader in time.

Figure 3 shows the same evolution as Figure 2 but with the more conservative assumption that cold starts are not possible during the main sequence phase. In this case the outer boundary of the HZ is constant and equal to its ZAMS value. The inner boundary radius moves outward in time as before so the HZ now becomes narrower in time. For Cases 1 and 2 the HZ disappears altogether by the end of the main sequence. Because of the significant differences in the HZ and CHZ in Figures 2 and 3, the question of whether planetary cold starts are possible is an important one for future work. It might be possible, for example, that a major climatic perturbation from equilibrium such as a massive impact or extensive vulcanism, could "jump start" a frozen planet.

Continuously habitable zones for any assumed time period can be read off Figures 2 and 3 for the three limiting cases. For our most conservative Case 1 fluxes without cold starts, the 4.5 Gyr CHZ extends from 0.95 to 1.15 AU, which is a factor of 5 larger than Hart's (1979) CHZ for the Sun. For our least conservative Case 3 the 4.5 Gyr CHZ extends from 0.75 to 1.9 AU, or a factor of 30 times larger than Hart's CHZ. For simplicity of discussion here and in the next section we shall focus on the intermediate Case 2. For this case the 4.5 Gyr CHZ extends from 0.84 to 1.77 AU if cold starts are possible and from 0.84 to 1.43 AU if they are not. In the next section we will see that the discrepancy between our CHZs and those of Hart are even greater for stars of lower mass than the Sun.

Figure 4 shows the evolution of the Sun's HZ throughout its entire lifetime, assuming (necessarily) that planetary cold starts are possible. The post–main sequence phases have a cumulative duration of only about 1 Gyr, almost all of which is in the ascension of the first red giant branch. By the end of this time, the HZ is in the vicinity of 10 to 20 AU according to the model of Iben (Fig. 4), or 25 to 55 AU according to the model of Sackmann et al. (1993). Although it is doubtful that Earthlike planets could form at these distances around a 1 $M_\odot$ star, this figure points out that one cannot categorically rule out even red giant systems as possible, although short-lived, abodes for life.

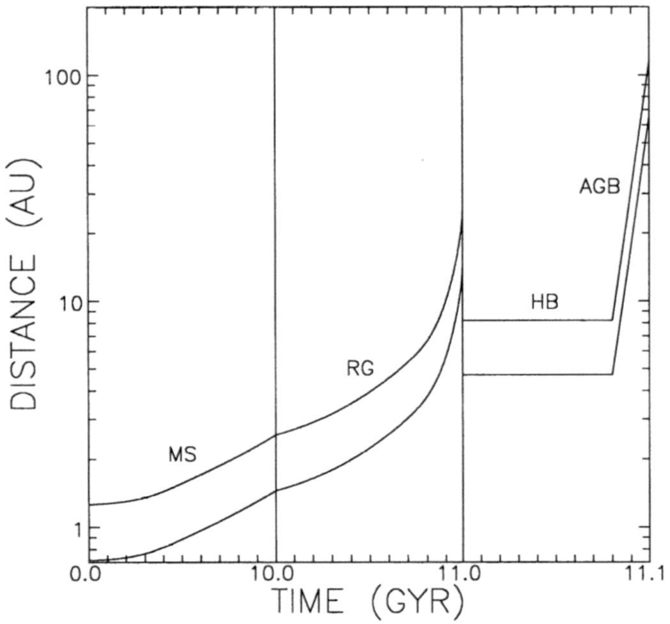

FIG. 4. *Evolution of the HZ during the entire lifetime of a 1 $M_\odot$ star. For the HZ to exist beyond 10 Gyr it is necessary to assume that cold starts are possible. Case 2 critical fluxes were used, and the phases shown are the same as those in Figure 1.*

EVOLUTION OF THE HABITABLE ZONE AROUND OTHER STARS

The same type of HZ calculations can be performed for stars with masses different from the Sun's. Main sequence stars much more massive than the Sun have large HZs on an absolute scale but their lifetimes are too short to be of interest as sites for the evolution of complex organisms (Huang 1960). For the massive O-type stars it is doubtful whether there is even sufficient time for planets to form during their ~ 10^6 yr stellar lifetimes (Bodenheimer 1989). The main sequence lifetime of a star, τ_{ms}, is proportional to M/L. Since $L \propto M^{4.75}$ over the mass range of most interest (Iben 1967a), $\tau_{ms} \propto M^{3.75}$. Here we restrict ourselves to stellar lifetimes greater than 2 Gyr, which corresponds to masses less than 1.5 $M_\odot$. At the other extreme, the low-mass main sequence M stars with masses $\leq 0.5\ M_\odot$ have lifetimes longer than the age of the Universe. Their evolution in 10 Gyr is negligible, so their 4.5 Gyr CHZ is identical to their ZAMS HZ. Although these are the most numerous stars in the Galaxy, several objections have been raised against the idea that they could support life-bearing planets. We discuss these points in the next section and conclude that it is premature to reject these stars as viable candidates for having habitable planets.

Figures 5 and 6 show the evolution of the HZ (for Case 2 fluxes) around stars of selected masses between 0.5 and 1.5 $M_\odot$, with and without cold starts, respectively. The effect of stellar temperature on the wavelength distribution of light and, hence, on planetary albedo has been taken into account in all figures. The fluxes were corrected for stellar temperature by quadratically fitting and extrapolating the three stellar temperatures calculated in the climate model of KWR. Cases 1 and 3 would yield smaller and larger HZs, respectively, as illustrated in Figures 2 and 3. Stellar luminosities and temperatures were taken directly from Iben (1967a,b) for masses 1.25 and 1.5 $M_\odot$. Since Iben's 1 $M_\odot$ star does not precisely reproduce the present Sun's luminosity, we have normalized $L(t)$ to be consistent with this and with the solar model of Gough (1981). Iben does not give the complete evolution for 0.5 and 0.75 $M_\odot$ stars. For these cases we used his ZAMS luminosity, temperature and lifetime to generate our own model. This is an adequate approximation since these stars do not evolve much in 10 Gyr.

For the stars in Figures 5 and 6, the widths of various CHZs can be read off for any assumed relevant time scale. Assuming no cold starts (Fig. 6), the CHZ for a given τ is equal to the HZ at time $t = \tau$. As in the case of the Sun, the 4.5 Gyr CHZs are much broader than those calculated by Hart (1979) for all stellar masses. Hart concluded that the 4.5 Gyr CHZ disappeared altogether for spectral types later than K0, corresponding to masses $\leq 0.85\ M_\odot$. Our results show that, even without cold starts, there is a significant 4.5 Gyr CHZ for all K and M stars.

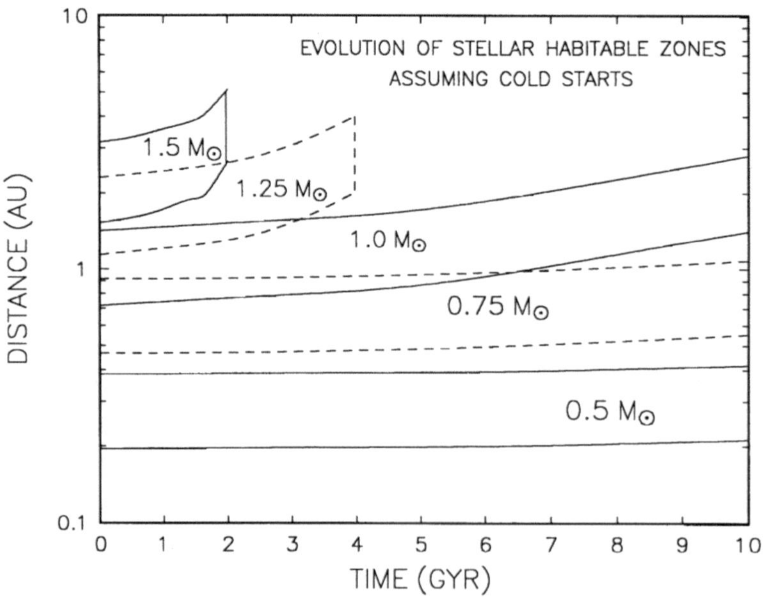

FIG. 5. *Evolution of the HZ around stars of different masses assuming that cold starts are possible. Case 2 critical fluxes were used and the evolution was truncated at the end of the main sequence phase.*

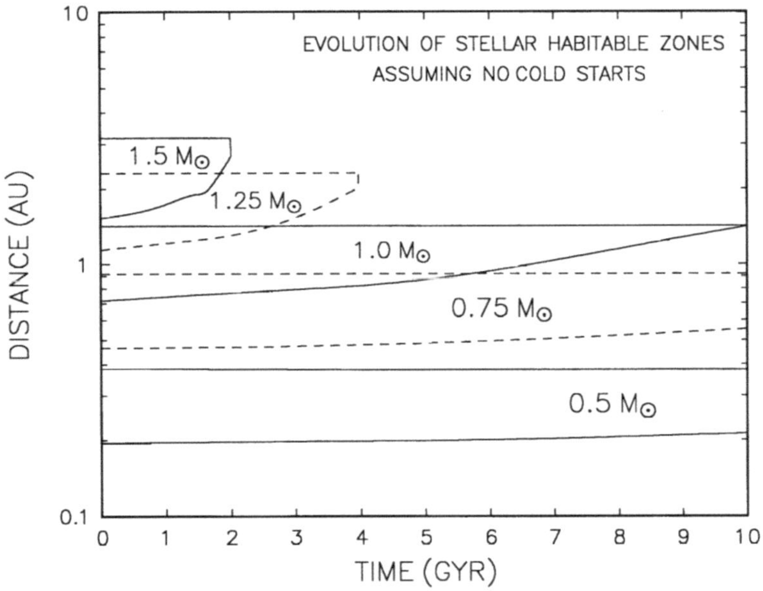

FIG. 6. *Evolution of the HZ around stars of different masses assuming that cold starts are not possible. Case 2 critical fluxes were used and the evolution was truncated at the end of the main sequence phase.*

The CHZs for masses between 0.1 and 0.5 $M_\odot$ are essentially the same as the ZAMS HZ and can be obtained from Figure 7 or 8. As noted earlier, the main reason our model gives different results from Hart's is that it includes climate stabilization by the carbonate-silicate cycle.

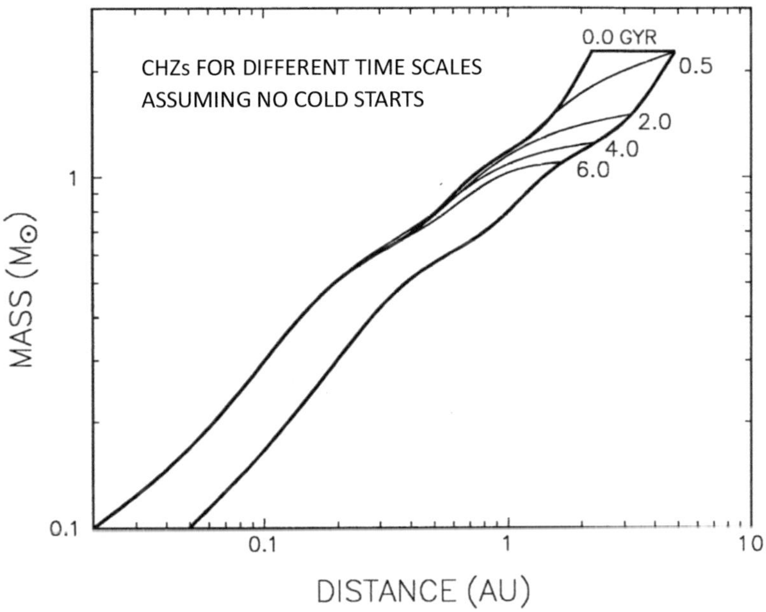

FIG. 7. *The CHZ for different assumed time scales, τ, for stellar masses in the range 0.1–2 M⊙, assuming cold starts are not possible. The τ = 0 outer envelope is the total HZ, from the inner radius at ZAMS to the outer radius at the end of the main sequence phase.*

The relevant CHZ time scale depends both on the type of life in which one is interested and on the rate of biological evolution. On Earth, intelligent life evolved in 4.5 Gyr, complex multicellular life in about 3.9 Gyr, and simple life within 1 Gyr after accretion. The extent to which these times depend on biological evolution rates as opposed to planetary evolution rates is an open question. One way of covering the range of possibilities is to plot CHZs for different assumed time scales, τ (Fig. 7).

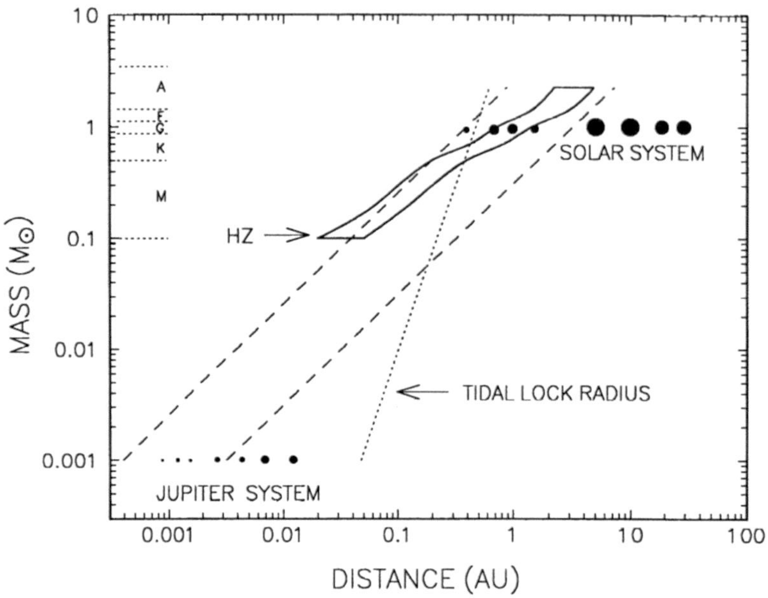

FIG. 8. *The ZAMS HZ (dark solid curve) as a function of stellar mass for Case 2 fluxes. The long-dash lines delineate the most probable terrestrial planet formation zone. The short-dash line is the radius for which an Earthlike planet in a circular orbit would be synchronously or slowly rotating as a result of tidal damping. Note that all such planets in the HZ around M stars are within this radius.*

Discussion

Habitable Zones and Planetary Accretion

In this section we consider where planets are most likely to accrete relative to the HZ. For the Solar System, numerical simulations of terrestrial planetesimal accretion (Wetherill 1991, 1996) resulted in an average of one to two terrestrial planets with masses $\geq 0.5\ M_\oplus$ forming per solar system between 0.6 and 1.4 AU, for a variety of initial conditions. On this basis, Wetherill suggested that the occurrence of Earthlike planets may be a common feature of solar-type planetary systems. A similar conclusion was reached in earlier simulations by Isaacman and Sagan (1977). Wetherill's results, together with our CHZ calculations, suggest that there is a high probability of at least one Earthlike planet forming in the (Case 2) 4.5 Gyr CHZ around solar mass stars. Even for our most conservative Case 1 evolution, the probability is ~ 0.5. Modeling of planetary accretion for stars of mass different from the Sun's is discussed by Wetherill (1996).

Evidence that protoplanetary nebulae are a common feature of low-mass star formation is found in observations of massive disks around young T Tauri stars and in observations of disks around many normal old main sequence stars, including G and later types (Aumann et al. 1984, Backman and Paresce 1993). A recent HST (Hubble Space Telescope) survey of over 100 young solar mass T Tauri stars in the Orion nebula found that fully half possessed observable protoplanetary disks (O'Dell and Wen 1994). Taking into account selection effects, it is estimated that at least half of all nearby main sequence stars have significant nonphotospheric IR excesses at IRAS wavelengths (10–100 μm). These excesses have been interpreted in terms of disks containing orbiting grains of sizes typically between one and 100 μm. In the case of β Pic the disk has been resolved optically (Smith and Terrile 1984). The lifetime of the observed grains due to various destruction mechanisms is generally less than the age of the star. Consequently, if the phenomenon is not transient, the observed grains must have larger sources, presumably planetesimals (Whitmire et al. 1988, Backman and Paresce 1993).

Extensive modeling of the IRAS data and observations at shorter and longer wavelengths have shown that the three prototype systems (α Lyr, α Psa, and β Pic) have gaps or holes in their inner disks of radii in the range ~ 20–80 AU. One possible explanation for the maintenance of these gaps against Poynting-Robertson and collisional diffusion is a planetary system. Roques et al. (1994) and Lazzaro et al. (1994) have recently investigated this possibility as an explanation for the gap and other disk structure in β Pic. Other explanations for the gap and the large-scale disk asymmetry are sublimation and a distant brown dwarf companion (Whitmire et al. 1988). To summarize, there is good evidence that, in general, protoplanetary nebulae are a common feature of star formation, but individual nebulae may differ significantly in various other properties such as mass, angular momentum, and magnetic structure, even around stars of similar mass.

We now wish to consider where terrestrial planets might form relative to the HZs of stars of different masses. The dark solid curve in Figure 8 shows the ZAMS HZ (for Case 2 fluxes) as a function of stellar mass. For stars with masses only slightly different from the Sun's, the zeroth order approximation is to assume Solar System planetary distances. Taking our own planetary system (which is typical of the Wetherill simulations) as an example, there would be two Earthlike planets, Venus and Earth, within the ZAMS HZ of an early K star and two planets, Earth and Mars, within the ZAMS HZ of a late F star. The Sun may also have had two planets in its ZAMS HZ, Venus and Earth. However, only for the early K star would both planets remain habitable for 4.5 Gyr. Thus, our own planetary system would have fared better, in terms of the a priori evolution of complex life, around an early K-type star. For middle (late) K stars, Venus (Mercury) would remain in the habitable zone for 4.5 Gyr. The late F star would have one planet, Mars, within its HZ for the main sequence lifetime. Stars more massive than 1.25 $M_\odot$ are relevant only for $\tau \leq 4.5$ Gyr. Similar results follow if we use the Wetherill (1991) simulations, rather than the actual Solar System. These considerations suggest that stars of spectral type between late F and early K (masses between approximately 1.25 and 0.85 $M_{\odot}$) are likely to have one or two planets within the $\tau \geq 4.5$ Gyr CHZ. In this zeroth order approximation the early K stars have a somewhat higher a priori probability of possessing long-term habitable planets.

For stars with masses that are significantly different from 1 $M_{\odot}$, we do not expect the zeroth order approximation to be adequate. If it were, it would predict that the most numerous stars in the Galaxy, the M stars, would not have any habitable planets, since the HZ would lie well inside the planet formation zone. In reality we expect that the planetary formation zone should depend on stellar mass and therefore that planets will form closer to lower-mass stars. We wish to approximate how the planetary formation region might scale with stellar mass in a statistical sense (i.e., other variables such as angular momentum and magnetic fields being neglected). Since the primary influence of a central star on the region of planet formation is directly or indirectly gravitational in nature (Bodenheimer 1989), an approximate mass scaling law might be expected. The more complex general problem of planet formation around stars of different masses is discussed by Lissauer (1989, 1995).

We consider first pre–main sequence (PMS) grain condensation in an optically thin nebula. In this case the local grain temperature is determined solely by the stellar flux. We assume that the distance at which various compounds condense is determined primarily by the local temperature. The distance at which a given temperature occurs, r, is proportional to $\sqrt{L_{PMS}}$. During the PMS phase the luminosity is the result of gravitational contraction, which is proportional to M^2. Thus, the distance at which grains of a given composition condense is proportional to M. This result is only approximate since L_{PMS} also depends on the contracting star's radius and on the contraction time scale. However, these effects are relatively small and they act in opposite senses. If the nebula is optically thick, the situation is more complicated: the local temperature then depends on the turbulent heat transfer properties of the disk in addition to the gravitational contraction energy. The trend should be the same, however, and the distance at which a given temperature is reached should still be proportional to mass. Using different physical arguments, Cameron (1963) has also suggested that planetary accretion distances scale with stellar mass.

We assume conservatively that, during the planet formation phase, the refractory silicate condensation radius, $a_1(M)$, for a 1 $M_{\odot}$ star lies inward of Mercury's orbit (0.39 AU) and that the condensation radius, $a_2(M)$, for water lies beyond the outer asteroid belt (3.2 AU). With these assumptions, we can generate two functions that should bound the most likely region for terrestrial planet formation by simply scaling linearly with stellar mass from these two Solar System values ($a_{1,2} \propto M$). These functions are plotted in Figure 8 as the two parallel dashed lines. Of course, as yet we do not know that terrestrial planets actually exist around other stars, but if they do, the dashed lines in Figure 8 should approximately bracket their accretion radii.

Our suggestion that planetary formation distances scale with stellar mass is consistent with two available examples—the Sun and Jupiter. The Galilean satellites may have formed from an optically thin Jovian nebula, analogous to the solar nebula (Pollack and Reynolds 1974). Jupiter's early luminosity would have then determined the temperature structure and thus the condensation region in the satellite nebula. Jupiter's innermost Galilean satellite, Io, is composed largely of rocky materials. The two large outer moons, Ganymede and Callisto, are H_2O rich. Europa has a significant amount of H_2O and represents the approximate transition region between rocky and icy bodies. Interestingly, it can be seen in Figure 8 that the lower dashed line, which represents the outer terrestrial planet formation boundary, passes through the Jupiter system near Europa, with Io (and several other smaller rocky satellites) lying well inside the zone. The Galilean moons, especially Io and Europa, may have undergone some orbital expansion since their formation, so their original orbits could have been somewhat farther inside the nominal terrestrial planet formation zone.

Except for the gap in the asteroid belt, the planets in the Solar System and the major satellites in the Jovian system are distributed approximately logarithmically with distance and are therefore spaced uniformly on a log scale as illustrated in Figure 8. The explanation for this spacing is at present unknown; it may simply represent a general tendency of accretion from a disk, or it may be coincidental. Assuming that the Solar System is not atypical we take the number of planets, dN, in the interval $d(\log r)$ to be a constant C,

$$\frac{dN}{d(\log r)} = C \qquad a_1(M) < r < a_2(M)$$

We assume, based on the mass scaling extrapolation in Figure 8, that C does not depend strongly on stellar mass. The precise value of C ($\sim$ 4 for the Solar System) is not important for the purpose of comparing different mass stars. Integrating dN between the inner and outer HZ boundaries, $r_1(M,t)$ and $r_2(M,t)$, gives the number of planets in the HZ at time t:

$$N_{HZ}(M,t) = C \int_{r_1}^{r_2} d(\log r) = C \log \left[\frac{r_2(M,t)}{r_1(M,t)}\right]$$

(For masses $\leq 0.3\ M_\odot$ a nontrivial part of the HZ lies interior to the terrestrial planet formation zone and so the lower integration limit should be a_1. Since our scaling of a_1 to the orbit of Mercury is probably too conservative and since this limit appears in the argument of a logarithm, we neglect this effect in the following discussion.) This result for $N_{HZ}(M,t)$ is readily interpretable in Figures 5 and 6 as the logarithmic width of the HZ of a star of mass M at time t, multiplied by a constant.

At $t = 0$ all spectral types plotted have the same ratio, $r_2/r_1 (\approx 2)$ and therefore the same number of planets in their HZs (~ 1).

For the case of no cold starts (Fig. 6), the HZ at a given time also corresponds to the CHZ for $\tau = t$. By inspection, the log CHZ for $\tau = 4.5$ Gyr is approximately the same for G, K, and M stars. Consequently, these stars should have comparable numbers of planets in their 4.5 Gyr CHZ. However, for $\tau \geq 4.5$ Gyr the K and M stars have nearly the same number of qualifying planets but the G stars' log CHZ diminishes, becoming zero at $\tau = 10$ Gyr. Overall, therefore, K stars and the abundant M stars are better candidates for having long-term habitable planets than are more massive stars if our assumptions of mass scaling and logarithmic spacing are valid. If M stars cannot be disqualified for other reasons (such as permanent glaciation due to synchronized planetary spins, as discussed in the next section) then they may possess most of the habitable planets in the Galaxy, notwithstanding their small (in absolute size) HZs.

Can Habitable Planets Exist Around M Stars?

The tidal interaction between an M star and any planets in its HZ will be much greater than that between the Earth and Sun. The inner HZ around a 0.1 $M_\odot$ star is at a distance of only 0.04 AU, or about 15 stellar radii. At this distance the orbital period is nine days. Since a mature M star is expected to rotate with a longer period, this corresponds to a tidally unstable configuration (Counselman 1973). However, using the analysis of Lecar et al. (1976) and Zahn (1977) we find that the time scale for a 1 $M_\oplus$ planet in the HZ to spiral into a fully convective star is greater than a Hubble time scale.

It was pointed out by Dole (1964) that planets around M stars might be synchronously rotating as a result of tidal damping. The dotted line in Figure 8 is our calculated tidal radius, r_T. It is the distance at which an Earthlike planet in a circular orbit would become tidally locked in a specified amount of time, t, taken here to be 4.5 Gyr. The tidal radius in cgs units is given by Peale (1977) as

$$r_T = 0.027 \left(\frac{P_0 t}{Q}\right)^{1/6} M^{1/3}$$

where P_0 is the original rotation period of the planet, Q^{-1} is the solid body plus ocean specific dissipation function, and M is the stellar mass. Today the Earth's Q is dominated by shallow seas and is 13. However, a value this small is inconsistent with the Earth-Moon system evolution, and the average solid body Q is presumably closer to 100 (Burns 1986). In calculating r_T in Figure 8 we used $Q = 100$, $t = 4.5$ Gyr, and $P_0 = 13.5$ h. Fortunately, r_T is relatively insensitive to the uncertainties in these parameters. Figure 8 shows that, after 4.5 Gyr of evolution, Earthlike planets in circular orbits lying within the HZ of stars of mass $\leq 0.5\ M_{\odot}$ will be synchronously rotating. This could result in the permanent freezing of all water and other volatiles on the dark hemisphere. Such freezing would certainly occur for synchronous rotation if the initial atmospheric pressure were sufficiently low. However, using a simple energy balance model, Haberle et al. (1996) find that only a modest surface pressure of ~ 100 mb, depending on stellar flux, is required to prevent the atmosphere from freezing out.

These results suggest that synchronously rotating planets may be habitable. However, M star planets may not all be synchronously rotating in any case. There are at least two alternative possibilities in the Solar System. One is that planets might be prevented from achieving synchronous rotation. The existence of orbital resonances, such as Mercury's 3 : 2 resonance with Venus, would tend to prevent synchronous rotation from occurring. A different example of nonsynchronous rotation is illustrated by Venus. The small amount of retrograde rotation of that planet should damp out in a small fraction of 1 Gyr. It is suspected that atmospheric tides have stabilized the rotation rate and prevented synchronous rotation from being achieved (Dobrovolskis 1980). Finally we note that, synchronized or not, planets around many M stars will be subjected to intensive flares, though it is not obvious that this would be a critical problem for habitability.

Planets and Habitable Zones in Multiple Star Systems

More than 65% of nearby solar-type stars are members of binary or multiple star systems (Duquennoy and Mayor 1991). These systems have sometimes been rejected as likely locations for habitable planets for two reasons—the suspected dynamical instability of planetary orbits, and the belief that, even granted orbital stability, planets would be unlikely to form at the required distances (Lissauer 1989).

The former objection was addressed by Graziani and Black (1981) and Pendleton and Black (1983), who found that stable, near-circular, planetary orbits existed when the planet's orbital radius was typically a factor of 5 or more larger than the binary separation (external case—planet orbits barycenter of the binary system), or vice versa, when the binary separation is a factor of 5 or more greater than the planet's distance from either star (internal case—planet orbits one star). More recently, Kubala et al. (1993) compared the Hill and Laplace stability criteria (e.g., Szebehely and McKenzie 1981) for the case of external planetary orbits. This analytical study supported the original Graziani-Black stability criterion for external orbits. However, other work (e.g., Dvorak 1986, Rabl and Dvorak 1988, Dvorak et al. 1989 and references therein) found a less restrictive limit. Rather than a factor-of-5 ratio, they obtained a factor-of-2 ratio, a result possibly at odds with observations of triple star systems. In these studies, numerical calculations for both internal and external orbits follow the systems for a relatively small number of stellar orbits (~ 500). For a review of other studies of the stability of planets in multiple star systems see Hale (1994, 1996).

Using the binary semimajor axes statistics of Duquennoy and Mayor (1991) and the Graziani-Black stability criteria and our HZ results, we estimate that ~ 50% of internal binaries and ~ 10% of external binaries could support habitable planets in at least short-term stable orbits. Numerical experiments have not as yet tested for stability on longer, more interesting, time scales.

With regard to the second objection it is generally expected that an accretion disk(s) will remain after binary star formation (Artymowicz et al. 1991). The time scale for planetesimal formation is likely to be short compared with the disk lifetime. Evidence that planetesimal accretion can occur is suggested by the fact that multiple star systems exhibit IR excesses with about the same frequency as do single stars (Backman and Paresce 1993). As discussed earlier in connection with single stars, these observations imply the presence of larger sources (planetesimals), indicating that accretion of planets is not necessarily ruled out dynamically.

According to Wetherill (1991) a perturbation capable of causing a relative velocity between planetesimals of 100 km/s would shut down runaway accretion and prevent embryo (Moon- to Mercury-sized bodies) formation. We have done preliminary numerical calculations that show this level of perturbation is readily achieved within the 10^4–10^5 yr time scale of runaway accretion in all but the widest (internal) binary systems. The typical solar-type binary has a separation of ≈ 38 AU and an eccentricity of ≈ 0.7 (Duquennoy and Mayor 1991). Assuming both stars formed at roughly the same time, this typical system and those with smaller semimajor axes may not accrete planets at and beyond one AU from the primary star. More separated systems will tend to have random eccentricities (mean = 0.7) and random relative inclinations (mean = 32 deg) between the expected planetesimal plane (= equatorial plane of primary star) and the binary orbital plane (Hale 1994). We find that many of these systems will produce the critical stirring velocity at one AU and that if a similar runaway accretion criterion is used at 5 AU only a small fraction of known binaries would accrete a Jovian planet. Jovian planets may be necessary to bring water and other volatiles and organics to terrestrial planets (Delsemme 1991) and/or shield habitable planets from excessive impacts (Wetherill 1994). It remains to be investigated under what conditions dynamical friction, collisions, or gas drag can relax relative velocities faster than a binary companion can accelerate them. In cases where a Jovian planet does not form but a terrestrial planet at one AU does, it is possible that under some range of parameters the companion star can simulate Jupiter's role in perturbing volatiles into the terrestrial region.

Conclusions and Future Work

Recently it has been proposed (Goldin 1994) that NASA adopt as a "new unifying mission" the search for a habitable planet around a nearby star. In addition to observations, such a mission would require (1) knowledge of the location and extent of habitable zones around specific nearby single and multiple star systems, (2) an assessment of the formation and long-term stability of planetary orbits in the HZs of these systems, and (3) an evaluation of the habitability of synchronously rotating planets around M dwarfs, which are the most numerous of the nearby stars.

In this paper we have focused primarily on the first of these objectives and reviewed results of recent HZ calculations that employed a modern climate model in conjunction with standard stellar evolution models to calculate the size and evolution of HZs around stars of various masses. Our main conclusions are that HZs and CHZs are an order of magnitude larger than suggested by an earlier model and that the average number of planets in the HZ may be independent of stellar type. This latter result, especially in conjunction with the results of Haberle et al. (1996) on the habitability of synchronously rotating planets, suggests that the nearby M stars should be considered candidates in the search for a habitable planet. Further progress in our knowledge of the location and evolution of HZs depends mainly on improvements in climate models—for example, by taking clouds into account directly and determining whether planets initially beyond the outer HZ radius can subsequently be "cold started" as the stellar luminosity increases.

Many of the astronomical problems that need to be addressed in the search for a habitable planet are dynamical in nature. As discussed in the previous section, a binary companion may shut down the runaway accretion process. A similar perturbation in a single star system by a Jupiter-mass planet might prevent the accretion of a terrestrial planet in the HZ of stars somewhat more massive than the Sun (i.e., HZ in an asteroid belt). Assuming that larger Jupiter-like planets will be discovered first, it will then be possible to assess the long-term dynamical stability of planetary orbits located within the HZ under the perturbative influence of the larger planet. If unstable, these systems could then be eliminated from the search. This may or may not be a problem for the numerous nearby M dwarfs (whose bolometric luminosities are typically two or three orders of magnitude less than the Sun's), since any Jupiter-like planets in these systems are likely to be much closer to the HZ in absolute distance. Prior to the anticipated discovery of Jupiter-like planets, similar long-term dynamical calculations could also be performed for HZ orbits in both accretion-modeled planetary systems (Wetherill 1996) and known nearby binary star systems. Disk accretion modeling of isolated M dwarf planetary systems might suggest whether these systems are sufficiently stable from large planet perturbations to be likely candidates in the search for a habitable planet. In previous three-body calculations of planetary orbit stability in binary systems, the planet was implicitly the most massive one in the system. This would likely be a Jovian planet rather than a potentially habitable planet. For the investigation of the stability of habitable planets in binary systems, four-body experiments are needed.

As previously noted, the results of Haberle et al. suggest that even a modest atmosphere may be sufficient to prevent water and other volatiles from freezing out on the dark hemisphere of a synchronously rotating planet in the HZ of a late K or M star. As the authors conclude, these results are encouraging but need to be substantiated by more sophisticated modeling of atmospheric heat transport.

REFERENCES

ARTYMOWICZ, P., C.J. CLARKE, S.H. LUBOW, AND J.E. PRINGLE (1991), "The Effect of an External Disk on the Orbital Elements of a Central Binary," *ApJ* **370:** L35–L38.

AUMANN, H.H., F.C. GILLETT, C.A. BEICHMAN, T. DE JONG, J.R. HOUCK, F.J. LOW, G. NEUGEBAUER, R.G. WALKER, AND P.R. WESSELIUS (1984), "Discovery of a Shell Around Alpha Lyrae," *ApJ* **278:** L23–L27.

BACKMAN, D.E., AND F. PARESCE (1993), "Main-Sequence Stars with Circumstellar Solid Material: The Vega Phenomenon," in *Protostars and Planets III*, E.H. Levy et al., eds. (Tucson: University of Arizona Press), pp. 1253–1304.

BARROW, J.D., AND F.J. TIPLER (1986), *The Anthropic Cosmological Principle* (Oxford: Oxford University Press).

BODENHEIMER, P. (1989), "The Impact of Stellar Evolution on Planetary System Development," in *The Formation and Evolution of Planetary Systems*, H.A. Weaver and L. Danly, eds. (Cambridge: Cambridge University Press), pp. 243–273.

BURNS, J.A. (1986), "The Evolution of Satellite Orbits," in *Satellites*, J.A. Burns and M.S. Matthews, eds. (Tucson: University of Arizona Press), pp. 117–158.

CALDEIRA, K., AND J.F. KASTING (1992), "Susceptibility of the Early Earth to Irreversible Glaciation Caused by Carbon Dioxide Clouds," *Nature* **359:** 226–228.

CAMERON, A.G.W. (1963), "Stellar Life Zones," in *Interstellar Communication*, A.G.W. Cameron, ed. (New York: W.A. Benjamin), pp. 107–114.

CARTER, B. (1983), "The Anthropic Principle and Its Implications for Biological Evolution," *Phil. Trans. R. Soc. Lond. A* **310:** 347–363.

COUNSELMAN, C. (1973), "Outcomes of Tidal Evolution," *ApJ* **180:** 307–314.

DELSEMME, A.H. (1991), "Origin of the Biosphere of the Earth," in *Bioastronomy: The Search for Extraterrestrial Life*, J. Heidmann and M.J. Klein, eds. (Berlin: Springer-Verlag), pp. 117–123.

DOBROVOLSKIS, A. (1980), "Atmospheric Tides and the Rotation of Venus. II. Spin Evolution," *Icarus* **41:** 18–35.

DOLE, S.H. (1964), *Habitable Planets for Man* (New York: Blaisdell Publishing Company).

DOYLE, L.R., R. VIKRAMSINGH, D.P. WHITMIRE, AND N.C. HEATHER (1995), "Circumstellar Habitable Zones and Mass Loss from Young Solar-type Stars. II. Observational Considerations," in *Progress in the Search for Extraterrestrial Life*, ASP Conf. Ser., S. Shostak, ed., **74:** 195–204.

DOYLE, L.R., N.C. HEATHER, R. VIKRAMSINGH, AND D.P. WHITMIRE (1996), "Stellar Mass Loss and the Young Solar System Habitable Zone," this volume.

DUQUENNOY, A., AND M. MAYOR (1991), "Multiplicity Among Solar-type Stars in the Solar Neighborhood. II. Distribution of the Orbital Elements in an Unbiased Sample," *A&A* **248:** 485–524.

DVORAK, R. (1986), "Critical Orbits in the Elliptic Three-Body Problem," *A&A* **167:** 379–386.

DVORAK, R., CH. FROESCHLÉ, AND CL. FROESCHLÉ (1989), "Stability of Outer Planetary Orbits (P-Types) in Binaries," *A&A* **226:** 335–342.

FOGG, M.J. (1992), "An Estimate of the Prevalence of Biocompatible and Habitable Planets," *J. Brit. Interplanet. Soc.* **45:** 3–12.

GOLDIN, D. (1994), "Address to the 75th Anniversary Meeting of the AGU, Washington D.C.," *Nature* **369:** 349.

GOUGH, D.O. (1981), "Solar Interior Structure and Luminosity Variations," *Solar Phys.* **74:** 21–34.

GRAEDEL, T., I. SACKMANN, AND A. BOOTHROYD (1991), "Early Solar Mass Loss: A Potential Solution to the Weak Sun Paradox," *Geophys. Res. Lett.* **18:** 1881–1884.

GRAZIANI, F., AND D.C. BLACK (1981), "Orbital Stability Constraints on the Nature of Planetary Systems," *ApJ* **251:** 337–341.

HABERLE, R.M., C.P. MCKAY, D. TYLER, AND R.T. REYNOLDS (1996), "Can Synchronously Rotating Planets Support an Atmosphere?," this volume.

HALE, A. (1994), "Orbital Coplanarity in Solar-type Binary Systems: Implications for Planetary System Formation and Detection," *AJ* **107:** 306–332.

HALE, A. (1996), "Determining Habitable Planetary Environments Within Binary and Multiple Systems," this volume.

HART, M.H. (1978), "The Evolution of the Atmosphere of the Earth," *Icarus* **33:** 23–39.

HART, M.H. (1979), "Habitable Zones Around Main Sequence Stars," *Icarus* **37:** 351–357.

HUANG, S.-S. (1960), "Life Outside the Solar System," *Sci. Am.* **202 (4):** 55–63.

IBEN, I. (1967a), "Stellar Evolution Within and Off the Main Sequence," *Ann. Rev. Astron. Astrophys.* **5:** 571–626.

IBEN, I. (1967b), "Stellar Evolution. VI. Evolution from the Main Sequence to the Red Giant Branch for Stars of Mass 1 $M_\odot$, 1.25 $M_\odot$ and 1.5 $M_\odot$," *ApJ* **147:** 624–649.

IBEN, I. (1974), "Post Main Sequence Evolution of Single Stars," *Ann. Rev. Astron. Astrophys.* **12:** 215–256.

IBEN, I., AND A. RENZINI (1983), "Asymptotic Giant Branch Evolution and Beyond," *Ann. Rev. Astron. Astrophys.* **21:** 271–342.

ISAACMAN, R., AND C. SAGAN (1977), "Computer Simulation of Planetary Accretion Dynamics: Sensitivity to Initial Conditions," *Icarus* **31:** 510–533.

KASTING, J.F. (1988), "Runaway and Moist Greenhouse Atmospheres and the Evolution of Earth and Venus," *Icarus* **74:** 472–494.

KASTING, J.F. (1991), "CO_2 Condensation and the Climate of Early Mars," *Icarus* **94:** 1–13.

KASTING, J.F. (1996), "Habitable Zones Around Stars: An Update," this volume.

KASTING, J.F., O.B. TOON, AND J.B. POLLACK (1988), "How Climate Evolved on the Terrestrial Planets," *Sci. Am.* **258 (2):** 90–97.

KASTING, J.F., D.P. WHITMIRE, AND R.T. REYNOLDS (1993), "Habitable Zones Around Main Sequence Stars," *Icarus* **101:** 108–128.

KUBALA, A., D. BLACK, AND V. SZEBEHELY (1993), "Stability of Outer Planetary Orbits Around Binary Stars: A Comparison of Hill's and Laplace's Stability Criteria," *Celest. Mech. Dyn. Astron.* **56:** 51–68.

LAZZARO, D., B. SICARDY, F. ROQUES, AND R. GREENBERG (1994), "Is There a Planet Around β Pictoris? Perturbations of a Planet on a Circumstellar Dust Disk. 2. The Analytical Model," *Icarus* **108:** 59–80.

LECAR, M., J.C. WHEELER, AND C.F. MCKEE (1976), "Tidal Circulation of the Binary X-Ray Sources Hercules X-1 and Centaurus X-3," *ApJ* **205:** 556–562.

LISSAUER, J.J. (1989), "Which Stars Have Planets?," in *The Formation and Evolution of Planetary Systems*, H.A. Weaver and L. Danly, eds. (Cambridge: Cambridge University Press), pp. 304–308.

LISSAUER, J.J. (1995), "Urey Prize Lecture: On the Diversity of Plausible Planetary Systems," *Icarus* **114:** 217–236.

LOVELOCK, J.E., AND A.J. WATSON (1982), "The Regulation of Carbon Dioxide and Climate," *Planet. Space Sci.* **30:** 795–802.

MCKAY, C.P., AND W.L. DAVIS (1991), "Duration of Liquid Water Habitats on Early Mars," *Icarus* **90:** 214–221.

O'DELL, R., AND Z. WEN (1994), As reported in *Sci. News* **145:** 391.

PEALE, S.J. (1977), "Rotation Histories of the Natural Satellites," in *Planetary Satellites*, J.A. Burns, ed. (Tucson: University of Arizona Press).

PENDLETON, Y.J., AND D.C. BLACK (1983), "Further Studies on the Criteria for the Onset of Dynamical Instability in General Three-Body Systems," *AJ* **88:** 1415–1419.

POLLACK, J.B., AND R.T. REYNOLDS (1974), "Implications of Jupiter's Early Contraction History for the Composition of the Galilean Satellites," *Icarus* **21:** 248–253.

POLLACK, J.B., J.F. KASTING, S.M. RICHARDSON, AND K. POLIAKOFF (1987), "The Case for a Wet, Warm Climate on Early Mars," *Icarus* **71:** 203–224.

POLLARD, W.G. (1979), "The Prevalence of Earthlike Planets," *Amer. Sci.* **67:** 653–659.

RABL, G., AND R. DVORAK (1988), "Satellite-Type Planetary Orbits in Double Stars: A Numerical Approach," *A&A* **191:** 385–391.

REYNOLDS, R.T., S.W. SQUYRES, D.S. COLBURN, AND C.P. MCKAY (1983), "On the Habitability of Europa," *Icarus* **56:** 246–254.

REYNOLDS, R.T., C.P. MCKAY, AND J.F. KASTING (1987), "Europa, Tidally Heated Oceans, and Habitable Zones Around Giant Planets," *Adv. Space Res.* **7:** 125–132.

ROOD, R.T., AND J.S. TREFIL (1981), "Are We Alone?," in *The Possibility of Extraterrestrial Civilizations* (New York: Scribner Publishing Company).

ROQUES, F., H. SCHOLL, B. SICARDY, AND B.A. SMITH (1994), "Is There a Planet Around β Pictoris? Perturbations of a Planet on a Circumstellar Dust Disk. 1. The Numerical Model," *Icarus* **108:** 37–58.

SACKMANN, I.-J., A.I. BOOTHROYD, AND K.E. KRAEMER (1993), "Our Sun. III. Present and Future," *ApJ* **418:** 457–468.

SAGAN, C., AND C. CHYBA (1996), "Organic Shielding of UV-Labile Greenhouse Gases on Early Earth and Mars," abstract, this volume.

SCHWARTZMAN, D.W., AND T. VOLK (1989), "Biotic Enhancement of Weathering and the Habitability of Earth," *Nature* **340:** 457–460.

SMITH, B.A., AND R.J. TERRILE (1984), "A Circumstellar Disk Around Beta Pictoris," *Science* **226:** 1421–1424.

SOLOMON, S.C., AND J.W. HEAD (1991), "Fundamental Issues in the Geology and Geophysics of Venus," *Science* **252:** 252–260.

SZEBEHELY, V., AND R. MCKENZIE (1981), "Stability of Outer Planetary Systems," *Celest. Mech.* **23:** 3–7.

WALKER, J.C.G., P.B. HAYS, AND J.F. KASTING (1981), "A Negative Feedback Mechanism for the Long-Term Stabilization of Earth's Surface Temperature," *JGR* **86:** 9776–9782.

WETHERILL, G.W. (1991), "Occurrence of Earth-like Bodies in Planetary Systems," *Science* **253:** 535–538.

WETHERILL, G.W. (1994), "Possible Consequences of Absence of 'Jupiters' in Planetary Systems," *A&SS* **212:** 23–32.

WETHERILL, G.W. (1996), "The Formation of Habitable Planetary Systems," this volume.

WHITMIRE, D.P., J.J. MATESE, AND L.J. TOMLEY (1988), "A Brown Dwarf Companion as an Explanation of the Asymmetry in the Beta Pictoris Disk," *A&A* **203:** L13–L15.

WHITMIRE, D.P., R.T. REYNOLDS, AND J.F. KASTING (1991), "Habitable Zones for Earth-like Planets Around Main Sequence Stars," in *Bioastronomy: The Search for Extraterrestrial Life*, J. Heidmann and M.J. Klein, eds. (Berlin: Springer-Verlag), pp. 173–178.

WHITMIRE, D.P., L.R. DOYLE, R.T. REYNOLDS, AND J.J. MATESE (1995a), "Circumstellar Habitable Zones and Mass Loss from Young Solar-type Stars. I. Theory," in *Progress in the Search for Extraterrestrial Life*, ASP Conf. Ser., S. Shostak, ed., **74:** 183–193.

WHITMIRE, D.P., L.R. DOYLE, R.T. REYNOLDS, AND J.J. MATESE (1995b), "A Slightly More Massive Young Sun as an Explanation for Warm Temperatures on Early Mars," *JGR* **100:** 5457–5464.

ZAHN, J.-P. (1977), "Tidal Friction in Close Binary Stars," *A&A* **57:** 383–394.

40 Eridani Triple-Star System

Photo Credit: © Alan Hale, Earthrise Institute/Las Cumbres Observatory/Global Sky Partners

(Note: After the Circumstellar Habitable Zones conference, Alan Hale went on to codiscover the largest comet ever discovered–Hale Bopp.)

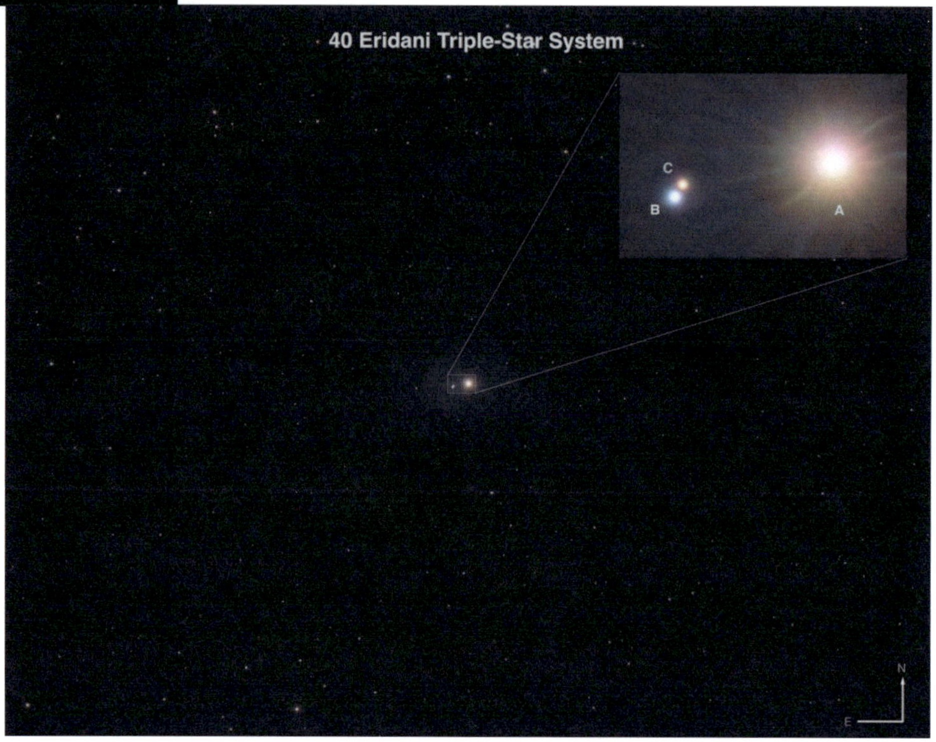

Photo Credit: © Pornthep Benyaapikul (https://app.astrobin.com/u/Pornthep.Benyaapikul) https://app.astrobin.com/i/hsbs8e

40 Eridani is about 16.3 light-years from Earth and is in fact a triple-star system in Eridanus. The brightest component is 40 Eridani A, an orange K-type star of about 4.4 magnitude, easily visible to the naked eye, and about 0.8 times the Sun's mass. The two more distant stars that accompany it are 40 Eridani B, a white dwarf, and 40 Eridani C, a red dwarf flare star. The two were discovered in January 1783 by William Herschel.

A planet orbiting 40 Eridani A would not have three Sunlike disks dominating its sky, but it would have two additional bright stellar companions orbiting in the distance, and slightly brighter than Venus seen from Earth in the evening sky.

In 2018, astronomers announced a possible super-Earth in the habitable zone of 40 Eridani A, but this was later rejected, as the signals were found to most likely be caused by the star itself. Incidentally, in the Star Trek franchise 40 Eridani was associated with Vulcan, Spock's home planet.

DETERMINING HABITABLE PLANETARY ENVIRONMENTS WITHIN BINARY AND MULTIPLE SYSTEMS

Alan Hale

Southwest Institute for Space Research
Cloudcroft, New Mexico 88317

Abstract

The presence of one or more stellar companions significantly complicates the issues involved in the discussion of suitable planetary environments for life, not the least of which is the question whether planets can even form in such a system. Given that they can, other items that must be addressed include stability of planetary orbits and the effects of varying insolation. Several theoretical studies have shown that stable planetary orbits can exist in binary systems, provided that the stellar components are either close together or remain relatively far apart. The effects of varying insolation upon the habitability of a planet due to the varying distances of the planet from the stellar companions are difficult to quantify, but simple calculations suggest that if orbital stability criteria are met, the flux variations received at the planet are minimal. However, in systems where the components are of dissimilar masses, the evolution of the more massive component(s) may well affect the habitability of planets orbiting the less massive component(s) over biologically long time scales.

INTRODUCTION

Most studies concerning the forming and maintaining of habitable planetary environments have concentrated on our Solar System or on stars like our Sun. The Sun, however, is a single star, and many of the stars within the solar neighborhood and the Galaxy are not. Recent work (Duquennoy and Mayor 1991) suggests that approximately two-thirds of the solar-type stars in the solar neighborhood are members of binary or multiple systems. Thus, the effects of nearby stellar companions on the habitability of planetary environments must be considered in estimating the number of potential life-bearing planets within the Galaxy.

In the discussion of planets within binary or multiple systems, two scenarios typically require examination: the so-called "outer planet" case (Fig. 1a), wherein the stellar components are close together and the planet orbits both of them; and the "inner planet" case (Fig. 1b), wherein the stellar components are more widely separated and the planet orbits one of them. For the purposes of discussing habitable planetary environments, "outer planet" scenarios would involve those systems where the stellar components are separated by $\approx 10^{-2}$ to 10^{-1} AU, and "inner planet" scenarios involve those systems where the stellar components are separated by $\approx 10^{1}$ to 10^{2} AU. Systems with intermediate separations, say 10^{0} AU, would be unlikely to have habitable planets in stable orbits due to the stability considerations discussed below, while wider systems, say 10^{3} to 10^{4} AU, are far enough apart that they may be treated as two separate single-star systems.

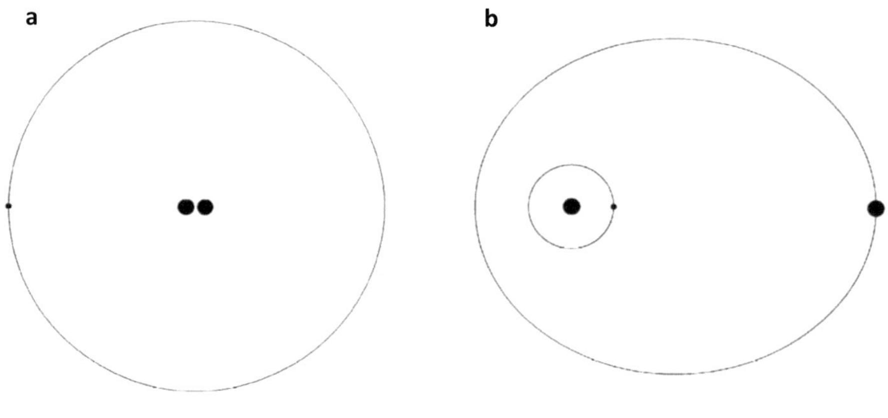

FIG. 1. *Possible configurations of a planetary orbit within a binary system. (a) "Outer planet" case, where planet orbits both of two closely separated stars; (b) "inner planet" case, where planet orbits one of two widely separated stars.*

The Existence of Planets Within Binary and Multiple Systems

The first issue that must be addressed within this discussion concerns the actual existence of such systems—can planets indeed form within binary or multiple systems? Most theoretical studies of planetary system formation are concerned with single-star systems like the Solar System; there are still enough unsolved problems in this endeavor that most researchers find it impractical to consider the perturbations induced by additional stellar companions. One exception to this has been early work by Heppenheimer (1974, 1978), who examined the problem of planetary formation within binaries in some detail and concluded that the existence of planets within binaries is unlikely unless the stellar components are separated by more than about 50 AU. Heppenheimer, however, based his calculations on the planetesimal accretion scenario devised by Goldreich and Ward (1973), which is now known to be incomplete (Cameron 1988). The effects that Heppenheimer's reliance on the Goldreich and Ward theory may have had on the validity of his calculations is difficult to judge. As the planetary formation process continues to become better understood, the complexities caused by the presence of additional stellar companions may become a fruitful and worthwhile area for future studies.

From an observational standpoint, the evidence is tantalizing but inconclusive. Backman and Paresce (1993) point out that, according to IRAS data, there is no lack of observed infrared excess among binary systems compared with single stars. Millimeter observations by Beckwith et al. (1990) of T Tauri stars and related objects in the Taurus-Auriga region indicate that at least the wider binaries appear to have circumstellar disks, although these appear to be less massive, on the average, than the disks accompanying single stars. More recently, Zuckerman and Becklin (1993) have found that the K5V binary HD 98800 is apparently surrounded by a massive zodiacal dust band, on the order of 10^6 times as massive as that in our Solar System. (There has also been a recent report by Vidal-Madjar et al. [1993] that the A2V binary 68 Ophiuchi is accompanied by a β-Pictoris-type disk, but this has apparently been withdrawn as erroneous; J. Schneider, pers. comm.) These observations suggest that the materials that eventually can accrete into planets can indeed coexist with binary systems; whether the planets could actually form from these materials is the unanswered question.

If planetary existence is in fact incompatible with the presence of additional stellar companions, any discussion of the effects of the stellar companions on the habitability of planets within the system is of course superfluous. For the sake of discussion, the remainder of this paper will assume that planets can indeed form within binary and multiple systems.

Stability of Planetary Orbits

From the standpoint of having a planetary environment suitable for the formation and evolution of life, a stable planetary orbit is a strong necessity, else any life-forms on a given planet would be subject to severe variations in received insolation, with resulting drastic effects on the planet's climate. Numerous studies have been published concerning the stability of planetary orbits within binary systems (e.g., Harrington 1977, Szebehely 1980, Pendleton and Black 1983, Schechter 1984, Rabl and Dvorak 1988, Donnison and Mikulskis 1992, Dobrovolskis et al., abstract, this volume) and all have essentially concluded that stable orbits can indeed exist provided that certain criteria concerning the separation between the stellar components (and between the planet and the stellar components) and the components' masses are met. The primary criterion is the ratio $\frac{p}{a}$, where p is the periapsis distance from the outer body (the planet in the outer planet scenario, the stellar companion in the inner planet scenario) to the barycenter of the inner system (star/star in outer planet scenario, star/planet in the inner planet scenario), and a is the semimajor axis of the inner system. Acceptable lower limit determinations for $\frac{p}{a}$ have ranged from about 3.3 (Donnison and Mikulskis 1992) to about 5.0 (Pendleton and Black 1983). Doyle (1993) has conservatively estimated that a lower limit to the $\frac{p}{a}$ ratio of 10 might be necessary for the formation of planets in stable orbits.

Most of the theoretical studies of binary/planetary orbital stability have assumed that the stellar components and the planets orbit in the same plane. At least two studies have briefly treated the effects of noncoplanar orbits upon stability—Harrington (1972), who found no significant effects on stability by noncoplanar orbits; and Pendleton and Black (1983), who found that in the inner planet scenario, instability begins to appear if the mutual inclination of the orbits is greater than about 50°. The total number of configurations examined by both of these studies is rather small, however. It is appropriate to ask whether such configurations might occur in nature. In an examination of equatorial vs. orbital coplanarity in solar-type binaries, Hale (1994) determined that noncoplanarity would not appear to be an issue in those binaries where the stellar components are separated by less than about 40 AU (and systems with component separations larger than this may be far enough apart that noncoplanarity may not matter much anyway). On the other hand, both Fekel (1981) and Hale (1994) determined that from one-third to one-half of hierarchical multiple systems appear to exhibit distinct noncoplanarity between the inner and outer systems. An interesting example is the ξ Ursae Majoris system (Fig. 2), where all three orbits within the system are observed at distinctly different inclinations. The existence of systems like this suggests that stable noncoplanar systems can exist in nature, although one might suspect that ξ Ursae Majoris may not be a suitable environment for habitable planets!

VARIATIONS IN RECEIVED INSOLATION

The presence of additional stellar components would in most cases subject planets within the system to varying levels of insolation. It is difficult to estimate how much variation in insolation over a periodic interval is "acceptable" from the standpoint of biological and climatological evolution. Dole (1970) estimated, somewhat arbitrarily, that received flux variations up to 10% would be acceptable for the maintenance of a habitable environment, and there is little reason to suspect that this figure is greatly in error.

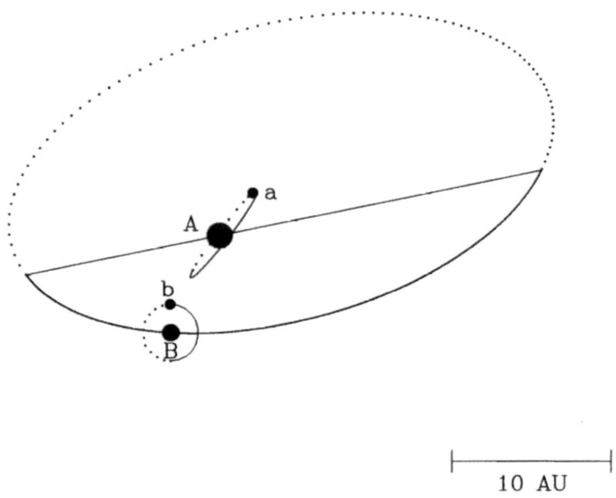

FIG. 2. *Orbital plot of ξ Ursae Majoris system, illustrating three distinctly different orbital inclinations within the same system. The* AB *orbit is retrograde, and inclined 58° to the plane of the sky; the* A**a** *orbit is also retrograde, and inclined 85° to the plane of the sky; the* B**b** *orbit is inclined 18° to the plane of the sky. Those portions of the orbits above the plane of the sky are indicated by a solid line, those portions below by dots. The line of nodes for the* AB *orbit is also indicated by a solid line. The indicated node crossings for the* B**b** *orbit are arbitrary. The positions of components* B *and* **a** *relative to* A *are for epoch 1994.05, the position of component* **b** *relative to* B *is arbitrary. The 10 AU scale in the lower right is for the* AB *orbit; the* A**a** *orbit has been magnified by a factor of 10, and the* B**b** *orbit has been arbitrarily magnified. (Orbital information is from Mason et al. 1995 for the* AB *and* A**a** *orbits; inclination for the* B**b** *orbit is from Hale 1994, assuming coplanarity between the stellar equatorial and systemic orbital planes.)*

Numerical simulations suggest that, if the stellar components' separations are such that stable planetary orbits can exist, in either the inner or outer planet scenarios, then variations in received stellar flux are not significant in their effects on planetary habitability. Harrington (1977) performed calculations for two scenarios, one with the Sun being replaced by a binary with both components having a mass of 0.5 $M_{\odot}$ and separated by 0.2 AU, the other with Jupiter being replaced with a 1 $M_{\odot}$ star (both configurations being stable by the $\frac{p}{a}$ criterion). The effects of these two configurations on the Earth's mean temperature are shown in Harrington's

Figure 3; in neither of these cases is the temperature variation (over an annual period) significantly greater than 10 K, not dissimilar to the current annual variation.

For a further example, it is instructive to consider the α Centauri system. This system comprises two stars of spectral types G2V and K1V (as well as a distant M companion, which can be ignored for this discussion); according to the orbital elements in the catalog by Worley and Heintz (1983), the orbital period is 79.9 years, the semimajor axis is 23.4 AU, and the eccentricity is 0.52, which causes the separation to vary from 11.3 AU at periastron to 35.4 AU at apastron. Stability analyses of this system by Benest (1988, 1991) suggest that stable planets out to 5 AU could exist around either component, thus planets in an Earthlike orbit should certainly be stable; the $\frac{p}{a}$ criterion is satisfied. The system is shown from a pole-on view in Figure 3; the habitable zones, the boundaries of which are here arbitrarily defined as being those radii that receive the same flux (from the respective parent sun) as Venus and Mars do in our Solar System, are shown for each component. These were calculated assuming effective temperatures of 5770 K and 5350 K for components A and B, respectively, and radii of 1.23 and 0.80 $R_\odot$ (Soderblom 1986). The corresponding orbital distances of an Earth analog are 1.23 and 0.69 AU.

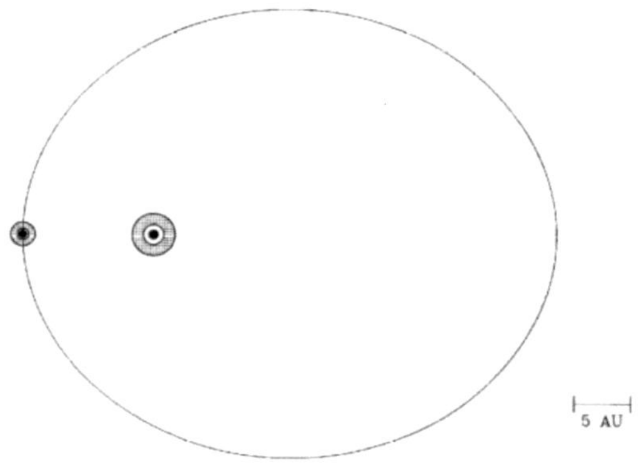

FIG. 3. *The α Centauri system, viewed pole-on; component* B *is shown orbiting a stationary component* A. *The habitable zones are indicated, via stippling, for each component; the definition of habitable zone used here is that given in the text.*

The mean annual fluxes received by an Earth analog around either component through one orbital revolution of the α Centauri system are shown in Figures 4a and b, normalized to the mean annual flux the Earth receives in our Solar System. The overall augmentation in flux received by either planet as a result of the other stellar component is minimal in both cases, being 3% for an Earth analog around component A and 5% for such a planet around component B. The variation in flux received throughout the system's orbital period amounts to 4% for the planet around component A and 7% for the planet around component B. Both of these are well within the 10% upper limit criterion suggested by Dole. Taking α Centauri as a "typical" binary where an inner planet scenario might exist, these results, together with those of Harrington, suggest that in any such system where the criteria for stable orbits for Earthlike planets are satisfied, any criteria regarding variation in received insolation are also satisfied.

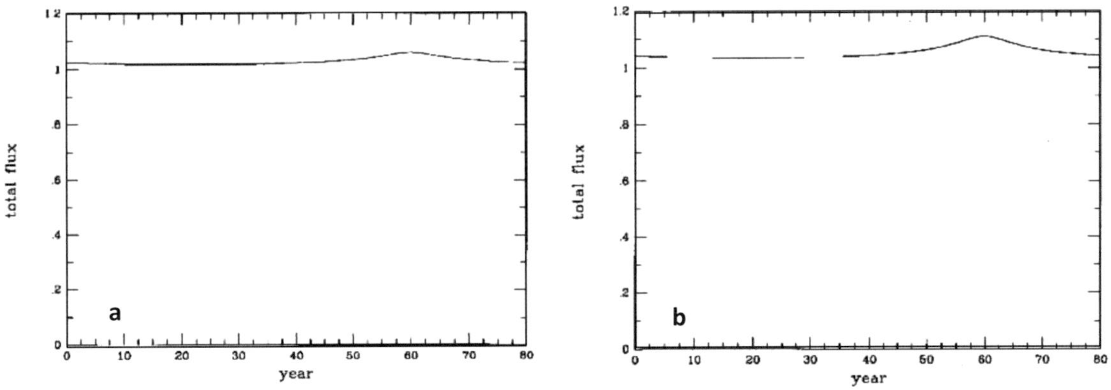

FIG. 4. *Average annual fluxes received by Earth analog planets in the α Centauri system over one orbital period of the stellar components. The flux is normalized to that received by the Earth in our Solar System. Time "0" is halfway between periastron and apastron; the time scale progresses through apastron, then through periastron, and back to time "0." (a) Earth analog around α Centauri A, (b) Earth analog around α Centauri B.*

An interesting problem in the issue of flux variation in an outer planet scenario was discussed by Doyle (1993), who noted that planets in orbits coplanar (or nearly so) with that of the stellar components would experience eclipses as the two stars repeatedly passed in front of and behind each other. While the magnitude and frequency of the flux variations would depend on the system's orbital characteristics, such as separation and eccentricity, and upon the components' respective masses, in a general sense any planets in such a system's habitable zone would experience significant drops in received insolation, for a duration of up to several hours, several times during its orbital period. While thermal inertia due to a planet's atmosphere and lithosphere should keep the temperature variations during eclipses from becoming too drastic, the effects of these regular and significant drops in received flux on a planet's biology are difficult to estimate. This is an issue that requires further investigation, ideally with a large number of binary system configurations.

EVOLUTION OF THE STELLAR COMPONENTS

The preceding discussion concerning the effects of varying insolation upon habitable environments within binary systems is, strictly speaking, valid only for a static system. However, stars evolve, and as they do the effects on their zone of habitability can be significant (see Kasting et al. 1993 for a recent thorough discussion). The evolution of the habitable zone for systems with an outer planet configuration will be driven by the evolution of the more massive stellar component; for the purposes of this discussion, such habitable zones may be treated as being around a single star—the one whose mass is largest. On the other hand, recent work by Mazeh et al. (1992) suggests that the stellar components of short period binaries, which would give rise to the outer planet configuration, tend toward similar masses. Thus the variation in stellar evolution of the components is perhaps not a significant issue in addressing the evolution of the habitable zones of such systems.

To test the effects of stellar evolution variation in the inner planet configuration, calculations for the α Centauri system similar to those described above were performed for an Earth analog around component B, as component A experiences its future evolution. For simplicity, the effects of the evolution of component B on its own habitable zone were ignored, and component A was replaced with our own Sun, whose evolution is better understood. The average annual fluxes received at the planet were computed for several future epochs using the data given in Table 2 of Sackmann et al. (1993); the results are shown in Table 1 (where column 1 gives elapsed time [T_f] from the present in Gyr, column 2 gives the average annual flux normalized to that received by the present Earth, column 3 the flux normalized to that currently received by an Earth analog around component B, and column 4 the periastron-to-apastron change in flux experienced during one orbital period of the binary system) and are shown graphically in Figure 5. These results suggest that as long as component A remains on the main sequence, there are no significant effects on the habitability of planets around component B; however, once component A begins to leave the main sequence the habitability of planets around B is essentially ended, both in terms of absolute flux received and in terms of flux variability, even though component B may have several Gyr of main sequence lifetime left. Thus, although it is less intuitive than is the case for outer planet systems, the evolution of habitable zones of inner planet binaries is also driven by the evolution of the more massive component.

TABLE 1. ***Flux Changes for Earth Analog Around Component B in Hypothetical Alpha Centauri System***

T_f *(Gyr)*	*Avg. Flux (Earth = 1)*	*Avg. Flux ($T_f = 0$)*	*Percent Change over Orbit*
0	1.043	–	5.9
3.0	1.050	1.006	6.8
4.8	1.056	1.012	7.5
6.4	1.086	1.041	11.4
7.1	1.071	1.027	9.6
7.5	1.179	1.130	22.5
7.6	1.252	1.200	30.4
7.7	3.059	2.933	81.07

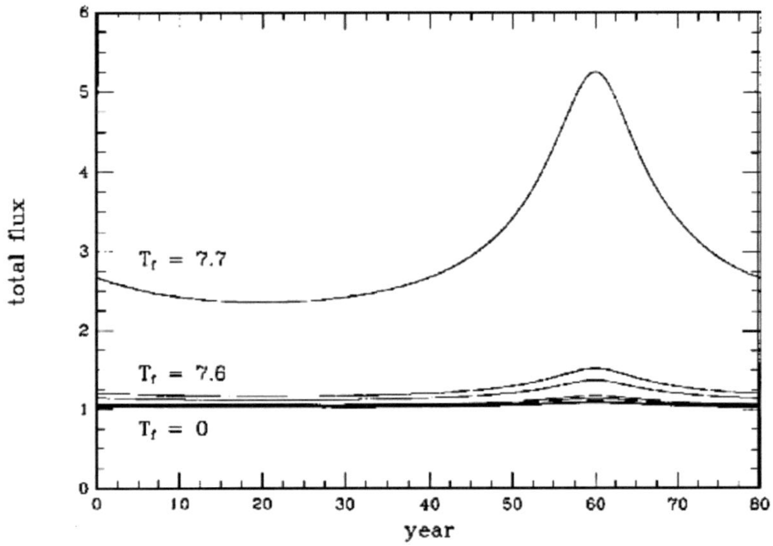

FIG. 5. *Average annual flux received by an Earth analog planet around component B in the α Centauri system over one orbital period of the stellar components, as component A (assumed to be identical to our Sun) evolves off the main sequence. (The evolution of component B is ignored.) The time* T_f *is in Gyr, with* $T_f = 0$ *being the present. Numerical results are given in Table 1.*

According to the recent study of solar-type binaries conducted by Duquennoy and Mayor (1991), the components of binaries with orbital periods representative of the inner planet configuration tend to exhibit dissimilar masses (with a peak perhaps occurring near a mass ratio of about 0.25; see their Fig. 10). When analyzing the habitability prospects of individual systems, the mass and age of the more massive component thus becomes a major factor for consideration. This can become especially important when time scales for biological evolution are considered. For example, the main sequence lifetime of an F3V star is approximately 4.4 Gyr (Allen 1976, §100; Lang 1980, Eqn. 5–72). However, in our Solar System, 4.6 Gyr were required to proceed from the formation of the Earth to the evolution of "intelligent" life. If this time scale is typical, then inner planet binary systems whose primary stellar component is of spectral type F3V or earlier would perhaps not be expected to have the requisite time to evolve intelligent life, regardless of the spectral type of the secondary component.

CONCLUSIONS

The presence of one or more stellar companions can have significant effects on the habitability of planets that might accompany the star(s) in the system. It is not at all certain that planets can even form in such systems; this question requires future theoretical investigation. (The unequivocal detection of planets within binary systems would certainly answer this question!) Given that planets can exist within binaries, the orbital stability of a planet within the system's habitable zone(s) becomes a significant issue in determining habitability; a stable orbit requires that the stellar components either be close together (i.e., $< \approx 0.2$ AU) or relatively far apart (i.e., $> \approx 10$ AU). If the criteria for a stable orbit are met, then the variation in flux received by a planet due to the presence of the other stellar component is, in most cases, not a significant factor in determining habitability. Certain cases, for example, mutual eclipses in a close stellar orbit, may create significant flux variations over short time scales (i.e., hours); the effects of these require further modeling. Over biological time scales, the evolution of the more massive component(s) in a binary or multiple system plays an important role in determining the habitability of all planets within the system, regardless of which star(s) the planets may orbit. Since there perhaps is no such thing as an "average" binary or multiple system, when it becomes possible to scrutinize the habitability of planets around stars in the solar neighborhood, the binary and multiple systems will almost certainly have to be treated on a case-by-case basis.

References

Allen, C.W. (1976), *Astrophysical Quantities*, 3d ed. (London: The Athlone Press).

Backman, D.E., and F. Paresce (1993), "Main-Sequence Stars with Circumstellar Solid Material: The Vega Phenomenon," in *Protostars and Planets III*, E.H. Levy and J.I. Lunine, eds. (Tucson: University of Arizona Press), pp. 1253–1304.

Beckwith, S.V.W., A.I. Sargent, R.S. Chini, and R. Güsten (1990), "A Survey for Circumstellar Disks Around Young Stellar Objects," *AJ* **99:** 924–945.

Benest, D. (1988), "Planetary Orbits in the Elliptic Restricted Problem. I. The α Centauri System," *A&A* **206:** 143–146.

Benest, D. (1991), "Habitable Planetary Orbits Around α Centauri and Other Binaries," in *Bioastronomy: The Search for Extraterrestrial Life*, J. Heidmann and M.J. Klein, eds. (Berlin: Springer-Verlag), pp. 44–47.

Cameron, A.G.W. (1988), "Origin of the Solar System," *Ann. Rev. Astron. Astrophys.* **26:** 441–472.

Dobrovolskis, A., C. Levit, J. Scargle, J. Cuzzi, L. Dones, M. Showalter, and K. Young (1996), "Orbital Stability of Planets of Binary Stars," abstract, this volume.

Dole, S.H. (1970), *Habitable Planets for Man*, 2d ed. (New York: American Elsevier).

Donnison, J.R., and D.F. Mikulskis (1992), "Three-Body Orbital Stability Criteria for Circular Orbits," *MNRAS* **254:** 21–26.

Doyle, L.R. (1993), "Binary Star Ecoshells," unpublished proposal.

Duquennoy, A., and M. Mayor (1991), "Multiplicity Among Solar-type Stars in the Solar Neighborhood. II. Distribution of the Orbital Elements in an Unbiased Sample," *A&A* **248:** 485–524.

FEKEL, F.C., JR. (1981), "The Properties of Close Multiple Stars," *ApJ* **246:** 879–898.

GOLDREICH, P., AND W.R. WARD (1973), "The Formation of Planetesimals," *ApJ* **183:** 1051–1061.

HALE, A. (1994), "Orbital Coplanarity in Solar-type Binary Systems: Implications for Planetary System Formation and Detection," *AJ* **107:** 306–332.

HARRINGTON, R.S. (1972), "Stability Criteria for Triple Stars," *Celest. Mech.* **6:** 322–327.

HARRINGTON, R.S. (1977), "Planetary Orbits in Binary Stars," *AJ* **82:** 753–756.

HEPPENHEIMER, T.A. (1974), "Outline of a Theory of Planet Formation in Binary Systems," *Icarus* **22:** 436–447.

HEPPENHEIMER, T.A. (1978), "On the Formation of Planets in Binary Star Systems," *A&A* **65:** 421–426.

KASTING, J.F., D.P. WHITMIRE, AND R.T. REYNOLDS (1993), "Habitable Zones Around Main Sequence Stars," *Icarus* **101:** 108–128.

LANG, K.R. (1980), *Astrophysical Formulae*, 2d ed. (Berlin: Springer-Verlag).

MASON, D., H.A. MCALISTER, W.I. HARTKOPF, AND M.M. SHARA (1995), "Binary Star Orbits from Speckle Interferometry. VII. The Multiple System Xi Ursae Majoris," *AJ* **109:** 332–340.

MAZEH, T., D. GOLDBERG, A. DUQUENNOY, AND M. MAYOR (1992), "On the Mass-Ratio Distribution of Spectroscopic Binaries with Solar-Type Primaries," *ApJ* **401:** 265–268.

PENDLETON, Y.J., AND D.C. BLACK (1983), "Further Studies on Criteria for the Onset of Dynamical Instability in General Three-Body Systems," *AJ* **88:** 1415–1419.

RABL, G., AND R. DVORAK (1988), "Satellite-Type Planetary Orbits in Double Stars: A Numerical Approach," *A&A* **191:** 385–391.

SACKMANN, I.-J., A.I. BOOTHROYD, AND K.E. KRAEMER (1993), "Our Sun. III. Present and Future," *ApJ* **418:** 457–468.

SCHECHTER, M. (1984), "Planets in Binary Star Systems," *Sky & Tel.* **68:** 394–395.

SODERBLOM, D.R. (1986), "The Temperatures of Alpha Centauri A and B," *A&A* **158:** 273–274.

SZEBEHELY, V. (1980), "Stability of Planetary Orbits in Binary Systems," *Celest. Mech.* **22:** 7–12.

VIDAL-MADJAR, A., R. FERLET, A. LECAVELIER DES ETANGS, G. PERRIN, F. SÈVRE, F. COLAS, F. DELADERRIÈRE, AND A.-M. LAGRANGE-HENRI (1993), "68 Ophiuchi," *IAU Circ. 5795.*

WORLEY, C.E., AND W.D. HEINTZ (1983), "Fourth Catalog of Orbits of Visual Binary Stars," *Publ. US Naval Obs.*, 2d ser. **24**, Part VII.

ZUCKERMAN, B., AND E.E. BECKLIN (1993), "Infrared Observations of the Remarkable Main-Sequence Star HD 98800," *ApJ* **406:** L25–L28.

Westerlund 2 in Gum 29

Photo Credit: NASA, ESA, the Hubble Heritage Team (STScI/AURA), A. Nota (ESA/STScI), and the Westerlund 2 Science Team

https://science.nasa.gov/asset/hubble/westerlund-2/ [retrieved 12-17-25]

The dazzling 'celestial fireworks display' is a giant cluster of thousands of stars called Westerlund 2 in what has been described as a raucous stellar breeding ground (Gum 29), located 20,000 light-years away from Earth in the constellation Carina. The photo was taken with Hubble's Wide Field Camera 3, which gave astronomers a clear view of the nebula and the dense concentration of the stars in the central cluster. The cluster is between 6 light-years and 13 light-years across.

This giant stellar grouping is only about 2 million years old and contains some of the Milky Way's hottest, brightest, and most massive stars. Some of its heftiest stars unleash torrents of ultraviolet light and hurricane-force winds of charged particles that etch away the enveloping hydrogen gas cloud. New Hubble observations are showing that lower-mass stars near the cluster's core do not have the large, dense clouds of dust that could become planets in a few million years. Hubble detected those planet-forming clouds embedded in disks encircling lower-mass stars farther away from the center.

The red dots scattered throughout the landscape are a rich population of newly forming stars still wrapped in their gas-and-dust cocoons. The brilliant blue stars seen throughout the image are mostly foreground stars.

STELLAR MASS LOSS AND THE YOUNG SOLAR SYSTEM HABITABLE ZONE

Laurance R. Doyle

SETI Institute
NASA Ames Research Center
Moffett Field, CA 94035

Neil C. Heather

Synernet Corp.
NASA Ames Research Center

Rajendra Vikramsingh

Palo Alto, CA 94303

Daniel P. Whitmire

Physics Department
University of Southwest Louisiana
Lafayette, LA 70504

Abstract

In defining the outer boundary of the solar habitable zone as the temperature at which CO_2 condenses, the lower luminosity of the zero age main sequence Sun was found to apparently be incompatible with evidence of liquid water on early Mars. One solution to this problem has been the suggestion that the *initial Sun was 4–7% more massive and lost this mass through a solar wind within the first 1–2 gigayears (Gyrs). Various pieces of evidence in the Solar System for this having taken place seem ambiguous at present. However, by analogy, this mass loss process should be taking place in other young solar-type stars. In this paper, in investigating the possibility that young solar-type stars lose a few percentage of their initial mass, we look for evidence in the radio continuum of moderate stellar winds from selected young solar-mass stars. Preliminary observational results are reported here.*

INTRODUCTION

In order to understand habitable zones around other stars we must first understand the history of the solar habitable zone. Here we will be addressing the *circumstellar* habitable zone (as opposed to other possible tidally or radioactively heated habitable zones; overview in Doyle and McKay 1992). We take the definition of a habitable zone to be the location where liquid water can exist for long periods of time, a constraint that is largely regulated on planetary surfaces by atmospheric greenhouse effects. The conservative inner boundary of the circumstellar habitable zone is the location where the moist runaway greenhouse effect occurs (0.95 AU in the Solar System; Kasting 1988). The outer boundary of the circumstellar habitable zone is conservatively defined as the location where CO_2 condensation begins—1.37 AU for an Earth-sized planet and 1.49 AU for a Mars-sized planet with a nitrogen atmospheric pressure of one bar (Kasting et al. 1993). (If the less conservative maximum-allowable greenhouse warming is taken as an outer limit, for example, the outer boundary of the solar habitable zone would then become about 1.67 AU.)

Geological evidence indicates that surface liquid water may have existed in large quantities on Mars during the first 1–2 Gyr of the Solar System (Carr 1987, McKay and Stoker 1989). However, standard solar evolution models put the initial solar luminosity at about 70% what it is today when it first arrived at the zero age main sequence (or ZAMS; Iben 1967). As first pointed out by Sagan and Mullen (1972), such a faint Sun would not have allowed liquid water on early Mars without substantial atmospheric greenhouse warming (Pollack et al. 1987). However, at Mars' distance from the young Sun, CO_2 condenses out at this lower initial solar luminosity, resulting in global freezing (Kasting 1991, 1996). This difficulty has been dubbed the "faint Sun paradox." Suggested solutions to this "paradox" have included the following: (1) extensive runoff channels on Mars may not have been caused by liquid water, or possibly were caused by liquid water upwelling from underground as a result of impacts or geothermal heat (Carr 1987, Squyers and Kasting 1994, as examples); (2) radical climate shifts caused by rapid obliquity changes may have caused widespread melting, resulting in runoff channels (Williams et al. 1996); (3) a modification on the changing obliquity suggestion in that significant changes in the orbital eccentricity of Mars (ranging as high as 0.2 in a matter of Gyrs; Laskar 1994) may also have caused periodic global flooding; (4) an organic haze produced from ammonia might have been able to provide the UV shielding necessary to preserve methane—with its lower-temperature greenhouse properties—in the early Martian atmosphere, allowing a hotter, wetter early Mars (Sagan and Chyba, abstract, this volume); and (5) the young Sun might have been slightly (4–7%) more massive, and therefore brighter than the standard solar models predict, when it first arrived at the ZAMS, consequently losing this mass over a time period on the order of 0.5–2 Gyr through a more extensive early solar wind (Whitmire et al. 1995, Doyle et al. 1995, Whitmire and Reynolds 1996). It is the investigation of this last suggested solution that is the primary subject of this paper.

Possible Direct and Indirect Evidence of a More Massive Early Sun

The luminosities of low-mass stars like the Sun are very sensitive to mass ($L_\odot \propto M_\odot^{4.75}$). Consequently a 4–7% increase in the ZAMS solar mass translates to a luminosity enhancement of 15–38% (Whitmire et al. 1995). The upper limit of 7% is dictated by the Earth's having to avoid a moist runaway greenhouse condition. The lower limit of 4% is dictated by the avoidance of condensation of CO_2 in the early Martian atmosphere. If the Sun arrived on the main sequence with slightly more mass, losing this mass through an initial solar wind three to four orders of magnitude higher than the present rate (which is about $10^{-14} M_\odot \mathrm{yr}^{-1}$), then the results may appear in the planetary record as well as in the present solar composition (see reviews in Whitmire et al. 1995 and Doyle et al. 1995). For example, one of the implications of a brighter early Sun would be an initial water inventory for Venus that was originally closer to that of Earth given the constraint of the present D^+/H^+ ratio there (estimated from processes outlined in Donahue and Hartle 1992). Similar impact cross sections on Earth and Venus might be expected if cometary delivery was the major source of the initial terrestrial-planet water inventory. Another implication is that on the early Earth, conditions under which the first biological organisms appeared would not have required as massive a CO_2 atmosphere to have avoided global glaciation if the initial Sun was just a few percent more massive (Caldeira and Kasting 1992). Possible direct evidence for an early, more massive solar wind may also exist in the irradiation record of selected early meteoritic samples (Caffee et al. 1987), as well as older lunar soils (Caffee et al. 1991). In highland lunar soils, the excess heavy component in lunar ilmenite, for example, may be explicable by an early solar wind several orders of magnitude greater than that today, and additional evidence for a sustained early solar wind in the lunar geological record would be worth investigating further.

Significant mass loss (on the order of $10^{-10}M_{\odot}yr^{-1}$) has apparently already been detected in several nearby M dwarf stars at microwave frequencies (Mullan et al. 1992). Radio continuum flux has also been detected at 3.5 cm for several G-type stars of varying ages (Güdel et al. 1994; see also Dale and Linsky 1981). Additional observations of these stars at other wavelengths would provide a spectral index that could be compared with that expected from an excess stellar wind (see next section). Small stellar winds can also be detected in the UV if there is a strong ultraviolet background so that ion absorption lines may be detected (Mullan et al. 1989, Ramsey 1990), but these instances may be rare (stellar winds in systems with eclipsing white dwarf components, for example). However, when they do occur, mass loss rates on the order of several times $10^{-11}M_{\odot}yr^{-1}$ can be detected.

Perhaps the most controversial *possible* evidence for an early solar wind is the lithium depletion detected for the Sun and numerous solar-type stars, generally attributed to either convective destruction in the stellar interior (Boesgaard 1990; see also Swenson and Faulkner 1992) or early stellar mass loss (Soderblom et al. 1990, Boothroyd et al. 1991, Zahn 1994, Pasquini et al. 1994, Cayrel de Strobel et al. 1994, Whitmire and Doyle 1995). There are problems with both approaches. The former theories need to better explain the general nondepletion of beryllium in the Sun and other stars necessitating a rather narrow temperature range for convective destruction of lithium in solar-type stars. The latter theories may lack a well-defined mechanism for angular momentum transport. At present, however, suggested mechanisms include stellar streamers coupled with magnetic activity (Brown et al. 1990, Badalyan and Livshits 1992, for example), which can cause mass loss rates of greater than $10^{-11}M_{\odot}yr^{-1}$ or direct meridian circulation coupled to an early massive stellar wind, allowing even higher mass loss rates (Zahn 1994). In this latter work, angular momentum in young tidally locked close binary stars is predicted to have been dissipated via shortening of the semimajor axis; more angular momentum is lost, but not from rotation (which could increase mixing). In single stars, angular momentum transport via meridian circulation is coupled to an early massive stellar wind, thereby slowing rotation (and thereby decreasing mixing). Thus very close (rotationally synchronized at an early age) solar-type binary stars should show less lithium depletion than single solar-type stars if angular momentum loss (i.e., a stellar wind) is the major mechanism for lithium depletion rather than subductive mixing. There is at present some supporting observational evidence for this underdepletion of several close solar-type binary systems compared with single solar-type field stars (Zahn 1994). (It may also be noted that if a large part of the interstellar lithium is thus due to stellar winds, some revision of estimates of the primordial nucleosynthesis of lithium would also be in order, perhaps significantly constraining a number of cosmological models.) The measurement of (or upper limits on) stellar winds from young solar-type stars could therefore assist the solution of a number of problems in planetary science as well as in some other fields.

As a consequence of solar mass loss, the planets would be expected to have migrated outward by the same proportion as the mass lost (Whitmire et al. 1995). The solar habitable zone would also have been initially farther from the Sun. In Figure 1 we show these results for an initial solar mass loss rate of $3.5 \times 10^{-11} M_{\odot} \mathrm{yr}^{-1}$ over 2 Gyr. The dashed lines represent the time evolution of planetary distances with and without this solar mass loss rate. The habitable zone boundaries (solid lines), which would increase slightly under normal solar evolution (after Whitmire et al. 1995), would initially migrate slightly inward as the Sun loses mass.

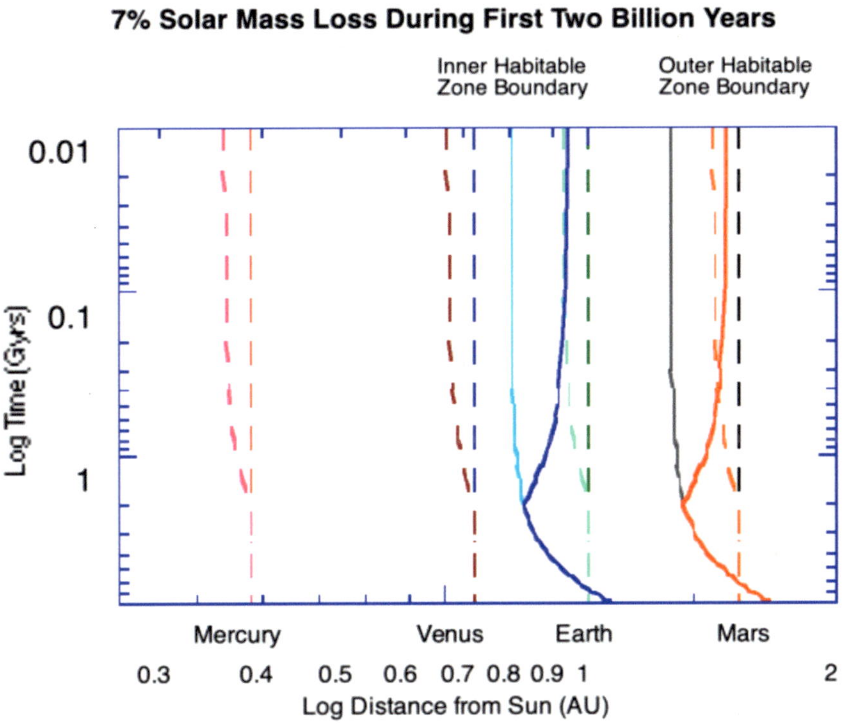

FIG. 1. *We show the evolution of the solar habitable zone for both an approximately constant solar mass and for a mass loss rate of* $7 \times 10^{-11} M_{\odot}$ *per year during the first 2 Gyr after the origin of the Solar System. The terrestrial planet positions are shown as dashes, while the habitable zone inner and outer boundaries are shown as solid lines. We use the most conservative definition of these boundaries (0.95 AU inner boundary for an Earthlike planet with a 1-bar carbon dioxide atmosphere, and 1.49 AU for the outer boundary for a Marslike planet with carbon dioxide in a 1-bar nitrogen atmosphere; after Kasting et al. 1993). We note the outward migration of the planets as well as the initial inward migration of the solar habitable zone boundaries as mass loss takes place. Mars is within the initial habitable zone only for a slightly more massive initial Sun while the Earth can be seen to just skirt the inner boundary of the habitable zone. Mars can be seen to eventually enter the habitable zone again in 1.4 Gyr from present while Earth leaves the habitable zone about 1 Gyr later.*

It should be noted that under the most conservative assumption of the habitable zone boundaries, Earth's orbit just aligns with the inner boundary of the moist runaway greenhouse effect, while Mars remains within the habitable zone for about 0.6 Gyr. Mars would remain well within the habitable zone for the full 2 Gyr under the less conservative boundary assumptions of Kasting et al. (1993). (We also note that standard stellar evolution will bring Mars back into the outer edge of the habitable zone in about 1.5 Gyr from present while the Earth would cross the inner edge of the solar habitable zone in about 2.4 Gyr from present.)

If the Sun lost some of its initial angular momentum through such a moderate stellar wind, then this process might be expected to be occurring in other very young solar-type stars. In the following section we outline the youth selection criteria we used to determine which nearby solar-type stars should be observed. We also present the results of our preliminary radio continuum observations of these stars in an effort to determine their spectral indexes to see if they correspond with those expected from moderate ($4 \times 10^{-11} M_\odot yr^{-1}$ to $7 \times 10^{-11} M_\odot yr^{-1}$) stellar winds.

Observing Method, Observed Stars, and Selection Criteria

One way to detect a stellar wind is to measure the excess radio continuum flux due to free-free electron transitions in the stellar wind (Wright and Barlow 1975). The distribution of this excess radio continuum will be such that $\log S_v \propto (2/3)\log v$, where S_v is the continuum flux at frequency v. This 2/3 spectral index of a stellar wind is fairly robust for various conditions of the wind including changes in velocity, temperature, and simplifying assumptions such as spherical symmetry (Leitherer and Robert 1991; see also Doyle et al. 1995). Radio spectral indexes more negative than about −0.5 are generally indicative of galaxies and some other extended sources. However, quasar spectra can also be flat or even slightly positive (Fomalont et al. 1984). Stellar flares also give, in general, a positive spectral index from 8.4 GHz to 4.9 GHz, but a negative spectral index from these frequencies toward 1.5 GHz (Stern et al. 1990, for example). Thus one may expect that a fairly strong positive spectral index indicative of a stellar wind would not generally be confused with most expected background sources.

The stellar wind will also have an upper frequency limit on its opacity depending on the velocity and temperature of the wind. For example, at the velocity and temperature of the solar wind at the Earth's orbit (214 solar radii), the solar wind could be detected at frequencies as high as 15 GHz. At its velocity and temperature at two solar radii, on the other hand, it might be detectable up to about 134 GHz. (Doyle et al. 1995 discuss the detectability of a five-segment wind model, although the solar wind is, of course, significantly more complicated than this.) For the limits of detectability calculated here we will take the stellar wind characteristics as uniform throughout (i.e., we will use a one-segment stellar wind model with a velocity and temperature equivalent to that of the present solar wind at the Earth's orbit; Doyle et al. 1995).

In Table 1 we list four nearby solar-type stars (HD number, rotation period, rotational velocity, separation in arcseconds if binaries, distance in parsecs, and total integration time in minutes at the given frequency) observed at the National Radio Astronomy Observatory's (NRAO's) Very Large Array (VLA) in Socorro, New Mexico.

All are G0 dwarfs thought to be quite young due primarily to their rapid rotation rates. HD39587 is a member star of the Ursa Majoris Cluster, whose color diagram puts its age at less than about 200 million years (Allen 1976). However, recent convective overshoot models (Mayor and Mermilliod 1991, for example) could double this stellar age determination, but this would not affect our basic conclusion of youth here (less than about 0.5 Gyr). Chromospheric activity is also an indication of stellar youth and was used here to determine the stellar rotation periods for these stars (Soderblom et al. 1991). The rotation period of HD206860 is not directly known but the projected rotational velocity, *V*sin*i* already indicates a very rapid rotation rate. All these stars are also within approximately 15 parsecs of the Sun, the present practical limit on detectability in the radio continuum at the mass loss levels we are trying to measure (Doyle et al. 1995).

TABLE 1. *Nearby Young G0 V Stars Observed at the VLA*

HD	P*(days)*	V*(k/s)*	*Sep(″)*	D*(pc)*	*4.8GHz*	*8.4GHz*	*15.0GHz*	*23.0GHz*
39587	5.13	10.7	0.6	9.6	108	23	–	–
72905	4.7	11.6	Single	13.7	604	156	88	–
115383	3.50	15.6	34.3	12.7	32	15	61	–
206860	Unkn.	≥10.2	Single	15.4	51	22	72	37

Total integration time at given frequency is in minutes; dashes indicate no observations at that frequency.

That the stellar rotational velocity can be related to the stellar age was first developed by Skumanich (1972) where it was thought that $t \propto V^{-2}$ (V is the stellar rotational velocity and t is the stellar age). It has recently been shown (Simon 1992) that such a linear relationship does not hold for stars less than about the Hyades age ($t < 5 \times 10^8$ years). In Figure 2 we show that for stars of known age a $t \propto V^{-1.8}$ is the best fit (Soderblom 1991). We have plotted the rotational velocities of the four observed stars in Figure 2 using this linear relationship. While the t-V curve will not remain linear toward the young age end of the diagram, there can nevertheless be little doubt that these four stars are very likely younger than the age of the Hyades (0.5–1 Gyr). The only star of these four whose rotation rate could be affected by a close companion would be HD39587 but its age, as mentioned, has also been independently determined by its membership in a young cluster. (As we shall see, this star does give significant ambiguity in interpreting its excess radio continuum due to its smaller companions.)

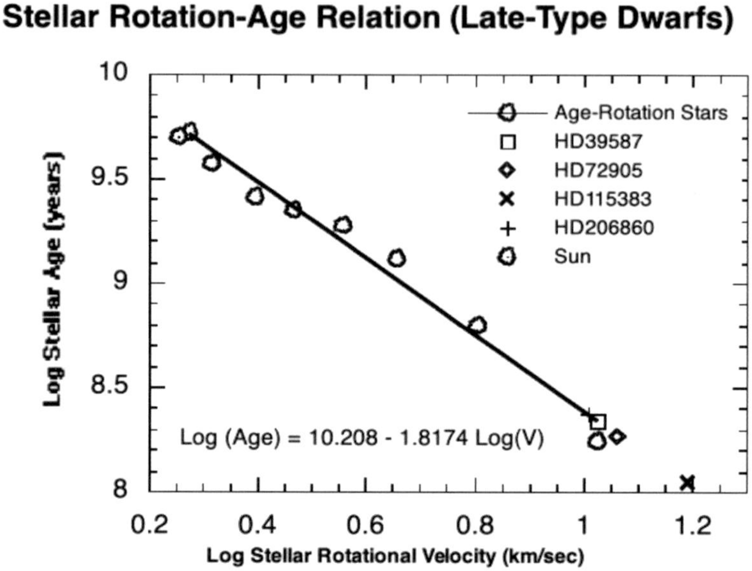

FIG. 2. *Stellar age–rotational velocity relationship for late-type dwarfs (after Soderblom 1991) with equation and line of best linear fit. The four stars in this study all give extrapolated rotation ages of less than about 250 Myr. The best fit equation overestimates the Sun's age by about 10%. The independent determination of the age of HD39587 from its Ursa Majoris cluster membership (158 Myr; Allen 1976) agrees with its parameterized age here, about 219 Myr). Rapid rotation is indicative of stellar youth even though the age-rotation relationship shown here is not truly linear for very young stars (Simon 1992).*

RADIO CONTINUUM OBSERVATIONS: PRELIMINARY RESULTS

Shown in Figures 3 through 6 are the radio continuum maps for HD39587 and HD72905 at 3.5 and 6.0 cm (8.4 and 4.8 GHz), respectively. Although the maps are very noisy in this preliminary analysis, significant flux at the stellar location is indicated at both frequencies—4.9 GHz (6.0 cm wavelength) and 8.4 GHz (3.5 cm wavelength) for HD39587 and at 8.4 GHz only for HD72905 at the level indicated in Table 2 (p. 157). (We plan to reduce the noise in these maps by further modeling and antenna phase processing.) These maps do, however, indicate that the radio continuum emission measurements needed to determine if nearby solar-type stars have stellar winds at the $4 \times 10^{-11} M_{\odot} yr^{-1}$ to $7 \times 10^{-11} M_{\odot} yr^{-1}$ level may be reached in reasonable observing times using the VLA.

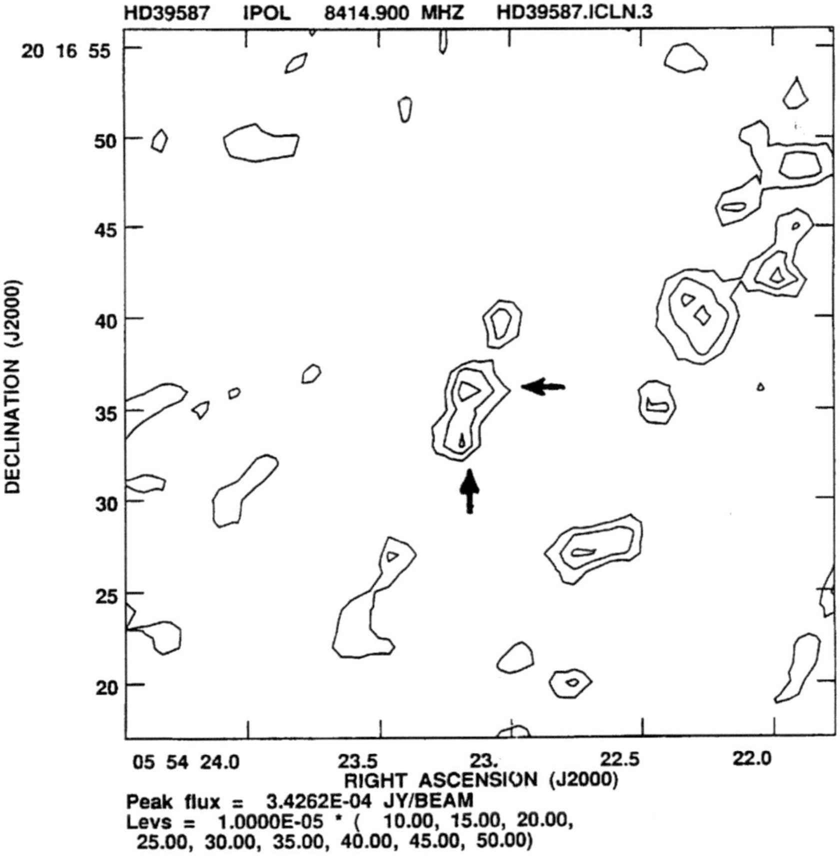

FIG. 3. *Radio map of the sky at a frequency of 8.4 GHz (wavelength of 3.6 cm) in the region of HD39587 (RA = $5^h54^m23^s$, Dec = $+20^\circ 16' 36''$) with a beam resolution of about 2.3 seconds of arc (0.15^s in right ascension) taken with the NRAO's Very Large Array. The flux contours at the arrows are at the position of HD39587 and represent a 2.6 sigma flux above the background.*

In Table 2 we show the radio continuum flux or flux limits at present reached by our VLA observations (only the 4.9 and 8.4 GHz data have so far been analyzed) along with the expected flux levels (one-segment models) for mass loss rates of $4 \times 10^{-11} M_\odot \mathrm{yr}^{-1}$ (4% per Gyr) and $7 \times 10^{-11} M_\odot \mathrm{yr}^{-1}$ (7% per Gyr). (HD206860 was also observed for 37 minutes at 23.0 GHz—data equivalent to reaching a flux limit of 0.22 mJy—as a preliminary test for excess radio continuum emission at higher frequencies, as shown in Table 1 [p. 151].) We see, for example, that significantly more continuum emission is being detected at both 8.4 GHz and 4.9 GHz for HD39587 than expected from the theoretical mass loss levels of 4–7% per Gyr, and that the levels detected have a negative spectral index (the 4.9 GHz emission is higher than the 8.4 GHz emission).

Considerable emission from HD72905 has also been detected at 8.4 GHz with an observational upper limit being placed on the 4.9 GHz emission, which is approaching the flux limit of interest to this investigation. A positive spectral index is possible but 4.9 GHz emission would have to be detected by about the 0.01 mJy level to be indicative of mass loss. The excess 8.4 GHz emission could also be caused by flare activity, and a few additional measurements are therefore needed.

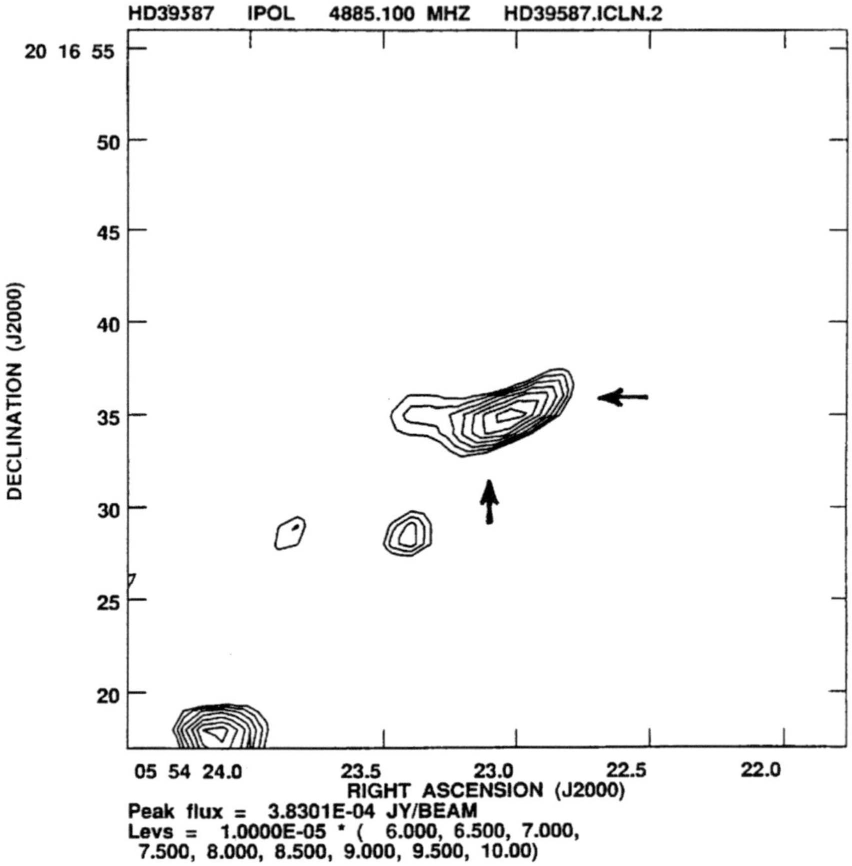

FIG. 4. *Radio map of the sky at a frequency of 4.9 GHz (wavelength of 6.1 cm) in the region of HD39587 with a beam resolution of about 3.9 seconds of arc (0.26^s in right ascension). The large flux contours indicated by the arrows are at the position of HD39587 and represent a 3.0 sigma flux above the background.*

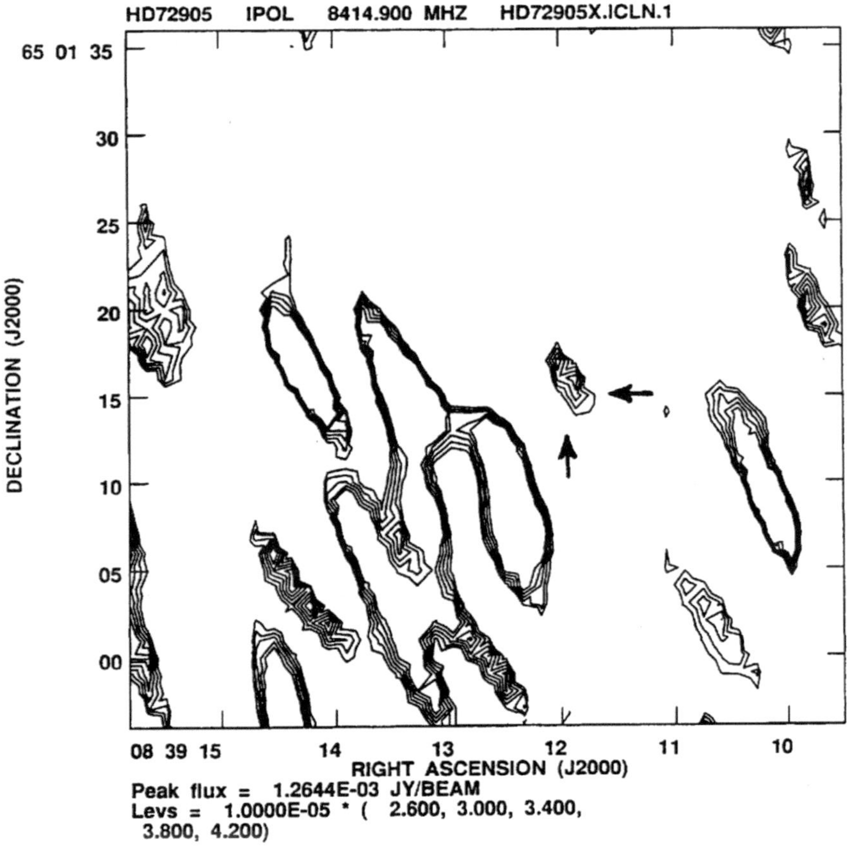

FIG. 5. *Radio map of the sky at 8.4 GHz in the region of HD72905 (RA = $8^h39^m12^s$, Dec = $+65^{\circ}01^m15^s$) with the beam resolution as given in Figure 3. The small, concentrated flux contours indicated by the arrows are at the position of HD72905 and represent a 2.2 sigma flux above the background.*

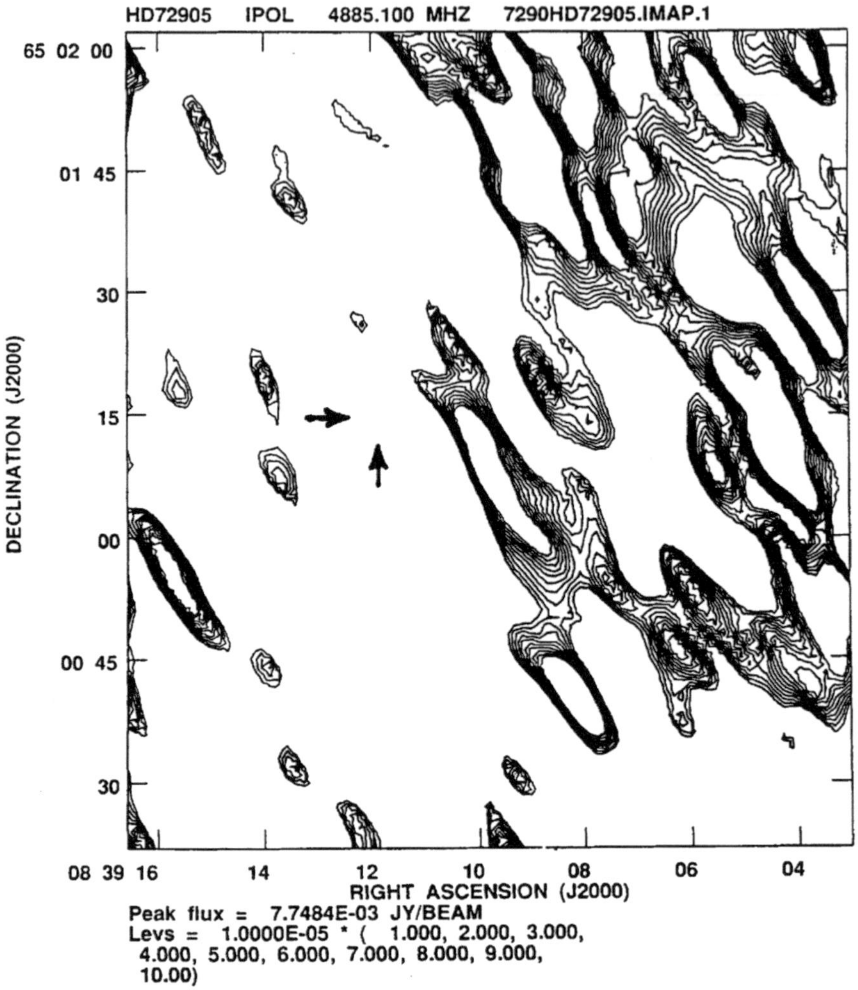

FIG. 6. *Radio map of the sky at 4.9 GHz in the region of HD72905 with the beam resolution as given in Figure 4. At the position of HD72905 indicated by arrows there was no appreciable emission detected down to a level of about 0.017 mJy (a limit equivalent to the flux that would be produced by a stellar wind mass loss rate of about 8.9 × $10^{-11}M_\odot$ per year).*

TABLE 2. ***VLA Stellar Radio Flux Detections or Detection Limits (in milliJanskys [mJy])***

HD	*4.9GHz*	*8.4GHz*	*15.0GHz*
39587	**.226**	**.093**	--
4%	.012	.018	.024
7%	.026	.037	.051
72905	**≤.017**	**0.041**	*≤.081*
4%	.006	.009	.012
7%	.013	.018	.025
115383	**≤.053**	**≤.055**	*≤.097*
4%	.007	.010	.014
7%	.015	.021	.029
206860	**≤.042**	**≤.045**	*≤.089*
4%	.005	.007	.009
7%	.010	.014	.020

Boldface indicates detections or limits, italics indicate observational time limits, dashes indicate no observations at that frequency.

Upper limits on the excess radio emission from the stars HD115383 and HD206860 reached so far are still significantly above the detection level of interest here and will also require additional observing time.

In Figures 7–10 we plot the Planck curve for HD39587, HD72905, HD115383, and HD206860, respectively, and their UBV (ultraviolet, blue, and visible) fluxes (from Hoffleit and Jaschek 1982) along with our detected radio continuum. As mentioned, significant emission has been detected for HD39587 but the spectral index is similar to one that might be produced by stellar flare activity rather than a stellar wind (which may therefore be masked). A previous continuum flux of about 0.6 mJy has been reported for this star at 6.0 cm (Dale and Linsky 1981), significantly above that detected here (0.23 mJy), indicating a variable source of 6.0 cm emission in this system. However, the negative slope in the continuum emission from HD39587 is difficult to explain as interference from galactic sources at the detection level we are working at. A survey at 1.5 GHz (20 cm wavelength) and 4.9 GHz (Fomalont et al. 1988) recorded only about one "background" detection at the 0.01 mJy level per square arcminute (most of these background sources also had large, negative spectral indexes, as expected from galaxies). It is therefore unlikely that background noise is interfering with the detection of stellar wind mass loss in HD39587. However, this system has three components (a G0 star, an M star, and a likely smaller component), and flare activity from any or all of these young stars could be interfering with our detection of stellar mass loss in this system. Flares in young dwarfs (RS CVn stars, for example; Stern et al. 1990, White et al. 1990) can have a significant negative spectral index over the time (hours) that we integrated at the 4.9 and 8.4 GHz frequencies, and large relative amplitudes, as well. Therefore detection of a stellar wind from the G0 component of HD39587 may not be possible without VLBI (very-large-baseline interferometry) techniques to resolve the individual components, also covering a larger frequency range.

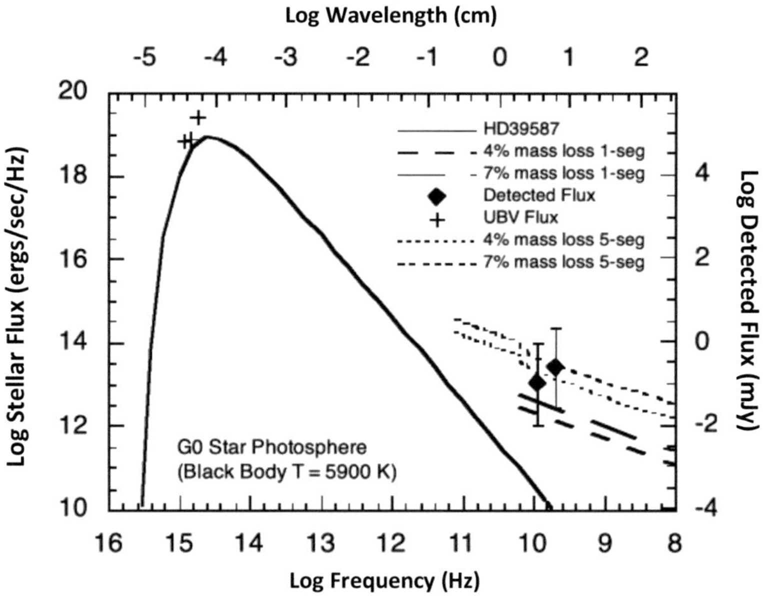

FIG. 7. *Radio flux detected for HD39587 where the Planck black body curve indicative of the stellar photosphere is shown along with measured U, B, and V filter stellar fluxes (the latter from Hoffleit and Jaschek 1982). The radio fluxes expected from four mass loss models (two one-segment and two five-segment solar wind models each with $4 \times 10^{-11} M_\odot$ per year and $7 \times 10^{-11} M_\odot$ per year mass loss rates, labeled 4% and 7%, respectively) are shown along with the detected radio fluxes at 3.6 and 6.1 centimeter wavelengths. A spectral index of about 2/3 is indicative of mass loss, as the models (dashed lines) indicate. The radio flux from this system is likely a convolution of flare and other activity from three stars—the G0 star, one M dwarf companion, and one possibly smaller-mass star—which may be masking evidence of stellar winds.*

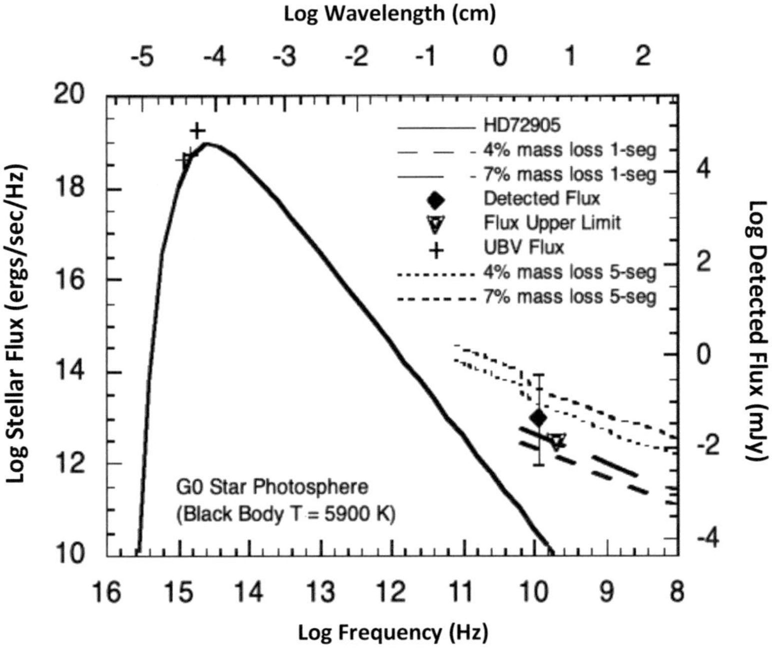

Fig. 8. *Radio flux detected for star HD72905. This figure is similar to Figure 7 with a detected flux at 3.6 centimeters. An upper limit on the flux at 6.1 centimeters of 0.017 mJy indicates that mass loss is not taking place at the 8.9 × $10^{-11} M_\odot$ per year level (for the conservative one-segment model) in this system. Once again flare activity may be taking place and a wider spectral sampling (with more sensitivity) is needed to determine if mass loss at the relevant 4 × $10^{-11} M_\odot$ to 7 × $10^{-11} M_\odot$ per year rate is taking place. We also note that if the five-segment model is correct (that is, there is a stellar wind with a significant temperature and velocity gradient), then observations at wavelengths of a few millimeters could also be useful in determining the spectral index.*

HD72905 is a single star, as far as is known, and we have preliminarily detected excess continuum flux at 3.5 cm. However, we have not detected any 6.0 cm emission down to the 0.017 mJy level (corresponding to a mass loss limit of about 8.9 × $10^{-11} M_\odot \mathrm{yr}^{-1}$). But the error bars must be further reduced and more data must be acquired before we can conclude that mass loss is not taking place in this system at the level of interest here. Continued observations of this system could be particularly fruitful. Our observations of HD115383 and HD206860 have just started, and significantly more time on these two stars will be required in order to be able to determine if they have significant stellar winds, as well. The upper limits on mass loss rates for HD115383 and HD206860 of about 1.8 × $10^{-10} M_\odot \mathrm{yr}^{-1}$ and 2.0 × $10^{-10} M_\odot \mathrm{yr}^{-1}$, respectively, are shown in Figures 9 and 10.

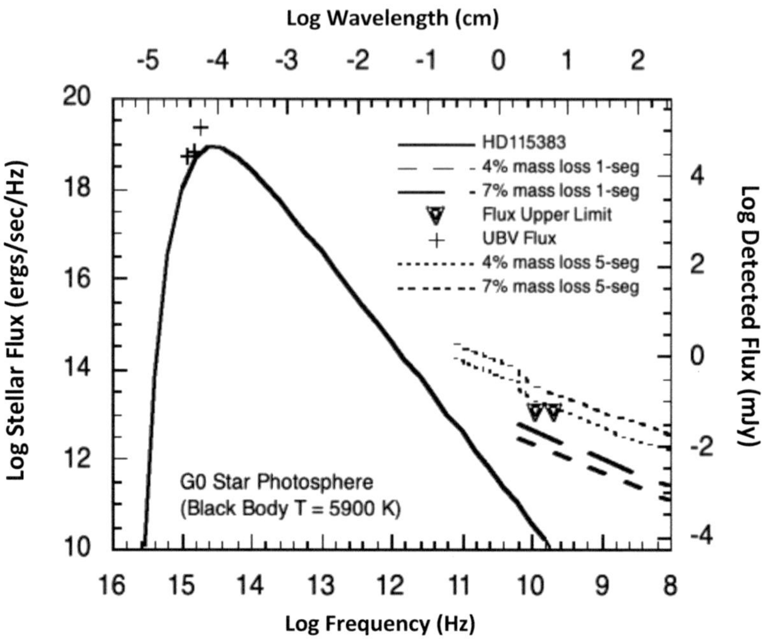

FIG. 9. *Preliminary observations of HD115383 show no radio flux at 3.6 or 6.1 centimeters at the 1.8 × $10^{-10}M_\odot$ per year mass loss level. More observations on this star are needed to determine if it is losing mass at the rate proposed for the young Solar System.*

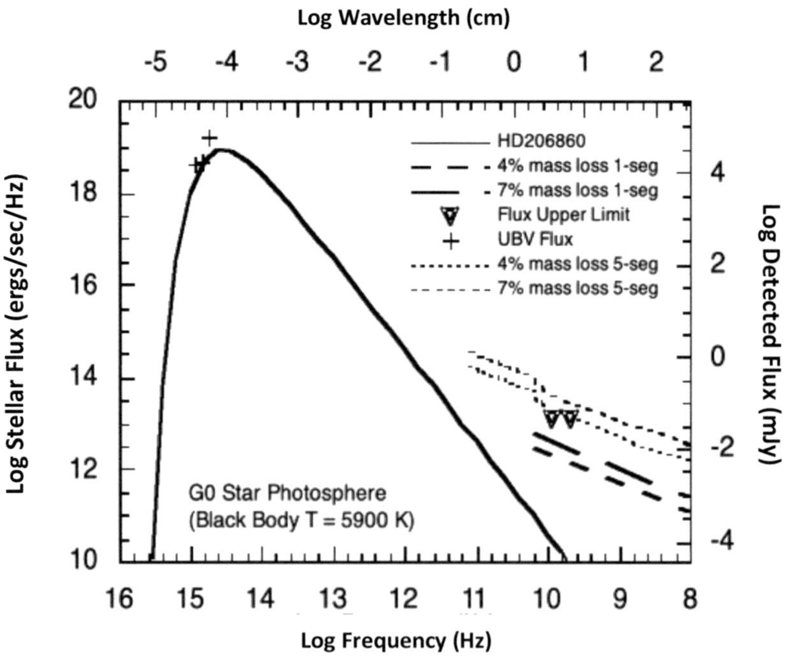

FIG. 10. *Observations of HD206860 show no radio flux at 3.6 or 6.1 centimeters at the 2.0 ×$10^{-10}M_\odot$ per year mass loss level. Additional observations are needed for conclusions relevant to the young Solar System mass loss rate to be reached.*

Conclusions

We have begun observations on four nearby young solar-type stars to ascertain if they could be losing mass via a moderate stellar wind at the $4 \times 10^{-11} M_\odot yr^{-1}$ to $7 \times 10^{-11} M_\odot yr^{-1}$ level. Radio emission from the G0 component of HD39587 will likely remain convoluted with emission from its smaller components so that small levels of mass loss will be difficult to measure. We have detected 8.4 GHz emission from the single star HD72905 but detected no emission at 4.9 GHz down to the 0.017 mJy level, corresponding to an upper limit on the mass loss rate of about 8.9% per Gyr, just above the range of interest. Clearly the spectral index, if there is 4.9 GHz emission, will be positive, but the error bars are still too large to conclude on the slope. The stars HD115383 and HD206860 require further substantial observations before any indexes can be implied. What we have shown, however, is that in general, the hypothesis that young solar-type stars lose mass at the moderate rate of a few percent per Gyr can be tested with radio continuum measurements for several young solar-type stars in reasonable observing times. The results of our further refinement of the existing data as well as acquisition of further radio continuum data should be able to answer a significant number of questions concerning the luminosity conditions extant in the early Solar System during the time of the first emergence of biological processes on the early Earth. In addition, our results could have repercussions for models of early runaway greenhouse effects on Venus, liquid water on early Mars, irradiation records on satellite surfaces as well as the meteoritic record, the irradiation history of comets (related to organics production as well as orbital evolution), angular momentum history of young stars, and cosmological models of primordial nucleosynthesis. If such mass loss did take place over the first 1–2 Gyr of the Solar System, and this process is general for solar-type stars, then the circumstellar habitable zones for solar-type stars would generally have been more stable for the emergence of early biological forms without requiring massive early neutral or oxidizing atmospheres, a result with implications for in situ production of the biogenic elements as well.

Note added in final proof—An update of more recent stellar atmospheric opacity tables and their consequences for lithium depletion can be found in Swenson, Faulkner, Rogers, and Iglesias ("The Hyades Lithium Problem Revisited," 1995, *ApJ* **425:** 286–302). Recent detection of radio emission from solar-type stars can be found in Lim and White ("Radio Detection of Late-Type [G-K] Dwarf Stars in the Pleiades," 1995, *ApJ* **453:** 207–213). And an especially interesting paper by Guzik and Cox ("Early Solar Mass Loss, Element Diffusion, and Solar Oscillation Frequencies," 1995, *ApJ* **448:** 905–914) indicates that the intermediate-degree modes of solar oscillations frequencies that probe the elemental composition at the bottom of the solar convection zone may not be compatible with mass-loss durations longer than about 0.2 Gyr. (Chaboyer,

Demarque, Guenther, and Pinsonneault, "Rotation, Diffusion, and Overshoot in the Sun: Effects on the Oscillation Frequencies and the Neutrino Flux," 1995, *ApJ* **446:** 435–444, include the effects of rotation on the diffusion rate, as well.) Such an impulsive mass-loss rate would be more easily detected in flux, but less in time, and its possible effects—such as those addressed here—on solar system bodies with regard to the habitable zone, would have to be studied in more detail.

Acknowledgments

This work was supported by the Exobiology Program Office at NASA Headquarters, Washington D.C., for which we specifically thank Dr. Michael Meyer. We also wish to thank the excellent staff of the NRAO in Charlottesville, Virginia and the Very Large Array in Socorro, New Mexico (and Dr. Barry Clark in particular) for the observing time and the continuous support in helping us to install and run the AIPS (Astronomy Image Processing System) at NASA Ames Research Center. We also thank Dr. David Goorvitch for the loan of the Sun Sparc station for this initial data reduction, and Dr. Max Bernstein for helpful discussions regarding the Venus atmosphere. Final and special thanks go to Dr. John Billingham, Dr. Bernard Oliver, Dr. Donald DeVincenzi, Dr. Frank Drake, and the staff of SETI Institute for encouragement, critiques, and support.

References

Allen, C.W. (1976), *Astrophysical Quantities*, 3d ed. (London: The Athlone Press).

Badalyan, O.G., and M.A. Livshits (1992), "Mass Loss by Active Late-Type Stars," *Sov. Astron.* **36:** 70–74.

Boesgaard, A.M. (1990), "Lithium and Beryllium in Stellar Atmospheres," in *Sixth Cambridge Workshop on Cool Stars, Stellar Systems, and the Sun*, ASP Conf. Ser., M.S. Giampapa and J.A. Bookbinder, eds., **9:** 317–327.

Boothroyd, A.I., I. Sackmann, and F.A. Fowler (1991), "Our Sun II. Early Mass Loss of $0.1M_\odot$ and the Case of the Missing Lithium," *ApJ* **377:** 318–329.

Brown, A., A. Vealé, P. Judge, J. Bookbinder, and I. Hubeny (1990), "Mass Loss Upper Limits for A and F Dwarfs," in *Sixth Cambridge Workshop on Cool Stars, Stellar Systems, and the Sun*, ASP Conf. Ser., M.S. Giampapa and J.A. Bookbinder, eds., **9:** 183–185.

Caffee, M.W., C.M. Hohenberg, T.D. Swindle, and J.N. Goswami (1987), "Evidence in Meteorites for an Active Early Sun," *ApJ* **313:** L31–L35.

Caffee, M.W., C.M. Hohenberg, R.H. Nichols, Jr., C.T. Olinger, R. Wieler, A. Pedroni, P. Signer, T.D. Swindle, and J.N. Goswami (1991), "Do Meteorites Contain Irradiation Records from Exposure to an Enhanced-Activity Sun?," in *The Sun in Time*, C.P. Sonett, M.S. Giampapa, and M.S. Matthews, eds. (Tucson: University of Arizona Press).

CALDEIRA, K., AND J.F. KASTING (1992), "Susceptibility of the Early Earth to Irreversible Glaciation Caused by Carbon Dioxide Clouds," *Nature* **359:** 226–228.

CARR, M. (1987), "Water on Mars," *Nature* **326:** 30–35.

CAYREL DE STROBEL, G., R. CAYREL, E. FRIEL, J.P. ZAHN, AND C. BENTOLILA (1994), "A Case of the Quadruple System ξUMa: Its Activity and Lithium Depletion," *A&A* **291:** 505–516.

DALE, G.E., AND J.L. LINSKY (1981), "First Detection of Nonflare Microwave Emission from the Coronae of Single Late-Type Stars," *ApJ* **250:** 284–292.

DONAHUE, T.M., AND R.E. HARTLE (1992), "Solar Cycle Variations in H^+ and D^+ Densities in the Venus Ionosphere: Implications for Escape," *Geophys. Res. Lett.* **19:** 2449–2452.

DOYLE, L.R., AND C.P. MCKAY (1992), "Exobiological Habitats: An Overview," in *Bioastronomy*, J. Heidmann and M.J. Klein, eds. (New York: Springer-Verlag), pp. 163–172.

DOYLE, L.R., R. VIKRAMSINGH, D.P. WHITMIRE, AND N.C. HEATHER (1995), "Circumstellar Habitable Zones and Mass Loss from Young Solar-type Stars. II. Observational Considerations," in *Progress in the Search for Extraterrestrial Life*, ASP Conf. Ser., S. Shostak, ed., **74:** 195–204.

FOMALONT, E.B., K.I. KELLERMANN, J.V. WALL, AND D. WEISTROP (1984), "A Deep 6-Centimeter Radio Source Survey," *Science* **225:** 23–28.

FOMALONT, E.B., K.I. KELLERMANN, M.C. ANDERSON, D. WEISTROP, J.V. WALL, R.A. WINDHOUST, AND J.A. KRISTIAN (1988), "New Limits to Fluctuations in the Cosmic Background Radiation at 4.86 GHz Between 12 and 60 Arcsecond Resolution," *AJ* **96:** 1187–1191.

GÜDEL, M., J.H.M.M. SCHMITT, AND A.O. BENZ (1994), "Discovery of Microwave Emission from Four Nearby Solar-Type Stars," *Science* **265:** 933–935.

HOFFLEIT, D., AND C. JASCHEK (1982), *The Bright Star Catalogue* (New Haven, CT: Yale University Observatory).

IBEN, I. (1967), "Stellar Evolution. VI. Evolution from the Main Sequence to the Red-Giant Branch for Stars of Mass $1_\odot$," *ApJ* **147:** 624–663.

KASTING, J.F. (1988), "Runaway and Moist Greenhouse Atmospheres and the Evolution of the Earth and Venus," *Icarus* **74:** 472–494.

KASTING, J.F. (1991), "CO_2 Condensation and the Climate of Early Mars," *Icarus* **94:** 1–13.

KASTING, J.F. (1996), "Habitable Zones Around Stars: An Update," this volume.

KASTING, J.F., D.P. WHITMIRE, AND R.T. REYNOLDS (1993), "Habitable Zones Around Main Sequence Stars," *Icarus* **101:** 108–128.

LASKAR, J. (1994), "Large Scale Chaos in the Solar System," *A&A* **287:** L9–L12.

LEITHERER, C., AND C. ROBERT (1991), "Observations of Stellar Winds from Hot Stars at 1.3 Millimeters," *ApJ* **377:** 629–638.

MAYOR, M., AND J.-C. MERMILLIOD (1991), "Rotational Velocities of Stars in Open Clusters: The Time Dependency Revisited," in *Angular Momentum Evolution of Young Stars*, S. Catalano and J.R. Stauffer, eds. (Dordrecht: Kluwer Academic Publishers), pp. 143–150.

McKay, C.P., and C.R. Stoker (1989), "The Early Environment and Its Evolution on Mars: Implications for Life," *Rev. Geophys. Space Phys.* **27:** 189–214.

Mullan, D.J., E.M. Sion, F.C. Bruhweiler, and K.G. Carpenter (1989), "Evidence for a Cool Wind from the K2 Dwarf in the Detached Binary V471 Tauri," *ApJ* **339:** L33–L36.

Mullan, D.J., J.G. Doyle, R.O. Redman, and M. Mathioudakis (1992), "Limits on Detectability of Mass Loss from Cool Dwarfs," *ApJ* **397:** 225–231.

Pasquini, L., Q. Lui, and R. Pallavicimi (1994), "Lithium Abundances of Nearby Solar-like Stars," *A&A* **287:** 191–205.

Pollack, J.B., J.F. Kasting, S.M. Richardson, and K. Poliakoff (1987), "The Case for a Wet, Warm Climate on Early Mars," *Icarus* **71:** 203–224.

Ramsey, L.W. (1990), "Optical and UV Spectra of RS CVn Stars," in *Sixth Cambridge Workshop on Cool Stars, Stellar Systems, and the Sun,* ASP Conf. Ser., M.S. Giampapa and J.A. Bookbinder, eds., **9:** 195–205.

Sagan, C., and C. Chyba (1996), "Organic Shielding of UV-Labile Greenhouse Gases on Early Earth and Mars," abstract, this volume.

Sagan, C., and G. Mullen (1972), "Earth and Mars: Evolution of Atmospheres and Surface Temperatures," *Science* **177:** 52–56.

Simon, T. (1992), "Bonas + 11: The Activity-Rotation-Age Relations for Solar-Type Stars," in *Seventh Cambridge Workshop on Cool Stars, Stellar Systems, and the Sun*, ASP Conf. Ser., M.S. Giampapa and J.A. Bookbinder, eds., **26:** 3–14.

Skumanich, A. (1972), "Time Scales for CaII Emission Decay, Rotational Braking and Lithium Depletion," *ApJ* **171:** 565–567.

Soderblom, D.R. (1991), "Main Sequence Angular Momentum Loss in Low-Mass Stars," in *Angular Momentum Evolution of Young Stars*, S. Catalano and J.R. Stauffer, eds. (Dordrecht: Kluwer Academic Publishers), pp. 151–165.

Soderblom, D.R., M.S. Oey, D.R.H. Johnson, and R.P.S. Stone (1990), "The Evolution of the Lithium Abundances of Solar-Type Stars. I. The Hyades and Coma Berenices Clusters," *AJ* **99:** 595–607.

Soderblom, D.R., D.K. Duncan, and D.R.H. Johnson (1991), "The Chromospheric Emission-Age Relation for Stars of the Lower Main Sequence and Its Implications for the Star Formation Rate," *ApJ* **375:** 722–739.

Squyres, S.W., and J.F. Kasting (1994), "Early Mars: How Warm and How Wet?," *Science* **265:** 744–749.

Stern, R.A., B.M. Haisch, Y. Uchida, F. Walter, A. Veale, A. Brown, O. Vilhu, and D. Hannikainen (1990), "Co-Ordinated *Ginga*, IUE and VLA Observations of Flaring Activity in σ^2 CR B," in *Sixth Cambridge Workshop on Cool Stars, Stellar Systems, and the Sun*, ASP Conf. Ser., M.S. Giampapa and J.A. Bookbinder, eds., **9:** 227–229.

Swenson, F.J., and J. Faulkner (1992), "Lithium Dilution Through Main-Sequence Mass Loss," *ApJ* **395:** 654–674.

White, S.M., M.R. Kundu, Y. Uchida, and N. Nitta (1990), "The Radio Spectrum of RS CVN Stars," in *Sixth Cambridge Workshop on Cool Stars, Stellar Systems, and the Sun*, ASP Conf. Ser., M.S. Giampapa and J.A. Bookbinder, eds., **9:** 239–242.

WHITMIRE, D.P., AND L.R. DOYLE (1995), "A Slightly More Massive Sun as an Explanation of Warm Temperatures on Early Mars," *JGR Planets* **100:** 5457–5464.

WHITMIRE, D.P., AND R.T. REYNOLDS (1996), "Circumstellar Habitable Zones: Astronomical Considerations," this volume.

WHITMIRE, D.P., L.R. DOYLE, R.T. REYNOLDS, AND J.J. MATESE (1995), "Circumstellar Habitable Zones and Mass Loss from Young Solar-type Stars. I. Theory," in *Progress in the Search for Extraterrestrial Life*, ASP Conf. Ser., S. Shostak, ed., **74:** 183–193.

WILLIAMS, D.M., J.F. KASTING, AND K. CALDEIRA (1996), "Chaotic Obliquity Variations and Planetary Habitability," this volume.

WRIGHT, A.E., AND M.F. BARLOW (1975), "The Radio and Infrared Spectrum of Early-type Stars Undergoing Mass Loss," *MNRAS* **170:** 41–51.

ZAHN, J.-P. (1994), "Rotation and Lithium Depletion in Late-Type Binaries," *A&A* **288:** 829–841.

Water Abundance in GK Tau Disk (MIRI emission spectrum)

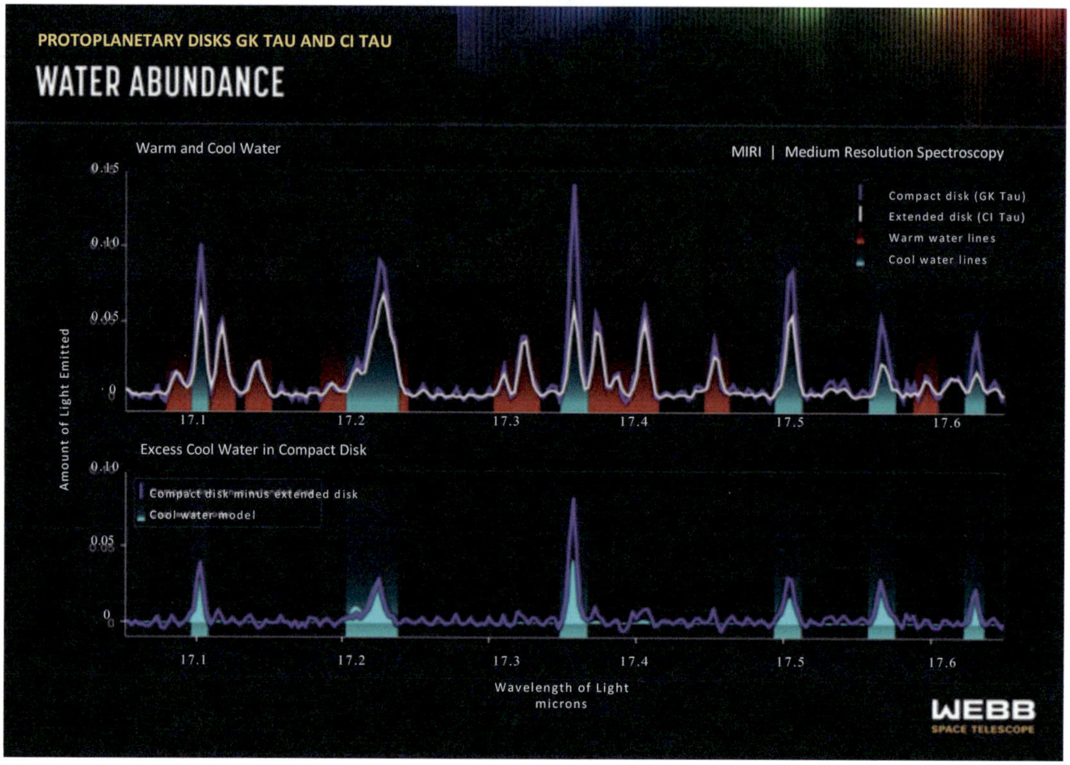

Credit: NASA, ESA, CSA, L. Hustak (STScI), A. Banzatti (Texas State University)
https://esawebb.org/images/driftingpebbles-ppdisk-1/ [retrieved 6-20-25]

Scientists using the NASA/ESA/CSA James Webb Space Telescope have made a *breakthrough discovery in revealing how planets are made.* By observing water vapour in protoplanetary disks, Webb confirmed a physical process involving the drifting of ice-coated solids from the outer regions of the disk into the rocky-planet zone.

Theories have long proposed that icy pebbles forming in the cold, outer regions of protoplanetary disks–the same area where comets originate in our solar system–should be the fundamental seeds of planet formation. The main requirement of these theories is that pebbles should drift inward toward the star due to friction in the gaseous disk, delivering both solids and water to planets. A fundamental prediction of this theory is that as icy pebbles enter into the warmer region within the "snowline"–where ice transitions to vapour–they should release large amounts of cold water vapour. This is exactly what Webb observed.

Webb observations were designed to determine whether compact disks have a higher water abundance in their inner, rocky planet region, as expected if pebble drift is more efficient and is delivering lots of solid mass and water to inner planets. The results confirmed expectations by revealing excess cool water in the compact disks, compared with the large disks.

This graphic compares the spectral data for warm and cool water in the GK Tau disk, which is a compact disk without rings, and extended CI Tau disk, which has at least three rings on different orbits. The science team employed the unprecedented resolving power of MIRI's MRS to separate the spectra into individual lines that probe water at different temperatures. These spectra, seen in the top graph, clearly reveal excess cool water in the compact GK Tau disk, compared with the large CI Tau disk.

The bottom graph shows the excess cool water data in the compact GK Tau disk minus the cool water data in the extended CI Tau disk. The actual data, in purple, are overlaid on a model spectrum of cool water. Note how closely they align. These results appeared in Andrea Banzatti et al., *Astrophysical Journal Letters,* November 8, 2023.

Recent Advances in Stellar Astronomy Relevant to Planet Formation

John R. Stauffer

Harvard-Smithsonian Center for Astrophysics
Cambridge, MA 02138

Abstract

The availability of large telescopes and greatly improved detectors have allowed stellar astronomers to investigate the environment in which stars and planets form in more detail than in the past. In this talk I will review recently published estimates of the fraction of stars formed in clusters, the fraction of stars born as binaries, the fraction of stars born with disks and the possible influence that pre–main sequence disks have on the main sequence evolution of low-mass stars.

Are Most Stars Born in Dense Clusters?

Until a few years ago, the standard answer to this question was that only about 10% of the stars are born in clusters and that the other 90% are born in relative isolation (i.e., in associations like Taurus-Auriga). This answer was obtained not by observing star-forming regions but by counting stars in nearby open clusters, correcting for the finite lifetimes of those clusters and then comparing the cluster star count with the total number of field stars in that volume of space (Roberts 1957).

With the development of infrared arrays, we now have the ability to peer into the interiors of molecular clouds and directly observe the birth environment of stars. This type of observation is still not easy, however, because the nearby star-forming regions often extend over several square degrees, whereas the infrared arrays are small and generally image an area on the order of only one square arc-minute at a time. Therefore, only a few molecular cloud complexes have been surveyed to date. The most provocative results come from the survey of the L1630 giant molecular cloud complex by Lada et al. (1991). Those authors identified four rich star clusters embedded within L1630, which they estimated to contain at least 58% and perhaps as high as 96% of the stars that have formed in this cloud. In order not to be producing more open clusters than are observed, Lada et al. postulated that most of these clusters become unbound when the molecular cloud dissipates and thus do not survive to be catalogued as discrete open clusters.

The L1641 cloud is the only other giant molecular cloud in the Orion star-forming region to be surveyed in this fashion (Strom et al. 1993). Strom et al. (1993) estimated that only about one-third of the stars in this cloud are currently in clusters or groups with the other approximately two-thirds spread out over the cloud as a distributed population. One possible conclusion from this is that the fraction of stars born in groups and clusters varies from one cloud complex to another. Evidence in favor of an alternative explanation is provided by the fact that the stars in the groups, in general, appear to be much younger than the stars in the distributed population. This suggests that even in L1641, most stars could have been formed in groups or clusters but that the groups that formed first have already been disrupted. Surveys of other molecular cloud complexes should allow us to decide between these two alternative explanations.

Even in Taurus-Auriga—the quintessential example of relatively isolated star formation—Gómez et al. (1993) have recently shown that formation in groups is more the rule than the exception. Gómez et al. first compiled a new and more complete membership list for Taurus combining well-known members from previous surveys with new members from proper motion surveys and analysis of IRAS data. Statistical analysis of this membership list showed that most of the stars have companions within a distance on the order of the size of a typical dense molecular cloud core and thus that "isolated star formation does not occur in low-density star-forming regions like Taurus." Figure 1 illustrates the nonrandom spatial distribution of star formation in Taurus.

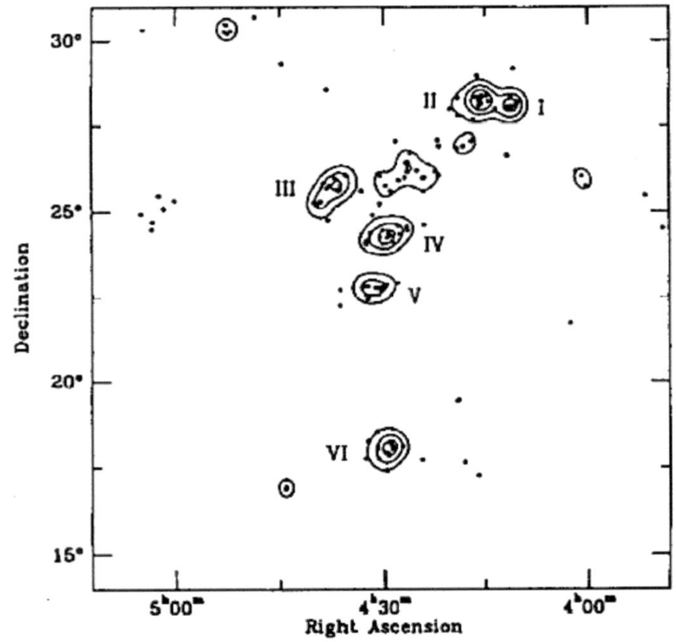

FIG. 1. *The isodensity contour map for the pre–main sequence star distribution in Taurus. The isolines are in units of stars per deg^2 and correspond to 5, 10, 15 and 20 stars/deg^2. Adapted from Gómez et al. (1993).*

ARE MOST STARS BORN IN BINARY SYSTEMS?

The "classical" study of the fraction of stars born as binaries was published by Abt and Levy (1976). Those authors combined their own radial velocity measurements with data in the literature to estimate that 58% of G dwarf primaries had at least one hydrogen burning companion, with the fraction possibly increasing to as high as 72% after correction for incompleteness. Abt and Levy claimed that the mass distribution of the secondary companions was different for short and long period binaries—binaries with periods less than 100 years tended to have secondaries with masses similar to that of the primary star, whereas longer period binaries have secondaries whose masses follow the distribution found for isolated field stars. This was thought to reflect the difference between "bifurcation" binaries and binaries whose components formed in relative isolation.

Duquennoy and Mayor (1991; D&M) have recently conducted a new survey of duplicity among solar mass stars, improving upon the Abt and Levy study by having a larger sample size and much more accurate radial velocities. D&M found a binary fraction of 57% after correction for incompleteness—comparable to, but slightly less than, the Abt and Levy result. D&M did not confirm the secondary mass–period relation found by Abt and Levy—their data were consistent with secondary masses following the field star distribution irrespective of orbital period.

Binary frequency does not appear to be a strong function of the mass of the primary star. Fischer and Marcy (1992) conducted a survey of the multiplicity of M dwarfs (mass ~ 0.4 $M_\odot$) and determined a companion frequency of 42%. They claimed that most of the difference relative to the G dwarf surveys was for the lowest-mass secondaries—M(sec)/M(prim) < 0.2; for M dwarf primaries these objects have $M < 0.08\ M_\odot$ and would thus be brown dwarfs. Because brown dwarfs are very faint unless they are quite young, the failure to find them as companions could simply be an observational bias; alternatively, nature may abhor brown dwarfs and the failure to detect companions in this mass range could reflect the general rarity of these objects.

There have been three recent surveys of multiplicity among low-mass pre–main sequence stars. Prosser et al. (1994) obtained HST (Hubble Space Telescope) planetary camera images of the Trapezium cluster—the densest known protocluster (age ~ 1 Myr) in Orion—and derived a binary frequency consistent with the D&M field star survey for the separation range to which they were sensitive—40 to 400 AU. If most stars are formed in dense clusters, this is the result that should have been expected. If most stars are born in looser aggregates, then it is perhaps surprising that the dense stellar environment in the Trapezium does not either affect the fraction of stars that are born as binaries or modify that fraction over time via gravitational encounters with other cluster members. The other two surveys have been directed toward determining the binary fraction in T Tauri associations (Ghez 1993, Reipurth and Zinnecker 1994). Ghez found a binary frequency three times higher than the D&M field star population in the separation range to which her observations were sensitive (15 to 225 AU); Reipurth and Zinnecker also found an enhanced binary frequency among the T Tauri stars. These results may indicate that the fraction of stars born as binaries, or at least the fraction that survive to ages ~ 1 Myr, is indeed a function of the local stellar density. An alternative explanation is that a significant fraction of the T Tauri binaries are disrupted by some process during pre–main sequence evolution.

In summary, most stars are probably formed in binary (or multiple) systems. Most of these binaries are probably formed within larger groups or clusters. Many of these clusters and probably some of these binaries are disrupted prior to the stars' reaching the main sequence. Therefore, one should not take the present epoch multiplicity of a field star as indicative of its birth environment—even stars that appear to be single now were likely born with one or many companions.

What Fraction of Stars Are Born with Disks?

Early models of star formation considered the spherical collapse of molecular cloud cores and ignored angular momentum and magnetic fields. Subsequent models included rotation and "predicted" the formation of protostellar disks. In some of these models, in fact, most of the mass that eventually resides in the star first passes through the disk. For this and other reasons, it has been of interest in recent years to attempt to determine the fraction of pre–main sequence stars that have observable disks and the time scale for those disks to become unobservable.

The "traditional" means of identifying disks around young stars is via the infrared emission from heated dust grains. Based on a survey of Taurus conducted at 10 μm, Skrutskie et al. (1990) found that approximately one-half of the stars with ages less than 3 Myr had detectable disks. Beckwith et al. (1990) carried out a similar survey at one mm, and also estimated that approximately one-half of the stars in Taurus have disks. There is a very good correlation between 10 μm emission strength and Hα emission: the stars with the strongest Hα emission (the classical T Tauri stars) have relatively large 10 μm excesses whereas the stars with relatively weak Hα emission (the "weak-lined" or naked T Tauri stars) have little or no 10 μm excess. This correlation is as expected given the current paradigm that the strong Hα emission from classical T Tauri stars arises from gas accreting from the disk onto the star.

One might expect that the formation and/or evolution of pre–main sequence (protoplanetary) disks might be different in dense clusters relative to that found in loose associations like Taurus. That is, clump-clump collisions, star-disk collisions or tidal interactions from close passages would seem likely to decrease the lifetimes of disks in dense clusters. Using Hα emission as a disk signature, Herbig and Terndrup (1986) suggested such an effect might be present in the Trapezium cluster because none of the ten Trapezium stars for which they obtained spectra showed the characteristic emission line profiles found in classical T Tauri stars. Recent images of the Trapezium cluster obtained with the Hubble Space Telescope by O'Dell et al. (1993) and Prosser et al. (1994) have serendipitously shown that protoplanetary disks apparently are present around many of the Trapezium cluster members. O'Dell et al. obtained their images primarily to study the structure of the HII region ionized by the high-mass stars in the Trapezium cluster; Prosser et al. obtained their images to derive photometry for the low-mass stars in the cluster. Both sets of images showed something unexpected though—stars encircled by rings or arcs of emission with sizes on the order of 200 AU, brightest in the direction facing θ^1 Ori C and θ^2 Ori A—the most luminous and hottest stars in the cluster. Figure 2 shows an example of one of these arcs (Stauffer et al. 1994). It is thought that these arcs arise when gas evaporated from circumstellar disks by the UV light from the O stars collides with the stellar wind from the O stars. Free-free emission from these compact nebulae is also detectable in the radio, and surveys with the VLA (Very Large Array in Socorro, New Mexico) have identified eighteen sources within 30 arcseconds of θ^1 Ori C as probable "lit up" disks. Stauffer et al. (1994) have estimated the fraction of stars with disks in the Trapezium by comparing the HST and VLA sources with a deep IR imaging survey of the stars in the cluster. Their estimate was that probably at least half, and possibly significantly more, of the Trapezium low-mass stars have circumstellar disks—that is, basically the same frequency as found in Taurus. Therefore, at least up to an age of 1 Myr, it appears that the fraction of stars with disks is not a strong function of the stellar environment.

Note that none of these studies has completely addressed the issue of what fraction of stars are born with disks—we are seeing stars at an age on the order of 1 Myr and not "at birth." Instead, we are determining what fraction of stars still have gas and dust disks observable with present day technology at a given age. That fraction is on the order of 50% at 1 Myr, decreasing to < 10% at 10 Myr (Strom et al. 1989). The stars that lose their disks earlier presumably either had less massive disks to start with or have more efficient ways of removing the disk. In either event, because there is a range in disk survival times, it will be useful in the future to consider whether or not those stars that lose their disks early are less likely to form planetary systems than those that retain their disks longer.

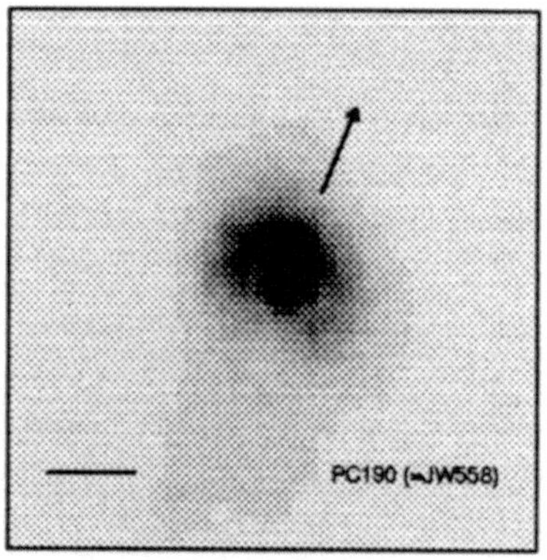

FIG. 2. *I band (~ 9000 Å) image from HST of JW 558 and its associated arc of emission. The arrow points toward θ^1 Ori C, and the bar in the lower left corresponds to a size at Orion's distance of 200 AU. Adapted from Stauffer et al. (1994); similar images can be found in O'Dell et al. (1993).*

JUST HOW DENSE IS THE TRAPEZIUM CLUSTER?

It is interesting to take a brief detour to quantify the difference in the star-forming environment experienced in, for example, Taurus relative to that experienced in the Trapezium cluster. The typical groups identified by Gómez et al. (1993) have on the order of ten to fifteen members and are located within a volume of a few cubic parsecs—thus indicating a spatial density on the order of five stars per cubic parsec. There are about 500 known probable members of the Trapezium cluster; within the central 25 arcseconds (0.053 parsec) from the center though,

McCaughrean and Stauffer (1994) estimate that there are 29 members, corresponding to a central spatial density of ~ 4.5×10^4 stars per cubic parsec. Inverting this density, one can estimate that the mean distance between stars in the core is ~ 0.027 pc or ~ 6000 AU. With this density, the one-dimensional velocity dispersion of stars, v, in the cluster (2.5 km-s^{-1}; Jones and Walker 1988), and an approximate cluster age of 1 Myr, one can estimate the collisional cross-section in the core of the cluster via $\sigma \sim 1/nv\tau$, and hence estimate the maximum disk radius that is allowed. With the above numbers, one derives r_{disk}(max) ~ 350 AU. Because the stars spend only a small fraction of their orbits in the core, this estimate is too pessimistic. However, even using a lower density corresponding to an average point in the cluster only reduces this radius to on the order of 1500 AU. Thus both disk-disk collisions and the photoevaporative effects of the O stars in the cluster may limit disk lifetimes in dense clusters like the Trapezium. Figure 3 shows an image of the central ~ 0.1 pc of the Trapezium, imaged at 2.2 μm (McCaughrean and Stauffer 1994).

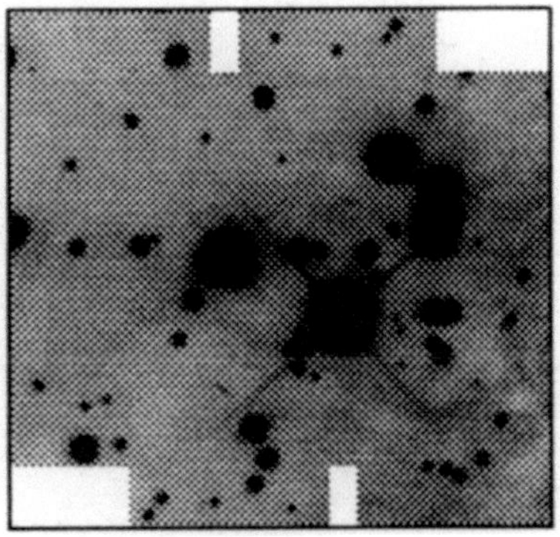

FIG. 3. *K band (2.2 μm) image of the central ~ 0.1 pc of the Trapezium cluster. The faintest stars visible probably have masses on the order of 0.2 $M_\odot$. The brightest star has a spectral type of O7.*

DO DISKS AFFECT THEIR "PARENT" STARS?

Prior to 1985, the working paradigm was that pre–main sequence stars with disks should have relatively large rotational velocities because disk matter was expected to accrete on the star, and the disk material brings with it relatively high specific angular momentum. However, observations of the rotational velocities of classical T Tauri stars and weak-lined T Tauri stars showed that this expectation was not fulfilled (cf. Hartmann and Stauffer 1989).

The new paradigm is that the rotational velocities of pre–main sequence stars are regulated by their disks. Magnetic fields threading the star and disk lock the rotational velocity of the star to a period on the order of five to ten days (Konigl 1991, Cameron and Campbell 1993). Left to its own devices, a pre–main sequence star should rotate faster as it contracts to the main sequence, due to conservation of angular momentum. The disk prevents this according to the model. Therefore, stars that lose their disks early should rotate more rapidly than their disked counterparts of the same age. Edwards et al. (1993) have presented observational evidence in support of this prediction—stars with infrared excesses in Taurus and Orion all have similar rotation periods on the order of eight days, whereas the stars without excesses rotate faster and have a range of periods (perhaps corresponding to a range of disk lifetimes). These data are shown in Figure 4.

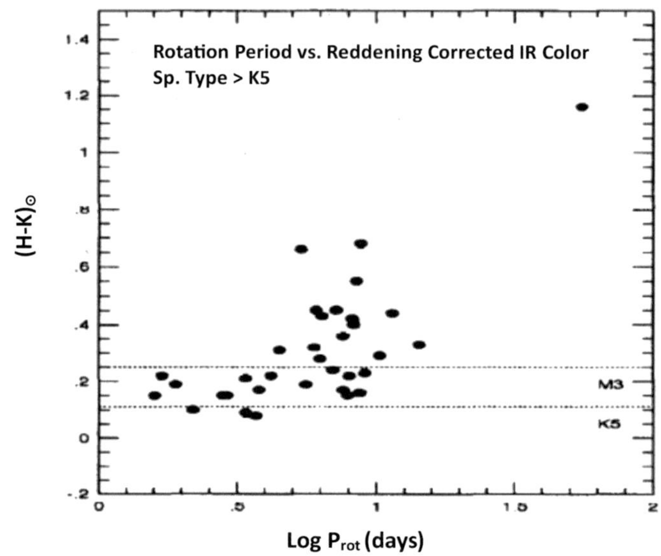

FIG. 4. *The correlation between rotation period and IR excess (the disk signature) for pre–main sequence stars in Taurus and Orion. Adapted from Edwards et al. (1993). The two dashed lines indicate the intrinsic H-K colors for stars of spectral type K5 and M3.*

Given the results shown in Figure 4, it should be expected that low-mass stars would arrive on the main sequence with a range of rotational velocities. This is indeed the case, as is illustrated in Figure 5, which shows the distribution of rotational velocities for stars in the Alpha Persei (age ≅ 50 Myr), Pleiades (age ≅ 70 Myr) and Hyades (age ≅ 600 Myr) clusters (Stauffer and Hartmann 1987, Stauffer et al. 1987, Stauffer et al. 1989, Prosser 1992). The Alpha Persei figure shows that G and K dwarfs arrive on the main sequence with rotational velocities ranging from less than 10 km-s^{-1} to on the order of 200 km-s^{-1}; between one-half and two-thirds of the G and K dwarfs are slow rotators ($v\sin i$ < 10 km-s^{-1}). By Pleiades age, angular momentum loss from winds has already significantly reduced the rotation rates of the rapid rotators so that the maximum rotational velocity is on the order of 100 km-s^{-1}. By the age of the Hyades, all of the G and K dwarfs rotate quite slowly (only two to three times faster than the present day Sun) and the dispersion in rotational velocity at a given effective temperature has essentially disappeared.

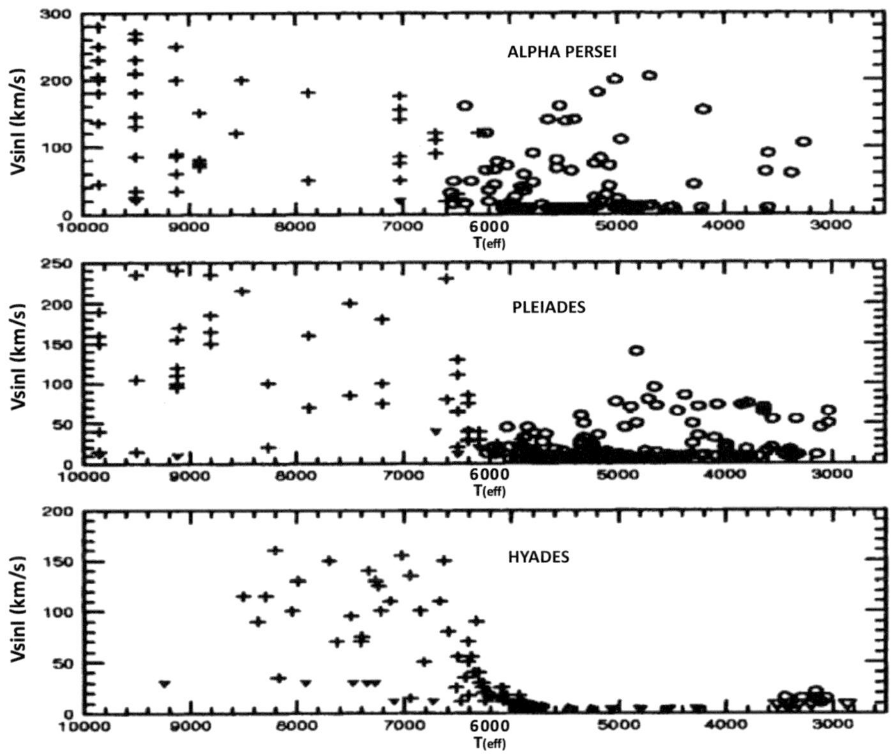

FIG. 5. *Spectroscopic rotational velocities versus effective temperature for stars in the three best-studied nearby open clusters.*

There are a few things about Figure 5 of particular interest to planetary astronomers. First, while it is often noted that > 90% of the angular momentum in the Solar System is in the planets, it should be clear from the Alpha Persei figure that this was not necessarily true when the Sun reached the main sequence. A significant fraction of G dwarfs arrive on the main sequence with rotational velocities on the order of 100 km-s^{-1}—that is, with angular momenta on the order of that of the planets in the Solar System. Second, whereas the Skumanich law ($v_{rot}\ \alpha\ t^{-1/2}$; Skumanich 1972) provides a good model of the expected rotational evolution for solar mass stars older than the Hyades, it is of quite limited value for young main sequence stars. That is, solar mass stars at Alpha Persei age range in rotation rate from a few times the Sun's rate to ~ 100 times the Sun's rate—clearly the dispersion in rotation is too large to make the Skumanich law prediction useful for individual stars. Lastly, from the Hyades figure it is also clear that once a solar mass star is > 500 Myr or 1 Gyr old, its rotational velocity is essentially uniquely determined by its mass—that is, one cannot determine whether the star arrived on the main sequence with a rotational velocity of 200 km-s^{-1} or 10 km-s^{-1}. That is, in particular, true for the Sun. However, if one adopts the disk-regulated pre–main sequence rotation paradigm, and assumes that stars with longer-lasting pre–main sequence disks are more likely to form planets, one would predict that the Sun more likely arrived on the main sequence as a slow rotator. With the currently available data, that conclusion is more a leap of faith than a sound deduction.

Final Thoughts

Stellar astronomers are beginning to have the tools to provide some answers to questions of interest to planetary astronomers and to pose additional questions. Planetary astronomers can pose provocative questions for stellar astronomers, such as raising the possibility of significant mass loss from the early Sun. (See Doyle et al., this volume.) We should talk more often.

References

ABT, H., AND S. LEVY (1976), "Multiplicity Among Solar-Type Stars," *ApJSupp* **30:** 273–306.

BECKWITH, S.V.W., A.I. SARGENT, R.S. CHINI, AND R. GÜSTEN (1990), "A Survey for Circumstellar Disks Around Young Stellar Objects," *AJ* **99:** 924–945.

CAMERON, A.C., AND C. CAMPBELL (1993), "Rotational Evolution of Magnetic T Tauri Stars with Accretion Disks," *A&A* **274:** 309–318.

DUQUENNOY, A., AND M. MAYOR (1991), "Multiplicity Among Solar-type Stars in the Solar Neighborhood. II. Distribution of the Orbital Elements in an Unbiased Sample," *A&A* **248:** 485–524.

EDWARDS, S., S. STROM, W. HERBST, J. ATTRIDGE, K. MERRILL, R. PROBST, AND I. GATLEY (1993), "Angular Momentum Regulation in Low-Mass Young Stars Surrounded by Accretion Disks," *AJ* **106:** 372–382.

FISCHER, D., AND G. MARCY (1992), "Multiplicity Among M Dwarfs," *ApJ* **396:** 178–194.

GHEZ, A. (1993), *The Multiplicity of T Tauri Stars in the Star Forming Regions Taurus-Auriga and Ophiucus-Scorpius: A 2.2 μm Speckle Imaging Survey*, PhD Thesis (Pasadena: California Institute of Technology).

GÓMEZ, M., L. HARTMANN, S. KENYON, AND R. HEWETT (1993), "On the Spatial Distribution of Pre–Main Sequence Stars in Taurus," *AJ* **105:** 1927–1937.

HARTMANN, L., AND J. STAUFFER (1989), "Additional Measurements of Pre-Main-Sequence Stellar Rotation," *AJ* **97:** 873–880.

HERBIG, G., AND D. TERNDRUP (1986), "The Trapezium Cluster of the Orion Nebula," *ApJ* **307:** 609–618.

JONES, B.F., AND M. WALKER (1988), "Proper Motions and Variabilities of Stars near the Orion Nebula," *AJ* **95:** 1755–1782.

KONIGL, A. (1991), "Disk Accretion onto Magnetic T Tauri Stars," *ApJ* **370:** L39–L43.

LADA, E., D. DEPOTY, N. EVANS, AND I. GATLEY (1991), "A 2.2 Micron Survey in the L1630 Molecular Cloud," *ApJ* **371:** 171–182.

MCCAUGHREAN, M., AND J. STAUFFER (1994), "High Resolution Near-Infrared Imaging of the Trapezium: A Stellar Census," *AJ* **108:** 1382–1397.

O'DELL, C., Z. WEN, AND X. HU (1993), "Discovery of New Objects in the Orion Nebula on HST Images—Shocks, Compact Sources, and Protoplanetary Disks," *ApJ* **410:** 696–700.

PROSSER, C. (1992), "Membership of Low-Mass Stars in the Open Cluster Alpha Persei," *AJ* **103:** 488–513.

PROSSER, C., J. STAUFFER, L. HARTMANN, D. SODERBLOM, B. JONES, M. WERNER, AND M. MCCAUGHREAN (1994), "HST Photometry of the Trapezium Cluster," *ApJ* **421:** 517–541.

REIPURTH, B., AND H. ZINNECKER (1994), "Visual Binaries Among Pre–Main Sequence Stars," *A&A* **278:** 81–108.

ROBERTS, M. (1957), "The Numbers of Early-Type Stars in the Galaxy and Their Relation to Galactic Clusters and Associations," *PASP* **99:** 141.

SKRUTSKIE, M., D. DUTKEVITCH, S. STROM, S. EDWARDS, AND K. STROM (1990), "A Sensitive 10-micron Search for Emission Arising from Circumstellar Dust Associated with Solar-Type Pre–Main Sequence Stars," *AJ* **99:** 1187–1195.

SKUMANICH, A. (1972), "Time Scales for CaII Emission Decay, Rotational Braking and Lithium Depletion," *ApJ* **171:** 565.

STAUFFER, J., AND L. HARTMANN (1987), "The Distribution of Rotational Velocities for Low-Mass Stars in the Pleiades," *ApJ* **318:** 337–355.

STAUFFER, J., L. HARTMANN, AND D. LATHAM (1987), "Rotational Velocities of Low-Mass Stars in the Hyades," *ApJ* **320:** L51–L55.

STAUFFER, J., L. HARTMANN, AND B. JONES (1989), "Rotational Velocities of Newly Discovered, Low-Mass Members of the Alpha Persei Cluster," *ApJ* **346:** 160–167.

STAUFFER, J., C. PROSSER, L. HARTMANN, AND M. MCCAUGHREAN (1994), "Additional Constraints on Circumstellar Disks in the Trapezium Cluster," *AJ* **108:** 1375–1381.

STROM, K., S. STROM, S. EDWARDS, S. CABRIT, AND M. SKRUTSKIE (1989), "Circumstellar Material Associated with Solar-Type Pre–Main Sequence Stars—A Possible Constraint on the Timescale for Planet Building," *AJ* **97:** 1451–1470.

STROM, K., S. STROM, AND M. MERRILL (1993), "Infrared Luminosity Functions for the Young Stellar Population Associated with the L1641 Molecular Cloud," *ApJ* **412:** 233–253.

Hydrocarbons in Protoplanetary Disk

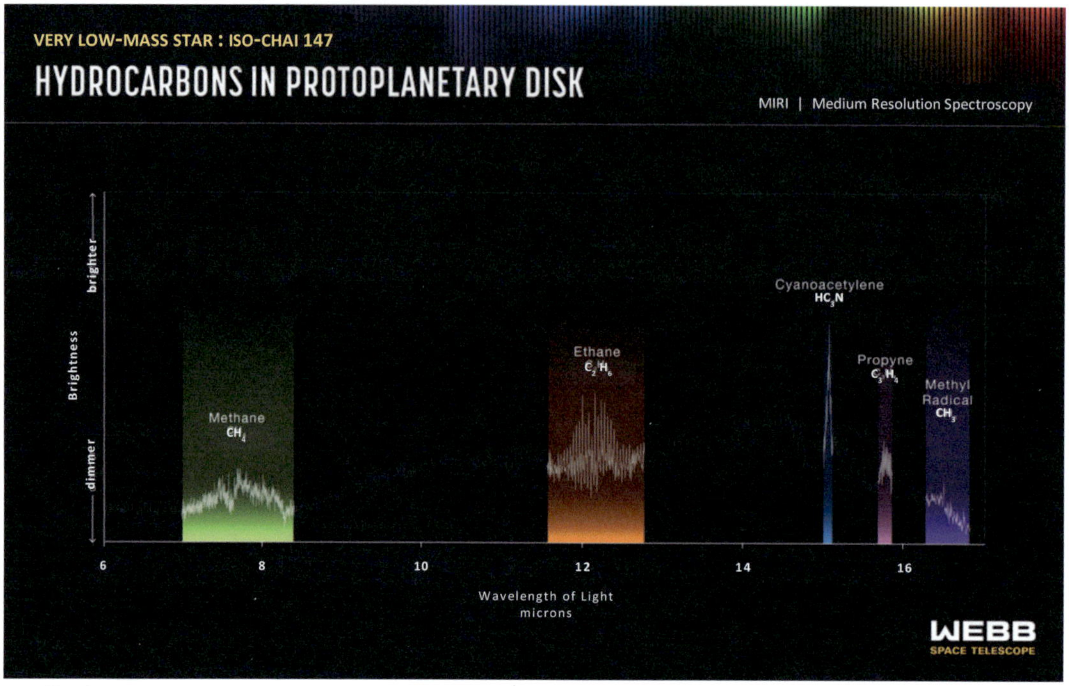

Credit: NASA, ESA, CSA, R. Crawford (STScI)
https://cdn.esawebb.org/archives/images/large/weic2416b.jpg [retrieved 6-20-25]

This graphic presents some of the results from the MIRI Mid-InfraRed Disk Survey (MINDS), which aims to build a bridge between the chemical inventory of disks and the properties of exoplanets. In a new study, the science team explored the region around a very low-mass star of 0.11 solar mass (known as ISO-Chal 147). These observations provide insights into the environment as well as basic ingredients for such planets to form. The team found that the gas in the planet-forming region around the star is rich in carbon. This could potentially be because carbon is removed from the solid material from which rocky planets can form, and could explain why Earth is relatively carbon-poor.

The spectrum revealed by Webb's Mid-InfraRed Instrument (MIRI) shows the richest hydrocarbon chemistry seen to date in a protoplanetary disk, consisting of 13 carbon-bearing molecules up to benzene. This includes the first extrasolar detection of ethane (C_2H_6), the largest fully saturated hydrocarbon detected outside our Solar System. Since fully saturated hydrocarbons are expected to form from more basic molecules, detecting them here gives researchers clues about the chemical environment. The team also successfully detected ethylene (C_2H_4), propyne (C_3H_4), and the methyl radical CH_3, for the first time in a protoplanetary disk.

This graphic highlights the detections of ethane (C_2H_6), methane (CH_4), propyne (C_3H_4), cyanoacetylene (HC_3N), and the methyl radical CH_3.

THE FORMATION OF HABITABLE PLANETARY SYSTEMS

G. W. Wetherill

Department of Terrestrial Magnetism
The Carnegie Institution of Washington
Washington, DC 20015

Abstract

The present state of development of a general theory of the formation of planetary systems from prestellar disks is described. Quantitative results have been found for several stages in the evolution of such systems. Significant progress has been made concerning development of terrestrial planet regions, which are also the most likely locales for the presence of liquid water and favorable conditions for the origin, evolution, and survival of living systems. According to these models, the source regions for terrestrial planets are not limited to local "feeding" zones, but include the present asteroidal regions, probably rich in volatile elements of biological importance, and possibly survivable organic compounds. These results suggest that habitable planets may be common throughout the Galaxy and Universe, but much remains to be understood before this suggestion can be considered to be at all well established.

INTRODUCTION

This document does not purport to be a proper scientific paper, but rather simply a description of the material that was presented orally at the Habitable Zones

Conference. It is hoped that this, together with some relevant references, will assist the reader in becoming familiar with the current state of understanding of this topic.

In recent years the Kant-Laplace model, in which our Solar System is assumed to have formed from a circumsolar disk of gas and dust, has received observational support by studies of circumstellar disks surrounding pre–main sequence stars (see reviews by Shu et al. [1993], Strom et al. [1993], Tscharnuter and Boss [1993]). One outcome of this recent work has been the discovery of the variable nature of these disks. It is probable that these observed disks are not yet the hypothesized "solar nebulae" from which planets form from a centrifugally supported disk containing the mass and angular momentum distribution that provide the initial conditions for theoretical calculations of planetary growth. Major advances in understanding the currently observationally and theoretically poorly constrained evolution of these disks are required before these initial conditions cease to be free parameters. Detailed evidence relevant to this stage of preplanetary evolution is preserved in meteorites (reviewed by Kerridge 1993). Utilization of this evidence requires development of models that are adequate to relate it to dynamical theory and astronomical observation.

Historically, primary attention in this field was directed toward the formation of our Solar System, rather than to the development of a general theory of planetary formation. This situation is now changing, inasmuch as the observational variety of circumstellar disks probably results in a variety of mass and angular momentum distributions even for stars of a given mass, and it is unlikely that a priori initial parameters for our Solar System will become available. Furthermore, theoretical models of the accumulation of these disks into planets emphasize the importance of close encounters between the growing bodies, a dynamical situation long recognized to be chaotic in nature (Öpik 1951). For these reasons the best approach is to attempt to construct a general theory of planet formation, in which our Solar System represents one of many possible systems, and to try to distinguish those features of these systems that are likely to be common, rather than improbable variations, some of which could be presented by our Solar System. General reviews of this subject have been given by Lissauer and Stewart (1993) and Lissauer (1993). A review directed toward the only currently known habitable planet, Earth, is that of Wetherill (1990).

It also does not seem likely that in the near future it will be possible to obtain either a numerical or theoretical model in which a swarm of planetesimals spontaneously evolves into a fully mature planetary system, containing terrestrial-like planets, asteroid belts, giant planets, and cometary source regions. A promising approach is to divide the problem into several stages, for which different theoretical mechanisms are appropriate, and then link these together into a complete quantitative model of planetary formation that can be tested by comparison with observations of our Solar System and with other observable planetary systems.

FORMATION OF PLANETS FROM INTERPLANETARY GRAINS

Formation of Planetesimals from Dust Grains

An important, but not well-understood, problem is the mechanisms by which micron-sized nebular dust grains agglomerate to form larger objects, including the chondrules and calcium-aluminum inclusions found in undifferentiated meteorites, as well as the ~ 1 km diameter planetesimals that are the initial planetary building blocks according to the currently fashionable planetary models for the growth of planets (Weidenschilling and Cuzzi 1993, Cuzzi et al. 1994). If planets indeed form from planetesimals, this is a process that Nature seems to be very good at, inasmuch as it must have occurred over a great range of physical and chemical conditions in our Solar System, and apparently around pulsars as well. For this reason it is probably safe to assume that sufficiently efficient conversion of dust to planetesimals does occur, even though it is not yet well understood.

Formation of ~ 10^{26}g Planetary Embryos from Small Planetesimals

The pioneering work on this problem is that of Safronov (1969). In this book, as well as in his earlier papers, the basic problems of formation of planets were clearly posed, and major contributions to their solution were made.

The initial number (~ 10^{11}) of small (~ 10 km) planetesimals in the terrestrial planet region and the asteroid belt is very large. It is not possible to follow the detailed evolution of the orbits of so many objects. For this reason, Safronov adopted the alternative of treating the simultaneous evolution of the mass and velocity distribution by the methods of the kinetic theory of gases.

An important extension of Safronov's work resulted from the demonstration by Stewart and Kaula (1980) that the positive-definite gravitational perturbations between planetesimals were accompanied by deceleration of the larger bodies of the system as a result of equipartition of energy between the larger and smaller bodies, resulting in so-called dynamical friction. These phenomena were incorporated into specific expressions by Stewart and Wetherill (1988) and used to demonstrate the rapid (10^5 years at one AU) growth of lunar- to Mercury-sized planetary embryos (Wetherill and Stewart 1989, 1993) (see also Kolvoord and Greenberg 1992). When the effects of fragmentation and gas drag are included along with advances in understanding of low relative velocity gravitational perturbations (Ida 1990), it is found that the runaway growth of these planetary embryos is self-regulated and generally in accord with the scaling laws described by Lissauer (1987), although subject to major stochastic fluctuations (Wetherill and Stewart 1993 and subsequent unpublished work). After about 10^5 years at one AU, and 10^6 years at

2.5 AU, the outcome of this stage of planetary growth is an assemblage of several hundred ~ 10^{26}g bodies extending nearly from the Sun to at least the outer asteroidal region, with a tendency for the mass and spacing of the embryos to increase with distance from the Sun. In addition, it is probable that a significant portion of the mass of the system remains in the form of smaller bodies—asteroid-sized residual objects and collision fragments. The larger bodies will be constrained by equipartition of energy to move in nearly circular, coplanar, noncrossing orbits. The results of a calculation of this stage of evolution are shown in Figure 1, taken from Wetherill and Stewart (1993).

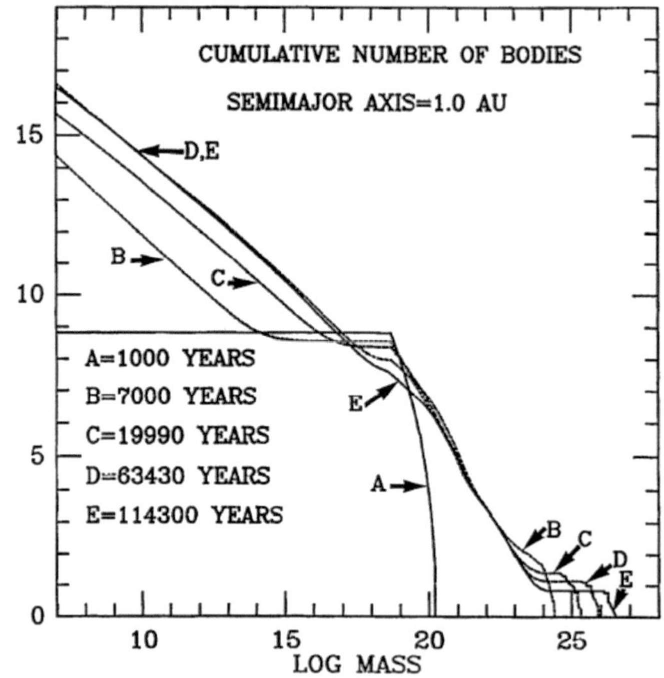

FIG. 1. *Evolution of the size distribution of a swarm of ~ 10 km planetesimals into ~ 10^{26}g "planetary embryos," originally published in Wetherill and Stewart (1993). Reprinted by permission of Academic Press, Inc.*

Curve A: As a result of combinatorial statistics an initial swarm of uniform-size bodies at 1.0 ± 0.08 AU forms about 50 bodies with about 30 times the initial mass within 1000 years.

Curve B: As a result of equipartition of energy, the velocities of the largest bodies decrease rapidly. This leads to their having an enhanced gravitational capture cross-section, producing > 1000 km diameter bodies in 7000 years by runaway growth.

Curve C (19990 years): Although runaway growth has continued, producing > 10^{25}g bodies, the gravitational cross-section enhancement tends to decrease because of the gravitational acceleration of the residual ~ 10^{18}–10^{20}g objects by the largest bodies.

Curves D, E: The higher velocities of the smaller bodies (~ 0.5 km/sec) leads to extensive fragmentation. Much of the mass of the system is transferred to small collision fragments. The velocities of these bodies are decreased by gas drag, leading to self-regulation of the runaway *process. By ~ 10^5 years, about one-half of the mass of the system is in large runaway "embryos" moving in nearly circular, coplanar orbits. Further evolution into full-size planets requires this system to be dynamically unstable with regard to mutual close encounters on an ~ 10^6-year time scale. Occurrence of this instability has not yet been established.*

Formation of Final Planets from Planetary Embryos

Some of these embryos will be comparable in mass to Mercury and Mars, but considerably smaller than Venus and Earth, and much smaller than the ~ 15 Earth mass cores of the giant outer planets. In the subsequent discussion, emphasis will be placed on the formation of planets in the potentially habitable terrestrial planet region, inasmuch as it is unlikely that distant objects will be warm enough to provide liquid surface water and the thermal conditions necessary to permit the formation and evolution of extensive living chemical systems. On the other hand, it is quite possible that in more subtle ways, the formation of large planets like Jupiter may have an important effect on the habitability of the terrestrial planets (Wetherill 1994a).

In order for planetary evolution to proceed beyond the planetary embryo stage, described in the previous section, it is necessary that the system of nearly circular, coplanar embryo orbits become dynamically unstable, permitting close encounters, strong mutual gravitational collision, and merger of these embryos to form Earth- and Venus-sized bodies. Although suggestions can easily be made regarding the establishment of this instability on an ~ 10^5–10^6-year time scale, this remains to be established quantitatively.

Assuming this instability does permit the orbits of the embryos to cross one another, a Monte Carlo technique has been developed to simulate the subsequent orbital evolution of the individual bodies of the system and their growth into planets. These techniques are ultimately based on the statistical expressions of Öpik (1951) for the orbital changes caused by close gravitational encounters and collisions with one another. The latest results of studies of this kind have been reported by Wetherill (1992) in which the results of 431 simulations under a wide range of parameters and assumptions are discussed. It is found that a variety of planetary systems are formed, but the range of this variety is insensitive to the variations in the choice of parameters and assumptions. The principal effect of the possible absence of gas giant planets in these systems is the presence of one or more terrestrial-like planets in the asteroid belt (Wetherill 1991), at distances unlikely to be habitable (Kasting et al. 1993). For a one solar mass star, with a surface density corresponding to our present Solar System, augmented by the material presumably initially present in the asteroid belt, the following features of our Solar System are explained in a natural way by this model.

(1) Evolution of the specific energy and angular momentum of the initial system into values similar to our Solar System.

(2) Over a significant range of variation in total surface density, production of a terrestrial planet system similar in mass to ours.

(3) Although subject to considerable stochastic variations, general agreement in the number, size, and position of the final planets with those of our Solar System.

(4) Spontaneous clearing of the asteroid belt of large embryos by the same processes that remove asteroids and meteorites from the asteroid belt today.

(5) Distribution of the final eccentricities and inclination of the terrestrial planets that includes those observed in our planetary system.

(6) Prevalence of giant impacts on the terrestrial planets of the size proposed to explain the formation of the moon and the loss of Mercury's silicate mantle.

(7) Natural evolution of the relative velocities of residual planetesimals in the asteroidal region to their present observed values.

(8) Natural evolution of ~ 1 AU wide "stratigraphy" in the asteroid belt.

The final distribution of the planets corresponding to the "nominal model" of Wetherill (1992) is shown in Figure 2. In this figure the results of 115 calculations are combined.

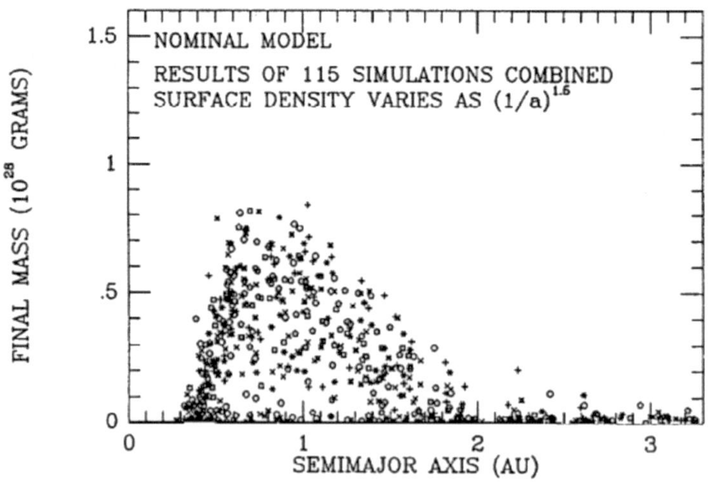

FIG. 2. *Superposition of the final planets formed from 115 calculations from the "nominal model" of Wetherill (1992), originally published in* Icarus **100:** *307–325. Reprinted by permission of Academic Press, Inc. This model assumes that the instability mentioned in the caption to Figure 1 occurs. As a consequence, close encounters between the embryos and resonant perturbations by the giant planets cause evolution of the system on an ~ 10^8-year time scale into systems resembling in a general way our Solar System, but in detail exhibiting differences resulting from the chaotic nature of the accumulation process. The different symbols for the points represent*

variations in the details of the model (Wetherill 1992). Similar results are found when the surface density varies as (1/a) (a = *semimajor axis*).

The principal feature demonstrated by this figure is the concentration of $> 10^{27}$g planets in the region interior to the present asteroid belt. The largest bodies are found between 0.5 and 1.2 AU. On the average, 4.0 final planets with masses greater than 10^{26}g are found interior to 2 AU. The average number of final planets of mass greater than 3.5×10^{27}g is 1.6 per simulation. Fifty percent of the simulations contain a "Mercury"—a final planet of mass $< 10^{27}$g interior to 0.48 AU. Most of these small bodies with small heliocentric distances are collision fragments. In 94% of the simulations a final planet larger than 10^{26}g appears in the position of "Mars." This final planet tends to be larger than our planet Mars. On the average, per simulation, 0.52 final bodies greater than twice the mass of Mars are found in the outer terrestrial planet region between 1.3 and 2.15 AU. No obvious differences are seen in the final planetary distribution resulting from the alternative versions of the nominal model.

In about 50% of the simulations a planet of mass larger than one-half Earth's mass is formed within the 0.95 to 1.37 AU habitable zone of a one solar mass star as given by Kasting et al. (1993). About one-third contain such a planet within their 4.6 Gyr continuously habitable zone.

Source Regions of the Terrestrial Planets

The availability and sources of the elements essential to life, particularly C, H, N, and O, have been widely discussed in the context of the origin of life on Earth. It is often supposed that an input of cometary material from the outer Solar System played an essential role in supplying this material. This may well be the case.

The recent developments in the formation of the terrestrial planets described in previous sections suggest, however, that other sources, particularly those from the asteroid belt, may also be of significance. In older models of the formation of the terrestrial planets that did not consider many phenomena now known to be important, it was usually supposed that each terrestrial planet had its own local "feeding zone," from which it accumulated planetesimals. Assumed variation of chemical composition with heliocentric distance thus primarily determined the chemical composition of these planets, except for the possibility of "late veneers" from more distant regions.

The quantitative model for the formation of the terrestrial planets and asteroid belt described above (Wetherill 1992) leads to a quite different concept regarding the source material from which the terrestrial planets were formed (Wetherill 1994b). A set of 59 new calculations have been made using the "nominal model" of Wetherill (1992). From these, thirteen cases were selected that resembled our terrestrial planets somewhat more than the others. The final masses of all these bodies are shown in Figure 3. These all include two large terrestrial planets of the general size and positions of Venus and Earth, and a wider range of distinctly smaller objects at larger and smaller heliocentric distances.

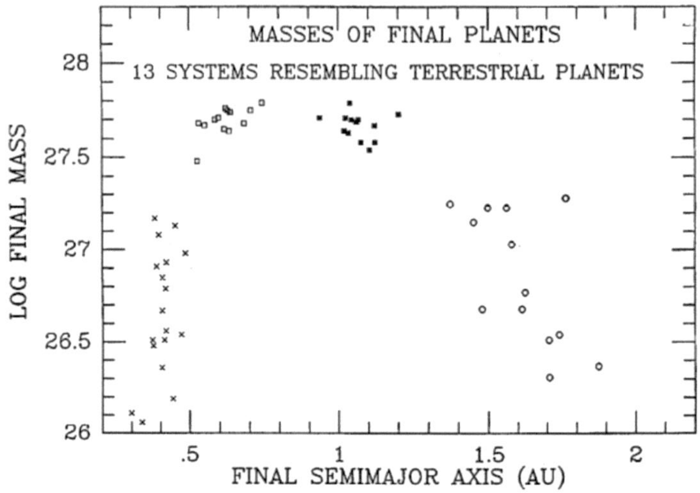

FIG. 3. *Final masses of 13 systems (out of a set of 56) that meet assumed criteria for "resemblance to terrestrial planets." Planets similar in mass to Venus (open squares) and Earth (solid squares) form between 0.5 and 1.2 AU. Small planets, with greater variation in mass, are found at the inner edge (crosses) and outer regions (open circles). The average results for all cases, including those less similar to the terrestrial planets (e.g., those containing three large planets) are described in the text. Reprinted with permission, from G.W. Wetherill, "Provenance of the Terrestrial Planets,"* Geochim. Cosmochim. Acta ***58:*** *4513–4520, 1994 (Elsevier Science Ltd., The Boulevard, Langford Lane, Kidlington 0X5 1GB, UK).*

The distribution of the initial semimajor axes of the sources of bodies that formed these final planets, weighted according to their mass, is shown in Figure 4 for a typical one of these simulations (solid squares). The final configuration for this particular calculation is shown in the upper left portion of the diagram.

It may be seen that the provenance of these four terrestrial planets ranges from ~ 0.5 to 2.5 AU. No planet has its own special feeding zone. In all cases a significant contribution is made by bodies in the region now occupied by asteroids. Furthermore, similar investigation of the evolution of the asteroid belt leads to the conclusion that these primordial asteroidal bodies, originally within 2.5 AU, are now represented in the outer asteroid belt, beyond 2.75 AU, in which dark, presumably light-element-rich asteroidal bodies predominate. It is also likely that the contribution of the asteroidal region to the terrestrial planets was augmented by capture by the terrestrial planets of small collisional debris brought from the primordial asteroid belt as collision debris by gas drag, and later, by Poynting-Robertson drag. For these reasons, an asteroidal contribution to Earth's inventory of elements essential to biological evolution should be expected, and possibly organic compounds as well.

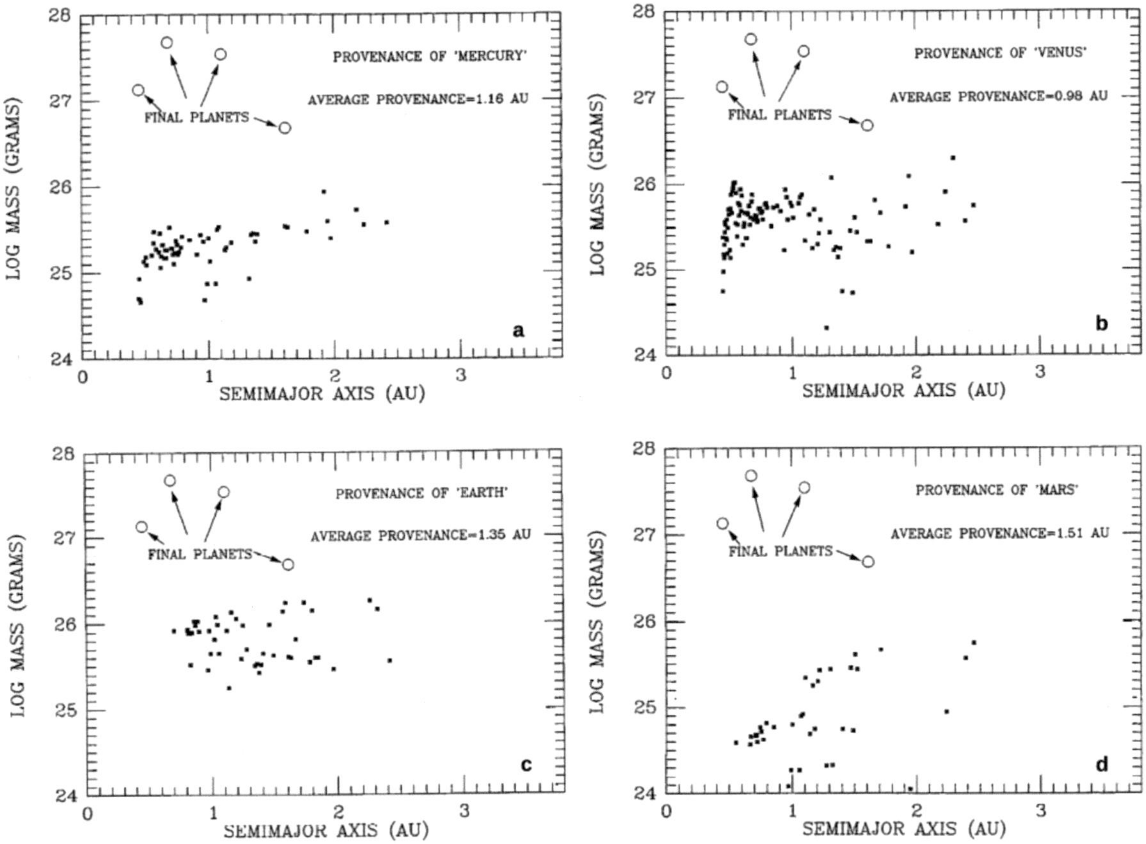

FIG. 4. *Distribution of source regions for a typical example of the systems illustrated by Figure 3. The final distribution of the planets is shown at the upper left-hand portion of the diagram. The positions of the initial bodies (weighted according to their mass contribution) for each of these four final planets are given separately in Figures 4a, b, c, and d. Despite their differences in final positions, every one of these four final planets receives significant contributions from bodies initially at heliocentric distances ranging from 0.5 to 2.5 AU. The weighted average provenance of each of the bodies is indicated on the figures. Reprinted with permission, from G.W. Wetherill, "Provenance of the Terrestrial Planets,"* Geochim. Cosmochim. Acta *58: 4513–4520, 1994 (Elsevier Science Ltd., The Boulevard, Langford Lane, Kidlington 0X5 1GB, UK).*

This wide provenance of terrestrial planet material is found for all of the systems included in Figure 3. It is also found for all 59 of the simulations, including those for which designations of surrogate planetary names were less appropriate. Despite this wide provenance of material for a given final terrestrial planet, there is a general correlation between provenance and final heliocentric distance (Fig. 5). Therefore, there is no reason to expect an exact agreement among the compositions of the terrestrial planets in other planetary systems, as a result not only of this correlation, but because of the stochastic variations that are evident in Figure 4.

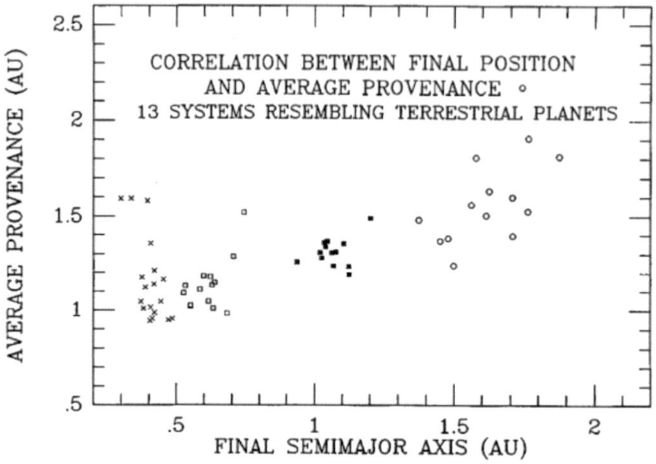

FIG. 5. *Distribution of the average provenance for each of the 13 systems. Symbols are the same as in Figure 3. (Planets similar in mass to Venus [open squares] and Earth [solid squares] form between 0.5 and 1.2 AU. Small planets, with greater variation in mass, are found at the inner edge [crosses] and outer regions [open circles]). Because of the chaotic nature of the evolution, wide variations in the average provenance of each individual planet are found. For this reason, there is no requirement that all of the terrestrial planets have nearly identical compositions. There is a general tendency for the more distant planets to be derived from more distant sources, but the extent of this correlation is at most comparable to the scatter in the average provenance of the individual bodies. Reprinted with permission, from G.W. Wetherill, "Provenance of the Terrestrial Planets,"* Geochim. Cosmochim. Acta ***58:*** *4513–4520, 1994 (Elsevier Science Ltd., The Boulevard, Langford Lane, Kidlington 0X5 1GB, UK).*

EFFECT OF VARIATION IN STELLAR MASS AND SURFACE DENSITY OF THE ASSEMBLAGE OF PLANETARY EMBRYOS

There is no reason to believe the models described in the previous sections are applicable only to our Solar System (Wetherill 1996). Preliminary results have been found in which the later stages of planetary growth, described under "Formation of ~ 10^{26}g Planetary Embryos from Small Planetesimals," are simulated including variation in the stellar mass and in the surface density of the preplanetary disk.

The results of calculations for ten simulations of a 1.5 $M_{\odot}$ star are shown in Figure 6a, with a proportionally higher initial surface density. The number of final planets per simulation is similar to that found previously. Some of them are considerably larger than Earth, and of sufficient mass to capture significant quantities of nebular gas. Except for several bodies intermediate in mass between Mars and Earth, at first glance most of these objects appear to be too hot according to the habitability calculations of Kasting et al. (1993). This may not be the case, however, because the clearing of the region beyond 2 AU is caused by the assumption of a "Jupiter" at the Solar System position. Variations of this assumption are being calculated at present.

A simulation for a 0.5 solar mass star is shown in Figure 6b, with a surface density equivalent to that of our Solar System. In this case, the larger planets tend to concentrate near one AU, beyond the habitable zones of Kasting et al. (1993). In this early work on this problem, there appears at first sight to be a concentration of planets near one AU that is relatively insensitive to stellar mass and surface density. If so, this leads to the speculation that stars similar to the mass of the Sun are more likely than others to be habitable. Such a conclusion would be premature at this stage of the investigation, however.

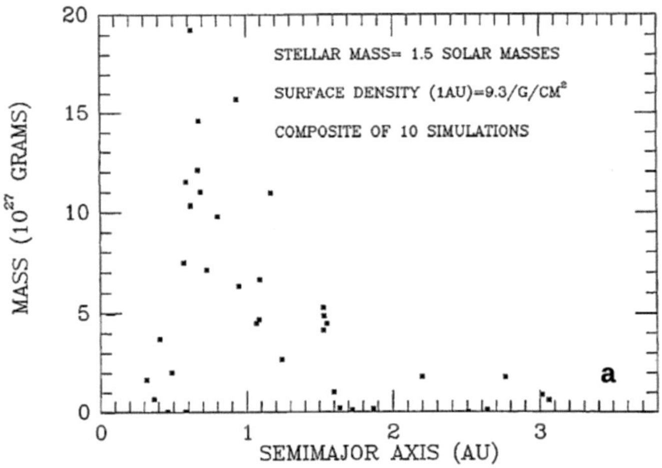

FIG. 6a and b. *Sample results of calculations for terrestrial planetary systems of different stellar mass and surface density. (a) Stellar mass = 1.5 M⊙, solid surface density = 9.3 g/cm (1.5 times value for nominal model of one M⊙ star). (b) Stellar mass = 0.5 M⊙, solid surface density = 6.2 g/cm^2 (same as that for nominal model of one M⊙ star).*

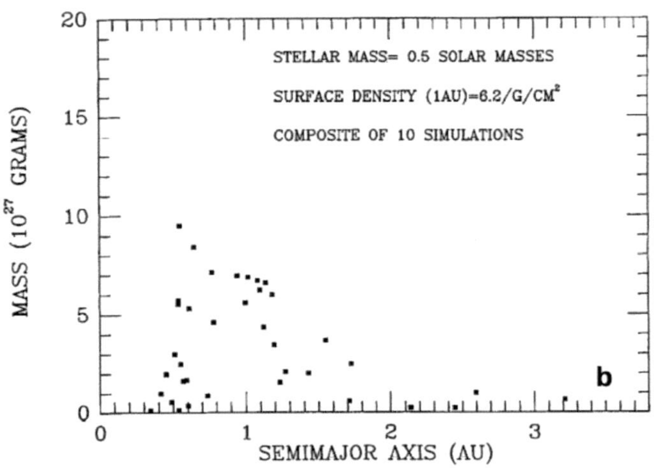

Acknowledgments

I wish to thank J. F. Kasting for valuable discussions and Jan Dunlap for her keyboard artistry. This work was supported by NASA grant NSG-1969.

References

Cuzzi, J.N., A.R. Dobrovolskis, and R.C. Hogan (1994), "What Initiated Planetesimal Formation?," *Lunar and Planetary Science XXV*, 307–308.

Ida, S. (1990), "Stirring and Dynamic Friction Rates of Planetesimals in the Solar Gravitational Field," *Icarus* **88:** 129–145.

Kasting, J.F., D.P. Whitmire, and R.T. Reynolds (1993), "Habitable Zones Around Main Sequence Stars," *Icarus* **101:** 108–128.

Kerridge, J.F. (1993), "What Can Meteorites Tell Us About Nebular Conditions and Processes During Planetesimal Accretion?," *Icarus* **106:** 135–150.

Kolvoord, R.A., and R. Greenberg (1992), "A Critical Reanalysis of Planetary Accretion Models," *Icarus* **98:** 2–19.

Lissauer, J.J. (1987), "Time Scales for Planetary Accretion and Structure of the Protoplanetary Disk," *Icarus* **69:** 245–265.

Lissauer, J.J. (1993), "Planet Formation," *Ann. Rev. Astron. Astrophys.* **31:** 129–174.

Lissauer, J.J., and G.R. Stewart (1993), "Growth of Planets from Planetesimals," in *Protostars and Planets III*, E. Levy, J.I. Lunine, and M.S. Matthews, eds. (Tucson: University of Arizona Press), pp. 1061–1088.

Öpik, E.J. (1951), "Collision Probabilities with the Planets and the Distribution of Interplanetary Matter," *Proc. R. Irish Acad.* **A54:** 165–199.

Safronov, V.S. (1969), *Evolution of the Protoplanetary Cloud and Formation of the Earth and Planets* (Moscow: Nauka Press). English translation: NASA TTF-677, 1972.

Shu, S., J. Najita, D. Galli, E. Ostriker, and S. Lizano (1993), "The Collapse of Clouds and the Formation of Stars and Disks," in *Protostars and Planets III*, E. Levy, J.I. Lunine, and M.S. Matthews, eds. (Tucson: University of Arizona Press), pp. 3–45.

Stewart, G.R., and W.M. Kaula (1980), "Gravitational Kinetic Theory for Planetesimals," *Icarus* **44:** 154–171.

Stewart, G.R., and G.W. Wetherill (1988), "Evolution of Planetesimal Velocities," *Icarus* **74:** 542–553.

Strom, S.E., S. Edwards, and M.F. Skrutskie (1993), "Evolution Time Scales for Circumstellar Disks Associated with Intermediate- and Solar-Type Stars," in *Protostars and Planets III*, E. Levy, J.I. Lunine, and M.S. Matthews, eds. (Tucson: University of Arizona Press), pp. 837–866.

Tscharnuter, W.M., and A.P. Boss (1993), "Formation of the Protosolar Nebula," in *Protostars and Planets III*, E. Levy, J.I. Lunine, and M.S. Matthews, eds. (Tucson: University of Arizona Press), pp. 921–938.

WEIDENSCHILLING, S.J., AND J.N. CUZZI (1993), "Formation of Planetesimals in the Solar Nebula," in *Protostars and Planets III*, E. Levy, J.I. Lunine, and M.S. Matthews, eds. (Tucson: University of Arizona Press), pp. 1031–1060.

WETHERILL, G.W. (1990), "Formation of the Earth," *Ann. Rev. Earth Planet. Sci.* **18:** 205–256.

WETHERILL, G.W. (1991), "Occurrence of Earth-like Bodies in Planetary Systems," *Science* **253:** 535–538.

WETHERILL, G.W. (1992), "An Alternative Model for the Formation of the Asteroid Belt," *Icarus* **100:** 307–325.

WETHERILL, G.W. (1994a), "Possible Consequences of Absence of 'Jupiters' in Planetary Systems," in *Planetary Systems: Formation, Evolution, and Detection*, B.F. Burke, J.H. Rahe and E. Roettger, eds. (Dordrecht: Kluwer Academic Publishers), pp. 23–32.

WETHERILL, G.W. (1994b), "Provenance of the Terrestrial Planets," *Geochim. Cosmochim. Acta* **58:** 4513–4520.

WETHERILL, G.W. (1996), "The Formation and Habitability of Extra-Solar Planets," *Icarus* **119:** 219–238.

WETHERILL, G.W., AND G.R. STEWART (1989), "Accumulation of a Swarm of Small Planetesimals," *Icarus* **77:** 350–357.

WETHERILL, G.W., AND G.R. STEWART (1993), "Formation of Planetary Embryos: Effects of Fragmentation, Low Relative Velocity, and Independent Variation of Eccentricity and Inclination," *Icarus* **106:** 190–209.

Globular Cluster M92 (NIRCam Detail)

Photo Credit: James Webb Space Telescope's NIRCam instrument
Image Processing: NASA, ESA, CSA, Alyssa Pagan (STScI)
https://science.nasa.gov/asset/webb/globular-cluster-m92-nircam-detail/
[retrieved 12-11-25]

Detail of the globular cluster Messier 92 (M92) captured by Webb's NIRCam instrument. This field of view captures the lower left quarter of the right half of the full image. Globular clusters are dense masses of tightly packed stars that all formed around the same time. In M92, there are about 300,000 stars packed into a ball about 100 light-years across. The night sky of a planet in the middle of M92 would shine with thousands of stars that appear brighter than those in our own sky. The image shows stars at different distances from the center, which helps astronomers understand the motion of stars in the cluster, and the physics of that motion.

ORIGIN AND EVOLUTION OF THE ROTATION OF TERRESTRIAL PLANETS

Luke Dones

San José State University Research Foundation
NASA Ames Research Center
Moffett Field, CA 94035

Abstract

The rapid rotation of Earth and Mars probably arose late in their formation from giant impacts by planetesimals with masses on the order of 20% of the planetary masses. Of the four terrestrial planets, only Mars may retain its primordial spin rate. Mercury and Venus have been despun by tides. Mutual planetary perturbations can cause large oscillations in the tilts of the planets' spin axes ("obliquities"); Mars' obliquity has probably varied by tens of degrees over the last 10 million years, while the Moon at present helps to stabilize the Earth's obliquity, which varies only by a few degrees. Billions of years from now, the Earth's obliquity will vary from 0 to 60 degrees or more. The effect of large obliquity variations on global climate may not be drastic, and stable obliquity is unlikely to be a requirement for the development of intelligent life on a planet.

"If everybody minded their own business," the Duchess said in a hoarse growl, "the world would go round a deal faster than it does."

Lewis Carroll, *Alice's Adventures in Wonderland*

In Stephen Dole's fascinating book, *Habitable Planets for Man*, he suggests that an Earthlike rotation rate favors the habitability of a planet. "Its period of rotation," Dole writes, "must be less than about 96 hours (4 Earth days) to prevent excessively high daytime temperatures and excessively low nighttime temperatures." In this paper we briefly review (1) how the planets are set spinning as they form, (2) how their rotation states change over time, and (3) the likely effects of spin on planetary habitability.

How Did the Terrestrial Planets Originally Acquire Their Spins?

Most of the planets rotate in the same direction that they revolve around the Sun. At the time Dole's book was published (1964), the rotation rates of seven planets were known. Of these, only Uranus rotates in a retrograde direction (i.e., opposite its direction of revolution). It is then natural to assume, as Dole does (p. 61), that the rapid prograde rotation of the planets is a logical consequence of their formation. However, both observation and theory suggest otherwise. Venus and Pluto are now known to spin retrograde. Recent studies of the planetary spin imparted by off-center impacts from bodies orbiting the Sun (Lissauer and Kary 1991, Dones and Tremaine 1993) show that a planet that forms from a uniform disk does not acquire rapid prograde spin. In fact, if the random velocities in the disk are very small, the planet spins rapidly, but retrograde! For large random velocities, the spin is prograde, but slow. Basically, rapid spin (i.e., once every 24 hours or faster) is impossible to obtain because of cancellation of prograde and retrograde angular momentum. The radii of curvature of the orbits of Earth impactors are typically hundreds of times the radius of the Earth. Thus incoming bodies have a hard time "knowing" which side of the planet to hit (thereby imparting forward or backward spin).

In response to this conundrum, two suggestions have been put forward. Lissauer and Kary (1991) suggested that a forming planet clears a gap around itself as it grows, so that the mass accreted at any given time tends to arise from the inner and outer edges of the planet's "accretion zone." Such material has, on the average, larger prograde angular momentum, and could produce rapid prograde rotation. Dones and Tremaine (1993) and Tremaine and Dones (1993) pointed out that the disk depletion required by this model has not been demonstrated in realistic models of planet formation (except at very early stages). In our view, planetary rotation is more likely to arise from one or two giant impacts toward the end of planet formation. Previous models of planetary obliquities (Lissauer and Safronov 1991) had generally assumed that most of a planet's spin arose through accretion of small bodies, with one or more late giant impacts providing the tilt. In our model, essentially all the spin comes from giant impacts. The typical spin period of a newly formed planet should be on the order of three to fifteen hours, the largest impactor has a mass on the order of 20% of the planet's mass, and the initial obliquity of the planet is equally likely to have any value in the range 0–180 degrees. Thus about half the planets should have formed with a prograde spin, while the other half should have been retrograde.

The models of both Lissauer and Kary, and Dones and Tremaine, make simplifying assumptions. These models posit (rather than calculate a priori) the orbits and masses of the bodies that impact the forming planets, and assume that the impactors are completely accreted by the planets. Final resolution of the origin of planetary spin must await future simulations of planet formation that simultaneously follow the evolution of the orbit and mass distributions throughout the course of planet formation, and realistically calculate the fate of the impactors. For the Earth, the Moon-forming impact likely gave the planet most of its spin angular momentum; Cameron and Benz (1991), however, argue that more than half of the angular momentum of the impactor may have been lost in escaping ejecta.

How Have the Spins of the Planets Changed Since Their Formation?

Since spin is a vector, it has both a magnitude ("spin rate") and a direction. The direction is measured by the obliquity, which is the angle between the planet's pole and the pole of the orbit (or, equivalently, the angle between the planet's equator and orbit plane). Obliquities less than 90 degrees are prograde; obliquities greater than 90 degrees are retrograde. Several processes can change the obliquity, spin rate, or both (Harris and Ward 1982, Williams 1993); these are discussed in the following sections.

Spin/Orbit Coupling Changes Obliquities Without Changing Spin Rates

The Sun exerts a torque on the equatorial bulge of a planet that makes the planet's spin axis precess like a top. (For the Earth, the Moon also exerts a torque that is twice as big as the solar torque.) The orbits of the planets also precess as a result of their mutual gravitational interactions. Thus a planet's obliquity can change because of changes in the orientation of either its spin axis or its orbit pole. If the spin axis precesses rapidly compared with the orbital precession, the spin axis is able to follow the instantaneous orbit pole, and the obliquity remains nearly constant. Mercury, Venus, and Earth currently are in this regime, with Earth's obliquity remaining in the range of 22.0–24.6 degrees over the last 18 million years

(Laskar et al. 1993a). On the other hand, if the spin axis precesses slowly, it follows the time-averaged orbit pole, and the obliquity varies in proportion to the planet's inclination to the invariable plane of the Solar System. The giant planets exhibit this type of variation.

The model of spin/orbit coupling just described assumes that the planetary orbits vary with time in a regular fashion. In the simplest "secular" theory, which treats a system of N planets as coupled harmonic oscillators, the orbital inclinations of the planets can be described as the sum of $(N-1)$ sinusoidal terms with periods ranging from roughly 50,000 years to 2 million years (Harris and Ward 1982). If the precession period of a planet's spin axis is close to one of these secular periods, a near-resonance results, and large obliquity variations can occur. It has been known since the work of Ward (1973) that a near-resonance causes Mars' obliquity to vary from 15 to 35 degrees over time scales of 1 Myr.

The actual variations in Mars' obliquity are likely to be even larger for two reasons. First, we now believe that the orbits of the planets are chaotic. For example, the orbits of the planets (and consequently, their obliquities) cannot be predicted 100 Myr into the past or future, because the uncertainties in the orbits grow exponentially with time due to their mutual gravitational interactions. (This chaos probably does not result in drastic changes in the sizes and shapes of the planetary orbits over the age of the Solar System, but this result has not been rigorously shown.) For our purposes, the implication of Solar System chaos is that the planetary orbits vary with a large number of frequencies; in fact, the planetary orbits cannot be well-represented over long time scales as a sum of sine waves. Thus there are many more opportunities for near-resonances than in the simple secular theory; in general, obliquity variations are larger when the full complexity of the planetary orbits is considered. Over long time scales, only statistical statements can be made about the likely variations of Mars' obliquity. Bills (1990) states that obliquities between 0 and 51 degrees may have occurred within the last 10 Myr, Touma and Wisdom (1993) infer a range of 11–49 degrees during the last 80 Myr, and Laskar and Robutel (1993) argue that obliquities can range from 0 to 60 degrees further in the past.

Second, obliquity variations can be amplified by changes in the period of precession of a planet's spin axis. This period is inversely proportional to the planet's departure from a spherical shape, which is measured by the gravitational harmonic J_2. The Tharsis Ridge of volcanoes on Mars increases the planet's J_2 by 6.5% (Ward et al. 1979, Ward 1992). Prior to the formation of the ridge, Mars' spin precession period would have been closer to resonance with one of the secular periods, resulting in obliquity variations from 9 to 46 degrees. Obliquity variations have been suggested as a possible cause of the formation of the layered polar deposits on Mars (Toon et al. 1980). Recent calculations with the NASA Ames

Mars General Circulation Model suggest that dust storms occur more frequently when Mars' obliquity is high (Haberle and Schaeffer 1995).

In contrast with Mars, the Earth at present experiences very small obliquity variations because its spin axis has a precession period of 26,000 years, shorter than any of the secular periods. The Moon is largely responsible for causing this rapid precession. "Without the Moon, the Earth's precession period would be ~ 81,000 years and its obliquity oscillations would exceed those of Mars" (Ward 1982). However, in the long term, the Moon will actually destabilize the Earth's obliquity (see p. 201).

Tidal Friction Changes Both Spin Rates and Obliquities

"Solid-body" tides in the planets driven by the Sun (and the Moon, in the case of the Earth) dissipate energy in the planets, and transfer angular momentum between rotational and orbital motion (Peale 1986, Goldreich 1972). For a planet whose rotation period is shorter than the orbital period of its "satellite" (the Sun or Moon), and whose spin is prograde, tidal friction slows the planet's rotation, and causes the satellite's orbit to spiral outward. Thus, in the past, the Earth spun faster, and the Moon was closer to the Earth; the original spin period of the Earth may have been only four or five hours. Tidal friction causes a planet's obliquity to evolve toward a specific prograde value for any initial rotation rate; this value is close to 0 for a slow rotator and close to 90 degrees for a fast rotator (Goldreich and Peale 1970). Since the Earth rotates rapidly, its obliquity is increasing slowly over geologic time, and may have been only 10 degrees when life began (Goldreich 1966, Touma and Wisdom 1994; cf. Williams 1993). However, Earth's spin state at very early times is uncertain: when the Moon was much closer to Earth, tidal interactions were complex. Before the Earth acquired the Moon, it may have undergone large, chaotic obliquity variations like the other terrestrial planets.

The usual end result of tidal evolution is synchronous rotation such as that of the Moon, the inner satellites of the giant planets, and the Pluto/Charon system, in which the spin period equals the orbital period. (Ultimately, the Earth's rotation period will likely also become synchronous at a value of about one month.) However, other end states are possible. Mercury's spin has been slowed by solar tides, but it does not rotate synchronously with its orbital period of 88 days; rather, Mercury spins three times for every two revolutions about the Sun. Capture into this state from an initially more rapid rotation is reasonably likely for a planet on an eccentric orbit (Goldreich and Peale 1968, Peale 1989). Tidal friction drives spins toward equilibria known as "Cassini states," in which the spin axis and orbit normal precess at the same rate. For instance, Mercury occupies Cassini state 1, which has an obliquity very close to 0 for a planet whose spin axis precesses rapidly (Harris and Ward 1982).

Core-Mantle Coupling Drives Obliquities Away from 90 Degrees

The liquid core and solid mantle of Earth do not rotate at exactly the same rate. Angular momentum transfer between the core and mantle appears to be the primary cause of the "decadal" variations in the length of day on the Earth (Hide and Dickey 1991). Viscous, electromagnetic, and "topographic" mechanisms for this core-mantle coupling have been proposed (Williams 1993). Core-mantle coupling drives prograde obliquities toward 0, and retrograde obliquities toward 180 degrees. It has little effect on the spin rate at the planet's surface, since the planet's total spin angular momentum is conserved. The importance of core-mantle coupling in the Earth's spin history is generally believed to be small, but Williams (1993) argues that core-mantle coupling reduced Earth's obliquity from an initially large prograde value (see p. 200). Core-mantle coupling could have driven Venus' obliquity to its present value near 180 degrees from any initial obliquity greater than 90 degrees (Dobrovolskis 1980, Yoder 1995).

Atmospheric Tides May Have Driven Venus' Obliquity Toward 180 Degrees

Planets have an asymmetric distribution of atmospheric mass because gas flows from the warm afternoon side to the cool morning side. In contrast with solid-body tides, which lag the tide-raising body, atmospheric tides lead the position of the Sun (Harris and Ward 1982). As a result, the effects of atmospheric tides are the inverse of solid-body tides; atmospheric tides speed up a planet's rotation rate and drive obliquities toward retrograde values. On Earth, the atmospheric tide is unimportant in the long-term evolution of the planet's spin. However, on Venus, atmospheric tides may balance the effect of the solar tide on the solid body of the planet and stabilize the planet's current slow retrograde spin (Dobrovolskis 1980). The relative importance of core-mantle coupling and atmospheric tides in determining the rotation state of Venus is unknown.

How Does a Planet's Rotation State Affect Its Habitability?

The seasons on Earth are caused primarily by the planet's nonzero obliquity, which causes the solar insolation to vary as each hemisphere tips toward and away from the Sun over the course of a year. The Earth's small orbital eccentricity, which causes the planet's distance from the Sun to vary during the year, plays a secondary role. For the present orientation of its spin axis, the Earth is closest to the Sun during winter in the northern hemisphere, so that the intensity of seasons is slightly decreased in the northern hemisphere and increased in the southern hemisphere. Ice ages on Earth are generally thought to be triggered by variations in the planet's orbital eccentricity, obliquity, and spin-axis orientation, which occur at dominant periods of ~ 100,000, 41,000, and 19,000–23,000 years, respectively (Pollack 1979, Imbrie and Imbrie 1980, Broecker and Denton 1990). Inspired by the apparently drastic effects produced by these modest astronomical variations, several authors have claimed that planets with rotation states unlike Earth's are inhospitable to life because of unfavorable climatic conditions.

Planets That Spin Too Slowly

Dole (1964, pp. 58–59) states that "For slow rotation rates, a limit would be reached when daytime temperatures became excessively high in the low latitudes below some critical latitude and when nighttime temperatures became excessively low poleward from this same latitude, or when the light-darkness cycle became too slow to enable plants to live through the long hot days and long cold nights." (Dole implicitly assumes that the planet's obliquity is small.) He estimates that four days might be a reasonable upper limit for the rotation period of a habitable planet. This estimate appears too pessimistic for two reasons. First, a planet's atmosphere and ocean buffer day-night temperature variations. Venus rotates once every 243 days, yet the planet's surface temperature varies by only a few degrees from day to night. Second, simple organisms can survive for months (or much longer) by forming spores when conditions are unfavorable, such as during a long Arctic winter; animals can hibernate. Long, hot days on a slowly rotating planet with an Earthlike atmosphere might pose a more severe problem, but one could live comfortably in a cave tens of meters below the surface. Whether intelligent life could evolve under these circumstances is uncertain.

Atmospheric Freeze-Out on Synchronously Rotating Planets

As a subcase of the above, Dole (1964, pp. 59–60, 68 ff.) argues that planets closer to a star than a certain critical distance will become locked into synchronous rotation by solar tides, possibly leading volatiles such as water and carbon dioxide to freeze out on the dark side of the planet. Kasting et al. (1993) calculate that planets within the habitable zones (defined by the stability of liquid water) of M stars and late K stars would be tidally locked and therefore lifeless. (In our Solar System, only Mercury lies in the nominal "synchronous zone.") Since most stars are of spectral type M, this effect is potentially important.

Two counter arguments can be made. The first, which Dole discusses, is that the planet's atmospheric circulation might prevent the precipitation of volatiles on the dark side. Haberle et al. (1996) estimate that the atmosphere (assumed CO_2-H_2O) of a synchronously rotating planet that receives an insolation more than about 30% that of present-day Earth will not freeze out, provided a modest amount of CO_2 (~ 0.1–1 bar) is available. Second, Mercury, whose synchronous rotation had been "confirmed" many times when Dole wrote his book in 1964, was shown the next year to rotate exactly 1.5 times during each orbit around the Sun (Goldreich and Peale 1968). Thus, as Kasting et al. state, Mercury rotates "nonsynchronously, albeit slowly."

Extreme Seasons on Planets with Large, Constant Obliquity

Williams (1993) has argued that the obliquity of the Earth's spin axis was greater than 54 degrees until about 600 Myr ago, then decreased rapidly to close to its present value over a period on the order of 200 Myr. For obliquities > 54 degrees, the annual insolation is greater at the poles than at the equator (Ward 1974). Williams uses the assumed high obliquity to explain the widespread equatorial glaciation (and lack of polar glaciation) inferred from paleomagnetic data in the late Precambrian period 600–800 Myr ago (Harland and Rudwick 1964). Williams further argues that large obliquities are not conducive to the development of advanced life: for instance, for obliquities > 45 degrees, the tropics and Arctic overlap, with the result that the "strongly seasonal climate experienced at all latitudes would likely be too stressful for all but relatively primitive organisms." Williams claims, in fact, that the Cambrian explosion of life ~ 550 Myr ago occurred *because* the Earth's obliquity finally became small around that time.

Williams addresses the evidence for and against his interesting hypothesis in great detail. However, while a large initial obliquity is indeed likely if the Earth's spin results from a giant impact, I still find Williams' scenario unconvincing. First, the tidal evolution of the Earth-Moon system causes the Earth's obliquity to *increase* with time; if tidal torques dominate the evolution, the axial tilt has likely been only 10–20 degrees for most of the last 4 billion years (Goldreich 1966, Touma and Wisdom 1994). Williams therefore invokes core-mantle coupling to reduce the Earth's obliquity from its assumed large value. However, the mechanism by which the strength of this coupling increases (to decrease the obliquity from > 54 degrees) and then decreases again (as required to prevent driving the obliquity to 0) is unclear. Second, the Earth's magnetic field might not have been a simple dipole during the late Precambrian, so that low magnetic latitudes would not necessarily imply low geographical latitudes (Embleton and Williams 1986, Williams 1993). If the glaciation was not preferentially equatorial, the need to assume high obliquities disappears. Third, there are other plausible explanations for the rise of the Metazoa, such as the attainment of a critical level of atmospheric oxygen on the order of 10% of the present level (Cloud 1983, Gould 1994).

Extreme Seasons on Planets with Chaotic Obliquity Variations

Laskar et al. (1993b), extending the arguments of Ward (1982), argued that if the Earth had no Moon, the planet's obliquity would exhibit large, chaotic variations, "reaching more than 50 degrees in a few million years and even, in the long term, more than 85 degrees. This would probably have drastically changed the climate on the Earth." In a year-in-review article in *Discover* magazine, this finding was interpreted thus: "astronomers think Earthlike planets with moons as big as ours are rare. If a large moon is a prerequisite for the evolution of intelligent life, beings like us might turn out to be exceedingly rare as well" (Naeye 1994).

While the Moon's stabilization of the Earth's obliquity is unquestionable, the implication for life is dubious. First, large impacts like the event which apparently formed the Moon are a natural consequence of planet formation (Tremaine and Dones 1993) so that, pending observations of other solar systems, the probability that a "terrestrial" planet has a large moon must be reckoned as 1 in 4 or better. (Earth is the only planet in the "continuously habitable zone" of the Sun [Kasting et al. 1993], so that in our Solar System, the chance that a *habitable* planet has a large moon is 1 in 1.) Second, large chaotic obliquity variations are not inevitable for a planet without a large moon; for some initial rotation rates and obliquities, the variations are regular (Laskar and Robutel 1993, Fig. 5). While Mercury was probably doomed to suffer chaotic variations as solar tides slowed its rotation, a Moonless Earth that formed with an initial rotation period less than about 12 hours and an obliquity less than 25 degrees would not yet have entered the chaotic zone. Third, even planets with large obliquities may be habitable. Williams et al. (1996) calculated climate models for planets with Earthlike land/sea ratios and large obliquities, and found that unless all the land is near one pole, the planet is habitable as defined by Dole, even for obliquities near 90 degrees, because the oceans damp the seasonal temperature cycles.

As an ironic sidelight, the actual role of the Moon in stabilizing the Earth's obliquity is double-edged; as Ward (1982) states, "It would be misleading . . . to claim we owe our dynamic and climatic stability to the Moon. During its lifetime, the Moon has drawn considerable angular momentum from the Earth's spin via tides. Had the Moon never been present, the Earth's day would have been shorter and the increased rotational flattening of the Earth . . . would have largely compensated for the lack of a lunar torque. . . ." As the Moon recedes in the future, the Earth's spin-axis precession rate will decrease, ultimately driving the Earth into spin-orbit resonances 1–5 Gyr from now. As a result, the Earth's obliquity will vary from 0 to as much as 60 degrees over time scales of 100,000 years to one million years (Ward 1982), or even 0–90 degrees (Neron de Surgy and Laskar 1995), so that our Martian cousins will no longer envy our steady climate.

What Does This All Mean?

The terrestrial planets probably formed with spin rates about as fast as, or faster than, the Earth's current rate. The spins of Mercury, Venus, and, to a lesser extent, Earth, have been slowed by tides. Mars probably spins at close to its primordial rate. All of the terrestrial planets likely underwent large, chaotic obliquity variations in the past; Mars still experiences large swings of its spin axis, and Earth will again do so as the ever-more-distant Moon's influence weakens. As this example illustrates, the Earth's spin evolution is intimately tied to the Moon; what angular momentum the Moon gave, it now takes away.

In the past, when the Moon was closer to Earth, the day was shorter, and the height of the tides would have been greater in proportion to the inverse third power of the Moon's distance. Walker and Zahnle (1986) interpret a 23-year periodicity in the Australian banded iron formation as the Moon's nodal period 2.5 Gyr ago. If this analysis is correct, the tides would have been, on the average, only ~ 55% higher at that time. Compared with this modest increase, the changing arrangement of the continents has a larger effect on the strength of the tides at a given locale. As the continents move, the rate of energy dissipation in the Earth's oceans varies, so that the Moon's orbital distance cannot be calculated a priori at earlier times. However, when the Moon was much closer to Earth, its orbital evolution was very rapid, so it is unlikely that tides were drastically higher when life gained a firm foothold on Earth some 3.8–4.0 Gyr ago.

The rotation rate and obliquity of a planet could, in principle, affect its habitability by their effects on atmosphere circulation and climate. Jenkins et al. (1993) find that for a given solar insolation, a more rapidly rotating Earth is warmer. This effect may help to counteract the "faint early Sun" problem. Calculations by Williams et al. (1996) suggest that a wide range of obliquities is compatible with a habitable Earth. It is therefore premature to claim that the Earth spins in the best of all possible states.

Acknowledgments

I thank Tony Dobrovolskis for comments on the manuscript, and Laurance Doyle and Ellen Blue for their encouragement and patience.

REFERENCES

BILLS, B.G. (1990), "The Rigid Body Obliquity History of Mars," *JGR* **95:** 14137–14153.

BROECKER, W.S., AND G.H. DENTON (1990), "What Drives Glacial Cycles?," *Sci. Am.* **262 (1):** 49–56.

CAMERON, A.G.W., AND W. BENZ (1991), "The Origin of the Moon and the Single Impact Hypothesis IV," *Icarus* **92:** 204–216.

CLOUD, P. (1983), "The Biosphere," *Sci. Am.* **249 (3):** 176–189, 202.

DOBROVOLSKIS, A. (1980), "Atmospheric Tides and the Rotation of Venus. II. Spin Evolution," *Icarus* **41:** 18–35.

DOLE, S.H. (1964), *Habitable Planets for Man* (New York: Blaisdell Publishing Company).

DONES, L., AND S. TREMAINE (1993), "Why Does the Earth Spin Forward?," *Science* **259:** 350–354.

EMBLETON, B.J.J., AND G.E. WILLIAMS (1986), "Low Palaeolatitude of Deposition for Late Precambrian Periglacial Varvites in South Australia: Implications for Palaeoclimatology," *Earth Planet. Sci. Lett.* **79:** 419–430.

GOLDREICH, P. (1966), "History of the Lunar Orbit," *Rev. Geophys.* **4:** 411–439.

GOLDREICH, P. (1972), "Tides and the Earth-Moon System," *Sci. Am.* **226 (4):** 43–52, 118.

GOLDREICH, P., AND S.J. PEALE (1968), "The Dynamics of Planetary Rotations," *Ann. Rev. Astron. Astrophys.* **6:** 287–320.

GOLDREICH, P., AND S.J. PEALE (1970), "The Obliquity of Venus," *AJ* **75:** 273–283.

GOULD, S.J. (1994), "The Evolution of Life on the Earth," *Sci. Am.* **271 (4):** 85–91.

HABERLE, R.M., AND J. SCHAEFFER (1995), "Obliquity Experiments with a Mars General Circulation Model," *Bull. Amer. Astron. Soc.* **27:** 1066.

HABERLE, R.M., C.P. MCKAY, D. TYLER, AND R.T. REYNOLDS (1996), "Can Synchronously Rotating Planets Support an Atmosphere?," this volume.

HARLAND, W.B., AND M.J.S. RUDWICK (1964), "The Great Infra-Cambrian Ice Age," *Sci. Am.* **211 (2):** 28–36.

HARRIS, A.W., AND W.R. WARD (1982), "Dynamical Constraints on the Formation and Evolution of Planetary Bodies," *Ann. Rev. Earth Planet. Sci.* **10:** 61–108.

HIDE, R., AND J.O. DICKEY (1991), "Earth's Variable Rotation," *Science* **253:** 629–637.

IMBRIE, J., AND J.Z. IMBRIE (1980), "Modeling the Climatic Response to Orbital Variations," *Science* **207:** 943–953.

JENKINS, G.S., H.G. MARSHALL, AND W.R. KUHN (1993), "Precambrian Climate: The Effects of Land Area and the Earth's Rotation Rate," *JGR* **98:** 8785–8791.

KASTING, J.F., D.P. WHITMIRE, AND R.T. REYNOLDS (1993), "Habitable Zones Around Main Sequence Stars," *Icarus* **101:** 108–128.

LASKAR, J., AND P. ROBUTEL (1993), "The Chaotic Obliquity of the Planets," *Nature* **361:** 608–612.

LASKAR, J., F. JOUTEL, AND F. BOUDIN (1993a), "Orbital, Precessional, and Insolation Quantities for the Earth from −20 Myr to +10 Myr," *A&A* **270:** 522–533.

LASKAR, J., F. JOUTEL, AND P. ROBUTEL (1993b), "Stabilization of the Earth's Obliquity by the Moon," *Nature* **361:** 615–617.

LISSAUER, J.J., AND D.M. KARY (1991), "The Origin of the Systematic Component of Planetary Rotation. I. Planet on a Circular Orbit," *Icarus* **94:** 126–159.

LISSAUER, J.J., AND V.S. SAFRONOV (1991), "The Random Component of Planetary Rotation," *Icarus* **93:** 288–297.

NAEYE, R. (1994), "Moon of Our Delight," *Discover* **15 (1):** 72–74.

NERON DE SURGY, O., AND J. LASKAR (1995), "On the Long Term Evolution of the Spin of the Earth," *Bull. Amer. Astron. Soc.* **27:** 1172.

PEALE, S.J. (1986), "Consequences of Tidal Evolution," in *The Solar System: Observations and Interpretations,* M.G. Kivelson, ed. (Englewood Cliffs, NJ: Prentice-Hall), pp. 275–288.

PEALE, S.J. (1989), "Some Unsolved Problems in the Evolutionary Dynamics of the Solar System," *Celest. Mech. Dyn. Astron.* **46:** 253–275.

POLLACK, J.B. (1979), "Climatic Change on the Terrestrial Planets," *Icarus* **37:** 479–553.

TOON, O.B., J.B. POLLACK, W. WARD, J.A. BURNS, AND K. BILSKI (1980), "The Astronomical Theory of Climate Change on Mars," *Icarus* **44:** 552–607.

TOUMA, J., AND J. WISDOM (1993), "The Chaotic Obliquity of Mars," *Science* **259:** 1294–1297.

TOUMA, J., AND J. WISDOM (1994), "Evolution of the Earth-Moon System," *AJ* **108:** 1943–1961.

TREMAINE, S., AND L. DONES (1993), "On the Statistical Distribution of Massive Impactors," *Icarus* **106:** 335–341.

WALKER, J.C.G., AND K.J. ZAHNLE (1986), "Lunar Nodal Tide and Distance to the Moon During the Precambrian," *Nature* **320:** 600–602.

WARD, W.R. (1973), "Large-Scale Variations in the Obliquity of Mars," *Science* **181:** 260–262.

WARD, W.R. (1974), "Climatic Variations on Mars. I. Astronomical Theory of Insolation," *JGR* **79:** 3375–3386.

WARD, W.R. (1982), "Comments on the Long-Term Stability of the Earth's Obliquity," *Icarus* **50:** 444–448.

WARD, W.R. (1992), "Long-Term Orbital and Spin Dynamics of Mars," in *Mars,* H. Kieffer, B. Jakosky, and C. Snyder, eds. (Tucson: University of Arizona Press), pp. 298–320.

WARD, W.R., J.A. BURNS, AND O.B. TOON (1979), "Past Obliquity Oscillations of Mars: The Role of the Tharsis Uplift," *JGR* **84:** 243–259.

WILLIAMS, D.M., J.F. KASTING, AND K. CALDEIRA (1996), "Chaotic Obliquity Variations and Planetary Habitability," this volume.

WILLIAMS, G.E. (1993), "History of the Earth's Obliquity," *Earth Science Reviews* **34:** 1–45.

YODER, C.F. (1995), "Venus' Free Obliquity," *Icarus* **117:** 250–286.

Spacing of Planetary Orbits About Stars of Differing Masses

Jack J. Lissauer

Astronomy Program
Department of Earth and Space Sciences
State University of New York
Stony Brook, NY 11794-2100

Abstract

The number of planets expected to orbit within the habitable zones of single stars can be estimated using a combination of models of planet formation and orbital stability requirements. The frequency of such planets generally increases slowly as the ratio of the masses of the planets to that of the star decreases.

The orbital spacing of planets is an important factor in determining how many planets are likely to exist within habitable zones. Although modern theories of planetary growth do not yield deterministic "Bode's Law" formulae for the orbits of planets (Wetherill 1990, 1996), characteristic orbital spacings do exist, and they depend on the masses of the star and planets in the system.

The nearly planar and circular orbits of the planets in our Solar System argue strongly for planetary formation within flattened circumstellar disks (Kant 1755, Laplace 1796). Terrestrial planets are believed to grow by pairwise accretion of solid bodies until the spacing of planetary orbits becomes large enough that the configuration is stable for the age of the system (Safronov 1969, Wetherill 1990).

The initial stages of the formation of gas giant planets probably also involved binary agglomeration of solid bodies, with accretion of significant quantities of gas initiated once the heavy element core had acquired enough mass to gravitationally trap hydrogen and helium, the two cosmically most abundant elements, which together are believed to make up ~ 99% of the mass of protoplanetary disks. Issues involving the ultimate sizes and spacings of gas giant planets are complex and poorly understood (Lissauer et al. 1995), but as described below, nearby planets are usually the strongest perturbers, so the details of giant planet sizes and spacings probably do not substantially affect the spacings of planets within habitable zones unless the giant planets themselves orbit within or near these zones.

The remainder of this article will concentrate on the spacing of the orbits of terrestrial-type planets. The criteria for orbital stability will be examined first, and then applied to derive scaling laws for the orbital spacing and masses of terrestrial-type planets as a function of stellar mass and the surface mass density of solids within a protoplanetary disk. See Lissauer (1993) for a general review of models of planet formation.

The simplest and best understood aspects of the stability of planetary orbits are those that can be studied using the planar circular restricted three-body problem, in which all bodies are assumed to orbit within the same plane, one planet is assumed to be massless, and the other planet orbits the star on a circular orbit. If the massless body is initially on a circular orbit far from the longitude of the massive planet, it can be shown that the bodies will never experience close approaches, provided that the ratio of the separation in semimajor axes, Δa, to the planet's orbital radius, a, satisfies

$$\frac{|\Delta a|}{a} \geq 2.4 \left(\frac{M_p}{M_\star}\right)^{1/3}, \tag{1}$$

where M_p and $M_\star$ are the masses of the planet and star, respectively. When both planets have nonzero mass, Equation 1 holds, provided M_p is taken to be the sum of the planetary masses (Gladman 1993).

Deterministic chaos also plays an important role in the stability of planetary orbits. Trajectories tend to be chaotic whenever resonances overlap. Strong resonances are more closely spaced near a planet, and resonances overlap over a range in semimajor axis centered on the perturbing planet and having a half-width given by

$$\frac{|\Delta a|}{a} \geq 1.5 \left(\frac{M_p}{M_\star}\right)^{2/7} \tag{2}$$

(Duncan et al. 1989).

Note that although the minimum separation for stability in two-planet systems increases with planetary mass, the growth rate is far less than linear. Thus, if orbital stability is determined primarily by perturbations from the nearest planet, then the merger of planets on adjacent orbits will usually make the system more stable.

Eccentricities and inclinations can also affect stability. As both eccentricity and inclination require excess energy, they both give the systems more freedom of motion, and thus allow for greater instability. However, although eccentricities generally destabilize systems because bodies approach each other more closely, inclinations more often than not increase (at least short-term) stability, because the bodies tend to remain farther apart. Retrograde orbits are known to be more stable than prograde ones under most circumstances (Huang and Innanen 1983). Resonances can either increase orbital stability (e.g., the Hilda asteroids, which are locked in a 3 : 2 mean motion commensurability with Jupiter), or destabilize orbits, such as at the 3 : 1 Kirkwood gap (Wisdom 1982).

The rapid advance in computing hardware and algorithms over the past few decades has led to increasingly realistic simulations of the dynamical stability of the planets and test particles within our Solar System and of artificial planetary systems. Planetary orbits appear to be chaotic, but no close approaches or orbit crossings have been found among the eight major planets (Sussman and Wisdom 1992). Test particles appear to be cleared from most regions between the planets (other than the asteroid belt) in time scales less than the age of the Solar System. Particles are rapidly depleted from some zones but remain much longer in other regions (Holman and Wisdom 1993); thus, stability considerations depend on the time scale with which one is concerned.

Stability criteria for systems with several planets are not fully understood. However, scaling laws based on generalizations of Jacobi-Hill stability and the resonance overlap criterion to many-body systems provide remarkably consistent fits when applied to the planets in our Solar System as well as to planetary satellite systems (Lissauer 1995).

Perturbations from several planets add in complex, nonlinear ways. Minimum separations for stability are larger than for two planet systems, but the dependence on approximately the 2/7 to 1/3 power of the mass ratio appears to be preserved (Lissauer 1995).

Equations 1 and 2 can thus be used to scale the spacing of the terrestrial planets in our Solar System to characteristic spacings expected for other planetary systems in which the stars and planets have different masses. Using the Jacobi-Hill stability criterion or the resonance overlap criterion, one obtains

$$\frac{|\Delta a|}{a} \propto \left(\frac{M_p}{M_\star}\right)^{1/3}, \tag{3}$$

or

$$\frac{|\Delta a|}{a} \propto \left(\frac{M_p}{M_\star}\right)^{2/7}, \tag{4}$$

respectively, where M_p is the characteristic mass of the planets in the region.

These scalings can be extended back to the properties of the disk out of which the planets formed. Although the latter stages of terrestrial planet growth involved mixing of planetesimals from various regions of the protoplanetary disk, for determining characteristic planetary masses it suffices to think of terrestrial planets as accreting all of the condensed material in an annulus centered on their orbits and having width equal to the separation between planetary orbits, Δa. Thus,

$$M_p \approx 2\pi a \Delta a \sigma, \tag{5}$$

where σ is the surface mass density of solids at the distance α from the star.

Combining Equation 5 with Equations 3 and 4, respectively, one obtains:

$$M_p \propto \frac{a^3 \sigma^{3/2}}{M_\star^{1/2}} \tag{6a}$$

$$\Delta\alpha \propto \frac{a^2 \sigma^{1/2}}{M_\star^{1/2}} \tag{6b}$$

and

$$M_p \propto \frac{a^{14/5} \sigma^{7/5}}{M_\star^{2/5}}, \tag{7a}$$

$$\Delta\alpha \propto \frac{a^{9/5} \sigma^{2/5}}{M_\star^{2/5}}. \tag{7b}$$

The above scalings can be used to estimate the number of planets in the habitable zones of single stars of differing masses from the distribution of planets observed in our Solar System. If the width of a habitable zone is determined by the total flux per unit area of stellar radiation impinging on a planet, then the fractional width, $\delta a/a$, of a habitable zone is independent of stellar luminosity (and thus also of stellar mass). Kasting (1996) estimates that the width of the habitable zone within our Solar System is approximately equal to the spacing between terrestrial planets; thus it is not surprising that one habitable planet orbits the Sun. For planets of a given mass, the number that can lie within a given logarithmic interval increases slowly with stellar mass (Eqns. 3 and 4). Smaller planets may be more closely spaced (Eqns. 3 and 4). Increasing the surface mass density of a protoplanetary disk results in more massive planets (Eqns. 6a and 7a), but they are separated by greater distances (Eqns. 6b and 7b).

The formation of planets in multiple star systems is not well understood (Lissauer 1989). Assuming they were able to accrete, planets could remain in stable orbits within habitable zones provided the binary orbit was either at least several times larger or at least several times smaller than those of the planets (Pendleton and Black 1983). However, perturbations from the binary would usually lead to more chaotic orbits and therefore to greater spacings between the planets, unless the scale of the binary orbit differed from that of the planets by a significantly larger amount.

ACKNOWLEDGMENTS

It is my pleasure to acknowledge valuable comments by Julia Espresate. This research was supported in part by the NASA Planetary Geology and Geophysics Program under grant NAGW-1107.

REFERENCES

DUNCAN, M., T. QUINN, AND S. TREMAINE (1989), "The Long-Term Evolution of Orbits in the Solar System: A Mapping Approach," *Icarus* **82:** 402–418.

GLADMAN, B. (1993), "Dynamics of Systems of Two Close Planets," *Icarus* **106:** 247–263.

HOLMAN, M.J., AND J. WISDOM (1993), "Dynamical Stability in the Outer Solar System and the Delivery of Short Period Comets," *AJ* **105:** 1987–1999.

HUANG, T.-Y., AND K.A. INNANEN (1983), "The Gravitational Escape/Capture of Planetary Satellites," *AJ* **88:** 1537–1548.

KANT, I. (1755), *Allegmeine Naturgeschichte und Theorie des Himmels* (Konigsburg und Leipzig: Johann Friederich Petersen). English translation: W. Hastie, "Universal Natural History and Theories of the Heavens," in *Kant's Cosmology* (New York: Greenwood Publishing), 1968.

KASTING, J.F. (1996), "Habitable Zones Around Stars: An Update," this volume.

LAPLACE, P.S. (1796), *Exposition du Système du Monde* (Paris: Cercle-Social). English translation: J. Pond, *The System of the World* (London: Richard Phillips), 1809.

LISSAUER, J.J. (1989), "Which Stars Have Planets?," in *The Formation and Evolution of Planetary Systems*, H.A. Weaver and L. Danly, eds. (Cambridge: Cambridge University Press), pp. 304–308.

LISSAUER, J.J. (1993), "Planet Formation," *Ann. Rev. Astron. Astrophys.* **31:** 129–174.

LISSAUER, J.J. (1995), "Urey Prize Lecture: On the Diversity of Plausible Planetary Systems," *Icarus* **114:** 217–236.

LISSAUER, J.J., J.B. POLLACK, G.W. WETHERILL, AND D.J. STEVENSON (1995), "Formation of the Neptune System," in *Neptune and Triton*, D. Cruikshank and M.S. Matthews, eds. (Tucson: University of Arizona Press), pp. 37–108.

PENDLETON, Y.J., AND D.C. BLACK (1983), "Further Studies on Criteria for the Onset of Dynamical Instability in General Three-Body Systems," *AJ* **88:** 1415–1419.

SAFRONOV, V.S. (1969), *Evolution of the Protoplanetary Cloud and Formation of the Earth and Planets* (Moscow: Nauka Press). English translation: NASA TTF-677, 1972.

SUSSMAN, G.J., AND J. WISDOM (1992), "Chaotic Evolution of the Solar System," *Science* **257:** 56–59.

WETHERILL, G.W. (1990), "Formation of the Earth," *Ann. Rev. Earth Planet. Sci.* **18:** 205–256.

WETHERILL, G.W. (1996), "The Formation of Habitable Planetary Systems," this volume.

WISDOM, J. (1982), "The Origin of the Kirkwood Gaps: A Mapping for Asteroidal Motion Near the 3 : 1 Commensurability," *AJ* **87:** 577–593.

A SEARCH FOR EARTH-SIZED PLANETS IN HABITABLE ZONES USING PHOTOMETRY

David Koch and William Borucki

NASA Ames Research Center
Moffett Field, CA 94035

Abstract

Several methods have been suggested for the detection of extrasolar planets. In the habitable zone, the radial velocity and astrometric methods are sensitive only to the detection of massive planets (greater than about ten to one hundred Earth masses). However, the photometric method is sensitive to Earth-sized planets in inner orbits about solarlike stars. The methodology for conducting a photometric space-based search is described. The information that can be directly and indirectly derived from the observations about any planetary systems discovered is discussed.

INTRODUCTION

Detection of extrasolar short period planets, particularly if they are in the liquid-water zone, would be one of the most exciting discoveries of our lifetime. Several methods for the detection of extrasolar planets have been suggested (TOPS 1992 and Schneider 1996). These methods include the use of astrometry, radial velocity measurements and photometry. The radial velocity and astrometric methods are sensitive only to the detection of massive planets (greater than about ten to one hundred Earth masses) in the habitable zone as shown in Figure 1. On the other hand, photometry is sensitive to Earth-sized planets (ESP) in inner orbits about solarlike stars (Rosenblatt 1971, Borucki and Summers 1984, Borucki et al. 1985, and Schneider and Chevreton 1990), as shown in Figure 2. These planets and their locations in the habitable zone are the ones of interest at this conference.

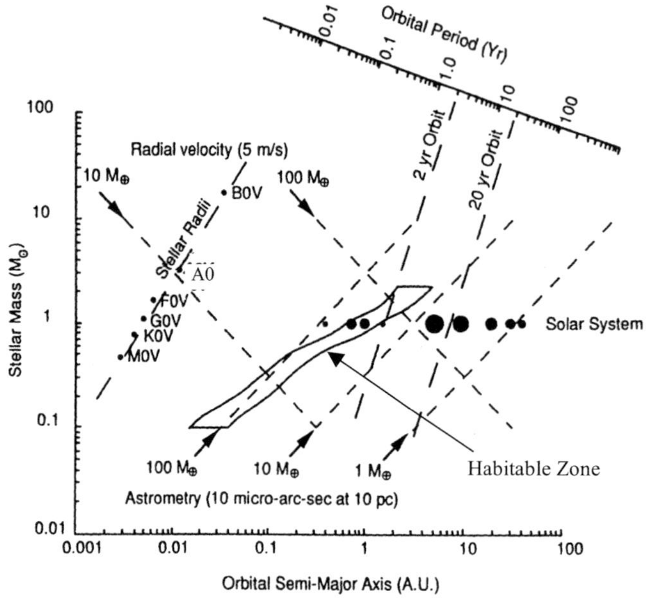

FIG. 1. *Minimum detectable planetary mass for the astrometric and radial velocity methods. The figure depicts the size and mass of various spectral-type dwarf stars. For these stars, any point on the plot defines a planetary mass, orbital semimajor axis and orbit period. The planets of the Solar System are illustrated. The habitable zone (see Kasting et al. 1993) is indicated by the outlined region. The contours for the minimum detectable planetary masses are indicated for the radial velocity and astrometric methods.*

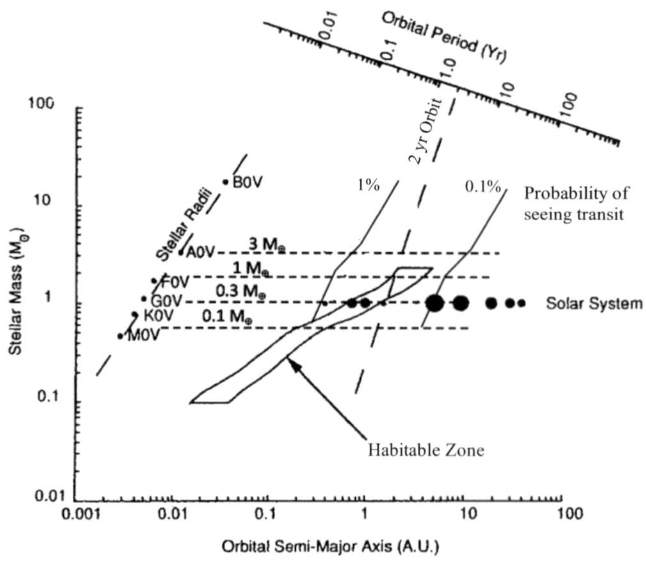

FIG. 2. *Minimum detectable planetary mass for the photometric method. The plot characteristics are the same as in Figure 1. The contours for the minimum detectable planetary masses are indicated for the photometric method. The sensitivity is based on the method's being capable of detecting a characteristic change of light of one part in 25,000 due to a planetary transit. The geometric random probability for detection is the stellar diameter divided by the orbital diameter. Contours for this probability are indicated.*

PROPERTIES OF TRANSITING PLANETS

Given a random orientation of orbital plane alignments with the line of sight to a star, the probability for proper alignment is the stellar diameter (D_*) divided by the planet's orbital diameter (D_p) as shown in Figure 3. Assuming our Solar System to be typical (having two ESP whose orbits are not coplanar to within an angle of $2D_*/D_p$), one would expect 1% of the stars monitored to exhibit planetary transits. To obtain a statistically significant result, several thousand stars need to be monitored, resulting in detection of about one transit per week throughout the mission. F, G, and K dwarfs are the most promising candidate stellar types.

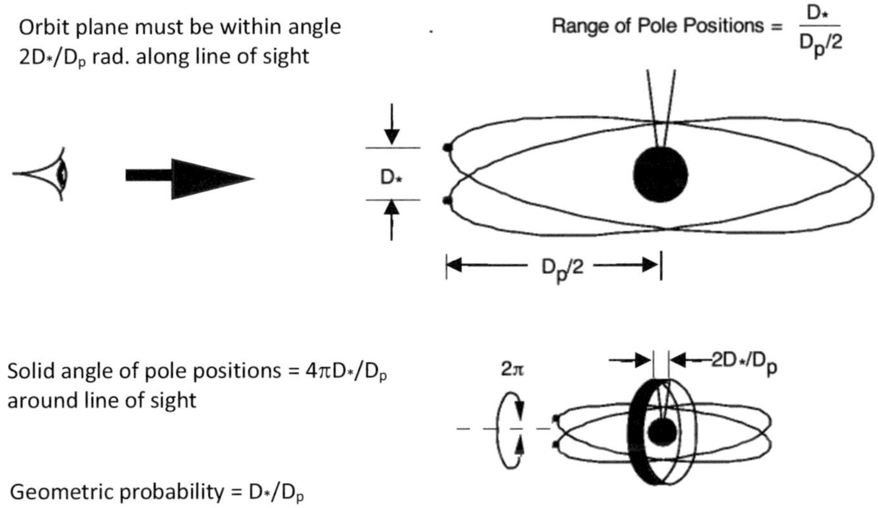

FIG. 3. *Geometric probability for transit sightings. For a transit to be observable, the planet must pass in front of the stellar disk; that is, the pole of the orbit must be within the angle* $D_*/(D_p/2)$, *where* D_* *is the stellar diameter and* $D_p/2$ *is the planetary orbital radius. This is true for all* 2π *angles of pole positions about the line of sight or* $4\pi D_*/D_p$ *steradians. Thus the random probability for seeing transits is* $p = D_*/D_p$.

ESP transiting solarlike stars will cause a decrease in the apparent luminosity of the star by one part in 12,000 with a duration of about thirteen hours for a nearly equatorial crossing. Planets as far out as Mars will have equatorial transit times of sixteen hours and those as close in as Mercury can be detected with a grazing transit time as short as four hours. The equatorial transit time is given by

$$\tau = D_* \sqrt{(R/GM_*)}, \tag{1}$$

where D_* is the stellar diameter, R is the planetary orbital radius and M_* is the stellar mass. For the Solar System, $\tau = 13.04\sqrt{R}$ hours, with R in AU. Not all transits will cross at the equator. However, 87% of the transits will have durations greater than half the equatorial transit time.

Discovery of a planet will require detection of multiple transits of the same amplitude, duration and correct period of occurrence. That is, after detection of two transits of the same amplitude and duration, the occurrence of a third and subsequent transits can be predicted to within a matter of hours. Planets in inner orbits will have periods of 0.3 to 2 years. Thus, a mission life of six years is required to detect a minimum of three transits by planets with orbits as large as that of Mars.

Mission Description

An instrument optimized for photometric observing of ESP around solarlike stars has the following features:

- It must have a photometric precision of ~ 1:36,000 (1 σ) per hour.
- It must be above the Earth's atmosphere, to avoid transparency variations that limit the photometric precision to less than 1:1000 over a large field of view.
- It must constantly view the same star field without interference from the Earth-limb or the Sun so as to obtain a continuous record on time scales from an hour to years. A 15,000 km 10° inclination orbit has been selected with the telescope pointed in a region close to the equatorial and ecliptic polar.
- It must simultaneously monitor several thousand stars. A one-meter-class telescope with a large field of view (on the order of 10° diameter) is required. A Schmidt telescope design is currently being evaluated and a focal plane array of CCDs to perform the simultaneous monitoring is being tested to demonstrate that CCDs can meet the photometric precision requirement.
- It must be able to efficiently extract the brightnesses for the selected set of stars for data processing to be manageable.

Specific Details from Each Planetary System Detected

Data from the photometric observations can provide information on the character of each planetary system detected. The basic information will be the orbital period, the depth of the transit and the transit duration, as well as the orbit plane inclination to a fraction of a degree.

The orbital period and stellar mass determine the orbital radius, where the stellar mass is determined from the spectral type and luminosity class of the star. From this one will know immediately if the planet is in the habitable zone for the given stellar type (see Fig. 2) (Kasting et al. 1993, Kasting 1996, Whitmire and Reynolds 1996).

From the transit depth the ratio of the planetary diameter to the stellar diameter is determined. The spectral type and luminosity class of the star define the stellar diameter and thereby determine the size of the planet detected. Presuming the planet to be terrestrial (density of about 5 g/cm^3) and given its size, a good approximation can be made for the mass of the planet and the value of its surface gravity.

From the stellar absolute magnitude and orbital radius, the stellar flux at the planet can be calculated. Given a range of albedos and assumptions on the amount and type of possible atmosphere, a range of surface temperatures can be estimated. Stellar activity data from this mission will provide further insight into fluctuations of the radiation that might affect the planetary climate.

If the planet is in a binary stellar system there are a number of other parameters to consider. From the timing of the transits with a close binary pair, the orbital eccentricity can be determined. Not only are the positions of the stars important in determining the amount of radiation reaching the planet, but also the variation in distance to each of the stars will affect the climate on the planet. The daylight variations can be calculated for the case of the planet in orbit around a close binary pair or the synodical variations can be calculated for an open binary system where the planet is in orbit around one of the pair. During part of the year in an open binary system when the planet is between the two stars, one can expect "night" to be as bright as fifty full Moons (stellar separation of 100 AU) to as much as one-tenth that of high noon (separation of 4 AU).

Results from the Ensemble of Systems

In addition to the information available about each specific planetary system detected, the ensemble of information from all of the planets detected will further our understanding of planetary formation theories. For the first time one will be able to derive the frequency of ESP for F, G and K dwarf stars. Prior to these measurements, it has commonly been assumed that this frequency is close to one. If only a few or no planets are detected then one must conclude that ESP are rare, the models of planetary system formation need to be substantially revised, that the origin of the Earth must be reconsidered, and that Earth itself might be unique. The value for the frequency of ESP can be used to predict the more general parameter for the frequency of planetary formation used in the Drake equation (Borucki and Koch 1995).

The ensemble of photometric observations should provide size and radial distribution information that can be used to test calculations of planetary formation, such as those of Wetherill (1991, 1996). Those calculations indicate that

- No planets much larger than Earth are expected for orbital radii between 0.3 and 2 AU.
- These planets can occupy any position within these radii.
- The distribution of planets will be affected by the presence or absence of a Jovian planet.
- Two ESP are expected within these radii.
- The orbits should be nearly coplanar.

The results on the size and radial distribution of the detected planets will address these hypotheses.

During the six-year mission, approximately one massive outer planetary transit should be seen. Ground-based radial velocity observations are capable of detecting other massive planets in the systems, provided they are in inner orbits. Follow-on ground-based photometry, even after the end of the mission, can confirm as well as search for massive planets in these systems. The depth of a Jovian transit is large, about 1%, and the chances for proper alignment along the line of sight are good, about 10%, since the orbit planes of the systems will be known to be close to the line of sight of the observer. These observations will help to further revise and refine models for planetary formation.

The coplanarity of orbits can be determined from the frequency of detection of multiple planetary systems (MPS). The probability for seeing multiple planets is given by

$$P_{21} = \frac{2}{\pi} \sin^{-1} \frac{D_*}{D_p \sin\phi_i}, \tag{2}$$

provided $D_*/D_p < \sin\phi_i$, where ϕ_i is the relative orbital plane inclinations and D_* and D_p are as before and as illustrated in Figure 4. When $\sin\phi_i < D_*/D_p$, the probability is near or equal to unity. For the Earth-Venus case, there would appear to be a 9.4% chance and for the Mars-Earth case there would be a 15% chance for an outside observer seeing transits by both planets. The frequency of MPS will provide a good measure of their coplanarity—that is, if the detection of MPS is high, then ϕ_i must typically be less than D_*/D_p. Whereas if the detection of MPS is low, then ϕ_i must typically be greater than D_*/D_p or few systems have more than one ESP.

About half of the stars being monitored will be in binary stellar systems. However, binaries do not necessarily preclude planetary formation and detection (Brandmeier and Doyle 1996). Theoretical modeling has demonstrated that long-lived planetary orbits exist when the planet's orbital radius is 3.3 times greater than the stellar separation or the stellar separation is 3.3 times greater than the planet's orbital radius (Pendleton and Black 1983). In the open binary case, planetary systems are even possible around both stars. Results from this mission should determine the frequency of ESP in binary systems.

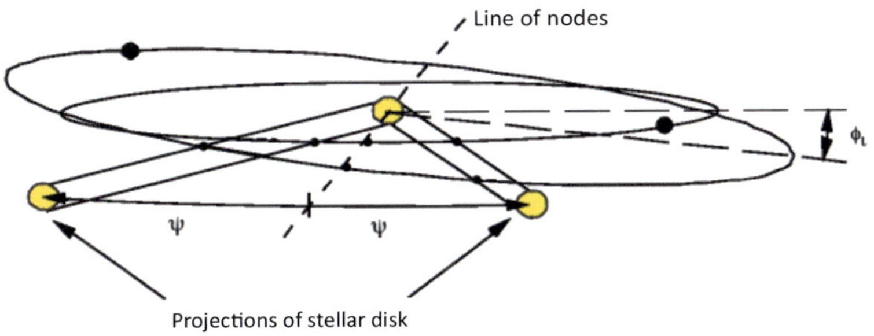

$\Psi = \sin^{-1} (D_*/D_p \sin\phi)$	Angle from Line of Nodes for which both transits can be seen D_*/D_p is the stellar angular radius viewed from a planet
$P = 4\Psi/2\pi$	Total probability for seeing two planets on either side of both nodes
$P_{EV} = 0.094$	Probability for Earth-Venus case
PME = 0.15	Probability for Mars-Earth case

FIG. 4. *Geometry for multiple planetary sightings. When the observer is viewing a transit near the line of nodes for two inclined planetary orbits, then transits for both planets will be seen. If the relative inclination is small compared with the apparent stellar angular diameter* ($2D_*/D_p < \phi_i$) *as viewed from a planet, then it is likely that transits from both can be seen.*

SUMMARY

The photometric method for planetary detection can provide for both the discovery of planetary systems and a direct and an indirect measurement of many of the properties of the planetary conditions. In particular, it can provide information about the frequency of ESP in the habitable zone about solarlike stars, orbital radius and period, stellar flux at the planet and stellar variability. Further, from the ensemble of measurements, one can test models for planetary system formation, including the cases for binary star systems. If few or no planets are detected, then theories of stellar and planetary formation need to be revised, the origin of the Earth must be reconsidered, the concept of the abundance of life in the Universe will have to be reexamined, and the Earth might be found to be unique.

References

BORUCKI, W.J., AND A.L. SUMMERS (1984), "The Photometric Method of Detecting Other Planetary Systems," *Icarus* **58:** 121–134.

BORUCKI, W., AND D. KOCH (1995), "Comparing the Expectations for Different Planetary Search Methods," in *Progress in the Search for Extraterrestrial Life*, ASP Conf. Ser., S. Shostak, ed., **74:** 173–180.

BORUCKI, W.J., J.D. SCARGLE, AND H.S. HUDSON (1985), "Detectability of Extrasolar Planetary Transits," *ApJ* **291:** 852–854.

BRANDMEIER, S., AND L.R. DOYLE (1996), "Simulation of Planetary Transits in Eclipsing Binary Star Systems," this volume.

KASTING, J.F. (1996), "Habitable Zones Around Stars: An Update," this volume.

KASTING, J.F., D.P. WHITMIRE, AND R.T. REYNOLDS (1993), "Habitable Zones Around Main Sequence Stars," *Icarus* **101:** 108–128.

PENDLETON, Y.J., AND D.C. BLACK (1983), "Further Studies on Criteria for the Onset of Dynamical Instability in General Three-Body Systems," *AJ* **88:** 1415–1419.

ROSENBLATT, F. (1971), "A Two-Color Photometric Method for Detection of Extra-Solar Planetary Systems," *Icarus* **14:** 71–93.

SCHNEIDER, J. (1996), "The Search for Habitable Extrasolar Planets: Current Programs and Long-Term Plans," this volume.

SCHNEIDER, J., AND M. CHEVRETON (1990), "The Photometric Search for Earth-Sized Extrasolar Planets by Occultation in Binary Systems," *A&A* **232:** 251–257.

TOPS (1992), *Toward Other Planetary Systems*, NASA.

WETHERILL, G.W. (1991), "Occurrence of Earth-like Bodies in Planetary Systems," *Science* **253:** 535–538.

WETHERILL, G.W. (1996), "The Formation of Habitable Planetary Systems," this volume.

WHITMIRE, D.P., AND R.T. REYNOLDS (1996), "Circumstellar Habitable Zones: Astronomical Considerations," this volume.

[Publisher's note: It was Bill Borucki's vision and persistence in pursuing funding for the search for extrasolar earth-sized planets that led to the Kepler mission and the discovery of Kepler 16b (the first planet ever discovered that orbits two stars), as well as many others. Our understanding of the frequency of planets around other stars increased dramatically after the launch of Kepler. Dave Koch and Bill Borucki headed the Kepler mission. *See also* Laurance R. Doyle et al., "Kepler 16: A Transiting Circumbinary Planet," *Science*, **333 (6049):** 1602–6, published in September 2011.]

Ignition of a Delta II 7925-10L Rocket with the Kepler Space Telescope on Board, March 6, 2009

Photo Credit: NASA/Sandra Joseph, Kevin O'Connell
https://commons.wikimedia.org/wiki/File:Ignition_of_Kepler%27s_Delta_II_7925-10L.jpg
[retrieved 10-31-25]

On March 6, 2009 at 10:49:57 p.m. EST, Kepler blasted off into the darkness aboard a Delta II rocket from Launch Complex 17-B at Cape Canaveral Air Force Station in Florida on a quest to find potentially habitable worlds.

The North America Nebula (NGC 7000) and the Pelican Nebula (IC 5070) in Cygnus

Photo Credit: Gil-Estel
https://commons.wikimedia.org/wiki/File:North_America_2007_09_08-14.jpg
[retrieved 12-18-25]

NGC 7000 or Caldwell 20 is a stunning example of an emission nebula. The clouds of gas that make up the nebula are ionized by a nearby star, causing the gas to glow as it emits energy. Colors emitted by emission nebulae depend on the chemical composition of the region. Sensitive cameras can pick up the reddish color that is characteristic of hydrogen, which dominates C20.

NASA's Hubble Wide Field and Planetary Camera 2 is part of a survey of T Tauri stars in nearby star-forming regions. T Tauri stars are young stars that vary in brightness over time. One reason this kind of star is important to astronomers is that a large portion of T Tauri stars have been found to host circumstellar disks. These disks of gas and dust may one day become solar systems.

J. Schneider runs the Encyclopaedia of Exoplanetary Systems database (https://exoplanet.eu/home/), which as of April 30, 2026 lists over 6,000 confirmed exoplanets and over 8,000 extrasolar objects.

THE SEARCH FOR HABITABLE EXTRASOLAR PLANETS: CURRENT PROGRAMS AND LONG-TERM PLANS

J. Schneider

Observatoire de Paris
92195 Meudon, France

Abstract

A comparative review is made of different techniques in the search for habitable extrasolar planets and of the planet parameters that can be derived in each method. Current searches and future projects are listed; special attention is paid to the possibility of detecting terrestrial planets from the ground.

INTRODUCTION

Theoretical discussions about circumstellar habitable zones are of great importance as they provide guidelines for their observational search. But it is also important to try to actually detect them, since their observation would certainly constrain the models. I will assume that habitable zones mean in fact habitable planets; one could

Note: Updates on this paper are accessible on the World Wide Web at the url: http://exoplanet.eu/. *See also* https://observatoiredeparis.psl.eu/jean-schneider-record-keeper-of-the-extrasolar-2770.html think of more exotic places like asteroid belts, but Huang (1960) has shown that a planetary body must be larger than the Moon to be habitable.

Until now only moderate efforts have been made in the search for habitable planets. In fact, the question of their existence has recently taken on very great importance: in a spectroscopic search of extrasolar planets around 21 solar-type stars, no single star was found to have a Jupiter-like companion (Walker et al. 1995). This result immediately leads to a further question—are planetary systems rare or are they on the contrary common but without Jupiters? This question is directly related to the present subject since, as will be addressed in the next section, habitable planets cannot be Jupiters.

The aim of this paper is to review all the conceivable methods of detection and to present the current efforts under way. In the following section all the possible detection methods that we can think of at present are discussed. Not all of them are practicably implementable. This is followed by an account of efforts being undertaken around the world.

What It Is Conceivable to Do

The potential success of a given detection method depends naturally on its technological limitations, but also on the different characteristics of the planets—their mass M, radius R, temperature T, and distance a from the parent star (I assume circular orbits). A preliminary question arises whether or not a habitable planet can lie anywhere in the (M, R, T, a) space. In fact there are regions in this space that are likely to be forbidden for the following reasons.

A General Constraint

Planets are divided into two categories: solid and gaseous. It is an empirical fact that all the solid planets or planetary cores in the Solar System have a mass of at most fifteen times the Earth's mass (Guillot et al. 1994). In the PSR 1257+12 system, the masses of the planets are three times the Earth's mass or less (Wolszczan 1994). In the absence of any other information, I will therefore take the conservative view that while a gaseous planet can have a Jupiter size and mass (or more), a solid extrasolar planet has at most fifteen Earth masses or ≈ 2.5 Earth radii.

On the other hand, a planet with an albedo A located at a distance a from a star with a radius R_* and a temperature T_* acquires a temperature T given by

$$T = \frac{(1-A)^{1/4}\, T_*}{\sqrt{2}} \left(\frac{R_*}{a}\right)^{1/2} \qquad (1)$$

Therefore, for a gaseous (Jupiter-like) planet, defined as having a temperature $T \leq$ 200 K to prevent hydrogen evaporation, one has:

$$a > a_{gas} = \frac{R_*}{2}\left(\frac{T_*}{\approx 200\ K}\right)^2 \quad (2)$$

This constraint leads to specification of two categories of planets—the massive planets ($M \geq 15\ M_\oplus$) are at the same time outer planets (defined by Eqn. 2), and therefore Earthlike planets ($M \leq 15\ M_\oplus$) are at the same time inner planets ($a < a_{gas}$). If one accepts the idea that a habitable planet is defined as having a temperature of 300 K to allow for liquid water, the inner planets are at the same time planets that can have a temperature appropriate to the development of a kind of biochemical activity.

The Observables in Each Detection Method

Consider a planetary system located at a distance D from the Solar System. As is well known, the presence of the planet, located at an angular distance $\Delta\theta$ from the star, can be revealed either directly by detecting its direct emission (by reflection of starlight with a magnitude difference Δm, or by its infrared excess F_{IR} peaking at λ_{max}), or indirectly by observing the reflex motion of the star; this reflex motion leads to a variation in angular position $\Delta\alpha$ and in radial velocity ΔV_R. In addition to these straightforward classical effects, the presence of the planet can be revealed by other means.

(1) If the star can be considered as a clock, like a pulsar, its reflex motion due to the presence of the planet modulates the time of arrival ΔT of the clock signals. Pulsars are generally considered hostile environments for life, but this point deserves further investigation. For instance, lethal radio, X and gamma ray radiations may be damped by a sufficiently thick atmosphere, as long as the pulsar beam does not directly hit the planet.

(2) If the star has modulations in its light curve (flares, etc.), the planet presents, by illumination, the same modulations with a delay DT and a relative amplitude R^2/a^2. An autocorrelation study of the star's light curve could reveal this echo (Argyle 1974).

(3) If the orbital plane of the planet is correctly oriented, it produces a drop η in the starlight during transits (duration $D_{occ} \approx$ 1–20 hrs). For a random orientation of the planet orbital plane, the geometric probability of the transit is $p = R_*/a$ for a single star. For an eclipsing binary this probability can amount to 100% since it is likely that the binary and planet orbital planes are identical. In the latter case the transit is double (Schneider and Chevreton 1990).

(4) The planet can produce a gravitational amplification A_G of the light of background stars with a duration T_G depending on its transverse velocity V (Mao and Paczynski 1991).

(5) Finally, the planet can, like Jupiter or Saturn, have an intrinsic radio emission with a flux F_R and could be searched for by looking for this emission (Lecacheux 1990). With the present state of knowledge, the latter is possible for any kind of planet and thus is not necessarily related to other planet characteristics such as a or M; for instance, in the Solar System, the Earth is brighter than Uranus and Neptune at low radio frequencies (Lecacheux 1994). It is not clear whether a large radio emission of the planet is incompatible with the presence of life.

Most of the observable quantities just mentioned are modulated by a phase factor $\phi(t)$ with a period P due to the orbital revolution of the planet. They are related to the intrinsic characteristics of the planetary system by the relations shown in Table 1.

The numbers in the right column are for an Earth-sized planet with an albedo one at a = one AU from its parent star with one solar mass at 5 pc from the Sun when the phase factor ϕ (t) is equal to one. For other planetary systems, these numbers scale in an obvious way with the planet and star parameters. They must be compared with the present or forthcoming performances of the instrumentation. We can use this comparison by classifying the methods of detection according to whether they are adapted to inner planets, outer planets or all planets. The result is shown in Tables 2 and 3 (where C is the composition of the planet's atmosphere). The third column in each table gives the physical parameters that can be derived from the observed quantities. The fourth column mentions the astrophysical artifacts that, independently of any systematic instrumental error, can mimic, at least before more extensive investigations, the presence of a planet.

The most common methods—imaging, astrometry, and accelerometry—listed in Table 2 are in fact sensitive only to the presence of outer planets (as far as current searches and sufficiently-close-to-completion projects are concerned). There are naturally ambitious projects to detect Earthlike inner planets with these methods (see for instance Angel 1989 and TOPS 1992), but they have no chance of becoming operational before at least a couple of decades.

A more exotic possibility is to search for the echo of the flares of a star on a nearby planet; this echo will have a delay proportional to a and an amplitude inversely proportional to a^2; thus, as shown by Argyle (1974) and Bromley (1992), this method can work only for very inner planets of Jupiter size, which, as discussed earlier, is a contradictory requirement.

TABLE 1. ***Relations Between the Observable Quantities and Intrinsic Characteristics of the Planetary System***

$\Delta\theta$	$=$	$\left(\frac{a}{D}\right) \times \phi(t)$	$=$	0.2 arc sec	
$\Delta\alpha$	$=$	$\left(\frac{M}{M_*}\right)\left(\frac{a}{D}\right) \times \phi(t)$	$=$	0.6μas	
Δm	$=$	$A\left(\frac{R}{a}\right)^2 \times \phi(t)$	$=$	25 mag	
ΔR_V	$=$	$\left(\frac{M\sin i}{M_*}\right)\left(\frac{GM_*}{a}\right)^{1/2} \times \phi(t)$	$=$	0.1 m s^{-1}	
ΔT	$=$	$\left(\frac{a}{c}\right)\left(\frac{M}{M_*}\right) \times \phi(t)$	$=$	3 msec	
η	$=$	$\left(\frac{R}{R_*}\right)^2$	$=$	10^{-4} mag	
D_{occ}	$=$	$\left(\frac{P}{\pi}\right)\left(\frac{R_*}{a}\right)$	$=$	15 hrs	
DT	$=$	$\left(\frac{a}{c}\right) \times \phi(t)$	$=$	8 min	
A_G	$\leq$	$4\left(\frac{\sqrt{GMD/c^2}}{R_*}\right)$	$\leq$	1 mag	(for $D = 1$ kpc)
T_G	$\approx$	$4\sqrt{GMD/c^2}\,/V$	$\approx$	4 hrs	(for $V = 100$ km s^{-1})

TABLE 2. ***Detection Methods Suited Only for Outer Planets***

Method	*Observables*	*Accessible Parameters*	*Physical Artifacts*
Imaging	$\Delta\theta$, Δm, P	AR^2, a, i	
Astrometry (optical)	$\Delta\alpha$, P, i	M, a, i	
Astrometry (radio)	$\Delta\alpha$, P, i	M, a, i	Activity (radio)
Accelerometry	ΔV_R, P	$M\sin i$, a	Chromospheric activity ? (Walker 1995)

TABLE 3. ***Detection Methods Suited for All Planets***

Method	*Observables*	*Accessible Parameters*	*Physical Artifacts*
Radio Emission	F_R, P ?	F_R, a ?	Other sources ?
IR Excess	F_{IR}, λ_{max}	R, a	Circumstellar ring
Occultations	η, D_{occ}, P	R, a, C (Schneider 1994), i	Star spots for single stars?
Gravitational Amplification	A_G, ΔT	M, a_{min} ?	
Timing	Δt, P	$M\sin i$, a	

As for the methods valid for all planets, the most efficient is the timing method. Unfortunately, it is valid only for pulsars and is thus helpless to detect main sequence stars. The gravitational amplification method has the disadvantage of being able to reveal the presence of a planet only once (during the single transit due to the proper motion of its parent star) and does not allow by itself for subsequent observations of a detected planet. But in very extensive surveys this method could provide statistical information on the number of planetary systems.

The occultation method is finally the only one that can detect inner planets around main sequence stars in the near future and can give some of their characteristics (R, a, i). It could even subsequently be used to study the planets' atmospheric composition (Schneider 1994).

What Is Being Done Around the World?

There are many projects and programs for the detection of extrasolar planets. Most of them are under development. Some of them have already started. Since we are concerned only with the detection of habitable planets, excluding therefore the outer planets, I will restrict myself in this paper to efforts of the former case. They are listed in Table 4.

TABLE 4. *Future Projects*

Method	*Group*	*Description*	*Status*
Imaging	Angel (1994)	Ground adaptive optics > 6 m	Design only
	Angel (1989)	Space 16 m interferometry	Design
Accelerometry	Connes (1994)	Laser calibrated, $\sigma V_R = 1$ m/s	Test
Occultations	FRESIP (Borucki and Koch 1994)	Dedicated satellite	NASA proposal
	DHEP	Dedicated satellite	ESA proposal
	AAT (Parry et al. 1994)	Ground	Proposal
Radio Emission	Paris-Kharkov (Lecacheux 1994, pers. comm.)	Direct radio emissions	Tests

Ideas, Projects and Developments

OCCULTATIONS

It is worthwhile to give additional details about the occultation method since it is the only one by which we can hope to detect habitable planets in the near future. There are two types of projects. In the space projects the aim is to monitor photometrically thousands of G stars to search for inner planets (Borucki and Koch 1994, Koch and Borucki 1996). From the ground it is possible to detect all kinds of planets, but because the photometric precision is then only $\approx 10^{-3}$, only dM stars are applicable (Schneider et al. 1990). One can increase the efficiency of a random search by choosing stars whose axis has been predetermined by the comparison of its rotation period, given by $V_{rot}R_*$, and its projected rotation velocity $V_{rot}\sin i$ (Doyle 1988).

The method finds some improvement when the target star is an eclipsing binary. There are then several favourable characteristics. The main advantage is that the chances for correct orientation of the planetary orbit are close to 100% (Schneider and Chevreton 1990). This advantage is strengthened by the precession, with a period $P_{pr} = 16 \cdot P/3\ (a/a_{sep})^2$, of the planet orbit around the binary (Schneider 1995)*: it forces an occultation to occur in any case at least once per half precession period.

Furthermore, the binary nature of the central star gives a very characteristic shape to the occultation curve: there is not only a double occultation (Schneider and Chevreton 1990), but it consists of a longer and a shorter occultation since the motion of the planet is in the same direction as the star for one of the components and in the opposite direction for the other one (Schneider and Chevreton 1990, Brandmeier and Doyle 1996). In addition, the ephemerides of the second occultation can be predicted with precision from the first one since the orbital motion of the binary is completely known. This makes a double occultation difficult to confuse with any artifact.

OTHER METHODS

Table 4 does not include the timing searches because there is no timing program specifically aimed at planet detection. Almost every timing campaign on pulsars has the potential to detect planets (see Phillips et al. 1993 for a review).

On the other hand, I should also mention the spectroscopic search, although no group has specific plans to search for terrestrial planets with this method. Indeed, while a terrestrial planet gives only a reflex velocity of 0.1 m/s for a solar-

* α_{sep} being the separation of the binary.

type star, this velocity is 0.5 m/s for a habitable planet orbiting a dM star. With a precision of one m/s as envisaged today on an $m_V = 3$ star (Connes 1994), the detection of Earthlike planets for these stars with a 10 m class telescope is not hopeless. For dM stars, terrestrial planets have orbital periods on the order of weeks and not of years as for Jupiters around Suns. This should render the detection easier. As for the habitability of planets around dM stars, it has first been argued that the tidal locking of the planet (which is very close to the star) precludes the presence of liquid water (Kasting et al. 1993). But more recently it has been shown that 3-D atmospheric circulation may prevent water evaporation (Hourdin 1994, pers. comm.; Haberle et al. 1996).

SEARCH FOR O_2 AND O_3

The final goal of detecting habitable planets is to detect signatures of life. One very promising way is the detection of molecular abundances that are compatible only with very out-of-equilibrium chemical reactions, similar to the high abundance of O_2 in the Earth's atmosphere due to the chlorophyllian photosynthesis (Lovelock 1975). It has been suggested to search for O_2 in the optical A band at 760 nm (Owen 1980) or for O_3 at 9.6 μm (Burke 1986).

The O_2 band in the visible can be searched for either by spectro-imaging of the planet or by pure spectroscopy during a planetary transit (Schneider 1994). In the first case a space mission would be necessary; in the second case it is worth investigating whether the observations could be made from the ground with a sufficiently large optical reflector (as no high angular resolution is necessary for spectroscopy).

In the case of the O_3 band at 9.6 μm, a satellite platform is required to prevent absorption by the Earth's atmosphere (Angel 1989). A more careful investigation, for instance the DARWIN project submitted to the European Space Agency (Léger et al. 1996), shows that it is even necessary to go to 3.5 AU or more from the Sun to suppress the IR background of the zodiacal light.

Current Observations

For about one year some groups have started effective searches for extrasolar planets with different methods. They are listed in the following table:

Table 5. *Observation Programs Currently Under Way*

Method	*Group*	*Description*	*Status*
Occultations	GNAT (Craine et al. 1992)	dM stars	?
	TEP Network (NASA Ames, IAC, Paris, Rochester, Crete, South Korea, Canaries)	Eclipsing binaries	Start 1994

Conclusion

The detection of habitable planets around other main sequence stars is still a great astrophysical and instrumental challenge. In fact it is one of the oldest scientific questions (Oates 1940 [quoting Epicurus]). In the present decade there is an increasing amount of effort to detect planets by several methods. While the search has up to now been unsuccessful because of instrumental limitations, the first clear answer to the question of their frequency has recently come up—Jupiters are not frequent (less than one for twenty stars), at least within 15 pc. This result should encourage astronomers to increase their efforts and especially to search for non-Jupiter-like (inner) planets, and makes the search for terrestrial planets more pertinent.

References

Angel, R. (1989), "Use of a 16 m Telescope to Detect Earth-like Planets," in *The Next Generation Space Telescope*, P. Bély, C. Burrows, and G. Illingworth, eds. (Baltimore, MD: Hubble Space Telescope Science Institute), p. 81.

Angel, R. (1994), "Ground-Based Imaging of Extrasolar Planets Using Adaptive Optics," *Nature* **368:** 203.

Argyle, E. (1974), "On the Observability of Extrasolar Planetary Systems," *Icarus* **21:** 199.

Borucki, W., and D. Koch (1994), "Multiplex Approach to the Photometric Detection of Planets," *A&SS* **212:** 293.

Brandmeier, S., and L.R. Doyle (1996), "Simulation of Planetary Transits in Eclipsing Binary Star Systems," this volume.

Bromley, B.C. (1992), "Detecting Faint Echoes in Stellar-Flare Light Curves," *PASP* **104:** 1049.

Burke, B.F. (1986), "Detection of Planetary Systems and the Search for Evidence of Life," *Nature* **322:** 340.

Connes, P. (1994), "Development of Absolute Accelerometry," *A&SS* **212:** 357.

CRAINE, E., E. ENGEL, AND R. TUCKER (1992), "A Cool Star Sample and Extra–Solar System Planetary Search Program," in *Seventh Cambridge Workshop on Cool Stars, Stellar Systems and the Sun*, ASP Conf. Ser., M.S. Giampapa and J.A. Bookbinder, eds., **26:** 588.

DOYLE, L. (1988), "Progress in Determining the Space Orientation of Stars," in *Bioastronomy—The Next Step* (Dordrecht: Kluwer Academic Publishers), p. 101.

GUILLOT, T., G. CHABRIER, P. MOREL, AND D. GAUTIER (1994), "Nonadiabatic Models of Jupiter and Saturn," *Icarus* **112:** 354.

HABERLE, R.M., C.P. MCKAY, D. TYLER, AND R.T. REYNOLDS (1996), "Can Synchronously Rotating Planets Support an Atmosphere?," this volume.

HOURDIN, F. (1994), personal communication.

HUANG, S.-S. (1960), "The Sizes of Habitable Planets," *PASP* **72:** 489.

HUYGENS, C. (1698), *The Celestial Worlds Discoverd*, Fac-simile: F. Cass and Co., 1968.

KASTING, J.F., D.P. WHITMIRE, AND R.T. REYNOLDS (1993), "Habitable Zones Around Main Sequence Stars," *Icarus* **101:** 108–128.

KOCH, D., AND W. BORUCKI (1996), "A Search for Earth-Sized Planets in Habitable Zones Using Photometry," this volume.

LECACHEUX, A. (1990), "On the Feasibility of Extra-Solar Planetary Detection at Very Low Radio Frequencies," in *Bioastronomy—The Search for Extraterrestrial Life*, J. Heidmann and M.J. Klein, eds. (Berlin: Springer-Verlag), p. 21.

LECACHEUX, A. (1994), "Solar System, Low Frequency Radio Astronomy from the Moon," *Adv. Space Res.* **14:** 193.

LECACHEUX, A. (1994), personal communication.

LÉGER, A., J.M. MARIOTTI, J.L. PUGET, D. ROUAN, AND J. SCHNEIDER (1996), "How to Evidence Primitive Life on an Exoplanet?—The DARWIN Project," this volume.

LOVELOCK, J.E. (1975), "Thermodynamics and the Recognition of Alien Biospheres," *Proc. R. Soc. Lond. B* **189:** 167.

MAO, S., AND B. PACZYNSKI (1991), "Gravitational Microlensing by Double Stars and Planetary Systems," *ApJ* **374:** L37.

OATES, W.J., ED. (1940), "Letter to Herodotus," in *The Stoic and Epicurean Philosophers: The Complete Extant Writings of Epicurus, Epictetus, Lucretius [and] Marcus Aurelius* (New York: Random House), pp. 3–15:5.

OWEN, T. (1980), "The Search for Early Forms of Life in Other Planetary Systems: Future Possibilities Afforded by Spectroscopic Techniques," in *Strategies for the Search for Life in the Universe*, M.D. Papagiannis, ed. (Dordrecht: D. Reidel Publishing Company).

PARRY, I.R., B. ESPERANZA CARRASCO, AND F.G. WATSON (1994), "Multi-Fibre Photometry—A Technique for Detecting Extra-Solar Planets," in *SPIE Astronomy Instrumentation Conference*, Hawaii.

PHILLIPS, J.A., S.E. THARSETT, AND S.R. KULKARNI, EDS. (1993), *Planets Around Pulsars*, ASP Conf. Ser. **36**.

SCHNEIDER, J. (1994), "On the Search for O_2 in Extrasolar Planets," *A&SS* **212:** 321.

SCHNEIDER, J. (1995), "On the Occultations of a Binary Star by a Circum-Orbiting Dark Companion," *Planet. Space Sci.* **42:** 539.

SCHNEIDER, J., AND M. CHEVRETON (1990), "The Photometric Search for Earth-Sized Extrasolar Planets by Occultation in Binary Systems," *A&A* **232:** 251–257.

SCHNEIDER, J., M. CHEVRETON, AND E. MARTIN (1990), "New Efforts in the Search for Extrasolar Planets," in *24th ESALB Symposium*, ESA SP-315, C. Battrick, ed. (Noordwijk: European Space Agency), p. 67.

TOPS (1992), *Toward Other Planetary Systems*, NASA.

WALKER, G. (1995?), "Spectroscopic Techniques for the Search for Other Planetary Systems," in *Proceedings of the Second TOPS Conference on Planetary Systems: Formation, Evolution and Detection.*

WALKER, G.A., A.R. WALKER, A.W. IRWIN, S.L.S. YANG, A.M. LARSON, AND D.C. RICHARDSON (1995), "A Search for Jupiter-Mass Companions to Near-by Stars," *Icarus* **116:** 359–375.

WOLSZCZAN, A. (1994), "Confirmation of Earth-Mass Planets Orbiting the Millisecond Pulsar PSR B1257+12," *Science* **264:** 538–542.

Kepler-16b: In the Light of Two Suns

Credit: NASA/JPL-Caltech/T. Pyle

https://www.nasa.gov/image-article/kepler-16b-light-of-two-suns/ [retrieved 7-22-25]

This artist's concept illustrates Kepler-16b, the first planet known to definitively orbit two stars—what's called a circumbinary planet. The planet, which can be seen in the foreground, was discovered by L. R. Doyle and colleagues (Doyle et al., 2011), of NASA's Kepler mission. The two orbiting stars regularly eclipse each other, as seen from our point of view on Earth.

The planet also eclipses, or transits, each star, and Kepler data from these planetary transits allowed the size, density and mass of the planet to be extremely well determined. The fact that the orbits of the stars and the planet align within a degree of each other indicate that the planet formed within the same circumbinary disk that the stars formed within, rather than being captured later by the two stars.

[Publisher's note: It was Bill Borucki's vision and persistence in pursuing funding for the search for extrasolar earth-sized planets that led to the Kepler mission and to the discovery of Kepler 16b (the first circumbinary planet ever discovered), in a collaboration with a team that for years had been focusing on looking for planets around binary stars. L. R. Doyle and colleagues published their results in Laurance R. Doyle et al., "Kepler 16: A Transiting Circumbinary Planet," *Science*, 333 (6049): 1602–6, published in September 2011.]

See also https://en.wikipedia.org/wiki/Kepler-16b

SIMULATION OF PLANETARY TRANSITS IN ECLIPSING BINARY STAR SYSTEMS

Stefan Brandmeier

Department of Computer Science
Polytechnic University of Konstanz
Konstanz, Germany

Laurance R. Doyle

SETI Institute
NASA Ames Research Center
Moffett Field, CA 94035

Abstract

The method of simulating planetary transits in eclipsing binary star systems is outlined. Some of the possible configurations of resultant planetary light curves have been generated for two very small mass systems (CM Draconis and XY Leo), with statistics of the transits compiled for the smallest system, CM Draconis. The transit profiles are of a quasi-periodic nature that differs significantly from the intrinsic noise source expected from the system. Cross-correlation of these models with actual photometric observations, therefore, will allow substantial improvements in the detectable signal-to-noise. For the CM Draconis system, the signal from Earth- to Neptune-sized planetary transits should be detectable.

INTRODUCTION

The detection of extrasolar planetary systems is one of the most challenging problems in observational astronomy today. Direct imaging of an extrasolar planet is at present possible only from orbit with either a chronographic telescope or an optical/infrared interferometer with tens of meters of baseline (reviewed in Black 1980, Schneider 1996). This is because, at best, the brightness ratio of a star to a planet is about four to five orders of magnitude (Jupiter-sized planets in the near infrared). The size and brightness are the important limiting factors in this detection method. Gravitational lensing is also an interesting technique, the requirement here being continuous observations of thousands of stars (TOPS 1992). Various indirect techniques for planetary detection include the wobbling of the star due to planetary motion in the system. The astrometric detection method requires milliarcsecond positional accuracy over decades for the detection of any planets Jupiter-like in size and orbit. However, it is not limited by the orientation of the planetary orbital plane. The radial velocity method requires meter-per-second measurement of Doppler shifts in the stellar spectra due to line-of-sight stellar wobbles caused by planets in the system. It also requires several years of observations but could likely detect Neptune-sized planets around solar-type stars. In these two methods the mass and distance of the planet from the star are the important parameters for detection. In the astrometric case, a larger planetary distance increases the magnitude of the astrometric wobble. In the radial velocity case, shorter planetary distances are advantageous since stellar offset velocity is what is being observed.

Finally, the photometric method, detecting planetary transits in front of their parent star, requires precise photometry (0.01% for the detection of an Earth-sized planet around a solar-type star). For random distributions of the planetary orbital plane, thousands of stars would have to be observed over years as only about 1–2% can be expected to have the planetary orbital plane edge-on to the observer (Borucki and Summers 1984, Doyle et al. 1984). In addition, there appears to be a practical limit, imposed by atmospheric scintillation, to the precision with which photometry can be performed of about 0.1%, that is, of about 0.001 in differential magnitude (Koch and Borucki 1996, and references therein).

One way to mitigate the necessity of observing so many stars is to preselect systems that are likely to have edge-on planetary orbits. If planets around binary stars form in the same plane, for example, as the binary orbital plane, then the edge-on eclipsing binary stars are natural candidates for observing planetary transits (Schneider and Chevreton 1990, Schneider and Doyle 1995). In the application of this method, simulations of the brightness variations due to planetary transits in various eclipsing binary systems must be performed in order to be able to recognize the presence of any planets in the photometric variations. Such simulations were first instituted by Bell and Borucki (1995), but we first introduce the light curves produced by such transits here.

SHORT REVIEW OF ECLIPSING BINARY LIGHT CURVE GENERATION

Brightness variations in binary systems due to the mutual eclipses of their components have been known for over two centuries. Such eclipses can be total or partial depending on the inclination i of the system (the orbital inclination of the binary orbital plane from the line of sight). *Primary minimum* refers to the lower-surface-brightness star passing in front of the higher-surface-brightness star thereby producing the deepest drop in the light curve brightness. This defines the epoch or zero point of the light curve. *Secondary minimum* refers to the reverse configuration of the two eclipsing stars. When the smaller star is passing *in front* of the larger star an annular eclipse occurs. When the smaller star is passing *behind* the larger star a total eclipse occurs. Partial eclipses occur when the inclination from the line of sight $i \neq 90°$. We will also use the term *transit* when a planet passes in front of either star. For main-sequence dwarf stars, primary minimum usually occurs during annular eclipses as the larger mass stars are usually the largest and have the brightest surface areas (the latter most relevant in the case of partial eclipses).

In simulating a system, in addition to the orbital parameters, the stellar radii, surface brightnesses (per area), and variations of brightness over the stellar disk (limb darkening) must be taken into account (e.g., Tsesevich 1973; some of these equations are corrected and given in Brandmeier 1994 as well as here). The total brightness of the stars is taken to be unity. A simplified expression for limb darkening can be taken as

$$I(\vartheta) = I(0)(1 - c_L + c_L \cos \vartheta)$$

where

$I(0)$ = the surface brightness at the center of the star from the line of sight

c_L = coefficient of limb darkening ($0 \leq c_L \leq 1$, $c_L = 0$ means uniform stellar surface brightness), and

ϑ = cosine of the angle between the line of sight and the normal to a point on the star's surface.

In this simulation the equations we use for the light reduction, ΔL, due to different configurations of the binary stars, are as follows.

For annular eclipses the distribution of light over the disk is given by (Tsesevich 1973)

$$I = I_0 \cos \vartheta = \frac{I_0}{R_1} \sqrt{{R_1}^2 - x^2 - y^2}$$

where x and y are rectangular coordinates with origin at the center of the larger star. The light reduction is then given by

$$\Delta L = \frac{I_0}{R_1} R_2^3 \int_{e-1}^{e+1} \left[\sqrt{2e(\xi - e + 1)(\xi - e - 1)(\xi - f)} + \left\{ \frac{1}{k^2} - \xi^2 \right\} \arcsin \sqrt{\frac{(1 + \xi - e)(1 - \xi + e)}{1/_{k^2} - \xi^2}} \right] d\xi$$

where R_1 and R_2 are the radii of the larger and smaller stars, respectively, and $e = \frac{\Delta}{R_2}$ (where Δ is the separation of the centers of the two stellar disks). Here the substitutions used are

$$k = \frac{R_2}{R_1}, \; f = \frac{\frac{1}{k^2} - 1 + e^2}{2e}, \text{ and } \xi = \frac{x}{R_2}.$$

The expression for the minimum luminosity at primary minimum ($\Delta = 0$, but $f \rightarrow \infty$ in the equation above) can be calculated from

$$L_{\min} = \frac{2\pi I_0 {R_2}^2}{3k^2} \left[1 - (1 - k^2)^{3/2} \right].$$

When the smaller star is behind the larger star, the light loss is simply equal to the luminosity of the secondary star:

$$L_2 = \frac{2\pi}{3} I_0 R_2^2.$$

Where the inclination of the system is not 90°, the expression for partial primary eclipses (corrected in Brandmeier 1994 from Tsesevich 1973) is

$$\Delta L = \frac{I_0}{R_1} R_2^3 \{ \int_{e-1}^{f} \left[\sqrt{2e(\xi - e + 1)(\xi - e - 1)(\xi - f)} + \left\{ \frac{1}{k^2} - \xi^2 \right\} arcsin \sqrt{\frac{(1 + \xi - e)(1 - \xi + e)}{1/_{k^2} - \xi^2}} \right] d\xi + \frac{\pi}{2k^3} \left(\frac{2}{3} - kf + \frac{k^3}{3} f^3 \right) \}$$

Finally, the expression for partial secondary eclipses (ibid.) is

$$\Delta L = 2I_0 R_2^2 \{ \tfrac{1}{2} \int_p^g \left[\sqrt{2e(\xi - p)(\xi - h)(\xi - g)} + (1 - \xi^2)\, arcsin \sqrt{\tfrac{(\xi - p)(h - \xi)}{1 - \xi^2}} \right] d\xi + \tfrac{\pi}{12}(2 - 3g + g^3) \}$$

where

$$g = \frac{1}{2e}(1 + e^2 - \frac{1}{k^2}),\ h = e + \frac{1}{k},\ \text{and}\ p = \frac{\Delta - R_1}{R_2}\ .$$

These are the expressions used to generate the eclipsing binary light curves. For most purposes here we assumed circular orbits and that the longitude of the ascending node was zero. Since the radii of the stars are expected to be much bigger than the radius of a transiting planet, we omitted any effects due to the planet's crossing the stellar limb. The light loss due to the transit of a planet is therefore calculated for the annular case, for example, by determining the limb darkening law that applies for that point and replacing the radius of the secondary star by the planetary radius and simply taking the ratio of the planet area to star area at that point.

The Simulation Model

We have developed a planetary transit model for eclipsing binary systems based on the classical simulations of eclipsing binary light curves where stellar radii, periods, eccentricities, inclinations, limb darkening and so forth, of the component stars can be well determined (Brandmeier 1994). We have also considered the photometric contribution from starspot activity (which, in general, will contribute to the high-frequency noise of the signal as a planet transits various starspots) and noise superposition (either various kinds of generated noise, noise directly from the Lacy 1977 data, or noise from the preliminary CCD data obtained by us on CM Draconis). In Figure 1a we plot (at very low resolution) the light curve for CM

Draconis (without noise) showing the general shape of the curve and the percentage drop in brightness due to the mutual eclipse (which fits well the data of Lacy 1977).

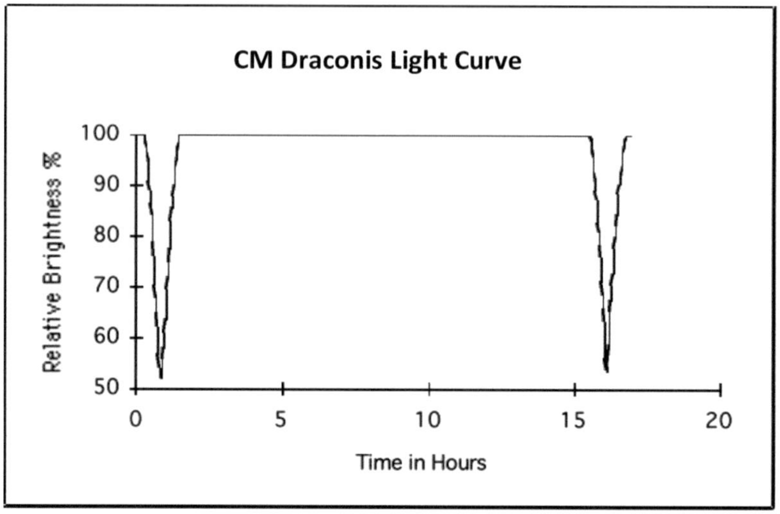

FIG. 1A.

Figures 1b through 1d show the relative percentage *drop* in brightness (note inverted vertical scale; figures after Brandmeier 1994) due to the transit of an extra-solar planet of two Earth-radii in an equivalent Earth orbit (i.e., 17.7-day period) (scaling to the equivalent flux from CM Draconis to equal one solar insolation at one AU). The effects of stellar eclipses have been removed so that some of the unusual planetary transit shapes may be better seen. In Figure 1b the stars are well separated and the durations of planetary transits differ only by the fact that one star was moving in the same direction as the planetary motion across the line of sight while the other was moving in the opposite direction at the time. Figure 1c starts out as a more normal transit configuration but in the second transit the planet is traveling in the same direction as the secondary star when the primary star (traveling in the opposite direction) moves past it (remember that the effects of the stellar eclipse itself are removed). The planet then travels a bit along with the secondary after the stellar eclipse, creating this strange variation in brightness (due mostly to changes in limb darkening as the planet transits various parts of the primary and secondary stellar disks as they pass each other).

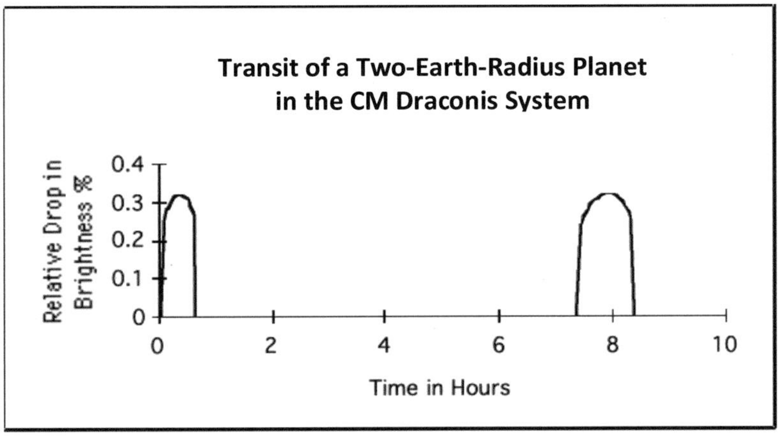

FIG. 1B.

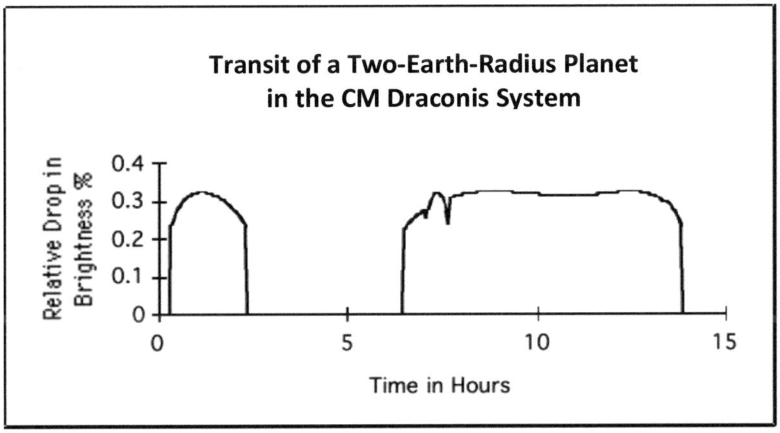

FIG. 1C.

In Figure 1d we have one of the somewhat rare triple transits where the planet transits the first star in the same direction, passes in front of the second star which is traveling in the opposite direction, which then catches up to it as it rounds elongation and starts traveling in the same direction as the planet, causing the third transit event. Statistics derived from the model of these transit events for 400 Earthlike orbital periods can be seen in Table 1.

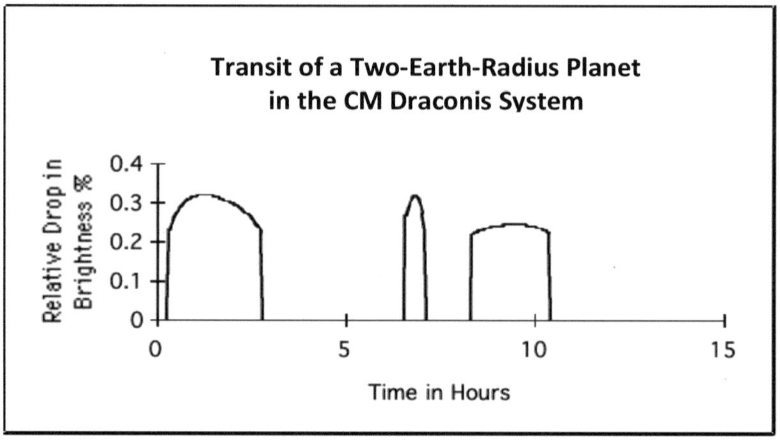

FIG. 1D.

TABLE 1. *Statistics of Two-Earth-Radii Transit Model (17.7-Day Circular Orbit) in the CM Draconis System*

Transit Events per Planet Period	
One	4%
Two	90%
Three	6%
Transit Occurrence	
Inside Eclipse	14%
Outside Eclipse	86%
Transit Depths (percentage)	
0.20–0.32	3%
0.32–0.34	97%
Transit Duration (hours)	
0.5–1.5	84%
1.5–13.7	16%
Hours Between Transit Events	
0.2–3	6%
3–4	3%
4–5	5%
5–6	15%
6–7	35%
7–8	36%

Cross-Correlation with Transit Models

Typical transit events for terrestrial planets in the CM Draconis system, then, should last about one hour each with a seven-hour average duration between transits. The depth of a two-Earth-radii planet should be about 0.3%. Of course, more points in transit are available during an eclipse event (sometimes over an order of magnitude more), and such a transit event (happening about 14% of the time for CM Draconis) has the same effect that a larger-aperture telescope would have on detection—allowing the collection of more transit points. These models can be run for various expected planetary distances and a cross-correlation can be done with real observational data (Jenkins and Doyle 1994). It has been shown that a signal-to-noise ratio (SNR) limit of about 0.67 can be reached for three such (double) transit events by cross-correlating these expected models with white noise–convolved data, which is characteristic of the observational data. During eclipse transits a simulated SNR of about 0.3 can be obtained, allowing the detection of two-Earth-radii planets with 1% photometry (Jenkins and Doyle 1994). The transit signals can be so well detected because, as can be seen, each transit provides an unusual signature that can be separated rather easily from rotational starspot modulation (a known period since most close binaries are tidally locked) as well as intrinsic stellar variability (which would not behave in such a quasi-periodic way).

Consequently (as suggested in Brandmeier 1994), multiple transit models can be generated, for example, for each possible terrestrial planetary orbital distance and cross-correlated with observational data. For a given temporal resolution for observations of about five minutes there would be 366 possible configurations of the two stars (the binary period of CM Draconis is 1.26838965 days). If we assume that Mercury formed as close as possible to the Sun (i.e., that any closer to the early Sun, and protoplanetary grains would have evaporated), then the equivalent distance from the present CM Draconis stars would give a planet with a 4.3-day period. Taking this out to the equivalent insolation received at Mars by the present Sun, then the farthest CM Draconis terrestrial planet would have a period of about 34 days. Equivalently, the innermost gas giant would be at a distance in this system of about seven months (Schneider 1996; but see also Wetherill 1996 for single star planet formation, and Boss 1995 for considerations of the thermal gradient of the protosolar nebula). Thus, various models could be cross-correlated over the observing time in order to detect evidence of a terrestrial-sized planet in the CM Draconis system. (In the event of a Jupiter-like planetary transit in this system, the photometric drop would be about 9%, easily detectable without cross-correlation.)

FURTHER APPLICATIONS OF THIS PLANETARY TRANSIT MODEL

In Figures 2a and b are shown some simulations of a possible two-Earth-radii planetary transit occurring in the XY Leonis system. The period of XY Leonis is 0.284110 days. We see that in this very close system, transits will more than likely occur while the star is in eclipse. As the planet transits, the two stellar components will continue to move behind it resulting in almost continuous transit events for several hours. Again, it is clear that the signatures of these transit events are such that they will not be confused with other likely sources of noise in the cross-correlation of the models with the observational data.

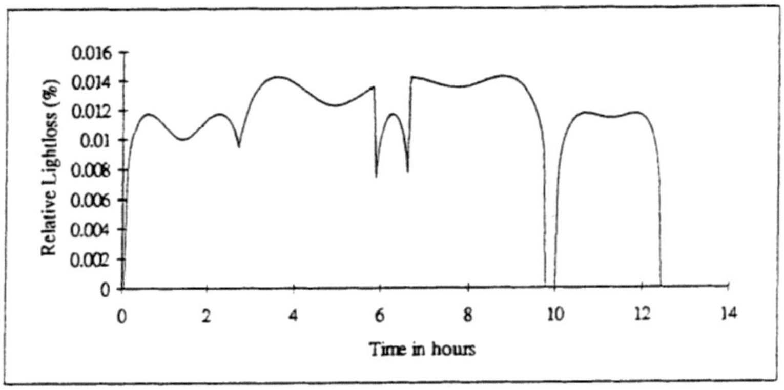

FIG. 2A

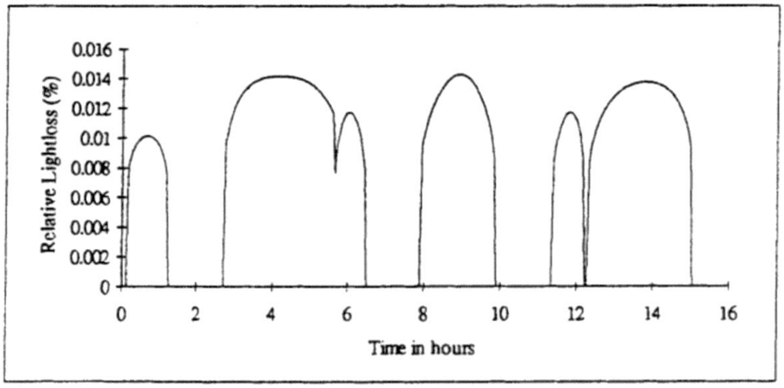

FIG. 2B

There are about six dozen other systems (Brancewicz and Dworak 1980) of dwarf spectral types G0 or later where terrestrial-sized planets may be detectable from space assuming the differential photometric precision of FRESIP (Frequency of Earth-Sized Inner Planets; see Koch and Borucki 1996). In this project, thousands of stars would be observed in order to detect planetary transits around 1% of them (assuming statistically random planetary orbital plane orientations). At least half of these systems can be expected to be edge-on (eclipsing) binaries and the analysis performed here would apply. Thus, if the photometric precision to detect Earth-sized planetary transits around solar-type stars can be achieved (0.01%), and long-duration (six-year) missions will allow an SNR of about 0.1 to be reached, then Earth-sized planetary transit detections for systems with equivalent areas as high as a dozen times that of the Sun could be reached, given that an accurate eclipsing binary model for each such system could be derived.

In conclusion then, the modeling of planetary transits in eclipsing binary star systems has broad application to the detection (or constraints on the formation) of extrasolar planets. Cross-correlation of such models with ground-based observations helps to mitigate the effects of scintillation and other sources of noise, which should allow detections at SNRs of less than one, since the signal produced by a planetary transit event is only quasi-periodic, and unlike any other noise sources that the system variability, atmosphere, or detection equipment may produce. This modeling cross-correlation method can be applied to several systems for ground-based detection of Earth- to Neptune-sized planets and could be extended to assist space-based missions as well. We look forward to its application to both types of observations in the near future.

Note added in final proof—In addition to the three pulsar planets (A. Wolszczan, "Confirmation of Earth-Mass Planets Orbiting the Millisecond Pulsar PSR B1257 + 12," 1994, *Science* **264:** 538–542), three giant planets around three different systems have been discovered. M. Mayor and D. Queloz have found an $M\sin i = 0.5$ Jupiter-mass planet at 0.05 AU from the solar-type star 51 Pegasus ("A Jupiter-Mass Companion to a Solar-Type Star," 1995, *Nature* **378:** 355–359). G. Marcy and P. Butler (A.P. Boss, "Giants and Dwarfs Meet in the Middle," 1996, *Nature* **379:** 397–398) have found an $M\sin i = 2.3$ Jupiter-mass planet at a distance from 47 Ursae Majoris equivalent to a position between Mars and Jupiter in the Solar System, and another planet of $M\sin i = 6.5$ Jupiter masses at a distance from 70 Virginis equivalent to a position between Mercury and Venus in the Solar System. Although none of these planets could be thought of as habitable, they are encouraging in their implications that planet formation may be widespread. M. McCaughrean and C. Robert O'Dell (to appear in *AJ* May 1996*), with imaging of six protoplanetary disks, have implied that half of the stars in the Galaxy may have formed planets.

* Mark J. McCaughrean and C. Robert O'Dell (1996), "Direct Imaging of Circumstellar Disks in the Orion Nebula," *AJ* **111:** p. 1977.

References

Bell, J.F., III, and W.J. Borucki (1995), "Characteristics of Transits by Earth-Sized Planets in Binary Star Systems," in *Progress in the Search for Extraterrestrial Life*, ASP Conf. Ser., S. Shostak, ed., **74:** 165–170.

Black, D.C. (1980), "In Search of Other Planetary Systems," *Space Sci. Rev.* **25:** 35–81.

Borucki, W.J., and A.L. Summers (1984), "The Photometric Method of Detecting Other Planetary Systems," *Icarus* **58:** 121–134.

Boss, A.P. (1995), "Proximity of Jupiter-like Planets to Low-Mass Stars," *Science* **267:** 360–362.

Brancewicz, H.K., and T.Z. Dworak (1980), "A Catalogue of Parameters for Eclipsing Binaries," *Acta Astronomica* **30:** 501–524.

Brandmeier, S. (1994), *Simulation of Planetary Transits in Eclipsing Binary Star Systems,* Master's Thesis in Computer Science at Polytechnic University of Konstanz, Germany.

Doyle, L.R., T.J. Wilcox, and J.J. Lorre (1984), "The Space Orientation of Stars," *ApJ* **287:** 307–314.

Jenkins, J.M., and L.R. Doyle (1994), "Detection of Extrasolar Planets in Eclipsing Binary Systems: A New Cross-Correlation Technique," abstract in *Bull. Amer. Astron. Soc.* **26:** 1089.

Koch, D., and W. Borucki (1996), "A Search for Earth-Sized Planets in Habitable Zones Using Photometry," this volume.

Lacy, C.H. (1977), "Absolute Dimensions and Masses of the Remarkable Spotted dM4e Eclipsing Binary Flare Star CM Draconis," *ApJ* **218:** 444–460.

Schneider, J. (1996), "The Search for Habitable Extrasolar Planets: Current Programs and Long-Term Plans," this volume.

Schneider, J., and M. Chevreton (1990), "The Photometric Search for Earth-Sized Extrasolar Planets by Occultation in Binary Systems," *A&A* **232:** 251–257.

Schneider, J., and L.R. Doyle (1995), "Ground-Based Detection of Terrestrial Extrasolar Planets by Photometry: The Case for CM Draconis," *Earth, Moon, and Planets* **71:** 153–173.

TOPS (1992), *Toward Other Planetary Systems,* NASA.

Tsesevich, V.P. (1973), "Photometric Phases of Eclipses," in *Eclipsing Variable Stars*, V.P. Tsesevich, ed., translated by R. Hardin (New York: Halsted-Wiley).

Wetherill, G.W. (1996), "The Formation of Habitable Planetary Systems," this volume.

How to Evidence Primitive Life on an Exoplanet? The DARWIN Project

A. Léger

IAS, bât. 121, Université Paris XI
91405 Orsay, France

J. M. Mariotti

Observatoire de Paris
92195 Meudon, France

J. L. Puget

IAS, bât. 121, Université Paris XI

D. Rouan

Observatoire de Paris

J. Schneider

Observatoire de Paris

Abstract

We report on the concept of an IR observatory with interferometric rejection to search for primitive life on an extrasolar planet which has been proposed to the European Space Agency.

Scientific Objectives

Motivation for the Search for Primitive Life

The search for life in the Universe will be a central item in the scientific activity of the twenty-first century. The TOPS report (1992) states: "Much of religion and philosophy has focused on attempts to understand how we and our world came to be." We "ask about the prevalence of planetary systems throughout the universe . . . and the likelihood that other planets have given birth to life, even advanced and intelligent forms of life. *We live in a remarkable time,* when human beings . . . have attained the possibility of finding *the first real answers* to some of these most meaningful questions."

As financial resources are scarce, it is important to distinguish which are the key steps. Current opinion states that we are to find exoplanets in the next ten to fifteen years (e.g., by astrometry with the European VLTI [Very Large Telescope Interferometer]). The knowledge of the frequency of appearance of primitive life on habitable planets would be a capital breakthrough, whereas the detection only of Jupiter-like exoplanets would not really answer our basic questions. The DARWIN concept is one of the very few possibilities we have to address this question.

How to Find the Evidence for It? Why a Space Mission?

The presence of abundant O_2, or O_3, in the atmosphere of an exoplanet has been shown to be a criterion for photosynthetic activity (Lovelock 1975, Owens 1980, Burke 1986, Léger et al. 1993). The main detectable signatures of these species are in the visible (O_2 : 760, 680 nm), the infrared (IR) (O_3 : 9.6 μm) and the radio (O_2 : 5 mm). The latter is not attractive because of the λ^{-2} dependence of the number of emitted photons for a 300 K thermal source.

The main challenge when searching for these signatures in a planetary atmosphere is the presence of the nearby, much brighter, star. In the Sun/Earth case the luminosity ratio is 7×10^9 in the visible and 10^7 at 10 μm, pointing out the advantage of observing in the IR (Bracewell 1978, Angel et al. 1984). After the starlight is rejected as much as possible, the signatures, if present, appear as dips in the spectrum. Considering the quantum noise alone, the feasibility of the detection can be evaluated by calculating the integration time (t) required for a 5σ detection of a 50% deep spectral band:

- **O_2 band in the visible (760 nm . . .).** With large ground-based interferometers (e.g., VLTI) in the visible, the maximum rejection factor (ρ) of the starlight with respect to the planet emission, if ever possible, will be 10^2. With a telescope-detector yield of 15%:

$$t = 3 \times 10^5 \, (L/10^{-2}L_S)(\rho/10^2)^{-1}(R_{pl}/R_\oplus)^{-4}(D/4 \text{ pc})^2 \text{ hours}$$

where L is the star luminosity, L_S the Sun one, R_{pl} the planet radius, $R_\oplus$ the Earth one, and D the distance of the system to us. Even for nearby M stars, this is *not possible.*

- **O_2 band (760 nm . . .) during a planet transit.** When a planet transits in front of a star (Schneider and Chevreton 1990) a small fraction of the star flux travels through the planet atmosphere and is specifically absorbed. For an M star system, this occurs during ≈ 30 h per year but the number of years needed for the band detection is prohibitive (~ 103).
- **O_3 band in the IR (9.6 μm).** A null interferometer in space can reject the starlight (black fringe) and transmit the planet's light (bright fringe) if it can resolve the system. For a planet at a suitable distance from the star to have a temperature of 300 K (Kasting et al. 1993), the condition to separate the star from the planet is

$$D < 20\left(\frac{L}{Ls}\right)^{1/2}\left(\frac{l}{20 \text{ m}}\right) \text{ parsec}$$

where l is interferometer base. Much higher rejection factors than on the ground are possible and, assuming that all other sources of noise are at levels below that due to stellar light—as explained below—the integration time becomes

$$t = 34\left(f\frac{L}{Ls}\right)\left(\frac{D}{6.3 \text{ pc}}\right)^2\left(\frac{\rho}{510^4}\right)^{-1}\left(\frac{Rpl}{R\oplus}\right)^{-4}\left(\frac{Res}{20}\right)\left(\frac{\varnothing 2}{1.3 \text{ m}}\right)^{-2} \text{ hours}$$

where $f\,L/L_S$ is the ratio of the star to Sun fluxes at 10 μm, f depends upon the star temperature (0.7–5.5 for stars of spectral types F-M), R_{pl} is the planet radius, *Res* is spectrometer resolution, $\varnothing_2$ is the diameter of the twin telescopes. Note that there are about 100 stars closer to us than 6.3 pc.

Clearly, among these possibilities, the last is the only one feasible. The ability of a space mission to accommodate a cold telescope and allow high interferometric contrasts makes it possible, although difficult. The detection cannot be done from the ground.

Conceptual Description

The concept is a null interferometer associated with an IR spectrometer. It has two versions—the "minimum" DARWIN mission, made of 2 × 1.3 m telescopes with a 10 m base, and the "full" DARWIN mission, with 4 × 0.9 m telescopes in a cross pattern, with 30 m bases. The spectral range is 5–25 μm or broader. At 9.6 μm, the number of photons N_{pl} received per hour in one spectral element from the planet is

$$N_{pl} = 9.610^2 \left(\frac{D}{6.3\ \mathrm{pc}}\right)^{-2} \left(\frac{\varnothing 2}{1.3\ \mathrm{m}}\right) \left(\frac{Res}{20}\right)^{-1} \left(\frac{R_{pl}}{R_\oplus}\right)^2 \ \text{photons} \cdot \text{hours}^{-1}$$

The problem is to extract such a tiny flux out of strong ones. Sources of noise and spurious signals are discussed below.

Starlight Leaks

At 9.6 μm, the ratio of the residual flux from the star F_{*res} to that of the planet F_{pl} is

$$\frac{F_{*res}}{F_{pl}} = 200 \left(f \frac{L}{L_S}\right) \left(\frac{\rho}{5104}\right)^{-1} \left(\frac{R_{pl}}{R_\oplus}\right)^{-2}$$

It critically depends upon the rejection factor ρ. The factors limiting ρ are the following.

Optics quality and microroughness—the amount of light scattered to the dark field by these optical defects is proportional to the variance of the optical path difference errors. Achieving an optical quality of $\lambda_{vis}/100$ in the visible (0.5 μm) will limit the scattered light at 10 μm to 1.10^{-5}. Other sources of stray light have to be smaller.

Optical path cancellation—an error δ in optical path adjustment will produce light leaks. $\delta = \lambda_{IR}/2000$ will result in a 10^{-5} level of light intensity. But if the control of the adjustment is done in visible light (0.5 μm), where photons are numerous, the requirement drops to $\lambda_{vis}/100$. A further requirement is that the optical path difference must be achromatic in the IR.

Guiding—the cancellation of the fields in the interferometer is perfect only if the object is on axis. If the interferometer axis is displaced by only 1/1600 of the angular size of the Airy disks formed by each telescope, the light leak is 10^{-5}. However, if the tracking is performed at 0.5 μm, the requirement is reduced to 1/80 of the Airy disk size (1.3 mas, for a one m telescope).

Star size—although the central star is not resolved by the interferometer, it is not a perfectly coherent source, resulting in a light intensity proportional to $R^2 \times \alpha^2$ where α is the angular size of the star, and R the angular resolution of the interferometer ($R = l/\lambda$).

The first three limitations are technological in nature and each can be kept within the quoted limits with available technologies used at best. Their total contribution will hence be less than 2×10^{-5}. The last one is more fundamental and results in a severe trade-off for the resolution. However, Angel (1989) has proposed an optical arrangement involving four telescopes in a diamond pattern that yields light leaks proportional to $R^4 \times \alpha^4$, keeping these leaks to levels much smaller than 10^{-5} at the expense of an increased complexity. This leads to the "full" DARWIN concept. A search for other clever optical concepts should be pursued but it is comforting to already have one concept that fulfills the requirements.

Zodiacal Light Likely To Be Associated with the Planetary System

At 10 μm, the solar integrated zodiacal light (*ZL*) luminosity is 3×10^2 larger than the Earth's flux. If the distant planetary system is similar to ours, its *ZL* should not generate too much noise because it would be on the order of the starlight leak but rejected by the interferometer because it is centrally peaked. However, it probably carries the silicate band at 9.7 μm in emission which is close to the O_3 line to be detected. The modulation of the planet signal during its orbital motion will need to be used to properly separate them.

Cirrus Background

According to our present understanding of the physics of the midinfrared emission from cirrus clouds found by IRAS, this galactic background gives a mean flux at 9.6 μm of

$$N_{cir} = 1.2 \times 10^4 \, (Res/20)^{-1} \qquad \text{photons} \cdot \text{hour}^{-1}$$

The requirement that the corresponding integration time is less than thirty hours limits the distance of observable planets to $D < 11.5$ pc.

Solar System Zodiacal Emission

From an earth orbit, the flux from zodiacal light (*ZL*),

$$N_{ZL} = 3.0 \times 10^6 \, (Res/20)^{-1} \qquad \text{photons} \cdot \text{hour}^{-1},$$

is *totally prohibitive.*

To avoid this difficulty the mission has to go farther out from the Sun than the Earth's orbit. The *ZL* background becomes weaker than the cirrus background at $d > 3.5$ AU (Jupiter is at 5.2 AU).

Telescope Temperature

A thermal emission by the mirrors inferior to the starlight leaks at 9.6 μm, but also at 15 μm (CO_2 band) implying $T < 30$ K for a comfortable detection (up to 17.5 μm). This should be possible with passive cooling as shown by the studies done for the EDISON project, taking into account the distance to the Sun of more than 3.5 AU.

Detector Noise

Currently available Si:As BIBs individual detectors already have a dark current comparable to the cirrus background ($3e^{-}s^{-1}$ at 10 K) in the 5–28 μm range. Provided the detectors are cooled to stable temperatures around 4 K, they will have the required low noise and stability.

SCIENTIFIC RETURN

Telluric Planet Detection

The currently planned ground-based searches will not find Earth-sized planets. The DARWIN concept can detect them spectroscopically, thanks to the CO_2 15 μm band. The band is expected whether life is present or not at their surface and is observed in the spectra of Venus, Earth, and Mars.

Photosynthesis Activity

The mission can detect the presence or absence of photosynthetic activity on a large scale on the terrestrial planets it will find. This detection does not depend on the specific molecule that reduces CO_2 using photons, contrary to an experiment that would search for the chlorophyll spectral signature in the visible. The number of star systems in which the minimum DARWIN mission (2 × 1.3 m telescopes, base = 10 m) can give these answers is 35 and with the full DARWIN mission (4 × 0.9 m telescopes, base = 30 m) it increases to 110 in a four-year mission (assuming two years of integration). Most of them are G stars (55%) or K stars (25%), which are good candidates in terms of the habitability of their telluric planets.

Important By-Products

Although dedicated to bioastronomy, the DARWIN concept would have many important by-products. Potential programmes, using the coronographic capability of the instrument, are

- planetary science of nonterrestrial planets, with spectroscopy
- studies of the central part of young stellar disks/circumstellar envelopes
- narrow line region of AGN (asymptotic giant nuclei), QSOs (quasi-stellar objects)

There is also the astrophysics that can be done with a 2.5 m^2 collecting area IR telescope without the disturbance of the zodiacal light during a long mission, although the instrument is not optimised for this purpose (e.g., small field of view).

Technology Status and Requirements

A key point is the achievement of the high rejection factor as described above—it should be possible with existing technology (e.g., the optical path length equalizing planned for the VLTI), but requires development. The deployable structure of the interferometer and the active cooling system also require development.

Present spacecraft technologies that are applicable to the DARWIN project include

Solar panels, telemetry from the ROSETTA mission

Passive cooling from studies for EDISON and FIRST (Far Infrared Space Telescope) missions

IR detectors from ISO (Infrared Space Observatory) and SIRTF developments

Pointing from HST (Hubble Space Telescope)

Active detector cooling developed for FIRST

Deployment mechanisms developed for FIRST in the deployable version

Relation to Existing or Planned Space and Ground Systems

The best candidates to be observed would be those for which giant planets have been found by astrometric/Doppler studies from the ground. Such studies are likely to give results in less than fifteen years and the DARWIN mission could be made the next step in the search for planets. For that to be possible developments and studies should be undertaken now.

Summary of the Main Mission Requirements

Orbit: more than 3.5 AU from the Sun, away from Solar System planets

Lifetime: typically ten years

Operational strategy: autonomous

Stability and pointing: 3 axis stabilisation; 1.3 marcsec in the focal plane achieved by locking on the star itself in the visible

Power: to be determined

Data rate: low, compatible with existing ones for planetary missions

Readiness: major development needed: deployable structure in space, balance between the intensity and optical paths from the different telescopes

Cost: cornerstone class

Conclusion

The mission proposed addresses one of the most fundamental problems that science can deal with. Most of the basic technology needed for it has been (or is being) developed for other missions and should be available in the "post–Horizon 2000" European Space Agency (ESA) programme era. Nevertheless, the concept is a very ambitious one and will require careful studies and some development of new space technology. It will require a thorough system analysis to make sure that the fundamental limits we aim at can be achieved. From that point of view it fits well to the cornerstone concept as developed in ESA's scientific programme.

References

ANGEL, J.R., A.Y. CHENG, AND N.J. WOOLF (1984), "A Space Telescope for IR Spectroscopy of Earth-Like Planets," *Nature* **322:** 341.

ANGEL, R. (1989), "Use of a 16 m Telescope to Detect Earth-like Planets," in *The Next Generation Space Telescope*, P. Bély, C. Burrows, and G. Illingworth, eds. (Baltimore, MD: Hubble Space Telescope Science Institute), p. 81.

BRACEWELL, R.N. (1978), "Detecting Non-Solar Planets by Spinning IR Interferometer," *Nature* **274:** 780.

BURKE, B.F. (1986), "Detection of Planetary Systems and the Search for Evidence of Life," *Nature* **322:** 340.

KASTING, J.F., D.P. WHITMIRE, AND R.T. REYNOLDS (1993), "Habitable Zones Around Main Sequence Stars," *Icarus* **101:** 108–128.

LÉGER, A., M. PIRRE, AND F.J. MARCEAU (1993), "Search for Primitive Life on a Distant Planet: Relevance of O_2 and O_3 Detections," *A&A* **227:** 309.

LOVELOCK, J.E. (1975), "Thermodynamics and the Recognition of Alien Biospheres," *Proc. R. Soc. Lond. B.* **189:** 167.

OWEN, T. (1980), "The Search for Early Forms of Life in Other Planetary Systems: Future Possibilities Afforded by Spectroscopic Techniques," in *Strategies for the Search for Life in the Universe*, M.D. Papagiannis, ed. (Dordrecht: D. Reidel Publishing Company).

SCHNEIDER, J., AND M. CHEVRETON (1990), "The Photometric Search for Earth-Sized Extra-solar Planets by Occultation in Binary Systems," *A&A* **232:** 251–257.

TOPS (1992), *Toward Other Planetary Systems*, NASA.

Enceladus Water Emission Spectrum

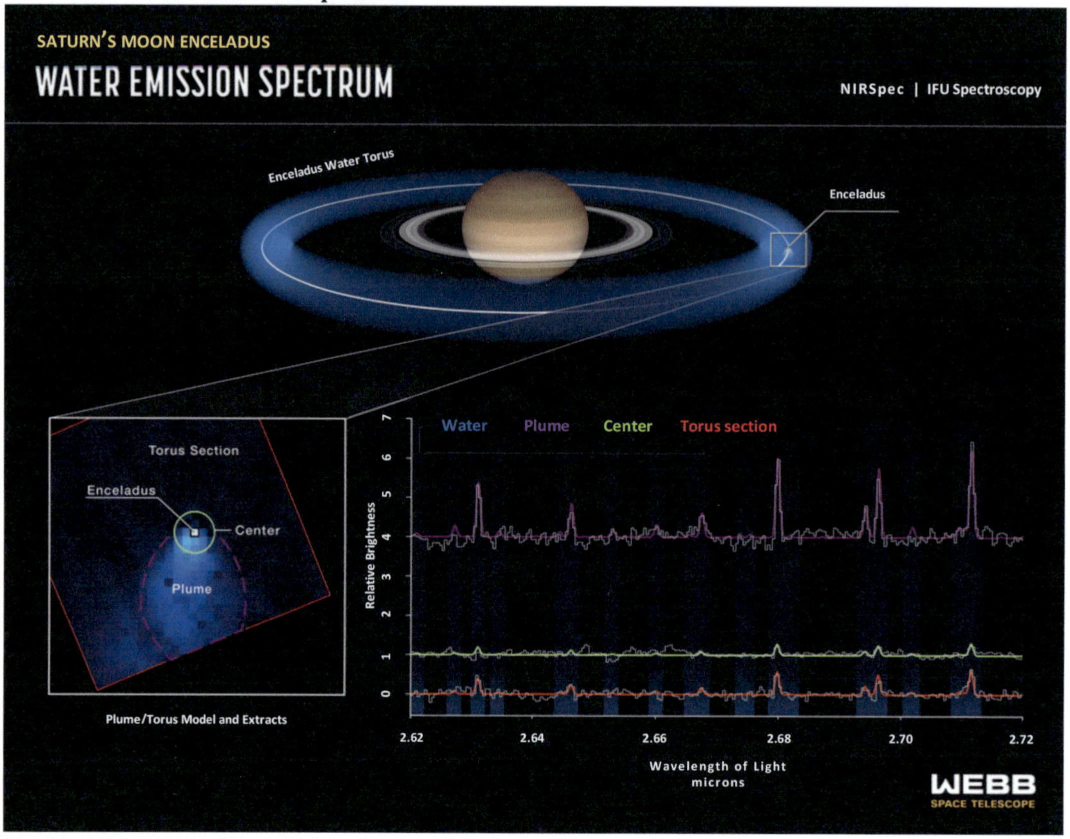

Credit: NASA, ESA, CSA, STScI, L. Hustak (STScI), G. Villanueva (NASA's Goddard Space Flight Center)
https://esawebb.org/images/weic2314b/ [retrieved 6-20-25]

NASA's James Webb Space Telescope's exquisite sensitivity and highly specialized instruments are revealing details into how one of Saturn's moons feeds the water supply for the entire system of the ringed planet. Enceladus, a prime candidate in the search for life elsewhere in our Solar System, is a small moon about four percent the size of Earth. New images from Webb's NIRCam (Near-Infrared Camera) have revealed a water vapour plume jetting from the south pole of Enceladus, extending out forty times the size of the moon itself. The Integral Field Unit (IFU) aboard the NIRSpec (Near-Infrared Spectrograph) instrument also provided insights into how the water from Enceladus feeds the rest of its surrounding environment.

Enceladus orbits around Saturn in just 33 hours, and as it does it sprays water and leaves behind a torus—or 'doughnut'—of material in its wake. This torus is depicted in the top diagram in light blue.

The unique sensitivity of Webb's IFU allowed researchers to detect many spectral features characteristic of water originating from the embedding torus around Enceladus and the plume itself. This simultaneous collection of spectra from the plume and the torus has allowed researchers to better understand their strong relationship. In this spectrum, the white lines are the data from Webb, and the best-fit models for water emission are overlaid in different colours—purple for the plume, green for the area central to the moon itself, and red for the surrounding torus.

This infographic shows a diagram of Saturn, Enceladus, and its torus at the top, the NIRCam image of Enceladus at the bottom left, and the spectra from the NIRSpec instrument at the bottom right.

Biological Aspects of Habitable Zones

Comet Ikeya-Seki

Photo Credit: James W. Young (TMO/JPL/NASA). Image taken on October 30, 1965.
https://commons.wikimedia.org/wiki/File:Ikeyaseki_tail_30Oct1965.jpg

[retrieved 12/12/25]

Ikeya-Seki is a member of the Kreutz sungrazers, a family of sungrazing comets characterized by orbits taking them extremely close to the Sun at perihelion. At the far extreme of their orbits, aphelion, Kreutz sungrazers can be a hundred times farther from the Sun than the Earth is, while their distance of closest approach can be less than twice the Sun's radius. They are believed to be fragments of one large comet that broke up several centuries ago and are named for German astronomer Heinrich Kreutz, who first demonstrated that they were related. These sungrazers make their way from the distant outer Solar System to the inner Solar System, to their perihelion point near the Sun, and then leave the inner Solar System in their return trip to their aphelion.

Several members of the Kreutz family have become great comets, occasionally visible near the Sun in the daytime sky. Comet Ikeya-Seki was the most recent of these, in 1965, and may have been one of the brightest comets in the last millennium. It has been suggested that another cluster of bright Kreutz system comets may begin to arrive in the inner Solar System in the next few decades.

More than 5,000 members of the family have been discovered since the launch of the SOHO satellite in 1995. None of these smaller comets has survived its perihelion passage. Larger sungrazers such as the Great Comet of 1843 and C/2011 W3 (Lovejoy) have survived their perihelion passage. Amateur astronomers have been successful at discovering Kreutz comets in the data available in real time via the internet.

Source: Wikimedia Commons

ARE COMETS NECESSARY FOR A HABITABLE ZONE?

Christopher F. Chyba

White House Fellow, National Security Council
Washington, DC 20506

The question posed by the title of this essay is at present impossible to answer. Are comets necessary for a habitable zone? We cannot yet even answer this question for our own Solar System. Were comets necessary for life on Earth? We do not know. But we have good reasons to think they may have been important.

The inner region of our Solar System, including the habitable zone, is extremely volatile-poor compared with the outer Solar System. McKay (1991) has compiled these data; for example, in units of carbon mass vs. the mass of all heavy atoms (atoms heavier than helium), Earth is depleted in carbon by several orders of magnitude compared with Uranus and Neptune. What is true for carbon holds for the other key "biogenic" volatiles as well, such as nitrogen, or for that matter, water.

So our Solar System presents us with a paradox: in close (but not too close) to the Sun, temperatures are right for liquid water to exist. Liquid water is the *sine qua non* of life as we know it. Yet the elements out of which life is made are extremely rare close to the Sun. Only in the outer Solar System, where temperatures are too low (barring uncertain mechanisms such as tidal and radionuclide heating sources)

Note: The ideas contained in this paper do not necessarily represent the views of the National Security Council or the United States Government.

for liquid water to exist for periods of time comparable to the age of the Solar System, are biogenic elements abundant.

An important question for understanding the origins of life around our own Sun is whether the low level of biogenic elements that is present on the terrestrial planets is coeval with these worlds, or was delivered to their surfaces by cometary or asteroidal bombardment subsequent to their bulk formation. Delsemme (1992) has summarized evidence that planetesimals at Earth's heliocentric distance would have accreted from dust grains too hot to have retained volatile elements. Nevertheless, detailed models of planetary accretion find that in the late stages of planetary formation, substantial radial mixing of planetesimals will occur (Wetherill 1990). Therefore, it is possible that the bulk Earth formed with its present complement of volatiles, brought to it within a number of larger planetesimals originating at greater heliocentric distances. This is why we cannot say with confidence that comets—or volatile-rich asteroids—were essential to the terrestrial origins of life. Perhaps even if there had been no comets, there would still have been enough volatiles present on Earth for life's origins.

There are other data that bear on this problem. Examination of the lunar cratering record suggests that the terrestrial complement of water and other volatiles could have a cometary origin if comets composed some tens of percentages of the impacting flux during the heavy bombardment (Chyba 1990). Moreover, this result is quantitatively consistent (though with admittedly large uncertainties) with siderophile abundances in the lunar crust (Sleep et al. 1989) and terrestrial mantle (Chyba et al. 1994). Finally, there is some evidence from noble gas abundances that appears to support a cometary origin of terrestrial planet volatiles (Owen and Bar-Nun 1993).

Beyond the question of volatile delivery looms the possibility that comets delivered not merely biogenic elements but intact organic molecules of prebiotic interest (Oró 1961, 1996; see also Deamer and Pashley 1989). It is extremely difficult for comets colliding with Earth to successfully deliver intact organics, because of the high temperatures reached in big impacts (Chyba et al. 1990); this problem is just beginning to be carefully explored experimentally (Bunch et al. 1996), although laboratory impact velocities are still too low to simulate actual terrestrial collisions. However, it is possible that a substantial quantity of organics might be synthesized in the cooling fireball subsequent to impact (Oró 1961, Chyba and Sagan 1992, Oberbeck and Aggarwal 1992). Moreover, cometary organics might also make a substantial contribution to the prebiotic organic inventory via arrival in interplanetary dust particles (IDPs) (Anders 1989, Chyba and Sagan 1992). Examination of polar ice samples should test whether IDPs are in fact currently delivering amino acids to the Earth (Bada et al. 1996).

The currently popular model for cometary formation in our own Solar System views comets as outer Solar System planetesimals that were not accreted into the outer planets (Fernández 1985, Weissman 1985). In this scenario, outer planets form through the collection of ice- and volatile-rich planetesimals (comets). However, during this process of formation, vast numbers of these planetesimals are scattered into interstellar space, the region of the Oort cloud, and the inner Solar System. Jupiter and Saturn, because of their large mass and their depth in the Sun's gravitational potential well, scatter most of their icy planetesimals on 10^6–10^7–year time scales into interstellar space. Uranus and Neptune, however, scatter most of their icy planetesimals into the Oort region, on time scales of ~ 10^9 years. Since the latter time scale is on the order of the age of the Solar System, it is likely that an inner cloud of comets continues to reside in the ecliptic, beginning around the orbit of Neptune. Comets that have aphelia raised to Oort cloud distances (10^4–10^5 AU; the nearest stellar system, α-Centauri, is about 4.3 light years or ~ 3×10^5 AU away) subsequently have perihelia raised and inclinations randomized by perturbations from passing stars and molecular clouds. The resulting spherical distribution of comets is the Oort cloud. Note that in the absence of Uranus and Neptune, our own inner Solar System would still have experienced an intense comet flux in its early history from the Jupiter-Saturn region (although on a short time scale, perhaps almost entirely prior to the putative Moon-forming impact), but little or no Oort cloud of comets would have formed. This is a point worth noting: there are readily imaginable solar systems in which cometary volatiles are delivered to the inner reaches of the system, but that nevertheless leave no Oort cloud as a vestige of this early scattering.

Tremaine (1993) has recently examined the formation of Oort clouds around other stars. He derives a series of requirements for a planet of given mass and orbital radius around a given star to be capable of contributing significantly to the formation of an Oort cloud. He finds that there is a rather small region of planetary mass–orbital radius space in which planets can reside and still contribute. It could therefore be that typical solar systems do not retain an Oort cloud of "leftover" icy planetesimals. Yet these systems could still once have scattered vast numbers of comets. The absence of an Oort cloud around a given star does not imply that the possible inner worlds of that system are devoid of volatiles or organics.

To the extent that comets formed out of the accumulation of unprocessed dust grains from the interstellar medium (ISM), the origins of life on Earth may be tied at the molecular level not only to the outer Solar System, but to interstellar space as well. Greenberg (e.g., 1982) is probably the individual most associated with this point of view regarding cometary formation. Whether dust grains from the ISM survived their fall into the primitive solar nebula more or less intact, to be eventually incorporated into comets, is an area of active research (Lunine et al. 1991). Nevertheless, apart from this modeling, we may ask what observational evidence tells us. While the details are tangled, spectroscopic evidence suggests that at least some meteoritic and cometary material is directly derived from the ISM (Pendleton 1996). Not only comets but also the organic chemistry taking place among the stars may have important ties to the origins of life on Earth.

References

Anders, E. (1989), "Pre-biotic Organic Matter from Comets and Asteroids," *Nature* **342:** 255–257.

Bada, J., K.L.F. Brinton, G.D. McDonald, and X. Wang (1996), "A Search for Extra-terrestrial Amino Acids in Polar Ice: A Progress Report," this volume.

Bunch, T.E., L. Becker, J. Bada, J.M. Paque, and J. Macklin (1996), "The Effects of Hypervelocity Impact on Carbonaceous Compounds: Experimental Insights," this volume.

Chyba, C.F. (1990), "Impact Delivery and Erosion of Planetary Oceans in the Early Inner Solar System," *Nature* **343:** 129–133.

Chyba, C., and C. Sagan (1992), "Endogenous Production, Exogenous Delivery and Impact-Shock Synthesis of Organic Molecules: An Inventory for the Origins of Life," *Nature* **355:** 125–132.

Chyba, C.F., P.J. Thomas, L. Brookshaw, and C. Sagan (1990), "Cometary Delivery of Organic Molecules to the Early Earth," *Science* **249:** 366–373.

Chyba, C.F., T.C. Owen, and W.-H. Ip (1994), "Impact Delivery of Volatiles and Organic Molecules to Earth," in *Hazards Due to Comets and Asteroids*, T. Gehrels, ed. (Tucson: University of Arizona Press), pp. 9–58.

Deamer, D.W., and R.M. Pashley (1989), "Amphiphilic Components of the Murchison Carbonaceous Chondrite: Surface Properties and Membrane Formation," *Origins of Life* **19:** 21–38.

Delsemme, A.H. (1992), "Cometary Origin of Carbon, Nitrogen and Water on the Earth," *Origins of Life* **21:** 279–298.

Fernández, J.A. (1985), "The Formation and Dynamical Survival of the Comet Cloud," in *Dynamics of Comets: Their Origin and Evolution*, A. Carusi and G.B. Valsecchi, eds. (Dordrecht: D. Reidel Publishing Company), pp. 45–70.

Greenberg, M. (1982), "What Are Comets Made Of? A Model Based on Interstellar Dust," in *Comets*, L.L. Wilkening, ed. (Tucson: University of Arizona Press), pp. 131–163.

Lunine, J.I., S. Engel, B. Rizk, and M. Horanyi (1991), "Sublimation and Reformation of Icy Grains in the Primitive Solar Nebula," *Icarus* **94:** 333–344.

McKay, C.P. (1991), "Urey Prize Lecture: Planetary Evolution and the Origin of Life," *Icarus* **91:** 93–100.

Oberbeck, V.R., and H. Aggarwal (1992), "Comet Impacts and Chemical Evolution on the Bombarded Earth," *Origins of Life* **21:** 317–338.

Oró, J. (1961), "Comets and the Formation of Biochemical Compounds on the Primitive Earth," *Nature* **190:** 389–390.

ORÓ, J. (1996), "Planetary Formation of Biochemical Compounds in Circumstellar Habitable Zones," this volume.

OWEN, T., AND A. BAR-NUN (1993), "Noble Gases in Atmospheres," *Nature* **361:** 693–694.

PENDLETON, Y.J. (1996), "Interstellar Organic Material: Comparisons with Laboratory and Solar System Spectra," this volume.

SLEEP, N.H., K.J. ZAHNLE, J.F. KASTING, AND H.J. MOROWITZ (1989), "Annihilation of Ecosystems by Large Asteroid Impacts on the Early Earth," *Nature* **342:** 139–142.

TREMAINE, S. (1993), "The Distribution of Comets Around Stars," in *Planets Around Pulsars*, J.A. Phillips, S.E. Thorsett, and S.R. Kulkarni, eds. (San Francisco: Astronomical Society of the Pacific), pp. 335–344.

WEISSMAN, P.R. (1985), "Cometary Dynamics," *Space Sci. Rev.* **41:** 299–349.

WETHERILL, G.W. (1990), "Formation of the Earth," *Ann. Rev. Earth Planet. Sci.* **18:** 205–256.

Super-Earth Exoplanet 55 Cancri e (emission spectrum)

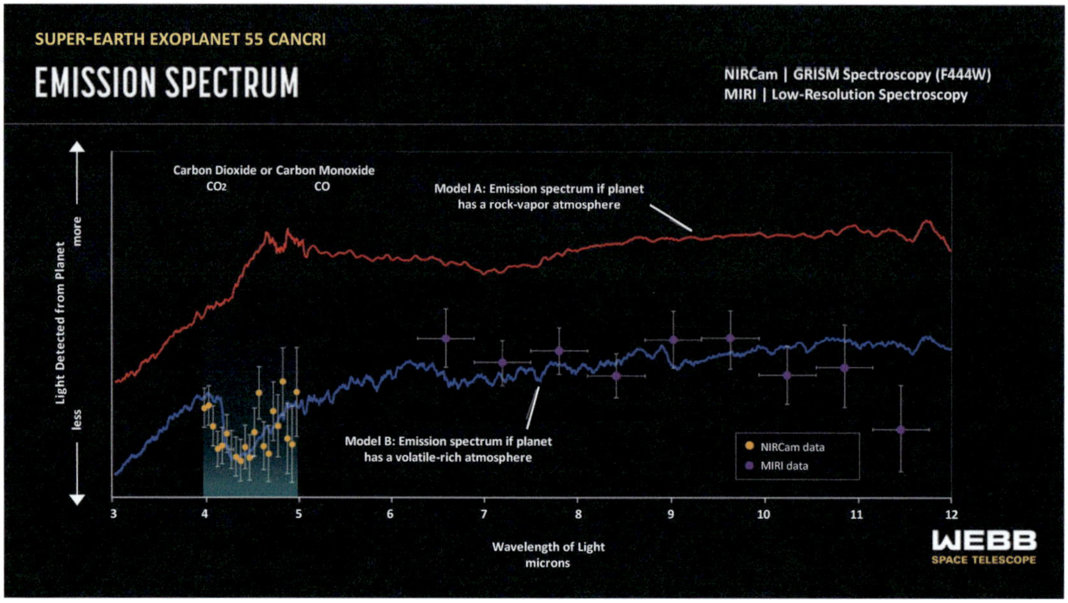

Credit: NASA, ESA, CSA, J. Olmsted (STScI), R. Hu (JPL), A. Bello-Arufe (JPL), M. Zhang (University of Chicago), M. Zilinskas (SRON Netherlands Institute for Space Research)
https://science.nasa.gov/asset/webb/super-earth-exoplanet-55-cancri-e-nircam-miri-emission-spectrum/ [retrieved 6-20-25]

A thermal emission spectrum captured by Webb's NIRCam (Near-Infrared Camera) in November 2022, and MIRI (Mid-Infrared Instrument) in March 2023, shows the brightness (y-axis) of different wavelengths of infrared light (x-axis) emitted by the super-Earth exoplanet 55 Cancri e. The spectrum shows that the planet may be surrounded by an atmosphere rich in carbon dioxide or carbon monoxide and other volatiles, not just vaporized rock.

The graph compares data collected by NIRCam (orange dots) and MIRI (purple dots) to two different models. Model A, in red, shows what the emission spectrum of 55 Cancri e should look like if it has an atmosphere made of vaporized rock. Model B, in blue, shows what the emission spectrum should look like if the planet has a volatile-rich atmosphere outgassed from a magma ocean that has a volatile content similar to Earth's mantle. Both MIRI and NIRCam data are consistent with the volatile-rich model.

The amount of midinfrared light emitted by the planet (MIRI) shows that its dayside temperature is significantly lower than it would be if it did not have an atmosphere to distribute heat from the dayside to the nightside. The dip in the spectrum between 4 and 5 microns (NIRCam data) can be explained by absorption of those wavelengths by carbon monoxide or carbon dioxide molecules in the atmosphere.

The spectrum was made by measuring the brightness of 4- to 5-micron light with Webb's NIRCam GRISM spectrometer, and 5- to 12-micron light with the MIRI low-resolution spectrometer, before, during and after the planet moved behind its star. The amount of each wavelength emitted by the planet (y-axis) was calculated by subtracting the brightness of the star alone (during the stellar occultation of the planet) from the brightness of the star and planet combined (before and after the occultation). Each observation lasted about eight hours. Note that the NIRCam data have been shifted vertically to align with Model B. Although the differences in brightness between each wavelength in the NIRCam band were derived from the observation (the data suggest a valley between 4 and 5 microns), the absolute brightness (the vertical position of that valley) could not be measured precisely because of noise in the data.

Planetary Formation of Biochemical Compounds in Circumstellar Habitable Zones

J. Oró

Department of Biochemical and Biophysical Sciences
University of Houston
Houston, TX 77204-5934

Abstract

The formation of the Earth-Moon system as a result of a catastrophic collision led to an Earth essentially devoid of water and other volatiles. These volatiles had to be acquired by late accretion processes from comets and other minor bodies of the Solar System. Comets contain large amounts of water and carbon compounds, a substantial fraction of which was captured by the primitive Earth. The captured carbon compounds from comets together with those synthesized on the primitive Earth by electrical discharges and other forms of energy, generated the biochemical molecules that eventually led to the emergence of life on our planet. The presence of comets in circumstellar disks, such as β-Pictoris, suggests that comets play a crucial role for the appearance of life in the habitable zones of other planetary systems.

Introduction

The Earth resides within the relatively narrow confines of the circumstellar habitable zone of the Solar System. The physical and chemical conditions necessary for life to appear and evolve on a planet such as ours have been described previously

(Oró et al. 1982, Oró 1988). Perhaps three of the most basic conditions or requirements are the presence of liquid water, sufficient concentrations of organic molecules, and a continuous supply of energy. These three conditions are satisfied by a planet such as the Earth with the right orbit and distance from the central star, and with the appropriate gravity to be able to retain water in the liquid state and volatile molecules of the biogenic elements in its atmosphere and hydrosphere.

The primary aim of this report is to summarize briefly the formation of the Earth-Moon system and the role that comets may have played in bringing to the primitive Earth water and simple precursors for the synthesis of biochemical compounds since it was first suggested (Oró 1961). Related studies from this laboratory on the planetary evolution aspects have been published (Oró and Mills 1989, Oró et al. 1992, Oró and Lazcano 1997). The prebiotic chemistry and other chemical evolutionary processes that may have been involved in the origin and early evolution of life on the Earth have been reviewed elsewhere (Miller and Orgel 1974, Oró et al. 1990, Deamer and Fleischaker 1994).

Formation of the Earth-Moon System

As pointed out by Cameron and other investigators (e.g., Cameron and Benz 1991), the formation of the Earth-Moon system resulted from a giant impact, namely, the collision of a large near-Mars-sized body with the protoearth, upon which all the iron of the impactor was injected into the Earth's core. Both the impactor and the Earth's mantle were melted, and a mass the size of the Moon was ejected into Earth's orbit, which subsequently coalesced into the Earth's only natural satellite. This theory not only explains the angular momentum evolution of the Earth-Moon system, but also the iron-poor nature of the Moon and the similar physical and chemical characteristics of the Earth's mantle and the Moon. The impact of the collision was such that both the Earth and the Moon became essentially devoid of any volatiles.

Late Accretion Capture of Cometary Matter

Once the Earth-Moon system was formed, it remained at relatively high temperature for some time. It took probably about 100 million years for the Earth to cool to a temperature of less than 100°C. Any collisions of comets and asteroids that may have occurred during this early period would have partially led to either melting, evaporation (generating an early atmosphere), or loss to interplanetary space of the in-falling matter. After the Earth was cooled, this bombardment process continued during the following 500 million years, allowing the Earth to acquire essentially all of the water of the oceans and carbon compounds. Several calculations of the cometary matter trapped by the Earth give the following values: 1×10^{23} g (Oró et al. 1980), $6 \times 10^{24-25}$ g (Ip and Fernández 1988), $1 \times 10^{23-26}$ (Chyba et al. 1990), and $1 \times 10^{24-25}$ g (Delsemme 1991, 1992).

This would have generated a substantial atmosphere made, to some extent, of internal terrestrial gases, but primarily of gases from comets and minor bodies impacting the Earth. The atmosphere must have contained neutral molecules such as CO_2, N_2, and H_2O vapor, and reduced molecules such as CH_4, NH_3 and small-molecular-weight carbon compounds. Such an atmosphere should have reflected the parent compounds present in comets, and may have had some similarities with the atmosphere of Titan, which is made of N_2, CH_4, HCN, HC_3N, and low-molecular-weight hydrocarbons (Sagan et al. 1992). However, there is an important difference. Due to the low temperature of Titan, the CO_2 and water are frozen in this satellite.

In support of the interaction of comets with the Earth, it is known that the D/H ratios of comet Halley and terrestrial oceanic water are identical (Bailey et al. 1990). Other studies (Owen et al. 1992) show that the terrestrial noble gas composition also reflects a cometary composition. The D/H ratio of Halley's comet is also identical to that of Titan (Bailey et al. 1990), the largest Saturn satellite, which has a thick atmosphere apparently also formed by late accretion from planetesimals and comets (Zanhle and Dones 1992).

During the end of the late accretion process, the in-falling cometary matter, either small comets or cometary dust, would have been slowly decelerated and eventually captured by the primitive atmosphere and hydrosphere without significant change (Anders 1989, Steel 1992). Even if the energy of a cometary interaction with the Earth's atmosphere was high enough to cause an explosion, the radicals and carbon compounds generated by the explosion, occurring in a non-oxidizing atmosphere, would be readily converted in the presence of water to precursors of biochemical compounds.

The 1908 Tunguska Event

The capture and preservation of some of the compounds from the in-falling body may occur even in the presence of an oxidized atmosphere, as indicated by some observations made on the atmospheric explosion that occurred about 10 kilometers above sea level in the Tunguska region of Siberia, June 30, 1908. For instance, it was established from investigations by the Meteoritic Committee of the USSR Academy of Sciences that trees in the periphery of the devastated forest (~ 100 km) grew six to seven times faster than normal after the explosion (Fesenkov 1971). Measurements by Russian botanists showed that the thickness of the tree rings increased dramatically during 1908 and 1909, but started to return to normal in

1910. Another pertinent observation made recently by French geochemists is that the analysis of Greenland ice cores shows the presence of a major ammonium formate spike (M. Legrand), and solid aerosols due to carbonaceous combustion (H. Cachier) in sections of the sediment corresponding to the year 1908 (Peel 1993). Furthermore, it is known that in the subsurface site of the Tunguska soil no rocky fragments were found. Only microscopic spherules of iron-nickel of cosmic composition and meteoritic dust have been found (Krinov 1960). It is also known that iron and nickel are two important constituents of comets.

A relatively simple explanation for these three observations is that the explosion was caused by the nucleus of a relatively small comet, as suggested by Astapovich and Whipple (Krinov 1960) and Fesenkov (1971), or a fragment of comet Encke, as calculated by Kresak (1978), measuring Encke's orbit. The alternative explanation of a stony object (Chyba 1993) may also be compatible with a highly dense solid fragment of a comet, or a carbonaceous object entering the atmosphere at about a 60° angle (Chyba et al. 1993).

At any rate, whatever the nature of the object was, it has to explain the spectacular growth of the trees, suddenly initiated in 1908. This is readily explained by a massive injection of ammonium salts and other cometary material (Fesenkov 1971) into the Tunguska soil by the shock waves of the atmospheric cometary explosion. The 1908 ammonium formate spike found in the Greenland ice cores stands in support of this explanation. The absence of any rocky fragments, except the cosmic iron-nickel microspherules buried in the soil, also indicates a likely cometary origin.

The K-T Boundary Event (65×10^6 y BP)

It would be of interest to examine the same 1908 Greenland ice core sections, in order to determine the possible presence of α-amino isobutyric acid (AIB) and isovaline, as was done in the case of the Cretaceous-Tertiary (K-T) boundary sediments from Denmark (Zhao and Bada 1989). If present, it would indicate that the 1908 Tunguska colliding body was similar in nature (not in size) to the one that generated the K-T boundary sediments (Alvarez et al. 1980), which also contain AIB as just indicated in addition to the extraterrestrial marker iridium. The K-T boundary sediments also contain amorphous carbon, presumably resulting from major forest fires.

If the results on the presence of AIB in the 1908 Tunguska explosion sediments were negative, it would not necessarily rule out the cometary nature of the extraterrestrial body, since apparently Halley's comet does not contain preformed amino acids (Kissel and Krueger 1987). The question that remains unanswered is whether the amino acid precursors in comets (HCN, aldehydes, NH_3 and H_2O)

would combine in the Earth's atmosphere after the atmospheric explosion of a comet. This may be possible as suggested by Zahnle and Grinspoon (1990). According to these authors, the AIB and other amino acids found in the sediments above and below the K-T boundary layer by Zhao and Bada (1989) were derived from the cometary dust falling into the Earth before and after the major impact of the extraterrestrial body. Finding more definite answers to questions concerned with the amino acid and organic compound formation (Bar-Nun et al. 1970) and capture (Steel 1992) from these types of extraterrestrial interactions, in which the explosion of an extraterrestrial object upon its encounter with the Earth is immediately followed by the generation of shock waves, will help us refine our knowledge on the nature of the two celestial bodies responsible for the Tunguska flattened forest and the K-T boundary impact. The latter impact presumably generated one of the largest known craters on Earth, the Chicxulub crater, at the Yucatán peninsula (Sharpton et al. 1993).

Reactions of Cometary Molecules

An overall summary of the abiotic synthesis of biochemical compounds that may be derived from simple interstellar or cometary molecules is given in Table 1.

Table 1. ***Biochemical Monomers and Properties That Can Be Derived from Interstellar and Cometary Molecules***

Cometary Molecules	*Formulae and Reactants*	*Biochemical Monomers and Properties*
1. Hydrogen	H_2	Reducing agent
2. Water	H_2O	Universal solvent
3. Ammonia	NH_3	Amination and catalysis
4. Carbon Monoxide	CO (+ H_2)	Fatty acids*
5. Formaldehyde	CH_2O	Ribose and glycerol
6. Aldehydes	RCHO (+ HCN and NH_3)	Amino acids
7. Hydrogen Sulfide	H_2S (+ other precursors)	Cysteine and methionine
8. Hydrogen Cyanide	HCN	Purines and amino acids
9. Cyanoacetylene	HC_3N (+ cyanate)	Pyrimidines
10. Phosphate (IDP) and Pyrophosphate	PO_4^{3-} and P_2O_5	Phosphates and nucleotides
11. Cyanamide	H_2CN	Oligonucleotides

Table 1. (cont.)

* *Fatty acids could also be derived from reduced and hydrolyzed cyano-oligo acetylenes, which have been detected in interstellar space.*

Hydrogen is the most abundant diatomic molecule in the Universe. It was the major component of the Solar Nebula, as it is now of the Sun and the Jovian planets. In comets, its elemental abundance (compared with solar abundances) is depleted, but it exists in substantial amounts combined in the form of water, ammonia, and organic compounds.

Water is the most abundant triatomic molecule in the Universe and also in comets. Its presence in liquid form is an absolute requirement for the emergence of life on a planet. Its absence in liquid form in comets also rules out the possibility of life emerging in these celestial bodies (Bar-Nun et al. 1981). It was obviously the solvent for the abiotic synthesis of biochemical compounds on the primitive Earth.

Ammonia was the first neutral molecule detected in interstellar space by microwave spectroscopy. NH_3 and CN are the dominant N carriers observed in the coma gas of Halley's comet (Wyckoff 1992). NH_3 provides the amino group of amino acids and acts as a moderate basic catalyst for the condensation of certain dipolar organic compounds into polymers—for instance, the condensation of hydrogen cyanide into the polymeric HCN (Völker 1960), which has been suggested to be present in comets (Matthews and Ludicky 1986, 1992).

Carbon monoxide alone cannot generate complex organic compounds. However, in the presence of molecular hydrogen in the Solar Nebula, it could have generated aliphatic hydrocarbons and fatty acids with the help of catalysts (Anders et al. 1974, Leach et al. 1978). Also, as shown by Löb (1913) and Miller (1974), carbon monoxide, ammonia, and water vapor could have produced amino acids by the action of electrical discharges in the terrestrial atmosphere. It should, however, be noted that unlike the work of Miller (1974), the experiments of Löb (1913) were not done within the context of the origin of life studies.

Formaldehyde and its homopolymer polyoxymethylene are abundant in Halley's comet (Huebner et al. 1989). It is well known that formaldehyde, by base catalysis, gives rise to the formation of monosaccharides (Butlerow 1861). Yet the formation of ribose is a problem that has not been completely solved since the reaction is not sufficiently selective (Shapiro 1988). There is a similar situation with the synthesis of 2-deoxyribose (Oró 1965).

Aliphatic aldehydes have been detected in interstellar space, and several are present in comets. They are also formed by the action of electrical discharges on methane, ammonia, and water, and are the precursors for the prebiotic synthesis of α-amino acids as shown by Miller in his pioneering experiments (Miller 1953, 1957).

Hydrogen sulfide and other sulfur compounds are present in comets (A'Hearn 1992). They are necessary for the formation of the sulfur-containing amino acids cysteine and methionine, which are present in proteins. Carbonaceous chondrites contain sulfur compounds. However, these two amino acids have not been found in the Murchison meteorite (Cronin 1989).

Hydrogen cyanide is prevalent in comets. It is necessary for the prebiotic synthesis of adenine (Oró 1960) and other purines and biochemical compounds (Oró and Kimball 1962). Hydrogen cyanide in aqueous ammonia solution can generate amino acids through the intermediate formation of HCN oligomers (Oró and Kamat 1961).

Cyanoacetylene is probably abundant in comets. It can condense with cyanate to form cytosine and uracil (Sánchez et al. 1966), which are the two pyrimidine bases present in RNA. Thymine (present in DNA) is more difficult to synthesize (Stephen-Sherwood et al. 1971).

Phosphates. The analysis of interplanetary dust particles (IDPs) of presumed cometary origin has given mass spectra corresponding to orthophosphate and pyrophosphate. These two compounds provide the phosphate necessary for the formation of mononucleotides.

Cyanamide is an interstellar molecule that in the laboratory has been shown to aid in the condensation of amino acids to peptides (Hawker and Oró 1981) and nucleotides to oligonucleotides (Sherwood et al. 1977). Comets may have other condensing agents, such as 2-methylimidazole, which forms activated nucleotides to yield polynucleotides (Bridson and Orgel 1980).

Stability of Organic Compounds

Whereas many biochemical compounds are labile to high temperatures, some of their chemical precursors are not. Even though it is not commonly recognized, some simple key carbon compounds such as CN, C_2 and CO, present in stellar atmospheres at about 6000 K, are more thermally stable than the most stable inorganic diatomic molecule, ZrO (Aller 1961). The simple cometary molecules listed in Table 1 are sufficiently stable to survive capture as in-falling matter by the primitive Earth. In the presence of liquid water they would eventually lead to the formation of many organic compounds. Thus, although possible, it is not necessary for the Earth to capture preformed biochemical compounds. In addition to their presence in comets, such compounds are probably also present in dark asteroids, interplanetary dust particles, and meteorites. It is well known that carbonaceous chondrites bring to the Earth amino acids, purines, and other biochemical compounds. It has also been suggested that cometary dust (Anders 1989, Steel 1992) and comets in grazing orbits (Clark 1988) may be more gently captured without significant change of their preformed biochemical compounds.

In summary, with the exception of cyanamide, all of the above molecules are found in comets, and most are quite stable, even at high temperatures. So, regardless of how cometary matter is captured, there should be no problem for these cometary precursors to be converted into the biochemical molecules essential for life.

PREFORMED BIOCHEMICAL COMPOUNDS IN COMETS

A number of organic compounds have been detected in the dust of comet Halley using the mass spectrometers on board the Vega and Giotto spacecrafts. These include low-molecular-weight hydrocarbons, aldehydes, nitriles, and heterocyclic compounds, including adenine (Kissel and Krueger 1987).

The polymerization of HCN can occur in the solid state and at low temperatures catalyzed by either ammonia or solar UV photons. Therefore, it would not be surprising if adenine were present in comets in significant amounts. In fact, HCN is the dominant parent molecule of CN (Wyckoff 1992) and the CN spectral line was one of the first lines observed in Halley's comet when it was still 4 AU from the Sun. Also, the surface albedo of comets is very low. So far the best explanation for this low albedo is the presence of black polymeric hydrogen cyanide (Matthews and Ludicky 1986, 1992).

Kissel and Krueger (1987) also observed the absence of any mass spectra corresponding to amino acids in comet Halley. At first, this would appear strange. However, one has to realize that even if the precursor molecules are there—namely aldehydes, ammonia, water and HCN—the water is frozen in the solid state, and is not in the liquid state, which would be required for the cyanohydrin-Strecker condensation (Miller 1957). Thus, whereas amino acids may not be present in comets, their precursors would be able to generate amino acids as soon as they were to encounter water in the liquid state (or vapor state) upon capture by the primitive Earth hydrosphere or atmosphere.

BIOCHEMICAL COMPOUNDS THAT COULD HAVE BEEN FORMED FROM HCN ON THE PRIMITIVE EARTH

As shown above, most of the biochemical compounds essential to life can be generated from precursor cometary molecules that are not destroyed by capture. Perhaps one of the most abundant and relevant of these precursors is hydrogen cyanide. In addition to adenine there are many other biochemical molecules that can be derived from HCN in aqueous solution in the presence of ammonia, for instance, simple C_1 compounds, aminoimidazole derivatives, and amino acids.

Among the first group we have formamidine, formamide, formic acid, and urea, which represent products of different reactions of ammonolysis, hydrolysis and rearrangement of HCN. Among the second group, we have primarily 4-amino-5-imidazole-carboxamidine and 4-amino-5-imidazole-carboxamide, which are the products of condensation and cyclization of HCN (Oró and Kimball 1962). The third group includes some α-amino acids. The amino acids are synthesized by reactions of HCN in aqueous ammonia. They involve first the condensation of HCN into dimer, trimer, and tetramer, followed by hydrolytic and degradative reactions. Thus the trimer aminomalonodinitrile can be converted into glycinamide and glycine by hydrolysis and decarboxylation. Similar reactions from the HCN tetramer give rise to the formation of alanine and aspartic acid (Oró and Kamat 1961).

It is of interest that these three amino acids can not only be obtained from aqueous ammonia solutions of HCN, but they are also three of the most abundant in experiments with electric discharges (Miller 1953). Other amino acids and biochemical compounds can be obtained from the hydrolysis of polymeric hydrogen cyanide. Briefly, they include hypoxanthine (Lowe et al. 1963), orotic acid (Ferris et al. 1978), and polyamidines, which could be converted to peptides by hydrolysis (Matthews and Ludicky 1992).

Concentration of Organic Compounds

If, as pointed out, the Earth-Moon system was formed by the catastrophic impact as described by the computer models of Cameron and Benz (1991), the Earth would be devoid of liquid water. Thus, the water had to be acquired from exogenous sources. The exogenous sources are planetesimals, comets, asteroids, and minor bodies of the Solar System, including carbonaceous chondrites and interplanetary dust particles.

It is obvious that when these bodies interact with the Earth and a fraction of their matter is captured, the water comes together with inorganic and organic compounds. Therefore, the concentration of organic compounds in water should be comparable to that present in the in-falling body. In comets it is about 20% and in the other minor bodies it should be higher since they have less water. So, the concentration of organic compounds immediately after their capture should have been relatively high. This would facilitate the processes of prebiotic synthesis and eventually molecular organization necessary to generate a protocell, or the simplest living entity (Lazcano 1994).

On the basis of the lunar cratering record, we know that the period from the initial moments of formation of the Solar System up until 3.9×10^9 years ago was a time of continuous bombardment and growth of planetary embryos by planetesimals. During the period of late accretion by the Earth, the rate of capture of water and carbon molecules as precursors of biochemical compounds must have been quite high (Delsemme 1991, 1992), and certainly much higher than the in-fall of meteorites, interplanetary dust particles and comets that occurred during the biological era. Therefore, it would not seem appropriate to extrapolate current values of organic matter in-fall to the first 500 million years of the Archean prebiological era during which time the process of late accretion was taking place.

Summary and Conclusions

Recent studies have given us more complete information on the formation of the Earth-Moon system and the key role that comets played in contributing water and a veneer of inorganic and organic matter to the primitive Earth. Thus it is of fundamental importance that the Rosetta or other future comet missions be undertaken to analyze in situ the composition of at least one comet. This would provide precise and much-needed knowledge pertaining to the inorganic and organic composition of these very interesting minor bodies of the Solar System.

On the basis of the observations of comets in the β-Pictoris disk (Beust et al. 1990, Oró et al. 1992), one could generalize the formation of comets to other planetary systems in our Galaxy. Assuming that other cosmic constraints are met (Oró et al. 1982), this would provide a universal mechanism for the appearance of life within circumstellar habitable zones. Detection of what may be the first planetary system with Earth-sized planets beyond our Solar System has been reported recently (Wolszczan 1994).

Acknowledgments

Support for this work has been received from NASA Grant NAGW 2788. I would like to acknowledge the continued cooperation of Thomas M. Mills and Antonio Lazcano. Thanks to Lucila Soria for help in preparing this manuscript.

References

A'Hearn, M.F. (1992), "Cometary Chemistry," in *Astrochemistry of Cosmic Phenomena*, P.D. Singh, ed. (Netherlands: IAU), pp. 415–420.

Aller, L.H. (1961), *The Abundance of the Elements* (New York: Interscience Publishers, Inc.), pp. 220–231.

Alvarez, L.W., W. Alvarez, F. Asaro, and H.V. Miche (1980), "Extraterrestrial Cause for the Cretaceous-Tertiary Extinction," *Science* **208:** 1095–1108.

ANDERS, E. (1989), "Pre-biotic Organic Matter from Comets and Asteroids," *Nature* **342:** 255–257.

ANDERS, E., R. HAYATSU, AND M.H. STUDIER (1974), "Catalytic Reactions in the Solar Nebula: Implications for Interstellar Molecules and Organic Compounds in Meteorites," in *Cosmochemical Evolution and the Origins of Life*, vol. 1, J. Oró, S.L. Miller, C. Ponnamperuma, and R.S. Young, eds. (Dordrecht: D. Reidel Publishing Company), pp. 57–67.

BAILEY, M.E., S.V.M. CLUBE, AND W.N. NAPIER (1990), *The Origin of Comets* (Oxford: Pergamon Press), pp. 451–454.

BAR-NUN, A., N. BAR-NUN, S.H. BAUER, AND C. SAGAN (1970), "Shock Synthesis of Amino Acids in Simulated Primitive Environments," *Science* **168:** 470–473.

BAR-NUN, A., A. LAZCANO-ARAUJO, AND J. ORÓ (1981), "Could Life Have Originated in Cometary Nuclei?," *Origins of Life* **11:** 387–394.

BEUST, H., A.M. LAGRANGE-HENRI, A. VIDAL-MAJDAR, AND R. FERLET (1990), "The β Pictoris Circumstellar Disk X. Numerical Simulations of Infalling Evaporating Bodies," *A&A* **236:** 202–216.

BRIDSON, P.K., AND L.E. ORGEL (1980), "Catalysis of Accurate poly(C)-Directed Synthesis of 3′-5′- Linked Oligoguanylates of Zn^{++}," *J. Mol. Biol.* **144:** 567–577.

BUTLEROW, A. (1861), "Formation Sintetique d'une Substance Sucreé," *Compte. Rend. Acad. Sci.* **53:** 145–147.

CAMERON, A.G.W., AND W. BENZ (1991), "The Origin of the Moon and the Single Impact Hypothesis IV," *Icarus* **92:** 204–216.

CHYBA, C.F. (1993), "Explosions of Small Spacewatch Objects in the Earth's Atmosphere," *Nature* **363:** 701–703.

CHYBA, C.F., P.J. THOMAS, L. BROOKSHAW, AND C. SAGAN (1990), "Cometary Delivery of Organic Molecules to the Early Earth," *Science* **249:** 366–373.

CHYBA, C.F., P.J. THOMAS, AND K.J. ZAHNLE (1993), "Atmospheric Disruption of Small Comets and Asteroids and the 1908 Tunguska Explosion," *Nature* **361:** 40–44.

CLARK, B.C. (1988), "Primeval Procreative Comet Pond," *Origins of Life* **18:** 209–238.

CRONIN, J.R. (1989), "Origin of Organic Compounds in Carbonaceous Chondrites," *Adv. Space Res.* **9:** 54–64.

DEAMER, D.W., AND G.R. FLEISCHAKER (1994), *Origins of Life: The Central Concepts* (Boston: Jones and Bartlett Publishers), p. 431.

DELSEMME, A.H. (1991), "Nature and History of the Organic Compounds in Comets: An Astrophysical View," in *Comets in the Post-Halley Era*, vol. 1, R.L. Newburn, Jr., M. Neugebauer, and J. Rahe, eds. (Dordrecht: Kluwer Academic Publishers), pp. 377–428.

DELSEMME, A.H. (1992), "Cometary Origin of Carbon, Nitrogen and Water on the Earth," *Origins of Life* **21:** 279–298.

FERRIS, J.P., P.C. JOSHI, E.H. EDELSON, AND J.G. LAWLESS (1978), "HCN: A Plausible Source of Purines, Pyrimidines and Amino Acids on the Primitive Earth," *J. Mol. Evol.* **11:** 293–311.

FESENKOV, V.G. (1971), "Organic Substances in the Universe," in *Chemical Evolution and the Origin of Life*, R. Buvet and C. Ponnamperuma, eds. (Amsterdam: North-Holland Publishing Company), pp. 495–498.

HAWKER, J.R., JR., AND J. ORÓ (1981), "Cyanamide Mediated Syntheses of Leu, Ala, and Phe Peptides Under Plausible Primitive Earth Conditions," in *Origins of Life*, Y. Wolfman, ed. (Dordrecht: D. Reidel Publishing Company), pp. 225–232.

HUEBNER, W.F., D.C. BOICE, AND A. KORTH (1989), "Halley Polymeric Organic Molecules," *Adv. Space Res.* **9:** 29–32.

IP, W.-H., AND J.A. FERNÁNDEZ (1988), "Exchange of Condensed Matter Among the Outer and Terrestrial Protoplanets and the Effect on Surface Impact and Atmospheric Accretion," *Icarus* **74:** 47–61.

KISSEL, J., AND F.R. KRUEGER (1987), "The Organic Component in Dust from Comet Halley as Measured by the PUMA Mass Spectrometer on Board Vega 1," *Nature* **326:** 755–760.

KRESAK, L. (1978), "The Tunguska Object: A Fragment of Comet Encke?," *Bull. Astron. Inst. Czechosl.* **29:** 129–134.

KRINOV, E.L. (1960), *Principles of Meteoritics* (Oxford: Pergamon Press), pp. 120–137.

LAZCANO, A. (1994), "The Transition from Non-Living to Living," in *Early Life on Earth*, Nobel Symposium no. 84, S. Bengtson, ed. (New York: Columbia University Press), pp. 60–69.

LEACH, W.W., D.W. NOONER, AND J. ORÓ (1978), "Abiotic Synthesis of Fatty Acids," in *Origin of Life*, H. Noda, ed. (Tokyo: Center for Academic Publications), pp. 113–122.

LÖB, W. (1913), "Uber das Verhalten des Formamids unter der Wirkung der stillen Entladung. Ein Beitrag zur Frage der Stickstoff-Assimilation," *Berichte der Deutschen Chemischen Gesellschaft* **46:** 684–697.

LOWE, C.U., M.W. REES, AND R. MARKHAM (1963), "Synthesis of Complex Organic Compounds from Simple Precursors: Formation of Amino-Acids, Amino-Acid Polymers, Fatty Acids and Purines from Ammonium Cyanide," *Nature* **199:** 219–222.

MATTHEWS, C.N., AND R.A. LUDICKY (1986), "The Dark Nucleus of Comet Halley: Hydrogen Cyanide Polymers," in *20th ESLAB Symposium on the Exploration of Halley's Comet*, ESA Report SP-250, vol. 2, B. Battrick, E.J. Rolfe, and R. Reinhard, eds. (Noordwijk: European Space Agency), pp. 273–277.

MATTHEWS, C.N., AND R. LUDICKY (1992), "Hydrogen Cyanide Polymers on Comets," *Adv. Space Res.* **12 (4):** 21–32.

MILLER, S.L. (1953), "A Production of Amino Acids Under Possible Primitive Earth Conditions," *Science* **117:** 528–529.

MILLER, S.L. (1957), "The Mechanism of Synthesis of Amino Acids by Electric Discharges," *Biochimica et Biophysica Acta* **23:** 480–489.

MILLER, S.L. (1974), "The First Laboratory Synthesis of Organic Compounds Under Primitive Earth Conditions," in *The Heritage of Copernicus: Theories "Pleasing to the Mind,"* J. Neyman, ed. (Cambridge, MA: MIT Press), pp. 228–242.

MILLER, S.L., AND L.E. ORGEL (1974), *The Origins of Life on Earth* (Englewood Cliffs, NJ: Prentice-Hall), p. 229.

ORÓ, J. (1960), "Synthesis of Adenine from Ammonium Cyanide," *Biochem. Biophys. Res. Comm.* **2:** 407–412.

ORÓ, J. (1961), "Comets and the Formation of Biochemical Compounds on the Primitive Earth," *Nature* **190:** 389–390.

ORÓ, J. (1965), "Stages and Mechanisms of Prebiological Organic Synthesis," in *The Origins of Prebiological Systems*, S.W. Fox, ed. (New York: Academic Press Inc.), pp. 137–171.

ORÓ, J. (1988), "Constraints Imposed by Cosmic Evolution Towards the Development of Life," in *Bioastronomy—The Next Steps*, G. Marx, ed. (Dordrecht: Kluwer Academic Publishers), pp. 161–165.

ORÓ, J., AND S.S. KAMAT (1961), "Amino Acid Synthesis from Hydrogen Cyanide Under Possible Primitive Earth Conditions," *Nature* **190:** 442–443.

ORÓ, J., AND A.P. KIMBALL (1962), "Synthesis of Purines Under Possible Primitive Earth Conditions. II. Purine Intermediates from Hydrogen Cyanide," *Arch. Biochem. Biophys.* **96:** 293–313.

ORÓ, J., AND T. MILLS (1989), "Chemical Evolution of Primitive Solar System Bodies," *Adv. Space Res.* **9:** 105–120.

ORÓ, J., AND A. LAZCANO (1997), "Comets and the Origin and Evolution of Life: A Scientific and Historical Introduction," in *Comets and the Origin and Evolution of Life*, P.J. Thomas, C.F. Chyba, and C.P. McKay, eds. (New York: Springer-Verlag), pp. 3–27.

ORÓ, J., G. HOLZER, AND A. LAZCANO-ARAUJO (1980), "The Contribution of Cometary Volatiles to the Primitive Earth," in *Cospar Life Sciences and Space Research*, vol. XVII, R. Holmquist, ed. (Oxford: Pergamon Press), pp. 67–82.

ORÓ, J., K. REWERS, AND D. ODOM (1982), "Criteria for the Emergence and Evolution of Life in the Solar System," *Origins of Life* **12:** 285–305.

ORÓ, J., S.L. MILLER, AND A. LAZCANO (1990), "The Origin and Early Evolution of Life on Earth," *Ann. Rev. Earth Planet. Sci.* **18:** 317–356.

ORÓ, J., T. MILLS, AND A. LAZCANO (1992), "Comets and the Formation of Biochemical Compounds—A Review," *Origins of Life* **21:** 267–277.

OWEN, T., A. BAR-NUN, AND I. KLEINFELD (1992), "Possible Cometary Origin of Heavy Noble Gases in the Atmospheres of Venus, Earth and Mars," *Nature* **358:** 43–46.

PEEL, D.A. (1993), "Cold Answers to Hot Issues," *Nature* **363:** 403–404.

SAGAN, C., W.R. THOMPSON, AND B.N. KHARE (1992), "Titan: A Laboratory for Prebiological Organic Chemistry," *Accounts of Chemical Research* **25**: 286–292.

SÁNCHEZ, R.A., J.P. FERRIS, AND L.E. ORGEL (1966), "Cyanoacetylene in Prebiotic Synthesis," *Science* **154:** 784–786.

SHAPIRO, R. (1988), "Prebiotic Ribose Synthesis: A Critical Analysis," *Origins of Life* **18:** 71–85.

SHARPTON, V.L., K. BURKE, A. CAMARGO-ZANOGUERA, S.A. HALL, D.S. LEE, L.E. MARIN, G. SUÁREZ-REYNOSO, J.M. QUEZADA-MUÑETON, P.D. SPUDIS, AND J. URRUTIA-FUCUGAUCHI (1993), "Chicxulub Multiring Impact Basin: Size and Other Characteristics Derived from Gravity Analysis," *Science* **261:** 1564–1567.

SHERWOOD, E., A. JOSHI, AND J. ORÓ (1977), "Cyanamide Mediated Synthesis Under Plausible Primitive Earth Conditions. II. The Polymerization of Deoxythymidine 5′-Triphosphate," *J. Mol. Evol.* **10:** 211–220.

STEEL, D. (1992), "Cometary Supply of Terrestrial Organics: Lessons from the K/T and the Present Epoch," *Origins of Life* **21**: 339–357.

STEPHEN-SHERWOOD, E., J. ORÓ, AND A.P. KIMBALL (1971), "Thymine: A Possible Prebiotic Synthesis," *Science* **173:** 446–447.

VÖLKER, T. (1960), "Polymere Blausäure," *Angew. Chem.* **72:** 379–384.

WOLSZCZAN, A. (1994), "Confirmation of Earth-Mass Planets Orbiting the Millisecond Pulsar PSR B1257+12," *Science* **264:** 538–542.

WYCKOFF, S. (1992), "Cometary Constraints on the Planet Forming Environment," *Adv. Space Res.* **12:** 13–20.

ZAHNLE, K., AND D. GRINSPOON (1990), "Comet Dust as a Source of Amino Acids at the Cretaceous/Tertiary Boundary," *Nature* **348:** 157–160.

ZAHNLE, K., AND L. DONES (1992), "Impact Origin of Titan's Atmosphere," in *Proceedings Symposium on Titan,* ESA SP-338, pp. 19–25.

ZHAO, M., AND J.L. BADA (1989), "Extraterrestrial Amino Acids in Cretaceous/Tertiary Boundary Sediments at Stevns Klint, Denmark," *Nature* **339:** 463–465.

A SEARCH FOR EXTRATERRESTRIAL AMINO ACIDS IN POLAR ICE: A PROGRESS REPORT

Jeffrey L. Bada

Scripps Institution of Oceanography
University of California at San Diego
La Jolla, CA 92093

Karen L. F. Brinton

Scripps Institution of Oceanography

Gene D. McDonald

Scripps Institution of Oceanography

Xueyun Wang

Scripps Institution of Oceanography

Abstract

Fifteen polar ice samples—fourteen from Greenland and one from Antarctica—have been analyzed for the extraterrestrial amino acid α-aminoisobutyric acid (AIB) in an effort to estimate the flux of interplanetary organic material to the Earth's surface. Only one sample (Greenland GISP II, 4270–4440 years old) contains detectable amounts of AIB, apparently the signature of a transient delivery event. The maximum oceanic concentration of AIB from such an event would be less than 10^{-9} M.

Introduction

Extraterrestrial sources have been suggested as an alternative for the Earth-based synthesis of the organic compounds necessary for the origin of life (Oró 1961, Chyba and Sagan 1992). Such an alternative source of organics may have been necessary if the atmosphere of the early Earth was composed mostly of nonreducing gases such as N_2 and CO_2 rather than CH_4 and NH_3 (Walker 1985, Kasting 1993, Kasting et al. 1993).

If extraterrestrial sources were important on the early Earth, they should still be supplying the Earth today with organic matter, although at a significantly reduced rate. Theoretical studies have indicated that organic molecules could be delivered to the Earth by cometary impactors (Chyba et al. 1990), but would be more likely to reach the Earth's surface intact if carried by micrometeorites and interplanetary dust particles (IDPs) with diameters in the tens to hundreds of microns (Anders 1989, Chyba and Sagan 1992). These particles can be collected in the stratosphere by aircraft, but the amount of material so far obtained by this method is much too low for detailed organic analyses. The only other alternative (besides in-orbit collection) is to attempt to detect exogenous organics at the surface of the Earth. Any extraterrestrial organic material that might reach the surface of the Earth, however, would be minuscule in comparison to terrestrial biogenic organics (see Table 1), and thus virtually impossible to detect in most terrestrial environments. The most pristine environment on Earth with respect to terrestrial organic matter is polar ice. Therefore, we have concentrated our search for extraterrestrial organic molecules—in particular extraterrestrial amino acids—on ice samples from Greenland and Antarctica.

Table 1. ***Amino Acids on the Earth***

	Carbon (10^{15} g)	*Amino Acid Content*	*Amino Acids (10^{15} g)*
Biosphere	600–800	< 10%	< 100
Polar Ice	3*	1 ppb†	0.03
Seawater Organics	2000–3000	1%	20–30
Soil Organics	1000	< 1%	< 10
Sediments			
Organics	3×10^7	0.02–1%	6000–300,000
Carbonates	10^8	20–300 ppm	1000–30,000

* *Total ice = 2.9×10^{19} kg; organic carbon = 100 μg C kg^{-1}.*

† *Total amino acid content of polar ice.*

We have analyzed several polar ice samples for the extraterrestrial amino acid α-aminoisobutyric acid (AIB). This molecule is the most abundant amino acid in carbonaceous meteorites (Cronin and Pizzarello 1983). Aside from a few fungal peptides (see, e.g., Mathew and Balaram 1983), it rarely occurs in the terrestrial biosphere. AIB has been found in association with the K-T boundary impactor (Zhao and Bada 1989), and routine and sensitive analytical techniques for this amino acid are available (Zhao and Bada 1995). AIB can therefore be used as a marker for the presence of extraterrestrial organics on the Earth.

METHODS

All ice core samples weighed between 3 and 7 kg. Each sample was melted and evaporated to dryness, and the residue redissolved in approximately one ml of double-distilled H_2O. Aliquots were then taken and derivatized by the o-phthaldialdehyde/N-acetyl cysteine (OPA/NAC) technique. The derivatives were separated by reverse-phase high-performance liquid chromatography (HPLC) with fluorescence detection and identified by comparison of retention times with authentic standards (Zhao and Bada 1995). Aliquots were also taken and hydrolyzed in double-distilled 6N HCl at 100°C for 24 hours. These aliquots were then dried and analyzed as above.

The sample believed to contain AIB was reacted with ninhydrin in pH 2.5 citrate buffer at 100°C for 5 minutes (Brinton and Bada 1994). The solution was then passed through an activated charcoal column, dried and analyzed as above.

RESULTS

We have analyzed one sample from the Allan Hills region of Antarctica, twelve samples from the Greenland Dye 3 site and two from the Greenland GISP II site. These ice samples range in age from approximately 100 to 100,000 years before present (BP). The levels of AIB present in these samples are shown in Table 2. AIB was not detected in the Antarctic sample, any of the twelve Dye 3 samples, or the 6000 year old GISP II sample. The levels in Table 2 are upper limits based on the detection limits of the OPA/NAC technique.

AIB was detected in one sample from the GISP II drilling site in Greenland (Table 2). The chromatographic retention time of the peak identified as AIB is consistent with that of an authentic standard. In addition, the yield of this peak increases between one minute and 15 minute derivatizations with the OPA/NAC reagent, as does the yield of authentic AIB but not the yields of protein amino acids (Zhao and Bada 1995). Reaction with ninhydrin results in approximately 25%

destruction of the putative AIB (again consistent with authentic standards; Brinton and Bada 1994), while the protein amino acids in the sample were almost completely destroyed. Since the behavior of the AIB peak is consistent with authentic AIB in these respects, we conclude that AIB is present in the 4270–4440 yr BP GISP II sample.

Although AIB is rare in the biosphere, contamination in ice samples cannot be completely dismissed. The amount of AIB detected in the sample does not increase with HCl hydrolysis, indicating that all the AIB present is in the free amino acid form rather than in peptide residues. This argues against terrestrial biological contamination as a source of the AIB detected in the GISP II sample. In addition, a second, older sample of GISP II ice does not contain detectable levels of AIB (see Table 2). Because both samples were drilled at the same time using the same techniques, AIB contamination from the drilling or sample handling procedures at the GISP II site can be ruled out.

TABLE 2. *Extraterrestrial Amino Acids in Polar Ices*

	Number of Samples	*Age*	*ppb AIB*
Antarctic		~ 10^5 yr BP	< 0.0005 (unhyd)
Allan Hills	1		< 0.002 (hydrol)
Greenland			
Dye 3	12	1860–1930 AD	< 0.001 (unhyd)
			< 0.001 (hydrol)
GISP II	2	4270–4440 yr BP	0.35 (unhyd)
			0.11 (hydrol)
		6010–6240 yr BP	< 0.001 (unhyd)

DISCUSSION

Anders (1989) has estimated the annual flux of intact organics from IDPs to the Earth as 3×10^8 g yr^{-1}. Assuming that IDP organic carbon is less than 1% extraterrestrial amino acids (AIB and isovaline), the expected concentration of extraterrestrial amino acids (ETAA) in Greenland ice would be $< 2 \times 10^{-11}$ g kg^{-1} ice, and the ETAA concentration in Antarctic ice would be $< 1 \times 10^{-10}$ g kg^{-1}. The measured upper limits for ETAA in Antarctic and Greenland Dye 3 ice, as well as for the 6000 yr BP GISP II sample, are consistent with and certainly no higher than the expected levels based on IDP flux. The 4300 yr BP GISP II sample, however, contains AIB at a concentration three orders of magnitude greater than the expected levels and that determined in the other ice samples analyzed.

If this AIB is extraterrestrial, it would seem to be a marker for a transient event that delivered extraterrestrial organics to the Earth around 4300 years ago. Failure to detect ETAA in the other samples listed in Table 2 suggests that the baseline flux of extraterrestrial organics at the Earth's surface is at or below the level estimated by Anders (1989). Studies of micrometeorite size distributions in Greenland ice (Maurette et al. 1987) indicate that the 3–7 kg samples analyzed for ETAA should contain an average of only 1–7 micrometeorites of around 100 μm diameter. The amount of AIB present in the GISP II sample is much too large to come from a few 100 μm particles, further suggesting that this AIB is the signature of a transient event.

If these data are the signature of a global ETAA delivery event around 4300 years ago, its astrophysical nature is not clear. The amount of AIB delivered globally by this event can be calculated from the measured AIB level in the GISP II sample, assuming a concentration of AIB similar to that of the Murchison meteorite (10 ppm), and assuming that the extraterrestrial material was deposited over the entire 170 year time span of the sample. Such calculations suggest an impactor with a diameter of around 7 km, large enough to have left lasting evidence of its arrival. If the extraterrestrial material was deposited completely in one year, the size of the impacting object is reduced to around one km. If the event delivered material only to the Greenland ice sheet rather than globally, the impactor size could be as small as 200 m. A collision between the Earth and a 200 m object would be expected to occur once every 10^3–10^4 years, while the impact of a one km object would occur once every 10^5 years (Chapman and Morrison 1994). A close approach of the Earth by a long-period comet, and the deposition on the Earth of cometary dust by a mechanism similar to that suggested for the K-T boundary amino acids (Zahnle and Grinspoon 1990) might be a plausible explanation. The ETAA flux in this case would be expected to gradually decrease to background levels over time.

It should be noted that the delivery event represented by the AIB in the GISP II sample, even if a global one, would result in a maximum oceanic AIB concentration of 3×10^{-11} M. Even if the AIB were initially concentrated in the upper 100 m of the ocean, the level of AIB would be only around 10^{-9} M. Such concentrations of extraterrestrial organic molecules would probably not be significant in prebiotic organic chemistry on the early Earth.

We hope to determine the areal extent of ETAA delivery from this event by analyzing ice core segments of the same age as the GISP II sample from other locations in either Greenland or Antarctica. Carbon isotope analysis may also be useful in confirming the extraterrestrial origin of the AIB detected, although the small amounts of amino acids involved may be at or below the limits of isotopic analysis.

ACKNOWLEDGMENTS

The authors thank Harmon Craig for the Dye 3 and Allan Hills samples, Tom Woiwode for help in collecting the GISP II samples, and the NASA Specialized Center for Research and Training in Exobiology at the University of California at San Diego for support.

REFERENCES

ANDERS, E. (1989), "Pre-biotic Organic Matter from Comets and Asteroids," *Nature* **342:** 255–257.

BRINTON, K.L.F., AND J.L. BADA (1994), "The Use of the Ninhydrin Reaction for the Elimination of Protein Amino Acids in Cosmogeochemical Samples," *Origins of Life* **24:** 125–126.

CHAPMAN, C.R., AND D. MORRISON (1994), "Impacts on the Earth by Asteroids and Comets: Assessing the Hazard," *Nature* **367:** 33–40.

CHYBA, C., AND C. SAGAN (1992), "Endogenous Production, Exogenous Delivery and Impact-Shock Synthesis of Organic Molecules: An Inventory for the Origins of Life," *Nature* **355:** 125–132.

CHYBA, C.F., P.J. THOMAS, L. BROOKSHAW, AND C. SAGAN (1990), "Cometary Delivery of Organic Molecules to the Early Earth," *Science* **249:** 366–373.

CRONIN, J.R., AND S. PIZZARELLO (1983), "Amino Acids in Meteorites," *Adv. Space Res.* **3:** 5–12.

KASTING, J.F. (1993), "Earth's Early Atmosphere," *Science* **259:** 920–926.

KASTING, J.F., D.H. EGGLER, AND S.P. RAEBURN (1993), "Mantle Redox Evolution and the Oxidation State of the Archean Atmosphere," *J. Geol.* **101:** 245–257.

MATHEW, M.K., AND P. BALARAM (1983), *Mol. Cell. Biochem.* **50:** 47–52.

MAURETTE, M., C. JEHANNO, E. ROBIN, AND C. HAMMER (1987), "Characteristics and Mass Distribution of Extraterrestrial Dust from the Greenland Ice Cap," *Nature* **328:** 699–702.

ORÓ, J. (1961), "Comets and the Formation of Biochemical Compounds on the Primitive Earth," *Nature* **190:** 389–390.

WALKER, J.C.G. (1985), "Carbon Dioxide Cycle on the Early Earth," *Origins of Life* **16:** 117–127.

ZAHNLE, K., AND D. GRINSPOON (1990), "Comet Dust as a Source of Amino Acids at the Cretaceous/Tertiary Boundary," *Nature* **348:** 157–160.

ZHAO, M., AND J.L. BADA (1989), "Extraterrestrial Amino Acids in Cretaceous/Tertiary Boundary Sediments at Stevns Klint, Denmark," *Nature* **339:** 463–465.

ZHAO, M., AND J. BADA (1995), "Determination of α-Dialkylamino Acids and Their Enantiomers in Geological Samples by High-Performance Liquid Chromatography after Derivatization with a Chiral Adduct of o-Phthaldialdehyde," *J. Chromatog. A* **690:** 55–63.

THE PHOTO PRODUCTION OF ORGANIC MOLECULES IN COMETARY AND INTERSTELLAR ICE ANALOGS

M. P. Bernstein

SETI Institute
NASA Ames Research Center
Moffett Field, CA 94035-1000

L. J. Allamandola

NASA Ames Research Center

S. A. Sandford

NASA Ames Research Center

S. Chang

NASA Ames Research Center

Well, I think that it is apparent to anyone who has perused today's schedule that there is now support for the notion that comets may have brought organic materials that were necessary for the formation of life on Earth (Chang 1979). It is therefore time for us to ask "what are the chemicals of which comets are composed?" It is imperative to answer this question because the organic molecules we find may profoundly affect our model of how life on this planet developed.

The first figure is just to remind us that out in the Oort cloud or the interstellar medium, where comets and interstellar ice grains are thought to form, it is so cold (10–50 K) that essentially anything hitting the grain, with the exception of He, sticks. So the comet or grain grows by a process of slow accretion, all the while bathed in background radiation (Greenberg 1977). When this radiation impinges upon the surface, bonds are broken, molecules are destroyed, and radicals and other fragments are formed. Under normal terrestrial conditions such species would be very short-lived but the low temperatures allow some of these to persist, trapped in a cage of ice. The process continues and more material builds, with molecular fragments throughout, until some event (such as entry into the Solar System) causes the ice to warm slightly, at which point the molecular fragments begin to move, react, and form new molecular species.

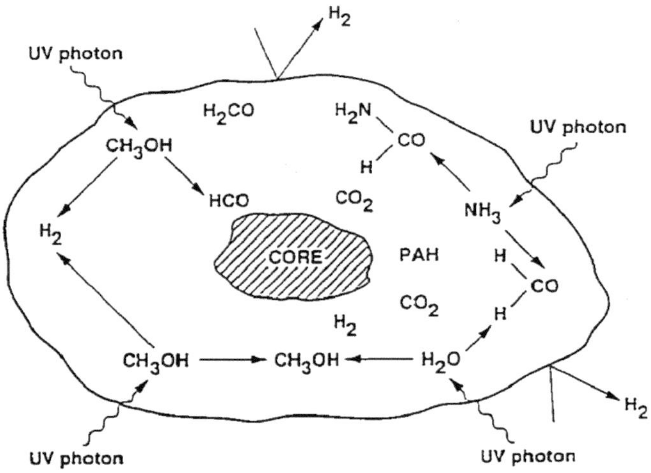

FIG. 1. *Typical ice grain formed by slow accretion.*

Our general approach is as follows. We begin with our best guess as to the bulk materials that compose interstellar ice and hence comets. If we mix these materials in astronomically relevant ratios and expose these mixtures to the conditions appropriate to the interstellar medium then we should produce some of the same compounds that form out there, but in the lab, where we can more easily identify trace organic species using infrared (IR) spectroscopy and gas chromatography–mass spectrometry (GC-MS). Having identified these organic molecules in the lab mixture we acquire IR data on the organic molecules individually, at low temperature and pressure, to facilitate the assignment of spectral bands caused by these organic species in the spectra of comets and dust in the interstellar medium.

Based on astronomical observations we are reasonably confident that cometary and interstellar ices contain water, methanol, carbon monoxide and ammonia

(Allamandola and Sandford 1988, Allamandola et al. 1988a, Hoban et al. 1991, Reuter 1992, Hoban et al. 1993).

In the interest of simulating the conditions of the interstellar medium, this gas mixture is deposited at 12 K in a vacuum chamber at a few times 10^{-8} Torr and ultraviolet (UV) radiation is used to simulate the background radiation. As depicted in Figure 2, the window is mounted on a rotating coldhead so the window can turn to face the deposition line or IR beam without breaking the vacuum. UV radiation was generated by a microwave-powered, hydrogen-flow, discharge lamp. The flux is evenly divided between the Lyman α line at 1216 Å and an ~ 200 Å wide collection of molecular transitions centered at about 1600 Å. The total flux produced by the lamp is ~ 2×10^{15} photons/cm^2 sec (Allamandola et al. 1988b). We monitor the effect of UV radiation via IR spectroscopy. In Figure 3a you see the spectrum of our starting mixture ($H_2O/CH_3OH/CO/NH_3$ = 20/10/1/1) at 12 K and then the changes caused by continued exposure to UV radiation in Figure 3b. The methanol (CH_3OH) is oxidized by the UV to formaldehyde (H_2CO), carbon monoxide (CO), and finally carbon dioxide (CO_2). After extended photolysis, methane (CH_4) is also formed as well as small amounts of the methyl radical ($CH_3\cdot$) and formyl radical ($\cdot CHO$). So at this stage (Figure 3b) we have simulated the effect of the radiation from the stars using UV and created some molecular fragments, embedded in our ice mixture.

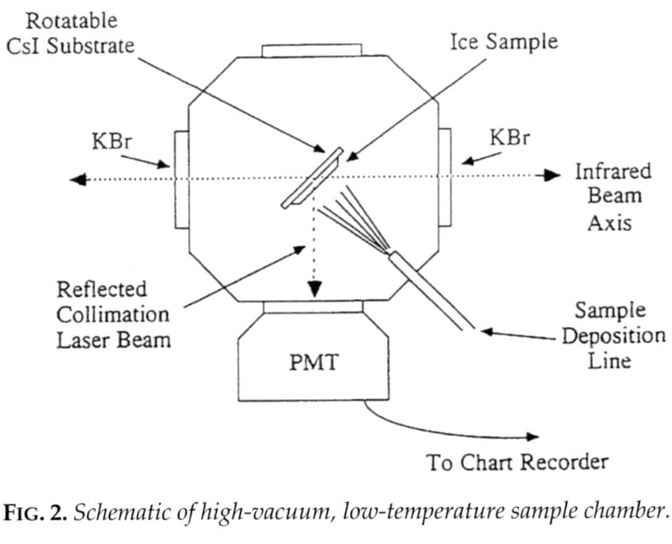

FIG. 2. *Schematic of high-vacuum, low-temperature sample chamber.*

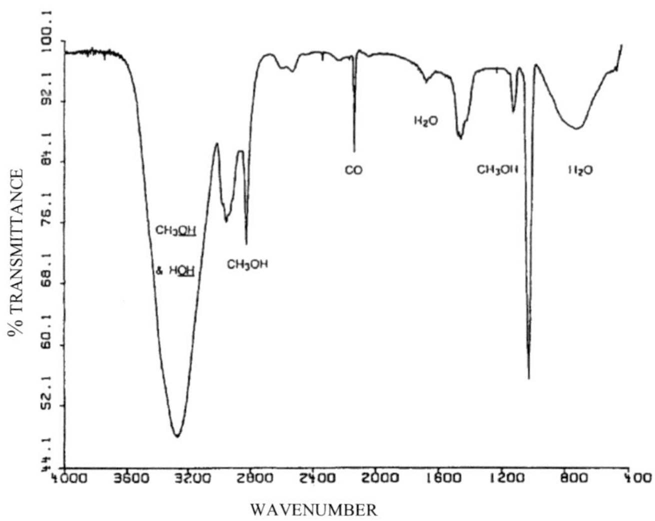

FIG. 3A. *Infrared (IR) spectrum of 20/10/1/1 $H_2O/CH_3OH/CO/NH_3$.*

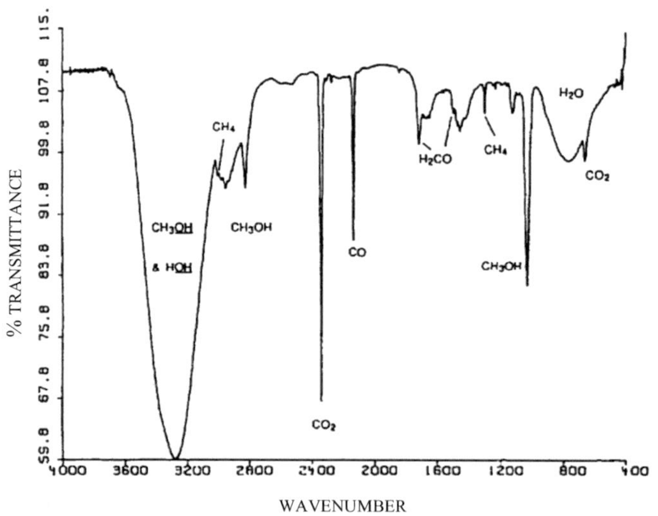

FIG. 3B. *IR spectrum of UV photolysed 20/10/1/1 $H_2O/CH_3OH/CO/NH_3$.*

Now this ice mixture must be fairly thin for the UV and IR to penetrate and the result is that when one warms the sample we see the IR spectrum shown in Figure 3c; that is to say, we see almost nothing at all. This is because most of the material evaporates, leaving only a trace residue behind, producing only ~ 1% features in the IR. Although it is possible to expand the scale and make statements about the general classes of compounds formed, this spectrum is not adequate if we wish to identify individual molecular species. So we proceed by putting down a second layer of starting mixture on top of the first after photolysis, in other words at the stage of Figure 3b we lay Figure 3a on top. After another dose of UV we put down another layer and so on. After twenty depositions and exposures of radiation we have a sample that is quite opaque to IR at 12 K but gives a nice spectrum at elevated temperatures after the volatiles have evaporated, and the result of this 160-hour experiment can be seen in Figure 4. At 300 K we have 20% features, which are much better than any such spectra taken before, and a series of functional groups can easily be identified based on this IR spectrum, including an alcohol, ketone, amide, at least one kind of CH, and a dominating 1000 cm^{-1} feature that would generally be assigned to a C–O single bond (Nakanishi and Solomon 1977), but we attribute to a C–N single bond.

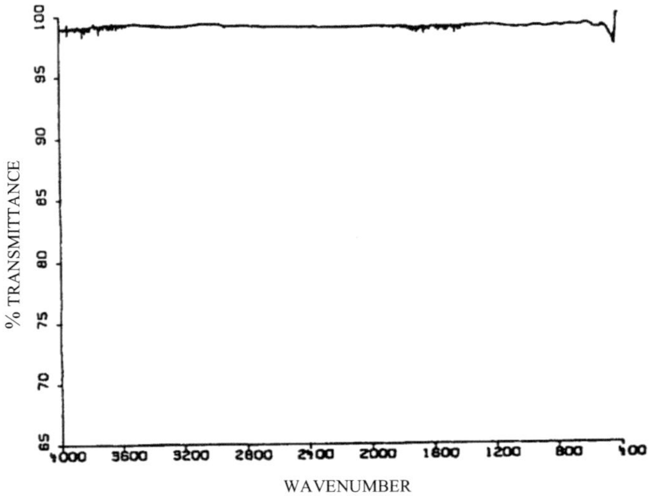

FIG. 3C (above). *IR spectrum of UV photolysed 20/10/1/1 $H_2O/CH_3OH/CO/NH_3$ warmed to 300 K.*

FIG. 4 (below). *300 K warm-up of 20/10/1/1 $H_2O/CH_3OH/CO/NH_3$*

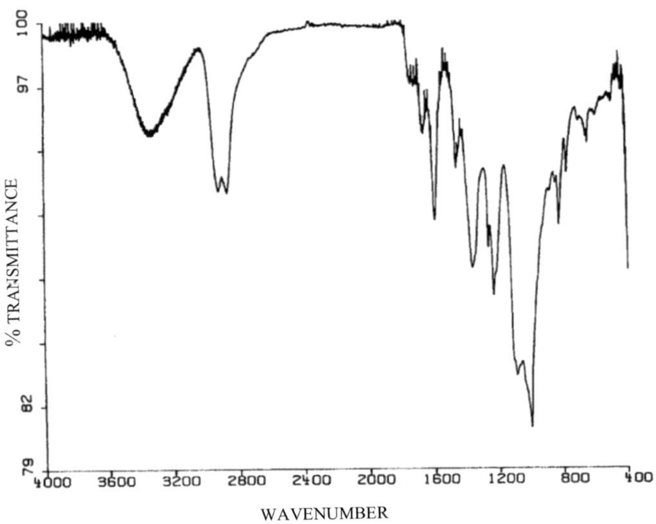

It was clear that we produced CO during the irradiation so its starting concentration may not be important, but changing the starting concentration of ammonia proved to be revealing. Increasing the concentration of ammonia by a factor of 4.5 (Fig. 5a and b) resulted in only small changes in the IR spectrum of the residue (for example the 3000 cm^{-1} CH stretch grows slightly with respect to that at 2900 cm^{-1} and the putative amide band at ~ 1660 cm^{-1} grows). This led us to suppose that very little of the ammonia was incorporated into the residue at room temperature, for if it were incorporated efficiently then the features caused by nitrogen-containing bonds should increase in magnitude in proportion to the concentration of ammonia in the starting mixture. Therefore, we fully expected that if we did not include the ammonia it would cause little change in the spectrum. To our great surprise we found that the spectrum was almost unrecognizable when ammonia was removed from the starting materials (compare Figure 5c to 5a and 5b). Even taking into account that some of the products could have been formed through catalysis by ammonia (and thus not actually incorporate it) this meant that a considerable amount of the residue contained nitrogen, although ammonia represents only a few percent of the starting mixture.

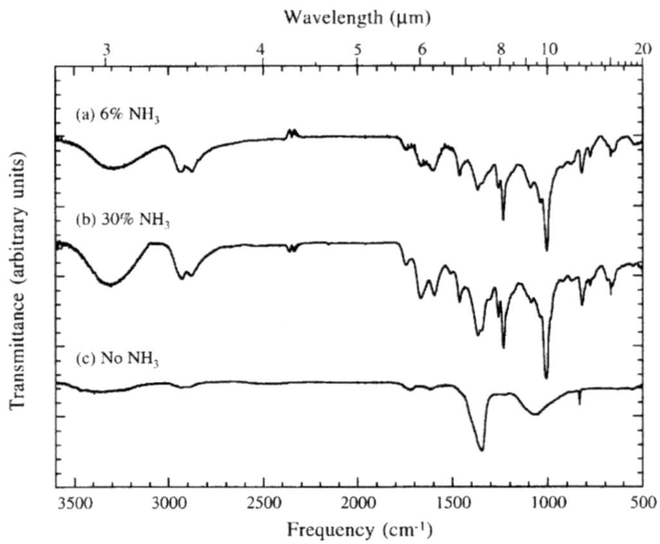

FIG. 5. *The dependence on NH_3 of the 3600–400 cm^{-1} (2.778–25μm) infrared spectra of 300 K residues produced by the UV photolysis of $H_2O/CH_3OH/CO/NH_3$ ice mixtures. The ice starting compositions were (a) $H_2O/CH_3OH/CO/NH_3$ = 100/50/10/10, (b) $H_2O/CH_3OH/CO/NH_3$ = 100/50/10/45, (c) $H_2O/CH_3OH/CO/NH_3$ = 100/50/10/0.*

In order to better characterize these organic residues we washed off the CsI window (using methanol in some cases and DMSO in others) under an atmosphere of dry argon and subjected the resulting solution to GC-MS analysis. Figure 6 shows these results for the astrophysical mix ($H_2O/CH_3OH/CO/NH_3$ =

20/10/2/2). Figure 6a shows the different components separated by chromatography, and 6b is the mass spectrum corresponding to the large peak seen at the left, in the chromatogram. The mass spectrum seemed to indicate the following—the atomic mass of the predominant component is 140 amu, there are no hydrocarbon moieties on the skeleton (based on the absence of a peak corresponding to loss of a methyl, ethyl, etc.), thus it probably has a ring or cage structure and it must have an even number of nitrogens (McLafferty 1973). Based on this information, and the IR spectrum, we were able to narrow this component down to just a handful of compounds. The known compound that best fits the major component of the residue was a nitrogen-rich organic cage molecule called hexamethylenetetramine (HMTA; Fig. 7), $C_6H_{12}N_4$, a white crystalline solid that sublimes at 280°C (Smolin and Rapoport 1959, Walker 1964). Our sample of HMTA was generously provided by B. N. Khare of Cornell University. The mass spectrum of HMTA bears a striking resemblance to that of our residue (compare Fig. 6b and d), and the comparison of the IR spectrum of our residue with that of HMTA (Fig. 8 [a] and [b]) makes it clear that HMTA composes a considerable portion of the residue at room temperature.

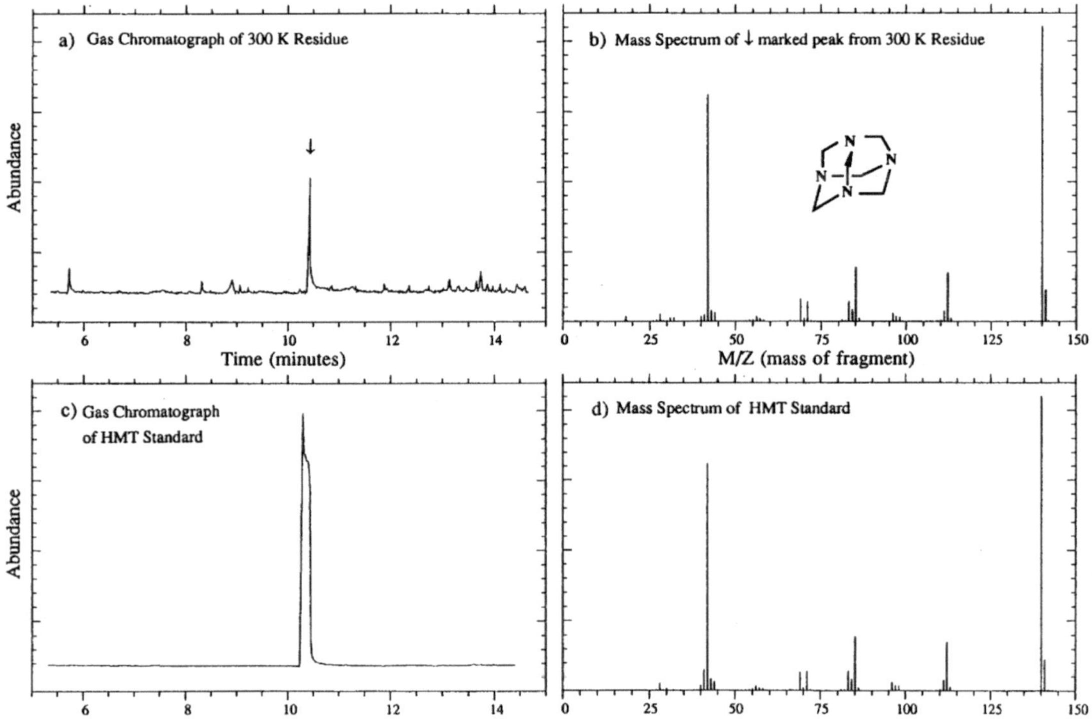

FIG. 6. *The gas chromatographs and mass spectra of the refractory residue produced by the irradiation of an $H_2O/CH_3OH/CO/NH_3$ = 100/50/10/10 ice followed by warming to 300 K compared with a sample of pure hexamethylenetetramine (HMTA). The gas chromatographs of the residue and HMTA standard are shown in (a) and (c), respectively, and the corresponding mass spectra of the most prominent peaks (retention time of about 10.2 minutes) are shown in (b) and (d).*

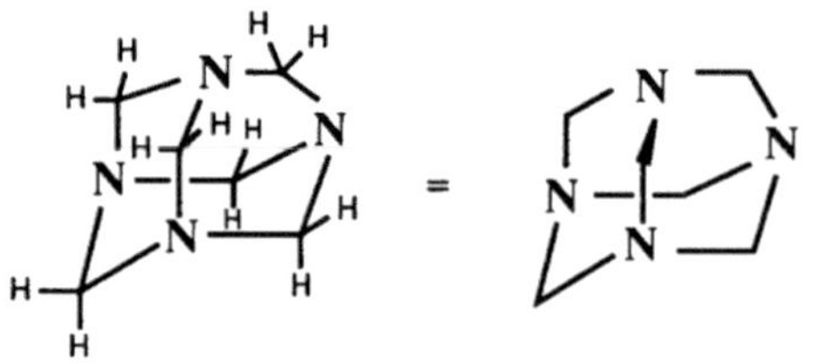

Fig. 7. *The chemical structure of hexamethylenetetramine, $C_6H_{12}N_4$ (the vertices are carbon atoms in an overlapping chair-type arrangement incorporating a tetrahedron of nitrogen atoms).*

This interesting molecule is not new. HMTA has been known to the chemistry community for over 130 years (Bulterov 1859, Dickinson and Raymond 1923) and it was suggested (based primarily on its ease of formation) to be a component of grains in 1972 by Khare and Sagan when, in the course of sample preparation, they noticed that HMTA formed spontaneously at room temperature when ammonia and formaldehyde were mixed (Sagan and Khare 1970, Khare and Sagan 1972). It has been noted that hydrolysis of HMTA by concentrated acid yields amino acids (Wolman et al. 1971), it was seen as a trace component in non-methanol-containing residues (Agarwal et al. 1985, Briggs et al. 1992) and its presence may explain the very low abundance of NH3 in comet Halley (Kargel 1992). Since it may be a component of the organic crust of comets and interstellar ice grains, the IR spectral properties and photochemistry of HMTA under astrophysical conditions (frozen in H_2O ice at low temperature) are of great interest and currently being studied in our laboratory. These latest results indicate that the UV photolysis of HMTA under astrophysical conditions produces, among other things, a CN stretch (2160 cm−1, 4.63 μm) in good agreement with that seen in the IR spectra of protostellar objects (Lacy et al. 1984, Larson et al. 1985, Tegler et al. 1993, Bernstein et al. 1994).

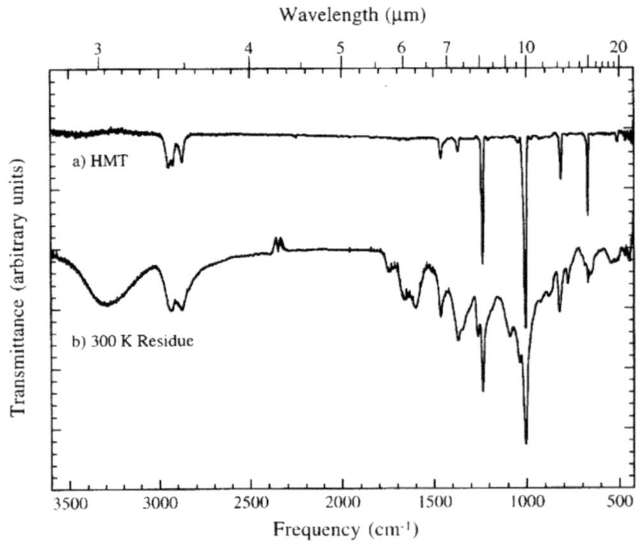

Fig. 8. *The 3600–400 cm^{-1} (2.778–25 μm) infrared spectrum of (a) a sample of pure HMTA frozen in H_2O at 12 K and subsequently warmed to 300 K, and (b) the residue generated from photolysis of an $H_2O/CH_3OH/CO/NH_3$ = 100/50/10/10 mixture that was subsequently warmed to 300 K (same as shown in Fig. 5a). The spectrum of HMTA is adapted from Bernstein et al. (1994).*

In conclusion, HMTA, which contains four nitrogens per molecule, composes 60% of the organic residue formed by photolysis and warming of ice analogs, although ammonia represented only 5% of the starting mixture. Based on the facility with which it forms in our experiments and its high nitrogen content, we join our colleagues in proposing that it may be present in interstellar and cometary ices (Sagan and Khare 1970, Khare and Sagan 1972, Agarwal et al. 1985, Briggs et al. 1992), and that much of the ammonia may have been tied up in this form (Kargel 1992). In addition, the observation of a nitrile stretch in the UV photolysis of HMTA suggests that HMTA could be the source of the CN observed in the tail of comets and the interstellar ice grains.

References

AGARWAL, V.K., W. SCHUTTE, J.M. GREENBERG, J.P. FERRIS, R. BRIGGS, S. CONNOR, C.P.E.M. VAN DE BULT, AND F. BAAS (1985), "Photochemical Reactions in Interstellar Grains. Photolysis of CO, NH_3 and H_2O," *Origins of Life* **16:** 21–40.

ALLAMANDOLA, L.J., AND S.A. SANDFORD (1988), "Laboratory Simulation of Dust Spectra," in *Dust in the Universe*, M.E. Bailey and D.A. Williams, eds. (Cambridge: Cambridge University Press), pp. 229–263.

ALLAMANDOLA, L.J., S.A. SANDFORD, A.G.G.M. TIELENS, AND T.M. HERBST (1988a), "Infrared Spectra of Dense Clouds in the C–H Stretch Region: Methanol and Diamonds," *ApJ* **399:** 134–146.

ALLAMANDOLA, L.J., S.A. SANDFORD, AND G.J. VALERO (1988b), "Photochemical and Thermal Evolution of Interstellar/Precometary Ice Analogs," *Icarus* **76:** 225–252.

BERNSTEIN, M.P., S.A. SANDFORD, L.J. ALLAMANDOLA, AND S. CHANG (1994), "The Infrared Spectrum of Matrix-Isolated Hexamethylenetetramine in Ar and H_2O at Cryogenic Temperatures," *J. Phys. Chem.* **98:** 12206.

BRIGGS, R., G. ERTEM, J.P. FERRIS, J.M. GREENBERG, J.P. MCCAIN, C.X. MENDOZA-GÓMEZ, AND W. SCHUTTE (1992), "Comet Halley as an Aggregate of Interstellar Dust and Further Evidence for the Photochemical Production of Organics in the Interstellar Medium," *Origins of Life* **22:** 287.

BULTEROV, A. (1859), "Hexamethylenetetramine," *Justus Liebigs Annalen Der Chemie* **11:** 250.

CHANG, S. (1979), "Comets, Cosmic Connections with Carbonaceous Meteorites and Interstellar Molecules," in *Space Missions to Comets*, NASA CP-2089, M. Neugebauer, D.K. Yeomans, J.C. Brandt, and R.W. Hobbs, eds., pp. 59–111.

DICKINSON, R.G., AND A.L. RAYMOND (1923), "The Crystal Structure of Hexamethylenetetramine," *J. Am. Chem. Soc.* **45:** 22.

GREENBERG, J.M. (1977), "From Interstellar Dust to Comets to Dust," in *Comets, Asteroids and Meteorites*, A.H. Delsemme, ed. (Toledo, OH: University of Toledo Press), p. 491.

HOBAN, S., M.J. MUMMA, D.C. REUTER, M.A. DISANTI, R.R. JOYCE, AND A. STORRS (1991), "A Tentative Identification of Methanol as the Progenitor of the 3.52-μm Emission Feature in Several Comets," *Icarus* **93:** 122–134.

HOBAN, S., D.C. REUTER, M.A. DISANTI, AND M.J. MUMMA (1993), "Comets," *Icarus* **105:** 548.

KARGEL, J.S. (1992), "Ammonia-Water Volcanism on Icy Satellites," *Icarus* **100:** 556–574.

KHARE, B.N., AND C. SAGAN (1972), "Experimental Interstellar Organic Chemistry: Preliminary Findings," in *Proceedings of the Symposium on Spectroscopic Studies of Astrophysical Interest* (Hyderabad), pp. 13–20.

LACY, J.H., F. BAAS, L.J. ALLAMANDOLA, S.E. PERSSON, P.J. MCGREGOR, C.J. LONSDALE, T.R. GEBALLE, AND C.P.E.M. VAN DE BULT (1984), "4.6 Micron Absorption Features Due to Solid Phase CO and Cyano Group Bearing Molecules Towards Compact Infrared Sources," *ApJ* **276:** 533–543.

LARSON, H.P., D.S. DAVIS, J.H. BLACK, AND U. FINK (1985), "Interstellar Absorption Features Towards the Compact Infrared Source W33A," *ApJ* **299:** 873.

MCLAFFERTY, F.W. (1973), *Interpretation of Mass Spectra* (Reading, MA: W.A. Benjamin).

NAKANISHI, K., AND P.H. SOLOMON (1977), *Infrared Absorption Spectroscopy* (San Francisco: Holden-Day).

REUTER, D.C. (1992), "The Contribution of Methanol to the 3.4 Micron Emission Feature in Comets," *ApJ* **386:** 330–335.

SAGAN, C., AND B.N. KHARE (1970), "Production of Organic Molecules in the Interstellar Medium," *Bull. Amer. Astron. Soc.* **2:** 340.

SMOLIN, E.M., AND L. RAPOPORT (1959), *S-Triazines and Derivatives* (New York: Interscience Publishers, Inc.), p. 545.

TEGLER, S.C., D.A. WEINTRAUB, L.J. ALLAMANDOLA, S.A. SANDFORD, T.W. RETTIG, AND H. CAMPINS (1993), "Detection of the 2165 Wavenumber (4.619 micron) XCN Band in the Spectrum of L1551 IRS5," *ApJ* **411:** 260.

WALKER, J.F. (1964), *Formaldehyde* (New York: Reinhold), p. 511.

WOLMAN, Y., S.L. MILLER, J. IBAÑEZ, AND J. ORÓ (1971), "Formaldehyde and Ammonia as Precursors to Prebiotic Amino Acids," *Science* **174:** 1039–1040.

CYANIDE POLYMERS IN CIRCUMSTELLAR HABITABLE ZONES

Clifford N. Matthews

Department of Chemistry
University of Illinois at Chicago
Chicago, IL 60607

Abstract

Hydrogen cyanide polymers—heterogeneous solids ranging in color from yellow to orange to red to black—may be among the most readily formed organic macromolecules within the Solar System. The nonvolatile black crust of comet Halley, for example, might consist largely of such polymers. It seems likely, too, that HCN polymers are a major constituent of the dark C≡N-bearing solids identified spectroscopically in the dust of some other comets. HCN polymerization could account also for the yellow-orange-red coloration of Jupiter and Saturn, as well as for the orange haze high in Titan's atmosphere.

Studies of these polymers show that a yellow-brown powder can be extracted by water and further hydrolyzed to yield α-amino acids. Several instrumental methods used for the separation and identification of these intriguing materials, including pyrolysis mass spectrometry, Fourier transform infrared photoacoustic spectroscopy, and supercritical fluid extraction chromatography, revealed fragmentation patterns and chemical functionalities consistent with the presence of polymeric peptide precursors—polyamidines—in HCN polymers.

Implications for prebiotic chemistry are profound. Primitive Earth may have been covered by HCN polymers through bolide bombardment or terrestrial synthesis, producing a proteinaceous matrix able to promote the molecular interactions leading to the emergence of life. Beyond Earth and the Solar System, on planetary bodies and satellites around other stars and in between, the existence of this preferred pathway must surely give rise to circumstellar habitable zones rich in the molecular precursors of life.

HYDROGEN CYANIDE POLYMERS: SYNTHESIS AND STRUCTURE

Liquid HCN (bp 25°C) polymerizes spontaneously to a dark brown or black solid at low temperatures in the presence of a base such as an amine or ammonia (Matthews and Moser 1967). Polymerization also occurs readily in nonaqueous solvents or in water (Völker 1960; Matthews and Moser 1966, 1967). Two types of structural units appear to be present in these solid materials. Most stable are the ladder polymers (Völker 1960) shown in Figure 1, formally derived from the olefinic tetramer of HCN, diaminomaleonitrile (A), which is usually found among the products. It seems probable, however, that polymerization to the substituted polymethylene (B) proceeds by way of an HCN dimer (Völker 1960, Kliss and Matthews 1962) for which several structures have been proposed (see Evans et al. 1991). Cyclization to C and D then leads to ladder structures possessing conjugated > C=N– bonds, as proposed by Völker (1960) on the basis of extensive physical and chemical investigations further supplemented by the work of Umemoto et al. (1987).

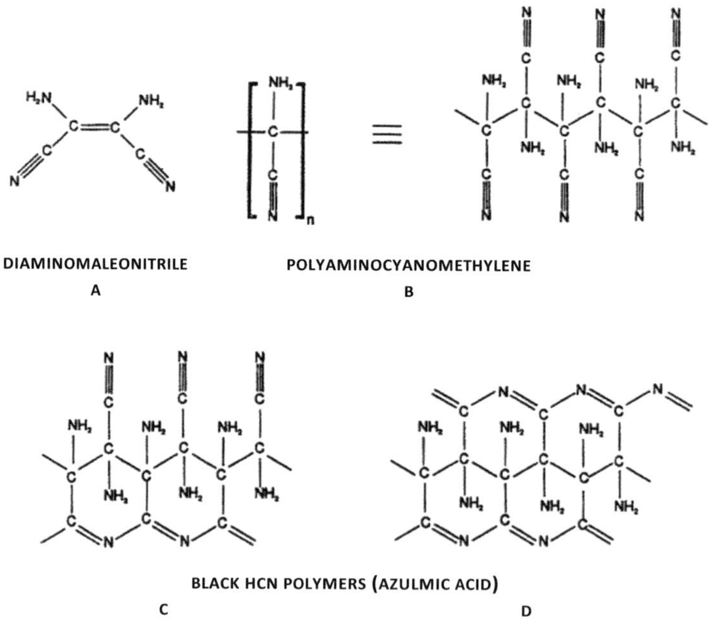

FIG. 1. *HCN polymers formally derived from HCN tetramer (diaminomaleonitrile).*

More controversial is the existence of the polyamidine structures shown in Figure 2. Polyaminomalononitrile (F) can be considered an addition polymer of the reactive trimer aminomalononitrile (E), though again it is possible that polymerization occurs through an HCN dimer (Kliss and Matthews 1962, Matthews and Moser 1967, Yang et al. 1976). Cumulative reactions of HCN on the highly activated nitrile groups of F then yield the heteropolyamidines (G), which are readily converted by water to heteropolypeptides (Matthews and Moser 1966, 1967) with release of ammonia and CO_2. Overall, this series of reactions constitutes a route for the direct synthesis of polypeptides without the intervening formation of α-amino acids (Matthews and Moser 1966, 1967; Matthews 1992a) (Fig. 3).

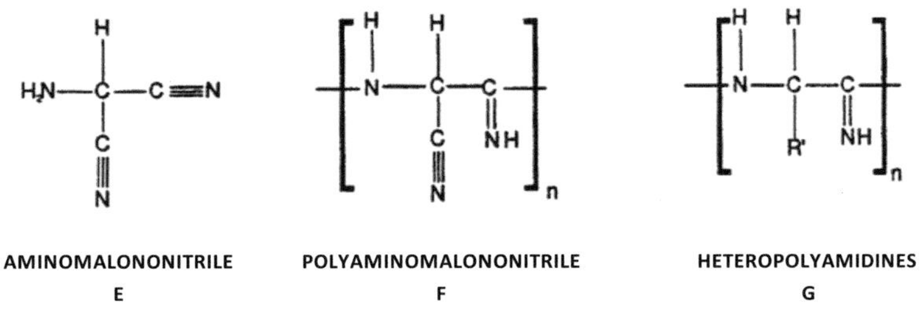

FIG. 2. *HCN polymers formally derived from HCN trimer (aminomalononitrile).*

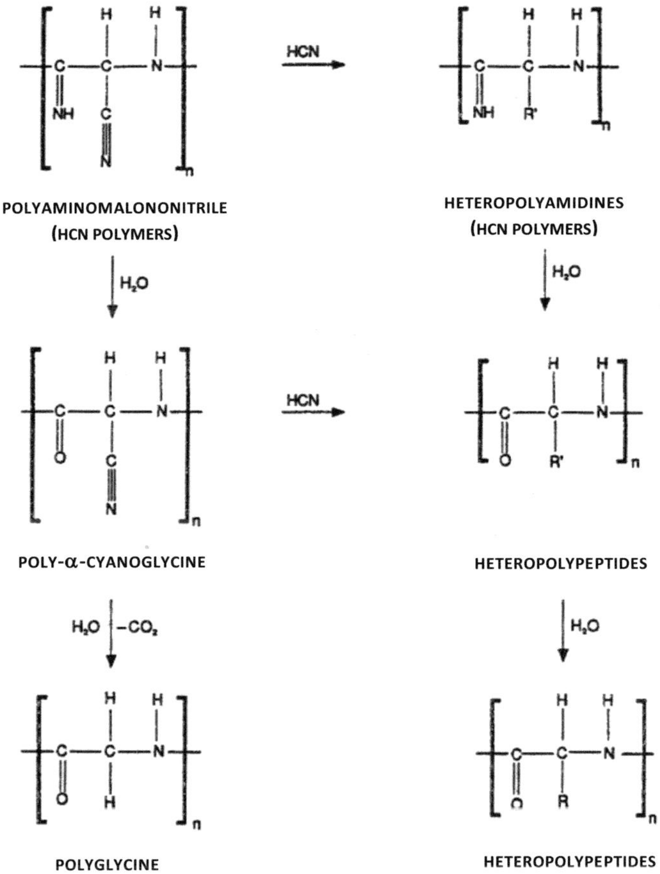

FIG. 3. *Polypeptides from polyamidines. Cumulative reactions of HCN on polyamino-malononitrile yield heteropolyamidines (with side chains* R′*), which are converted stepwise by water to heteropolypeptides (with side chains* R*).*

Several kinds of experiments have provided results consistent with this polyamidine model. In general, water-soluble, yellow-brown solids can be extracted from the products of each of the following types of reactions:
(1) Base-catalyzed polymerization of liquid HCN, alone, in water or in nonaqueous solvents (Matthews and Moser 1967);

(2) Electric discharge experiments producing HCN from methane-ammonia mixtures (Matthews and Moser 1966);

(3) Alkaline hydrolysis of aminoacetonitrile (Moser and Matthews 1968), aminomalononitrile (E; HCN trimer) (Moser et al. 1968), and diaminomaleonitrile (A; HCN tetramer) (Moser et al. 1968), all of which are ready sources of HCN at high pH;

(4) HCN modification of the reactive nitrile side chains of poly-α-cyanoglycine (see Fig. 3), a polyamide analog of polyaminomalononitrile (F) synthesized from the N-carboxyanhydride related to α-cyanoglycine (Warren et al. 1974, Minard et al. 1975).

Acid hydrolysis of these yellow-brown polymers yields not just glycine, the major product, but other α-amino acids as well, such as alanine, aspartic acid, glutamic acid, serine and threonine, together with some α-amino acids not found in proteins. Further GC-MS (gas chromatography–mass spectrometry) studies show that the glycine is perdeuterated when D_2O/DCl is used for hydrolysis instead of H_2O/HCl (Matthews et al. 1977), an indication of the highly acidic nature of the methine carbons of F and of the great reactivity of its nitrile groups brought about by its delocalized structure.

Nondestructive analysis of the total solid product obtained from HCN, as well as of separate components, became possible with the advent of cross-polarization magic-angle spinning solid-state NMR (nuclear magnetic resonance) spectroscopy (^{13}C and ^{15}N). In particular, the unambiguous presence of secondary amide groups, as in peptides, has been established by double-cross-polarization studies on polymers synthesized from equimolar amounts of $H^{13}CN$ and $HC^{15}N$ (Schaefer et al. 1982, Matthews et al. 1984, McKay et al. 1984, Garbow et al. 1987). Several instrumental methods are now being used for the separation and identification of these intriguing polymeric materials, including photoacoustic Fourier transform infrared spectroscopy, supercritical fluid extraction chromatography and pyrolysis mass spectrometry. Our integrated analytical approach has already revealed fragmentation patterns and chemical functionalities consistent with the presence of polymeric peptide precursors both in HCN polymers and in the Murchison meteorite (Liebman et al. 1995).

Taken together, these various analytical studies point to the presence of both peptide and ladder structures in HCN polymers exposed to water (see Table 1). Hybrid polymers with peptides attached to ladders through their free amino groups may be major components.

TABLE 1. ***Summary of Evidence Reported by Matthews et al. for the Presence of Peptidelike Segments in Hydrogen Cyanide Polymers and the Murchison Meteorite***

Analytical Mode	*Previous Studies* **(see Matthews 1992a)**	*Current Studies* **(see Liebman et al. 1995)**
1 Nondestructive Identification	*Solid-state (^{13}C, ^{15}N) NMR spectroscopy* HCN polymer extracts (H_2O) and insoluble residues contain polypeptide segments	*Photoacoustic Fourier transform infrared spectroscopy (FTIR-PAS)* Aqueous extracts of HCN polymer and the Murchison meteorite possess polypeptide segments
2 Separation by Fractionation	*Liquid-phase (Sephadex) column chromatography* HCN polymer extracts (H_2O) contain heteropolypeptide segments	*Supercritical fluid extraction chromatography (SFE–Transcap FTIR–Microscope)* Aqueous extracts of HCN polymer and the Murchison meteorite possess polypeptide segments
3 Analysis by Fragmentation	*Hydrolysis (GC-MS)* HCN polymers and the Murchison meteorite yield - amino acids (also α-hydroxy acids)	*Analytical pyrolysis by direct insertion pyrorobe–mass spectrometry (DIP-MS)* Hydrocarbons and polypeptides are present in Murchison meteorite; HCN polymers yield polypeptides

HYDROGEN CYANIDE POLYMERS IN THE SOLAR SYSTEM

Comparable polymers are synthesized when methane and ammonia mixtures are converted to hydrogen cyanide by electric discharges (Matthews and Moser 1967, Woeller and Ponnamperuma 1969). The original presence on cometary nuclei of frozen volatiles such as methane, ammonia, and water subjected to high energy sources (Whipple 1974) thus makes them possible sites for the formation and condensed-phase polymerization of HCN (Matthews and Ludicky 1986, 1992). Dust emanating from the nucleus, contributing to the coma and tail, would also arise partly from the polymer. Results from the recent Halley missions support this view. Following Vega observations of a dark surface largely masked by clouds (Sagdeev et al. 1986), the Giotto multicolor camera showed the presence of a potato-shaped core (≈ 15 × 10 km) with a very low albedo (2–4%) (Reinhard 1986, Keller et al. 1986). We propose that this nonvolatile black crust is composed largely of hydrogen cyanide polymers and related compounds (Matthews and Ludicky 1986). Continuing polymerization of HCN—a strongly exothermic process—may even be the cause of the dramatic cometary outbursts that have been noted by observers for many years (Rettig et al. 1992). The recognition that cyanide chemistry could be proceeding on Halley and other comets suggests that HCN polymerization is widespread within the Solar System and beyond. Most significantly, through spectroscopic detection of the 2.2 μm overtone of the C≡N stretching fundamental mode, Cruikshank et al. (1991) have recently identified molecules containing cyano groups in the very dark solids in the dust of the "new" comets Panther and Bowell. The infrared reflectance spectra of these bodies resemble the comparable spectra of HCN polymers in the same 0–3 μm region.

Carbonaceous chondrites from the asteroid belt might therefore also possess HCN products. Studies of the Murchison meteorite have shown that free α-amino acids are present (Kvenvolden et al. 1973) together with acid-labile amino acid precursors of undefined structure (Cronin 1976). Deuterolysis of these water-soluble yellow-brown solid extracts yields perdeuterated glycine, as would be expected from peptide segments in hybrid polymers derived from HCN (Matthews et al. 1977, 1980). Further evidence is being obtained by the integrated, organic analysis approach of Liebman et al. (1995) (see Table 1). Hydrogen cyanide polymerization could account, too, for much of the yellow-brown-orange coloration of Jupiter and Saturn (Matthews and Moser 1966, Woeller and Ponnamperuma 1969, Matthews 1982), especially since HCN has been found in Jupiter's reducing atmosphere and in the atmosphere of Titan, the largest moon of Saturn. Most intriguing is an orange haze high in Titan's stratosphere that may consist of organic polymers (Owen 1982). These could be polycyanides formed directly from HCN, some of which would be converted by water to heteropolypeptides after settling on the frozen surface or the hydrocarbon oceans of the satellite (Matthews 1982). Laboratory simulations of Jupiter and Titan chemistry—reactions of methane, ammonia (or nitrogen), and water subjected to high energy sources—have yielded "tholins" (Sagan and Khare 1979, Khare et al. 1984, McDonald et al. 1991), dark materials of undefined structure whose infrared reflectance spectra are remarkably similar to the spectra of HCN polymers (Matthews and Ludicky 1986, Cruikshank et al. 1991). Further analysis will show if polymers of hydrogen cyanide and of acetylene are indeed significant components of these "tholins," which on hydrolysis also yield α-amino acids.

Hydrogen Cyanide Polymers in the Milky Way

Since HCN is one of the more abundant interstellar molecules found outside the Solar System (Irvine et al. 1987, Greenberg 1989), the presence of HCN polymers on interstellar grains would be expected. Solid-phase cyano group molecules have indeed been detected by IR absorption spectra (4.6 μm) in the vicinity of protostars within molecular clouds of the Milky Way (Lacy et al. 1984). Further detections by Tegler et al. (1995) suggest that these X(CN) molecules may result from chemical processing of dust grains dominated by nonpolar icy mantles in the local environments of pre–main sequence stars.

Hydrogen Cyanide Polymers and the Origin of Life

Which came first, amino acids or their polymers? The ubiquitous presence of HCN in reducing environments invites the reexamination and possible reinterpretation of almost all previous research concerned with the origin of α-amino acids, including simulations of the chemistry of primitive atmospheres (Miller and Orgel 1974), studies of aqueous cyanide chemistry (Oró and Lazcano-Araujo 1980, Ferris and Hagan 1984), and meteorite analysis (Kerridge 1991). In my view, these investigations of reactions ostensibly yielding α-amino acids actually supply evidence for the abundant prebiotic and extraterrestrial existence of polymeric protein ancestors—heteropolypeptides synthesized directly from hydrogen cyanide and water (Matthews and Moser 1966, 1967; Matthews 1984, 1985). The detection of amino acid polymers in some of these experiments (Lowe et al. 1963; Matthews and Moser 1966, 1967; Woeller and Ponnamperuma 1969; Su et al. 1989; Khare et al. 1989; McDonald et al. 1991) adds to the plausibility of this conclusion.

Implications for prebiotic chemistry are profound. Primitive Earth may have been covered by HCN polymers, either through cometary deposition or by photochemical reactions in a reducing atmosphere. As polyamidines settled onto land and sea together with other organic products, a proteinaceous matrix developed able to take part in and promote the interactions leading to the emergence of life. These polyamidines could have been the original condensing agents directing the synthesis of nucleosides and nucleotides from available sugars, phosphates, and nitrogen bases. Most significant would have been the parallel synthesis of polypeptides and polynucleotides arising from the dehydrating action of polyamidines on nucleotides. On our dynamic planet this polypeptide-polynucleotide symbiosis mediated by polyamidines may have set the pattern for the evolution of protein–nucleic acid systems controlled by enzymes, the mode characteristic of life today (Matthews 1986, 1988, 1990, 1992b).

Beyond Earth and the Solar System, on planetary bodies around other stars and satellites and in between, the existence of this preferred pathway—HCN polymerization—could surely give rise to circumstellar habitable zones rich in the molecular precursors of life.

REFERENCES

CRONIN, J.R. (1976), "Acid-Labile Amino Acid Precursors in the Murchison Meteorite," *Origins of Life* **7:** 337–348.

CRUIKSHANK, D.P., W.K. HARTMANN, D.J. THOLEN, L.J. ALLAMANDOLA, R.H. BROWN, C.N. MATTHEWS, AND J.F. BELL (1991), "Solid C≡N Bearing Material on Outer Solar System Bodies," *Icarus* **94:** 345–353.

EVANS, R.A., P. LORENCAK, T.-K. HA, AND C. WENTRUP (1991), "HCN Dimers: Iminoacetonitrile and N-Cyanomethanimine," *J. Am. Chem. Soc.* **113:** 7261–7276.

FERRIS, J.P., AND W.J. HAGAN (1984), "HCN and Chemical Evolution: The Possible Role of Cyano Compounds in Prebiotic Synthesis," *Tetrahedron* **40:** 1093–1120.

GARBOW, J.R., J. SCHAEFER, R.A. LUDICKY, AND C.N. MATTHEWS (1987), "Detection of Secondary Amides in HCN Polymers by Dipolar Rotational Spin-Echo ^{15}N NMR," *Macromolecules* **20:** 305–309.

GREENBERG, J.M. (1989), "Interstellar Dust: An Overview of Physical and Chemical Evolution," in *Evolution of Interstellar Dust and Related Topics*, A. Bonetti, J.M. Greenberg, and S. Aeillo, eds. (Amsterdam: Elsevier North-Holland), pp. 7–52.

IRVINE, W.M., P.F. GOLDSMITH, AND A. HJALMARSON (1987), "Chemical Abundances in Molecular Clouds," in *Interstellar Processes*, D.J. Hollenbach and H.A. Thronson, Jr., eds. (Dordrecht: D. Reidel Publishing Company), pp. 561–609.

KELLER, H.U., C. ARPIGNY, C. BARBIERI, R.M. BONNET, S. CAZES, M. CORADINI, C.B. COSMOVICI, W.A. DELAMERE, W.F. HUEBNER, D.W. HUGHES, C. JAMAR, D. MALAISE, H.J. REITSEMA, W.K.H. SCHMIDT, P. SEIGE, F.L. WHIPPLE, AND K. WILHELM (1986), "First Halley Multicolour Camera Imaging Results from Giotto," *Nature* **321:** 320–326.

KERRIDGE, J.F. (1991), "A Note on the Prebiotic Synthesis of Organic Acids in Carbonaceous Meteorites," *Origins of Life* **21:** 19–29.

KHARE, B.N., C. SAGAN, W.R. THOMPSON, E.T. ARAKAWA, F. SUITS, T.A. CALCOTT, M.W. WILLIAMS, S. SHRADER, H. OGINO, T.O. WILLINGHAM, AND B. NAGY (1984), "The Organic Aerosols of Titan," *Adv. Space Res.* **4:** 59–68.

KHARE, B.N., C. SAGAN, W.R. THOMPSON, L. FLYNN, AND M.A. MORRISON (1989), "Amino Acids and Their Polymers in the Lower Clouds of Jupiter?—Preliminary Findings," *Origins of Life* **19:** 495–496.

KLISS, R.M., AND C.N. MATTHEWS (1962), "Hydrogen Cyanide Dimer and Chemical Evolution," *Proc. Natl. Acad. Sci. USA* **48:** 1300–1306.

KVENVOLDEN, K.A., J.G. LAWLESS, AND C.E. FOLSOME (1973), "Organic Matter in Meteorites," *Sci. Am.* **227 (6):** 38–54.

LACY, J.H., F. BAAS, L.J. ALLAMANDOLA, S.E. PERSSON, P.J. MCGREGOR, C.J. LONSDALE, T.R. GEBALLE, AND C.P.E.M. VAN DE BULT (1984), "4.6 Micron Absorption Features Due to

Solid Phase CO and Cyano Group Bearing Molecules Towards Compact Infrared Sources," *ApJ* **276:** 533–543.

LIEBMAN, S.A., R.A. PESCE-RODRÍGUEZ, AND C.N. MATTHEWS (1995), "Organic Analysis of Hydrogen Cyanide Polymers: Prebiotic and Extraterrestrial Chemistry," *Adv. Space Res.* **15 (3):** 71–80.

LOWE, C.U., M.W. REES, AND R. MARKHAM (1963), "Synthesis of Complex Organic Compounds from Simple Precursors: Formation of Amino-Acids, Amino-Acid Polymers, Fatty Acids and Purines from Ammonium Cyanide," *Nature* **199:** 219–222.

MATTHEWS, C.N. (1982), "Heteropolypeptides on Titan?," *Origins of Life* **12:** 281–283.

MATTHEWS, C.N. (1984), "Chemical Evolution: Protons to Proteins," *Proc. Roy. Inst. Gt. Britain* **55:** 199–206.

MATTHEWS, C.N. (1985), "Universal Protein Ancestors from Hydrogen Cyanide and Water," in *The Search for Extraterrestrial Life: Recent Developments*, M. Papagiannis, ed. (Dordrecht: D. Reidel Publishing Company), pp. 151–156.

MATTHEWS, C.N. (1986), "Simultaneous Synthesis of Polypeptides and Polynucleotides? Hydrogen Cyanide Polymers as Prebiotic Condensing Agents," *Origins of Life* **16:** 500.

MATTHEWS, C.N. (1988), "Cosmic Metabolism: The Origin of Macromolecules," in *Bioastronomy—The Next Steps*, G. Marx, ed. (Dordrecht: Kluwer Academic Publishers), pp. 167–178.

MATTHEWS, C.N. (1990), "Simultaneous Synthesis of Polypeptides and Polynucleotides. Hydrogen Cyanide Polymers as Prebiotic Condensing Agents," in *Prebiological Self Organization of Matter*, C. Ponnamperuma and F.R. Eirich, eds. (Hampton, VA: A. Deepak Publishing), pp. 171–178.

MATTHEWS, C.N. (1992a), "Dark Matter in the Solar System: Hydrogen Cyanide Polymers," *Origins of Life* **21:** 421–434.

MATTHEWS, C.N. (1992b), "Origin of Life: Polymers Before Monomers?," in *Environmental Evolution*, L. Margulis and L. Olendzenski, eds. (Cambridge, MA: MIT Press), pp. 29–38.

MATTHEWS, C.N., AND R.E. MOSER (1966), "Prebiological Protein Synthesis," *Proc. Natl. Acad. Sci. USA* **56:** 1087–1094.

MATTHEWS, C.N., AND R.E. MOSER (1967), "Peptide Synthesis from Hydrogen Cyanide and Water," *Nature* **215:** 1230–1234.

MATTHEWS, C.N., AND R.A. LUDICKY (1986), "The Dark Nucleus of Comet Halley: Hydrogen Cyanide Polymers," in *20th ESLAB Symposium on the Exploration of Halley's Comet*, ESA Report SP-250, vol. 2, B. Battrick, E.J. Rolfe, and R. Reinhard, eds. (Noordwijk: European Space Agency), pp. 273–277.

MATTHEWS, C.N., AND R.A. LUDICKY (1992), "Hydrogen Cyanide Polymers on Comets," *Adv. Space Res.* **12 (4):** 21–32.

MATTHEWS, C.N., J.E. NELSON, P. VARMA, AND R.D. MINARD (1977), "Deuterolysis of Amino Acid Precursors: Evidence for Hydrogen Cyanide Polymers as Protein Ancestors," *Science* **198:** 622–624.

MATTHEWS, C.N., J.E. NELSON, AND R.D. MINARD (1980), "Deuterolysis of Amino Acid Precursors from the Murchison Meteorite," *Abstracts 6th International Conference on the Origins of Life*, Jerusalem, June 22–27, p. 100.

MATTHEWS, C.N., R.A. LUDICKY, J. SCHAEFER, E.O. STEJSKAL, AND R.A. MCKAY (1984), "Heteropolypeptides from Hydrogen Cyanide and Water: Solid State ^{15}N NMR Investigations," *Origins of Life* **14:** 243–249.

MCDONALD, G.D., B.N. KHARE, W.R. THOMPSON, AND C. SAGAN (1991), "$CH_4/NH_3/H_2O$ Spark Tholin: Chemical Analysis and Interaction with Jovian Aqueous Clouds," *Icarus* **94:** 354–367.

MCKAY, R.A., J. SCHAEFER, E.O. STEJSKAL, R.A. LUDICKY, AND C.N. MATTHEWS (1984), "Double-Cross-Polarization Detection of Labelled Chemical Bonds in HCN Polymerization," *Macromolecules* **17:** 1124–1130.

MILLER, S.L., AND L.E. ORGEL (1974), *The Origins of Life on Earth* (Englewood Cliffs, NJ: Prentice-Hall).

MINARD, R.D., W. YANG, P. VARMA, J. NELSON, AND C.N. MATTHEWS (1975), "Heteropolypeptides from poly-α-Cyanoglycine and Hydrogen Cyanide: A Model for the Origin of Proteins," *Science* **190:** 387–389.

MOSER, R.E., AND C.N. MATTHEWS (1968), "Hydrolysis of Aminoacetonitrile: Peptide Formation," *Experientia* **24:** 658–659.

MOSER, R.E., A.R. CLAGGETT, AND C.N. MATTHEWS (1968), "Peptide Formation from Diaminomaleonitrile (HCN Tetramer) and from Aminomalononitrile (HCN Trimer)," *Tetrahedron Lett.*, **9:** 1599–1603.

ORÓ, J., AND A. LAZCANO-ARAUJO (1980), "The Role of HCN and Its Derivatives in Prebiotic Evolution," in *Cyanide in Biology*, B. Vennesland, E.E. Conn, C.J. Knowles, J. Westley, and F. Wissing, eds. (New York: Academic Press), pp. 517–541.

OWEN, T. (1982), "Titan," *Sci. Am.* **246 (2):** 98–109.

REINHARD, R. (1986), "The Giotto Encounter with Comet Halley," *Nature* **321:** 313–318.

RETTIG, T.W., S.C. TEGLER, D.J. PASTO, AND M.J. MUMMA (1992), "Comet Outbursts and Polymers of HCN," *ApJ* **398:** 293–298.

SAGAN, C., AND B.N. KHARE (1979), "Tholins: Organic Chemistry of Interstellar Grains and Gas," *Nature* **277:** 102–107.

SAGDEEV, R.Z., J. BLAMONT, A.A. GALEEV, V.I. MOROZ, V.D. SHAPIRO, V.I. SHEVCHENKO, AND K. SZEGÖ (1986), "Vega Spacecraft Encounters with Comet Halley," *Nature* **321:** 259–262.

SCHAEFER, J., E.O. STEJSKAL, G.S. JACOB, AND R.M. MCKAY (1982), "Natural Abundance N-15 NMR of the Solids from the Reaction of HCN and Ammonia," *Appl. Spectr.* **36:** 179–182.

SU, Y.-L., Y. HONDA, P.E. HARE, AND C. PONNAMPERUMA (1989), "Search of Peptide-like Materials in Electric Discharge Experiments," *Origins of Life* **19:** 237–238.

TEGLER, S.C., D.A. WEINTRAUB, T.O. RETTIG, Y.J. PENDLETON, C.B. WHITTET, AND C.A. KULESA (1995), "Evidence for Chemical Processing of Precometary Icy Grains in Circumstellar Environments of Pre-Main-Sequence Stars," *ApJ* **439:** 279–287.

UMEMOTO, K., M. TAKAHASHI, AND K. YOKOTA (1987), "Studies on the Structure of HCN Oligomers," *Origins of Life* **17:** 283–293.

VÖLKER, T.A. (1960), "Polymeric Hydrogen Cyanide," *Angew. Chem.* **72:** 379–384.

WARREN, C.B., R.D. MINARD, AND C.N. MATTHEWS (1974), "Synthesis of α-Cyanoglycine N-Carboxyanhydride and α-Cyanoglycine," *J. Org. Chem.* **39:** 3375–3378.

WHIPPLE, F.L. (1974), "The Nature of Comets," *Sci. Am.* **230 (2):** 48–57.

WOELLER, F., AND C. PONNAMPERUMA (1969), "Organic Synthesis in a Simulated Jovian Atmosphere," *Icarus* **10:** 386–392.

YANG, W., R.D. MINARD, AND C.N. MATTHEWS (1976), "Azacyclopropenylidenimine: A Low Energy Dimer of Hydrogen Cyanide," *J. Theor. Biol.* **56:** 111–123.

Irradiated Protoplanetary Disk

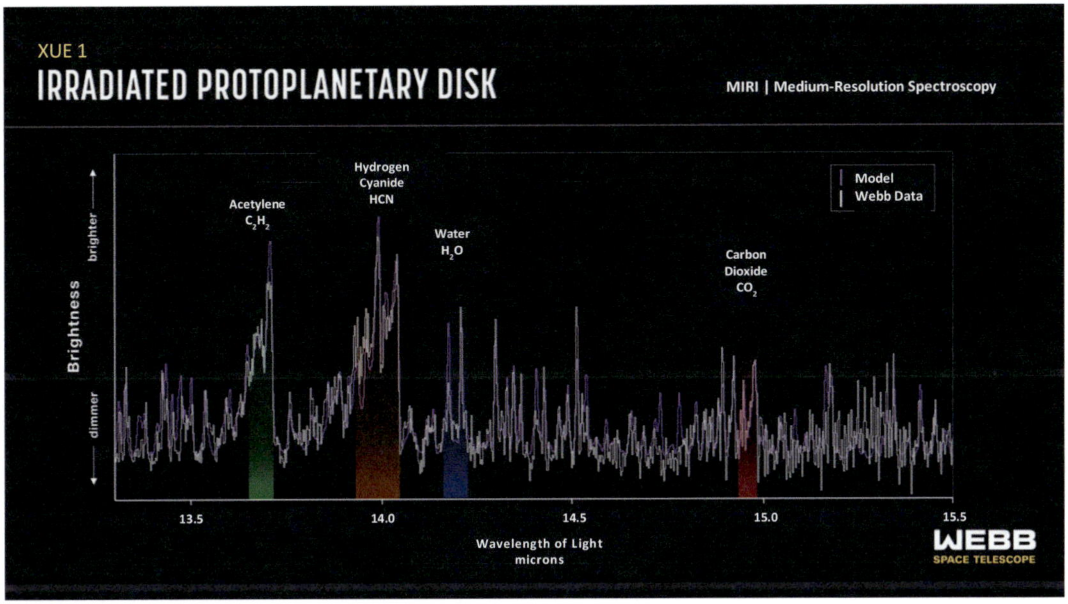

Credit: NASA, ESA, CSA, STScI, J. Olmsted (STScI), M. C. Ramírez-Tannus (Max Planck Institute for Astronomy)
https://esawebb.org/images/weic2329b/ [retrieved 6-20-25]

This graphic presents some of the first results from the eXtreme UV Environments (XUE) James Webb Space Telescope program. These results suggest that the conditions for rocky-planet formation, typically found in the disks of low-mass star-forming regions, can also occur in massive-star-forming regions and possibly a broader range of environments.

Astronomers focused on rocky-planet-forming regions of disks in the Lobster Nebula using Webb's Medium Resolution Spectrometer (MRS) of the Mid-InfraRed Instrument (MIRI). This first result focuses on the protoplanetary disk XUE 1, which is located in the star cluster Pismis 24.

The inner disk around XUE 1 revealed signatures of water (highlighted here in blue; around 14.2 microns), as well as acetylene (C_2H_2, highlighted in green; around 13.7 microns), hydrogen cyanide (HCN, highlighted in brown; around 14.0 microns), and carbon dioxide (CO_2, highlighted in red; around 14.95 microns). As indicated, some of the emission detected was weaker than some of the predicted models, which might imply a small outer disk radius.

Seismic Map of the Silverpit Crater

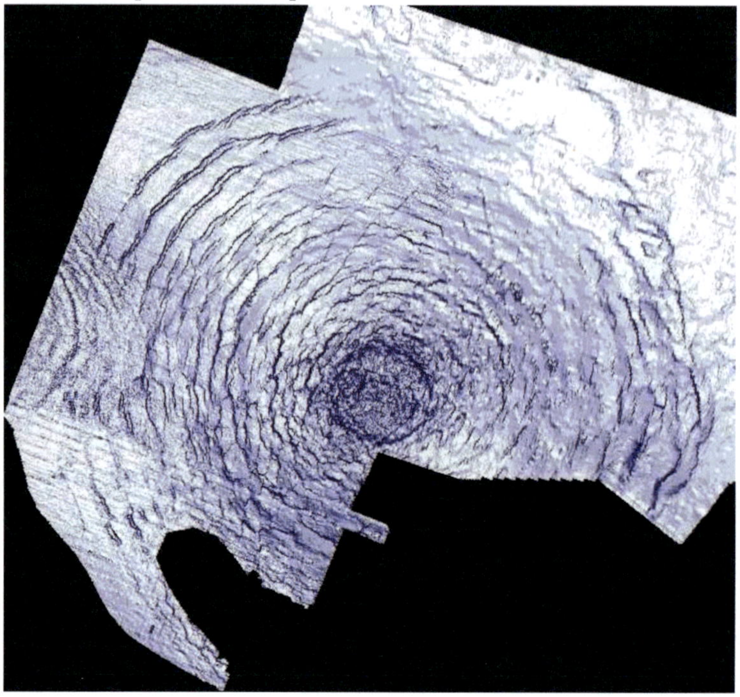

Photo Credit: Phil Allen (Production Geoscience Ltd.) and Simon Stewart (BP)
https://en.wikipedia.org/wiki/Silverpit_crater [retrieved 11-19-25]

After two decades of debate, researchers have announced that they finally have definitive proof that the Silverpit Crater, a buried structure in the North Sea, was created by an asteroid strike more than 40 million years ago.

The crater, discovered in 2002 about 80 miles (129 kilometers) off the Yorkshire coast, is 1.8 miles (3 km) wide and lies some 766 yards (700 meters) below the seafloor. Its circular shape and central peak look like hallmarks of a cosmic impact–but in the absence of definitive evidence, alternative explanations emerged. For instance, there were theories that shifting underground salt deposits or ancient volcanic activity could be responsible for the structure.

Now, a team led by Uisdean Nicholson of Heriot-Watt University in Scotland has solved the mystery at last. Using modern 3D seismic imaging and drill cuttings from a 1980s oil well, the researchers pieced together the clearest picture yet of the crater, confirming its asteroid impact origin.

"Silverpit is a rare and exceptionally preserved hypervelocity impact crater," Nicholson said. "We can use these findings to understand how asteroid impacts shaped our planet throughout history, as well as predict what could happen should we have an asteroid collision in future."

The new seismic data provided an "unprecedented look" at the structure, revealing unmistakable impact features, including a central uplift, an encircling moat, shattered rock zones and even smaller "secondary craters" carved by falling debris. The pattern of faults around the crater–with rocks pulled apart on the west side and compressed on the east–suggests the asteroid slammed in from the west at a shallow angle.

Microscopic analyses of drill samples were the smoking gun: rare grains of quartz and feldspar etched with microscopic scars that form only under the extreme pressures of a hypervelocity impact, not by any Earth-bound process. The presence of shocked minerals–tiny structural units within crystals found only in the high pressures of an asteroid impact–was a compelling indicator.

Source: Space.com

THE EFFECTS OF HYPERVELOCITY IMPACT ON CARBONACEOUS COMPOUNDS: EXPERIMENTAL INSIGHTS

T. E. Bunch

NASA Ames Research Center
Moffett Field, CA 94035-1000

Luann Becker and J. Bada

Scripps Institution of Oceanography
University of California at San Diego
La Jolla, CA 92093

Julie M. Paque

SETI Institute
Mountain View, CA 94043

J. Macklin

Department of Chemistry
University of Washington
Seattle, WA 98195

Abstract

Hypervelocity impact experiments were performed to test the survivability of carbonaceous impactors and to determine potential products that may have been synthesized during impact. Diamonds were launched by the Ames two-stage light gas gun into Al plate at velocities of 2.75 and 3.1 km s^{-1}. FESEM imagery *confirms that diamond fragments survived in both experiments. Earlier experiments found that diamonds were destroyed on impact above 4.3 km* s^{-1}. *Thus, the upper stability limit for diamond on impact into Al, as determined from our experimental conditions, is between 3.1 and 4.3 km* s^{-1}. *Impacts above this limiting velocity would probably destroy pertinent isotopic and noble gas information. Because diamonds in chondritic meteorites do contain such information, they must have accreted at very low velocities.*

Particles of the carbonaceous chondrite Nogoya were also launched into Al at a velocity of 6.2 km s^{-1}. *Laser desorption (*L^2MS*) analyses of the impactor residues indicate that the lowest- and highest-mass polycyclic aromatic hydrocarbons (PAHs) were largely destroyed on impact; those of intermediate mass (202–220 amu) remained at the same level or increased in abundance. In addition, alkyl-substituted homologs of the most abundant preimpacted PAHs were synthesized during impact. These results suggest that an unknown fraction of some organic compounds can survive low to moderate impact velocities and that synthesized products can be expected to form up to velocities of at least 6.5 km* s^{-1}.

Using information from natural terrestrial impact craters and these experiments, we conclude that the delivery of organic compounds by incoming projectiles to an evolving planetary body, together with organic synthesis on impact, were viable contributors to prebiotic evolution.

INTRODUCTION

The experimental results and interpretations presented here are part of a continuing research program aimed at determining the characteristics of carbonaceous matter that has sustained hypervelocity impact. Information on the behavior of carbonaceous materials on impact is rather sparse (e.g., Bunch et al. 1992, 1993), although Peterson et al. (1991) performed shock experiments on amino acid survivability. Two of the three crystalline forms of carbon (diamond and graphite) are known to occur in meteorites (Lewis et al. 1987), and diamond occurs in the interstellar medium (ISM) (Allamandola et al. 1993). Amorphous carbon and poorly crystallized graphite (PCG) in carbonaceous chondrites form the bulk of their carbon inventories, together with many organic compounds including polycyclic aromatic hydrocarbons (PAHs) that are also found in the ISM (Allamandola et al. 1987). These and other organic compounds may occur in comets (e.g., Langevin et al. 1987). The information contained within carbonaceous impactors is vital to understanding their origin and significance with respect to precursor materials and impact modifications.

In order to characterize and interpret carbonaceous residues, the following questions are being addressed. (1) Can carbon crystalline phases and organic compounds survive low-velocity ($\leq$ 7 km s^{-1}) impact, and if they do survive, what are their characteristics? (2) If they do not survive impact, what are their breakdown products or shock-synthesized products, if any? Light gas gun hypervelocity experiments have been conducted to constrain, within our experimental capabilities, these and other issues. Experiments to further refine techniques and establish additional impactor (projectile) criteria are continuing.

The original objective of this research was to build an experimental data base for use in the interpretation of the many hundreds of small (< 5 mm), carbonaceous-bearing craters on LDEF (Long Duration Exposure Facility). The experimental information can also be useful in ascertaining the survivability and potential syntheses of carbonaceous matter on impact into evolving planetary bodies. In the case of small impacting particles (less than a few mm), the experimental data would be useful only for bodies that have little or no atmosphere, as the impacting particles would suffer destructive thermal ablation and mask the precursor carbonaceous components, or be totally lost during travel through the atmosphere. The results presented herein can be extended to the interpretation of carbonaceous behavior in much larger impact events on bodies with atmospheres, although some scaling problems are involved. The survivability of carbonaceous matter is of critical importance in any discussion of circumstellar habitable zones.

Light Gas Gun Experiments: Sample Selection, Experimental and Analytical Techniques

Sample Selection and Experimental Conditions

Diamond, coarse-grained graphite, and Murchison and Nogoya carbonaceous chondrites were selected for study in impact experiments. Diamond was impacted to determine its stability under our experimental conditions; graphite was used to see if any other carbon forms (diamond/fullerenes) could be produced on impact; Murchison and Nogoya were used to test shock effects on actual meteoritic material.

Coarse-grained graphite was accelerated into Al targets under vacuum (0.5 mm Hg) at 5.92 km s^{-1}, Murchison particles at 4.8 and 5.89 km s^{-1}, Nogoya at 6.2 km s^{-1}, and diamond at 2.75, 3.1, 4.8, and 5.5 km s^{-1}. The prelaunch impactor particle size for Murchison, Nogoya, and graphite was ≈ 0.2 mm (minimum diameter) and diamond was ≈ 0.3–0.4 mm. For experiments of this type, the Ames light gas gun is limited to acceleration velocities of < 6.5 km s^{-1}. Peak pressures and temperatures cannot be directly measured.

Experimental Methodology—Two-Stage Light Gas Gun Experiments

Projectile grains were loaded into a small (3 mm cavity) Al carrier cup, capped with an Al plate and fitted into a sabot for launching. The two-stage light gas gun accelerates the sabot down a rifled barrel (1.2 m long; bore diameter = 9 mm) to velocities of ≈ 2–6.5 km s^{-1} depending on the amount of the powder charge (first stage), which in turn determines the speed of the deformable ram that compresses hydrogen gas (second stage). When the gas reaches a certain critical pressure, a diaphragm ruptures and the gas propels the sabot down the barrel. At the end of the barrel, the sabot, carrier, and particles are separated in a "blast" chamber, particle velocities are electronically recorded, the in-flight particles are photographed, and the particles continue on to impact with the target plate (at 90° to the target) in an evacuated chamber (vacuum pressures $\approx$ 0.5 mm of Hg). The impacted plate also serves as a witness plate with a 2.5 cm diameter hole through which the carrier travels. This technique allows only particles to impact, the alignment of the launch can be measured, and the range in diameters of the launched particles at the impact point can be measured from the holes in the mylar covering, which is attached beneath the hole in the impacted plate. This cluster shot approach is necessary for projectiles $<$ 1.0 mm in diameter as smaller grains cannot be individually launched.

Observational and Analytical Techniques

Samples were dry cut from the target plate. Craters and retained impactor debris were first observed by an optical light microscope, then by field emission scanning electron microscopy (FESEM). Samples were then analyzed for impactor residue by light element equipped energy dispersive spectroscopy (EDS).

MICRO RAMAN SPECTROSCOPY

Raman spectra were obtained by using the 488.0 nm line of a coherent radiation Innova 90 argon ion laser. The power of the laser radiation used here was between 20 and 40 mW and focused to a spot size of 5 microns. The scattered radiation was dispersed with a SPEX Industries 1477B Triplemate spectrometer equipped with a SPEX Micromate microsampling system that includes a modified Zeiss microscope and a Reticon intensified diode array detector. The spectrometer slit width was kept at 300 microns.

Integration times on the diode array ranged from 5 to 10 seconds and 10 or 20 acquisitions were averaged before data collection.

LASER IONIZATION MASS SPECTROMETRY (L^2MS)

The two-step laser methodology has been described by Hahn et al. (1987). In the first step, the pulsed output of a CO_2 laser (10.6 μm; 20 mJ/pulse; 10 μsec pulse width; 5-Hz repetition rate) is focused onto a small stainless steel disk (~ 1 mm diameter) containing the sample. The infrared (IR) radiation is readily absorbed by the meteorite minerals and causes the ejection of intact neutral molecules from their surfaces in a rapid, laser-induced thermal desorption process. The fact that desorption dominates over decomposition in rapid laser heating processes is well documented (Hahn et al. 1987, Engelke et al. 1991). The sample can be rotated manually in order to expose a fresh surface to the desorption laser. After an appropriate time delay (~ 130 μsec), the fourth harmonic of a Nd:YAG laser (266 nm; 1.5–2.0 mJ/pulse; 10-nsec pulse width; 5-Hz repetition rate) is used to induce 1 + 1 resonance-enhanced multiphoton ionization (REMPI) of the desorbed molecules in an interaction region about 5 mm from the surface. REMPI causes soft ionization so that the parent ions of the desorbed aromatic compounds almost exclusively dominate the spectrum. Total ionization efficiency is about a factor of 100 to 1000 greater than that of methods where ions are directly produced on a surface. One of the advantages to the L^2MS system is the spatial and temporal separation of the desorption and ionization, which results in more control than in one-step desorption/ionization processes. The laser-generated ions are mass separated in a linear time-of-flight (TOF) system (mass resolution = 500) and detected with a microchannel plate array. Data for the meteorite samples were averaged over 100 laser shots, although a complete mass spectrum can be obtained from a single shot.

Samples were prepared using MALDI (matrix-assisted laser desorption ionization). Previous reports have shown that laser desorption of neutral molecules can be improved by spraying a fine layer of sample on top of a matrix that absorbs at the wavelength of the laser (Hillenkamp et al. 1991). For our L^2MS system, the organic substrate sinapinic acid was used as the matrix. The matrix is sprayed directly onto the stainless steel disk (100 ng/mm^2) ensuring that the substrate is evenly dispersed over the entire surface of the disk. The impacted meteorite sample (sonicated in toluene) is then sprayed on top of the sinapinic acid film. The sample disk is mounted on a 7 mm diameter teflon probe tip and is introduced to the TOF mass spectrometer through a separate antechamber pumped down to zero millitorr before introducing it to the high vacuum (10^{-7} torr) of the system. Sample introduction takes about two minutes and the spectrum can be recorded immediately thereafter.

RESULTS

Graphite

These craters contain large areas of shiny, meltlike carbon, although crater liners are less lumpy compared with submicron graphite craters. Raman spectra of these craters indicate that intact, well-ordered graphite did survive impact, together with partially disordered graphite in the crater bottom centers; the most disordered graphite is found as ejecta outside of the craters (Fig. 1). Thus, the degree of disorder increases away from the crater centers. For reference, the Raman spectrum of crystalline graphite consists of a single band (below 2000 cm^{-1}) at 1580 cm^{-1}.

Raman spectra of poorly crystallized graphite (PCG) and glassy or vitreous carbon have additional bands in this spectral region at 1360 and 1620 cm^{-1}. The intensity of the band at 1360 cm^{-1}, relative to the band at 1580 cm^{-1} (height and width), is taken to indicate the degree of disorder in PCG and vitreous carbon.

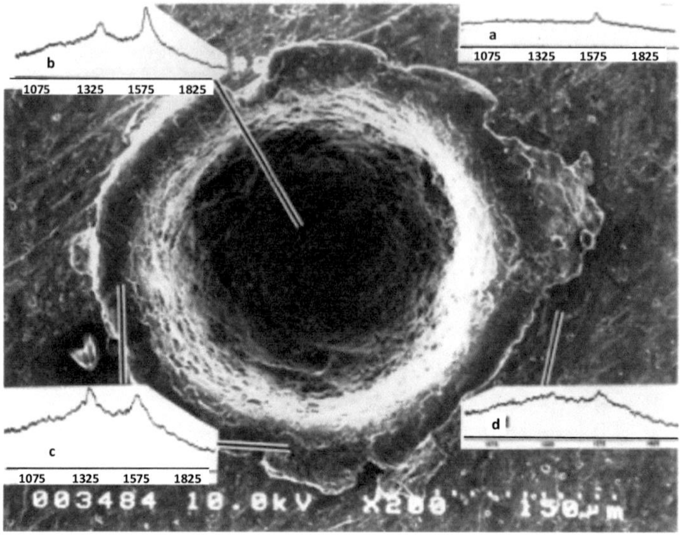

FIG. 1. *Graphite impact crater in Al formed at 5.92 km s^{-1}. Raman spectra show that graphite is still, in part, well crystallized in the crater bottom (b) compared with carbon on the rim (c) and ejecta (d) that have degraded into PCG and amorphous carbon, respectively. All are compared with (a), which is the reference graphite standard.*

The Raman spectrum of prelaunched graphite is shown in Figure 1 (a), together with three graphite spectra from the crater bottom (b), the rim (c), and ejecta outside the crater (d). The spectrum from the bottom is that of mostly crystalline graphite with some disordered carbon. These features were found only in the centers of the craters. The spectrum on the rim is similar to that obtained for typical vitreous carbon and is analogous to most carbon from the Allende meteorite (Macklin et al. 1987). In addition, it is also similar to PCG when it is heated in argon above 1200°C for up to 0.5 hour. The spectrum for ejecta carbon is indicative of even greater disorder in PCG material. It is comparable to the spectrum obtained from most carbon in the Murchison meteorite (Macklin et al. 1987).

Diamond

Typical craters formed from diamond impactors at 5.5 km s^{-1} are shown in Figure 2. Regardless of their size, diamond craters contain much thicker, continuous-melt liners (Fig. 3) compared with the other experimental carbon impactor craters.

Figure 4 shows a large melt droplet on the crater wall. EDS analysis indicates a weak Al signal along with a strong carbon signal; thus, we conclude that the melt liner and the droplet are carbon. Raman spectra of diamond impact craters in Al are the same as those obtained for PCG and vitreous carbon (Fig. 1c) and are similar to spectra from carbon in the Allende meteorite. This melt liner is covered by a thin layer of dusty carbon. Raman analyses show that the layer is a mixture of amorphous carbon and highly disordered PCG. The crater bottom consists of submicron grains and tiny "melt" spheres. The spheres may be the amorphous and vitreous carbon in the Raman analyses.

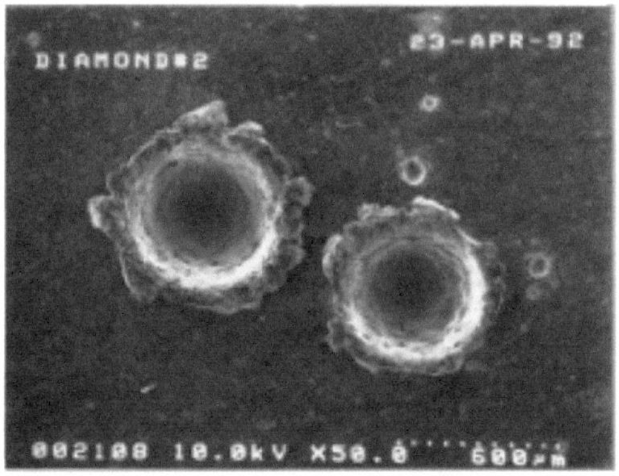

FIG. 2. *FESEM image. Diamond impact craters in Al.*

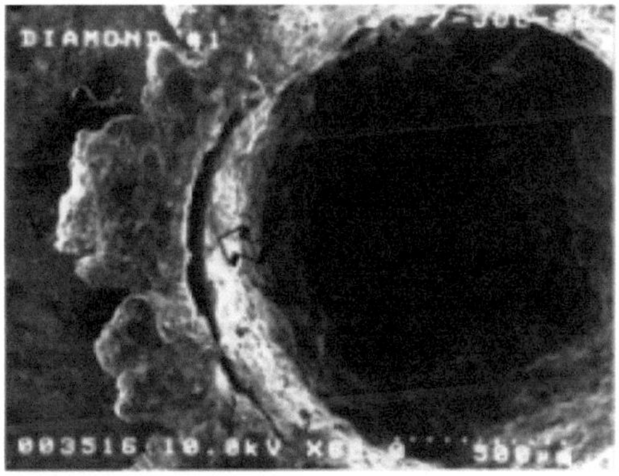

FIG. 3. *FESEM image. Arrow points to thick, carbonaceous cavity liner that formed from diamond degradation during impact at 5.5 km s^{-1}.*

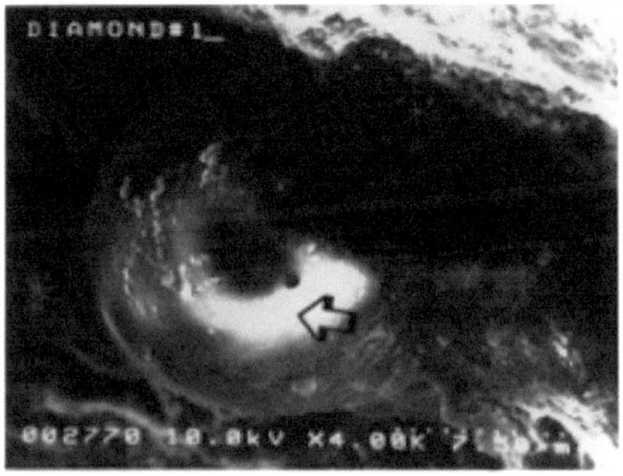

FIG. 4. *Melt droplet of carbon formed by the melting of diamond on impact.*

There is variability in the carbon distribution between the large craters (> 0.1 mm) and small craters (< 0.1 mm). Raman spectra of highly disordered (amorphous) carbon were obtained from the raised rim of small craters but not from the center of the crater, whereas no spectrum of amorphous carbon was obtained from the raised rim of the large craters, and measurements could not be obtained from the small crater interiors (bottoms).

Craters formed by launching diamond into Al at 2.75 and 3.1 km s^{-1} (Fig. 5) show morphological features quite different from those formed at higher velocities (5.5 km s^{-1}; Fig. 2) in that they have rough, irregular multiple crater bottoms, without melt features. In addition, diamond impactor fragments are visibly present (Fig. 6) and Raman spectra indicate that crystalline diamond is present in much greater abundance than PCG.

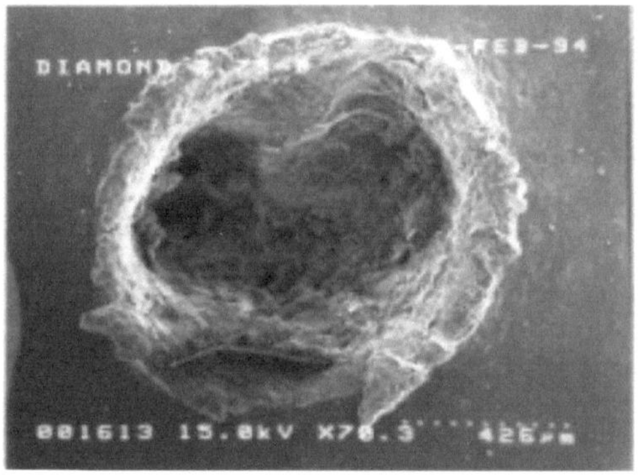

FIG. 5. *Diamond impacted into Al at 2.75 km s^{-1}. Note the rough, irregular crater features.*

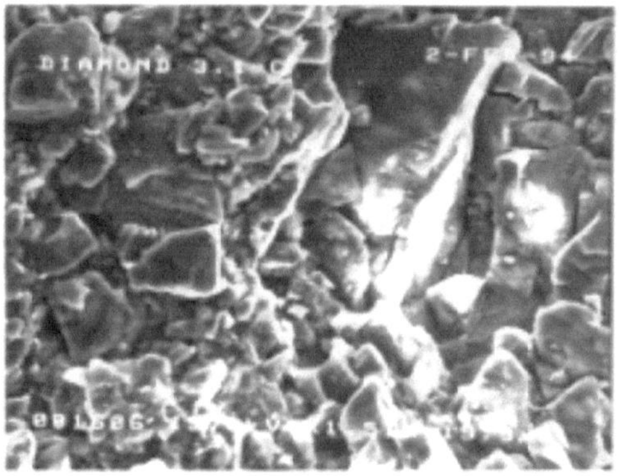

FIG. 6. *Residual, fragmented diamond impactor in crater of Fig. 5.*

Murchison and Nogoya

These craters have small amounts of retained impactor and impact alteration features that are unique among those described in this work (Fig. 7). Crater bottoms commonly have melt beads that grade into peculiar spongelike structures at the lower portions of the wall (Figs. 7 and 8). The upper walls and rims have large amounts of irregularly shaped melted to partially melted impactor.

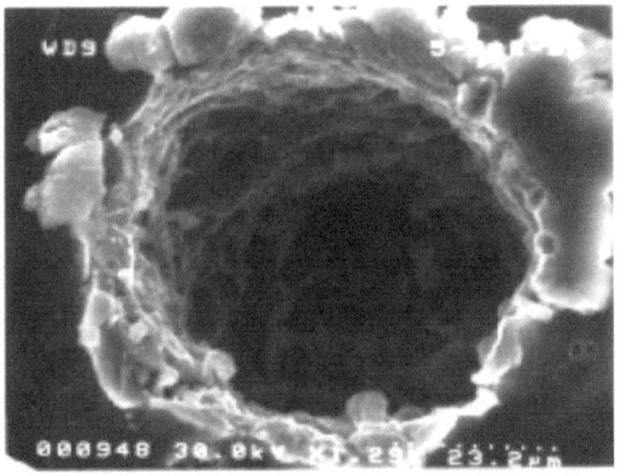

FIG. 7. *Small Murchison impact crater in Al formed at 5.89 km s^{-1}. Note spongy residue within crater.*

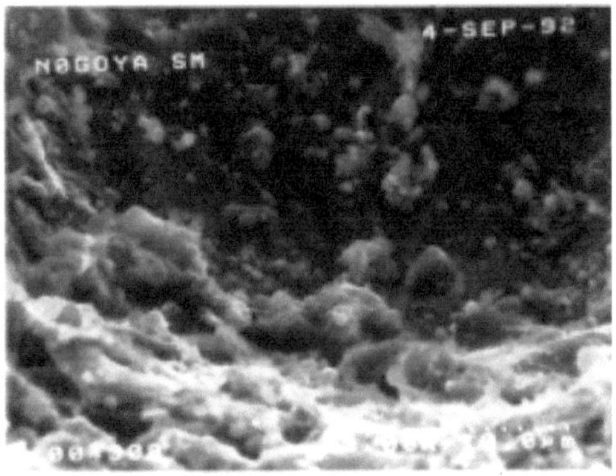

FIG. 8. *Intact and partially melted impactor debris in the bottom of a Nogoya crater.*

The prelaunch PAH composition from L^2MS analyses for Murchison is shown in Figure 9a and Table 1, and for Nogoya in Figure 10a and Table 1. The two impacted meteorite samples were then analyzed and compared with the non-impacted compositions (Fig. 9b for Murchison and Fig. 10b for Nogoya). Table 2 summarizes the PAH compositions after impact. In both samples the lower masses are no longer present but those of intermediate masses 184, 202, 206, 220 are preserved in part. This is consistent with what one might expect to see since the lower-molecular-weight PAHs would likely be the first to decompose/evaporate with the temperatures incurred at impact. There also appears to be an increase in alkylation compared with the nonimpacted composition, which would indicate that some of these PAHs were derived from more extensively heated precursors (probably due to impact) since the high extent of alkylation of PAHs has been attributed to the thermal cracking of the "organic polymer" (Ishiwatari and Fukushima 1979).

All three samples were scanned for higher-mass compounds both in the porphyrin and fullerene ranges, but none were observed.

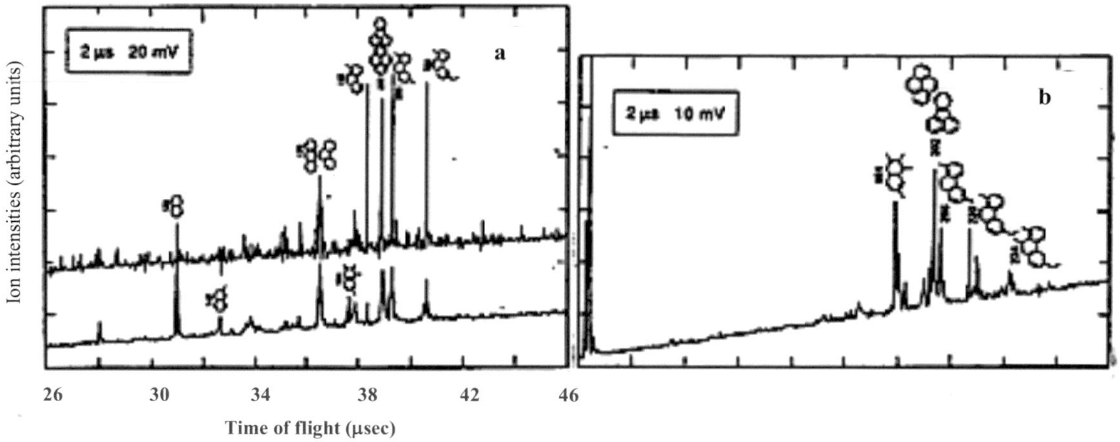

FIG. 9. *(a) Laser desorption/multiphoton ionization TOF mass spectra of nonimpacted Murchison meteorite. Top spectrum is a single shot, lower spectrum is an average of 100 shots. (b) 100-shot averaged spectrum for 4.8 km* s^{-1} *impact.*

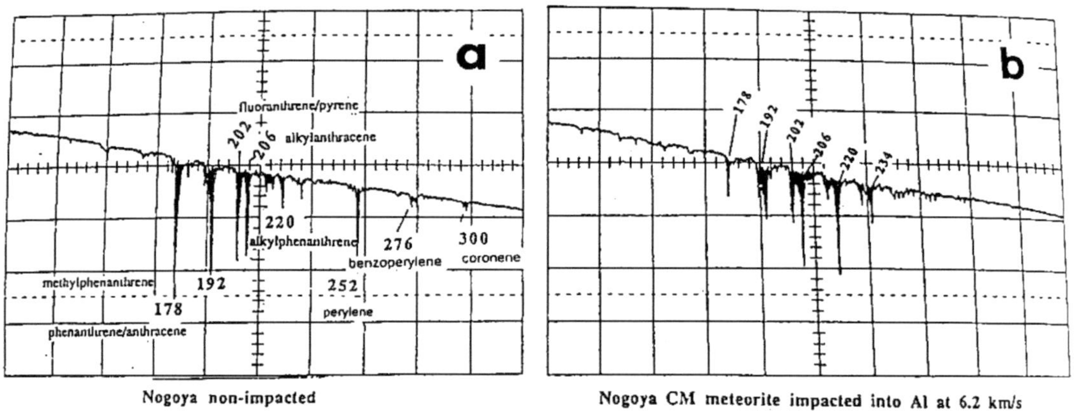

FIG. 10. *(a)* L^2*MS spectrum of nonimpacted Nogoya. (b) After impact into Al at 6.2 km* sec^{-1}.

Nogoya differs from Murchison in its PAH content by not having naphthalene (mass 128) or any of the alkyl-substituted species (homologs) of naphthalene (mass 142, 154, 170, 198), for example, 2-methyl naphthalene (mass 142). Note: a single alkyl substitution is a univalent aliphatic radical attached to the PAH structure. Alkylation comes from the "cracking of the polymer" or the loss of CH_2 from a polymer (Ishiwatari and Fukushima 1979) that is concentrated in constituent kerogens of the meteorite and attaches itself to the PAH benzene ring structure during a thermal event. Generally speaking, the greater the alkylation of PAHs, the lower the formation temperature (Wing and Bada 1991). For example, PAHs formed at 2000°C have no alkyl substitutions; a few are formed at 400°C, and up to six alkyl carbons at 150°C (Blumer 1976).

TABLE 1. ***PAHs in Prelaunched Murchison and Nogoya Carbonaceous Chondrites***

Murchison		*Nogoya*	
Mass	*Compound*	*Mass (amu)*	*Compound*
128	naphthalene		
142	2-methylnaphthalene		
178	phenanthrene / anthracene*	178	phenanthrene / anthracene*
192	2 methyl-phenanthrene / anthracene*	192	2 methyl-phenanthrene / anthracene*
202	pyrene / fluoranthene*	202	pyrene / fluoranthene*
206	C_{16}-alkylphenanthrene / anthracene*	206	C_{16}-alkylphenanthrene / anthracene*
220	C_{17}-alkylphenanthrene / anthracene*	220	C_{17}-alkylphenanthrene / anthracene*
		252	perylene†
		276	benzoperylene†
		300	coronene†

* *These are possible assignments of the observed peaks; different isomers of PAHs cannot be distinguished by L^2MS.*

† *These PAHs have also been found in Murchison (De Vries et al. 1993) by L^2MS technique at the ≈ 1 ppm level. They are either not in our sample or they are below our detection limits.*

TABLE 2. ***PAH Content in Murchison and Nogoya after Impact at 5.8 and 6.2 km s^{-1}, Respectively***

Mass	*Compound*	*Murchison*	*Nogoya*
Missing			
128	naphthalene	√	
142	2-methylnaphthalene	√	
252	perylene		√
276	benzoperylene		√
300	coronene		√
Greatly Reduced in Content			
178	phenanthrene / anthracene	√	√
192	methyl-phenanthrene / anthracene	√	√
202	pyrene / fluoranthene	√	√
About the Same			
206	C_{16}-alkylphenanthrene / anthracene	√	√
Increased Content/ Synthesized?			
220	C_{17}-alkylphenanthrene / anthracene	√	√
Synthesized?			
234	C_{18}-alkylphenanthrene / anthracene	√	√

Discussion

Graphite

Raman spectra indicate that some intact, well-ordered graphite did survive impact, together with partially disordered graphite in the crater bottom centers. The degree of disorder increases away from the crater centers. This suggests that carbonaceous residues found in the centers probably represent material that experienced the least amount of shock-induced damage. Impactor material outside of the craters, which shows the highest degree of disorder, was probably ejected early in the crater formation stages and was exposed to higher pressures and temperatures, in addition to having cooled more rapidly. In fact, all residues cooled exceedingly rapidly as the entire crater-forming event probably took no longer than nano- to microseconds (Hörz et al. 1983, Chyba et al. 1990), although cooling continued after the pressures dropped to the ground state. The cooling rate is unknown, but sufficiently high to prevent annealing/ordering to occur. Thus, the graphite that was disordered by very high temperatures was not annealed. Rims and the upper walls captured some of the late-stage ejecta and melt splash before the crater cavity formation was completed; this residue material is intermediate in the degree of disorder between that of the centers and the outside ejecta.

Diamond

From our impact experiments, we find that diamond can partially survive in impacts > 3.1 to < 4.3 km s^{-1}. At 4.3 km s^{-1} diamond completely decomposes to PCG with some melting, and clear evidence for melting is developed at impact velocities of 5.5 km s^{-1}. We cannot directly measure peak pressures and temperatures of the experiments, although peak pressures can be extrapolated from equation-of-state information for Al and diamond and from the thermodynamic model that solves for peak stress generated by impactors and their targets (Cintala 1993). For diamond impacts of 2.75, 3.1, 4.3, and 5.5 km s^{-1}, peak pressures are calculated as ≈ 35, 39, 60, and 84 GPa, respectively (M. Cintala, pers. comm.). The highest two pressures are sufficient to produce melting of anorthosite on impact into Al (Bernhard et al. 1992); diamond, which is 20% more dense than anorthosite, should partially melt at these peak pressures. The time duration of peak pressures and temperatures for these impact events is probably on the order of 10^{-6} to 10^{-9} sec and the total crater formation time is probably less than a few msec (extrapolated from data of Hörz et al. 1983). Although diamonds are not expected to be present as diamond IDPs (interplanetary dust particles) from the interstellar medium, they are expected to be present as small grains (up to tens of nanometers in diameter) in chondritic meteorites (Huss 1990). The apparent low thermal stability of diamond under shock as shown in these experiments makes the study of any extracted intact meteoritic debris from impact craters impractical for diamond. On the other hand, our experiments establish a lower limit on diamond stability with respect to impacts into silicates during planetary accretion of > 3 km s^{-1}, which would probably destroy pertinent isotopic and noble gas information. Diamonds in chondritic meteorites must have accreted at very low velocities.

Murchison and Nogoya

The consistent absence of the low-mass PAHs in all impactor residues and the very good reproducibility of the analyses, coupled with an average of 100 analyses for each impactor residue where the low-mass PAHs are missing, strongly suggests that the missing lower-mass PAHs, which have lower boiling points (T_b K), vaporized on impact. The lower-mass PAHs, naphthalene and alkyl-substituted naphthalenes (homologs), are absent in the impacted Murchison meteorite sample (5.9 km s–1). This absence is attributed to the higher volatility of these lower-molecular-weight PAHs, which presumably volatilized during impact. The T_b K values for the missing PAHs are < 400, whereas T_b K values for the surviving PAHs are > 400 (White 1986), which could imply an upper effective evaporation temperature limit sustained during impact for the surviving PAHs. Thus, it appears that at least some of the higher-mass PAHs common in meteorites like Murchison can survive hypervelocity impact under the experimental conditions presented here. In contrast, unshocked Nogoya does not contain PAHs below 178 amu and, in addition, contains PAHs above 252 amu. Figure 10b (shocked Nogoya) shows that the unsubstituted PAHs are either absent or greatly reduced by impact at 6.2 km s–1. This is expected for those below 202 amu, because of their higher volatility, but not for those above 252. For example, coronene has the highest boiling point of the common PAHs and is the most stable of aromatic compounds (Aihara 1992).

The absence of coronene and benzoperylene in impacted Nogoya may be due to their low abundance in Nogoya, which is estimated from the L^2MS spectrum (Fig. 10b) to be 1–5 ppm. The low concentration together with the known highly heterogeneous distributions of meteoritic organics may be due to sampling bias. However, this does not explain the absence of abundant perylene (252 amu) in shocked Nogoya. Radicati di Brozolo et al. (1994) suggested that the fullerenes found in an impact crater on LDEF where the impactor residue has a "chondritic" or extraterrestrial signature may have formed on impact. They also noted PAHs in their L^2MS spectrum of the carbonaceous residue; perylene is the most notable. Perhaps some of the precursor PAHs in the extraterrestrial impactor gave rise to fullerenes on impact in a way analogous to the formation of fullerenes from high-temperature combustion of naphthalene (Taylor et al. 1993). The lack of the highest-mass PAHs in impacted Nogoya may be due to their synthesis into fullerenes on impact. Although we have not found fullerenes in the impacted meteorite samples on cursory examination, the experiments needed to test this idea can be easily done. We suggest that impact-formed fullerenes are likely to be found in carbonaceous residues of large-scale impact craters.

As in the case for shocked Murchison, the alkylated homologs (206 and 220 amu) are the most abundant PAH species. In shocked samples of both meteorites, C_{18}-alkyl-phenanthrene/anthracene at amu 234 is present, possibly as an impact-synthesized PAH from further cracking of intrinsic kerogens in the meteorites. We can conclude that, within our experimental and analytical constraints, some PAHs do survive impact up to 6.2 km s^{-1}, some are destroyed, and a few appear to be synthesized (see Table 2, p. 315).

Significance for Circumstellar Habitable Zones

There has been considerable interest over the last few years in the concept of the delivery of organic compounds to Earth via comets, meteorites, and IDPs, and the survivability and possible syntheses of organics on impact. Whereas much of the total carbonaceous mass that was delivered to Earth during the Hadean period (4.5–3.8 Gyr ago) is thought to have been contained in IDPs compared with much larger bodies (Anders 1989), some of this carbonaceous material would have sustained thermal alteration/pyrolysis from atmospheric ablation on entry, if entry velocities were sufficiently high. Only particles $<\approx 0.02$ mm in diameter would have escaped serious alteration (Love and Brownlee 1991). The behavior of carbonaceous matter during ablation is largely unknown.

For evolving bodies with limited or no early atmospheres, small particle delivery of organics would seem to be a viable means if the effects of impact are not totally destructive. Our simple experiments under low impact velocities ($<$ 6.3 km s^{-1}) suggest that at least some PAHs may survive impacts at temperatures well above their stability limits and others may be synthesized. The reasons for these behaviors are not clear. Perhaps the very short times available for bond breaking and reactions to occur are insufficient for complete transformations. From our own experimental research and from a theoretical model of Chyba et al. (1990) we can suggest that an unknown fraction of the impactor escapes the intense and destructive pressures and temperatures of the peak shock (possibly material on the backside of the impactor) and is saved from destruction. The graphite experiment shows that the degree of impactor disorder is correlated with its position within the cratered area; that is, at least a layer of moderately ordered graphite occupies the crater bottom, consistent with the idea that the last portion of the impactor to reach the target is the least affected. Organic synthesis of precursor impactor components may occur in the much lower post-peak-temperature regime and may also be dependent on its position within the impact plume. Radicati di Brozolo et al. (1994) found that carbonaceous residue in the naturally occurring fullerene-bearing LDEF crater, in addition to fullerenes, consists of highly disordered carbon, fragmented masses of unknown carbonaceous materials, and very small amounts of typical meteoritic PAHs. So even at impact velocities approaching 13 km s^{-1}, a few PAHs appear to survive together with synthesized fullerenes and possibly other compounds.

It remains to be determined if we can extrapolate from our experiments to megaimpact events in terms of predicting survivability and synthesis of carbonaceous matter. In terrestrial cratering events, few impactor remains are found outside of the crater with time because of rapid weathering, erosion and/or burial. The interiors of craters, because of steep-sided walls, fill in rapidly, thus impactor evidence is quickly lost. However, it is widely known that heavily to lightly shocked metal meteorite fragments of the Canyon Diablo meteorite have been found on the rim and flanks of Meteor Crater, Arizona (age $\approx 5 \times 10^3$ years old) together with nodules (up to decimeters in size) of well-ordered graphite (Bunch, unpubl. data). Clearly, some lightly altered impactors do survive large cratering events. Moreover, Becker et al. (1994) have found large amounts of fullerenes in an impact debris layer of the ancient Sudbury Crater and conclude that the fullerenes were synthesized from impactor carbonaceous material. From these examples, it appears very likely that some portions of impactors do survive intact and synthesis of carbonaceous matter does occur. We conclude that the delivery of organic compounds by incoming projectiles to an evolving planetary body together with impact synthesis probably made viable contributions to prebiotic evolution.

Summary

The experimental results discussed here suggest that some impacting organic compounds (PAHs) can survive impact temperatures well above their stability limits while others are apparently synthesized. Table 3 below summarizes the experimental results of this work. The impact process provides a potential means for delivery of organic material to planetary bodies.

TABLE 3. ***Summary of Hypervelocity Impact Experiments Involving Carbonaceous Impactors***

Samples	*Velocity (km s^{-1})*	*Remarks*
Graphite	3–5.95	Graphite decomposed to various degrees of poorly crystallized graphite (PCG). Correlation of highest disorder with distance away from crater bottom center. Diamonds and fullerenes not formed.
Diamond	2.75–5.95	Diamond decomposed to PCG and vitreous/amorphous carbon above 4 km s^{-1} impact velocities.
Murchison	4.35–5.98	PAHs above molecular mass 178 mostly survived impact. Naphthalene and alkyl-substituted naphthalenes were destroyed. Additional alkyl homologs of phenanthrene/anthracene were synthesized, especially at 220 and 234 amu.
Nogoya	4.3–6.1	Carbonaceous chondrite that contains more water, bulk C, and somewhat different organic chemistry than Murchison. PAHs between mass 178 and 206 amu partially survived impact, although those above 240 amu were also destroyed. Additional alkyl homologs of phenanthrene/anthracene were synthesized, especially at 220 and 234 amu.

ACKNOWLEDGMENTS

We thank J. Erlichman for assistance in FESEM imagery and B. Langedyk, W. Logsdon, and J. Vongrey for their assistance during the impact experiments. We also thank the NASA Exobiology Program and the LDEF Science Office for partially supporting this work.

REFERENCES

AIHARA, J. (1992), "Why Aromatic Compounds Are Stable," *Sci. Am.* **240 (3):** 62–68.

ALLAMANDOLA, L.J., S.A. SANDFORD, AND B. WOPENKA (1987), "Interstellar Polycyclic Aromatic Hydrocarbons and Carbon in Interplanetary Dust Particles and Meteorites," *Science* **237:** 56–59.

ALLAMANDOLA, L.J., S.A. SANDFORD, A.G.G.M. TIELENS, AND T.M. HERBST (1993), "Diamonds in Dense Molecular Clouds: A Challenge to the Standard Interstellar Paradigm," *Science* **260:** 64–66.

ANDERS, E. (1989), "Pre-biotic Organic Matter from Comets and Asteroids," *Nature* **342:** 255–257.

BECKER, L.A., J. BADA, R.E. WINANS, J.E. HUNT, T.E. BUNCH, AND B.M. FRENCH (1994), "Fullerenes in the 1.8 byr Old Sudbury Impact Structure," *Science* **265:** 642–645.

BERNHARD, R.P., T.H. SEE, AND F. HÖRZ (1992), "Projectile Compositions and Modal Frequencies on the 'Chemistry of Micrometeoroids' LDEF Experiment," in *Second LDEF Symp.*, NASA CP-3194, pp. 551–573.

BLUMER, M. (1976), "Polycyclic Aromatic Compounds in Nature," *Sci. Am.* **234 (3):** 34–38.

BUNCH, T.E., F. RADICATI DI BROZOLO, R.H. FLEMING, D.W. HARRIS, D.E. BROWNLEE, AND T. REILLY (1992), "LDEF Impact Craters Formed by Carbon-Rich Impactors: A Preliminary Report," in *First LDEF Symp.*, NASA CP-3134, pp. 549–564.

BUNCH, T.E., L. BECKER, J. BADA, J. MACKLIN, F. RADICATI DI BROZOLO, R.H. FLEMING, AND J. ERLICHMAN (1993), "Hypervelocity Impact Survivability Experiments for Carbonaceous Impactors," in *Second LDEF Symp.*, NASA CP-3194, pp. 453–477.

CHYBA, C.F., P.J. THOMAS, L. BROOKSHAW, AND C. SAGAN (1990), "Cometary Delivery of Organic Molecules to the Early Earth," *Science* **249:** 366–373.

CINTALA, M.J. (1993), "Impact-Induced Thermal Effects in the Lunar and Mercurian Regolith," *JGR* **97:** 947–973.

DE VRIES, M.S., K. REIHS, H.R. WENDT, W.G. GOLDEN, AND H.E. HUNZIKER (1993), "A Search for C_{60} in Carbonaceous Chondrites," *Geochim. Cosmochim. Acta* **57:** 933–938.

ENGELKE, F., J.H. HAHN, W. HENKE, AND R.N. ZARE (1991), "Determination of Phenylthiohydantoin–Amino Acids by Two-Step Laser Desorption/Multiphoton Ionization," *Anal. Chem.* **59:** 909–912.

HAHN, J.H., R. ZENNOBI, AND R.N. ZARE (1987), "Subfemtomole Quantitation of Molecular Absorbates by Two-Step Laser Mass Spectrometry," *J. Am. Chem. Soc.* **109:** 2842–2843.

HILLENKAMP, F., M. KARAS, R.C. BEAVIS, AND B.T. CHAIT (1991), "Matrix-Assisted Laser Desorption/Ionization Mass Spectrometry of Biopolymolecules," *Anal. Chem.* **63:** 24–37.

HÖRZ, F., H. FECHTIG, AND J. JANICKE (1983), "Morphology and Chemistry of Projectile Residues in Small Experimental Craters," *Proc. 14th LPSC, JGR* **88:** B353–363.

HUSS, G.R. (1990), "Ubiquitous Interstellar Diamond and SiC in Primitive Chondrites: Abundances Reflect Metamorphism," *Nature* **347:** 159–162.

ISHIWATARI, R., AND K. FUKUSHIMA (1979), "Generation of Saturated and Aromatic Hydrocarbons by Thermal Alteration of Young Kerogen," *Geochim. Cosmochim. Acta* **43:** 1343–1349.

LANGEVIN, Y., J. KISSEL, J.-L. BERTAUX, AND E. CHASSEFIÉRE (1987), "First Statistical Analysis of 5000 Mass Spectra of Cometary Grains Obtained by PUMA 1 (Vega 1) and PIA (Giotto) Impact Ionization Mass Spectrometers in the Compressed Modes," *A&A* **187:** 761–766.

LEWIS, R.S., M. TANG, J.F. WACKER, E. ANDERS, AND E. STEEL (1987), "Interstellar Diamonds in Meteorites," *Nature* **326:** 160–162.

LOVE, S.G., AND D.E. BROWNLEE (1991), "Heating and Thermal Transformation of Micrometeoroids Entering the Earth's Atmosphere," *Icarus* **89:** 26–43.

MACKLIN, J., D.E. BROWNLEE, S. CHANG, AND T.E. BUNCH (1987), "Micro-Raman Spectroscopic Measurement of Carbon in Meteorites and Interplanetary Dust Particles," *Microbeam* **87:** 211–212.

PETERSON, E., F. HÖRZ, G. HAYNES, AND T. SEE (1991), "Modification of Amino Acids at Shock Pressures of 3 to 30 GPa," *Meteoritics* **26:** 384.

RADICATI DI BROZOLO, F., T.E. BUNCH, R.H. FLEMING, AND J. MACKLIN (1994), "Fullerenes in an Impact Crater on the LDEF Spacecraft," *Nature* **369:** 37–40.

TAYLOR, R., G.J. LANGLEY, H.W. KROTO, AND R.M. WALTON (1993), "Formation of C_{60} by Pyrolysis of Naphthalene," *Nature* **366:** 728–731.

WHITE, C.M. (1986), "Predictions of the Boiling Point, Heat of Vaporization, and Vapor Pressure at Various Temperatures for PAHs," *J. Chem. Eng. Data* **31:** 198–203.

WING, M.R., AND J.L. BADA (1991), "Geochromatography on the Parent Body of the Carbonaceous Chondrite Ivuna," *Geochim. Cosmochim. Acta* **55:** 2937–2942.

INTERSTELLAR ORGANIC MATERIAL: COMPARISONS WITH LABORATORY AND SOLAR SYSTEM SPECTRA

Yvonne J. Pendleton

NASA Ames Research Center
Moffett Field, CA 94035

Abstract

Spectra of objects that lie along several lines of sight through the diffuse interstellar medium (DISM) reveal an absorption feature near 3.4 μm, which has been attributed to saturated aliphatic hydrocarbons on interstellar grains. The similarity of the absorption bands near 3.4 μm (2950 cm^{-1}) along different lines of sight indicates that the carrier of this band lies in the diffuse dust. Several materials have been proposed as "fits" to the 3.4 μm feature over the years. A comparison of these identifications is presented. These comparisons illustrate the need for high-resolution, high signal-to-noise observational data as a means of distinguishing among laboratory organics as matches to the interstellar material. Although any material containing hydrocarbons will produce features in the 3.4 μm region, the proposed "matches" to the DISM do differ in detail. These differences may help in the analyses of the chemical composition and physical processes that led to the production of the DISM organics. A remarkable similarity between the spectrum of the diffuse dust and an organic extract from the Murchison meteorite suggests that some of the interstellar organic material may be preserved in primitive Solar System bodies. This feature has recently been detected in other galaxies (Bridger et al. 1993, Pendleton 1995, Wright et al. 1996) indicating the widespread availability of organic material for incorporation into planetary systems.

INTRODUCTION

Infrared spectral studies provide insight into the chemical identity of the organic component of the interstellar medium because the fundamental vibrational frequencies of the common chemical bonds between the most chemically abundant elements occur in the mid-infrared (5000–400 cm^{-1}, 2–25 μm). Infrared spectroscopic techniques have been used to detect organic material in the interstellar medium (Wickramasinghe and Allen 1980, Butchart et al. 1986, Adamson et al. 1990, Sandford et al. 1991, Pendleton 1993, Pendleton et al. 1994) as well as in primitive Solar System bodies such as comets, meteorites, and asteroids (Cruikshank and Brown 1987, Cronin and Pizzarello 1990, Tokunaga and Brooke 1990, Brooke et al. 1991, Ehrenfreund et al. 1991, Mueller et al. 1992, Pendleton and Cruikshank 1994).

Unequivocal identification of the specific molecules responsible for absorption and/or emission features observed in the ISM has met with limited success, in part due to technical constraints on the quality and completeness of astronomical spectra. Until fairly recently, adequate wavelength coverage at sufficiently high resolution and good signal-to-noise ratios were not available to distinguish among several possible "fits" to the 3.4 μm feature. Consequently, the existing data through the C–H region were "fit" with a variety of organic material, including plausible materials such as organic grain mantles (Greenberg 1982, Schutte and Greenberg 1988) and hydrogenated amorphous carbon (HAC) (Jones et al. 1987), and less plausible materials such as *Escherichia coli* bacteria (Hoyle and Wickramasinghe 1962, Hoyle et al. 1982).

Over the years the instrumentation has improved dramatically, so that we are now able to see structure previously hidden in the 3.4 μm region. The 3.4 μm absorption band is actually composed of several bands that correspond to the stretching modes of C–H in saturated aliphatic hydrocarbons (Sandford et al. 1991; Pendleton 1993, 1994; Pendleton et al. 1994; Sandford et al. 1995). In this paper, several comparisons are made of the diffuse dust line of sight toward galactic center IRS 6E and laboratory materials. The astronomical data were obtained at the NASA Infrared Telescope on Mauna Kea using a 32 channel near-infrared grating spectrometer.

DISCUSSION

Several materials have been suggested as candidate carriers of interstellar carbon. These include HAC (cf. Ogmen and Duley 1988, Adamson et al. 1990), quenched carbonaceous composite (QCC) (cf. Sakata et al. 1987), biological materials (cf. Hoyle et al. 1982), and residues produced by the irradiation of ices (cf. Allamandola et al. 1988, Schutte 1988, Sandford et al. 1991).

HAC can be produced in the laboratory using one of two standard methods: (1) through inefficient hydrocarbon burning in air and (2) by striking an arc in a controlled Ar atmosphere between two amorphous carbon electrodes (cf. Borghesi et al. 1987, Ogmen and Duley 1988). QCC is synthesized from hydrocarbon—principally methane—plasmas (Sakata and Wada 1989). Figure 1a compares the astronomical spectrum of the line of sight toward galactic center source GC IRS 6E with the laboratory spectrum of an HAC sample taken from Borghesi et al. (1987). Figure 1b contains a similar comparison for QCC that has been heated to 450°C. The QCC spectrum was taken from Sakata and Wada (1989). A comparison of GC IRS 6E spectrum to that of QCC at room temperature (not shown) showed that the better match occurred with the heated QCC shown here. Since heating at these temperatures tends to aromaticize the mixture, the better match between the heated QCC and the DISM is consistent with our conclusion that the DISM dust contains carbonaceous material other than aliphatics (Sandford et al. 1991, Pendleton et al. 1994). It is clear that both HAC and QCC provide a rough match to the interstellar C–H stretching feature, but that both materials produce a feature that differs from the interstellar feature in some details. It is not surprising that both HAC and QCC fit the interstellar data to first order since both materials contain the $-CH_3$ and $-CH_2-$ functional groups thought to be responsible for the interstellar absorption feature. The detailed mismatches in the positions and relative strengths of the various subfeatures are largely a reflection of differences in the relative abundances of $-CH_3$, $-CH_2-$, and other chemical groups in the various materials. For example, the weakness of the subfeature near 2960 cm^{-1} relative to the subfeature near 2925 cm^{-1} in the spectra of HAC and QCC when compared with the same bands in the interstellar feature suggests that the interstellar carrier is richer in $-CH_3$ groups than the particular laboratory samples of HAC and QCC shown here. Other materials (not shown here) that provide equally good comparisons with the interstellar data are the ice tholins (Khare et al. 1993) and ion-bombarded methane (Strazzulla et al. 1995).

Hoyle et al. (1982) have suggested that the C–H stretching band of diffuse dust may be due to organic materials of biological origin. In particular, they claimed that the spectrum of *E. coli* provided a good match to the interstellar C–H stretching feature. While we consider such an identification extremely unlikely, for completeness we have included a comparison of the spectrum of an *E. coli* sample and the spectrum of GC IRS 6E (Fig. 1c). Our spectrum of *E. coli* was taken at a resolution of one cm^{-1} using a Nicolet 740 Fourier transform infrared spectrometer. The *E. coli* sample was prepared using a standard KBr pellet technique (see Sandford 1984 for a detailed discussion of the sample preparation technique used). As with the spectra of HAC and QCC, the spectrum of *E. coli* is generally similar to the interstellar feature, just as would be expected for any material containing abundant aliphatic $-CH_3$ and $-CH_2-$ groups, but differs in detail.

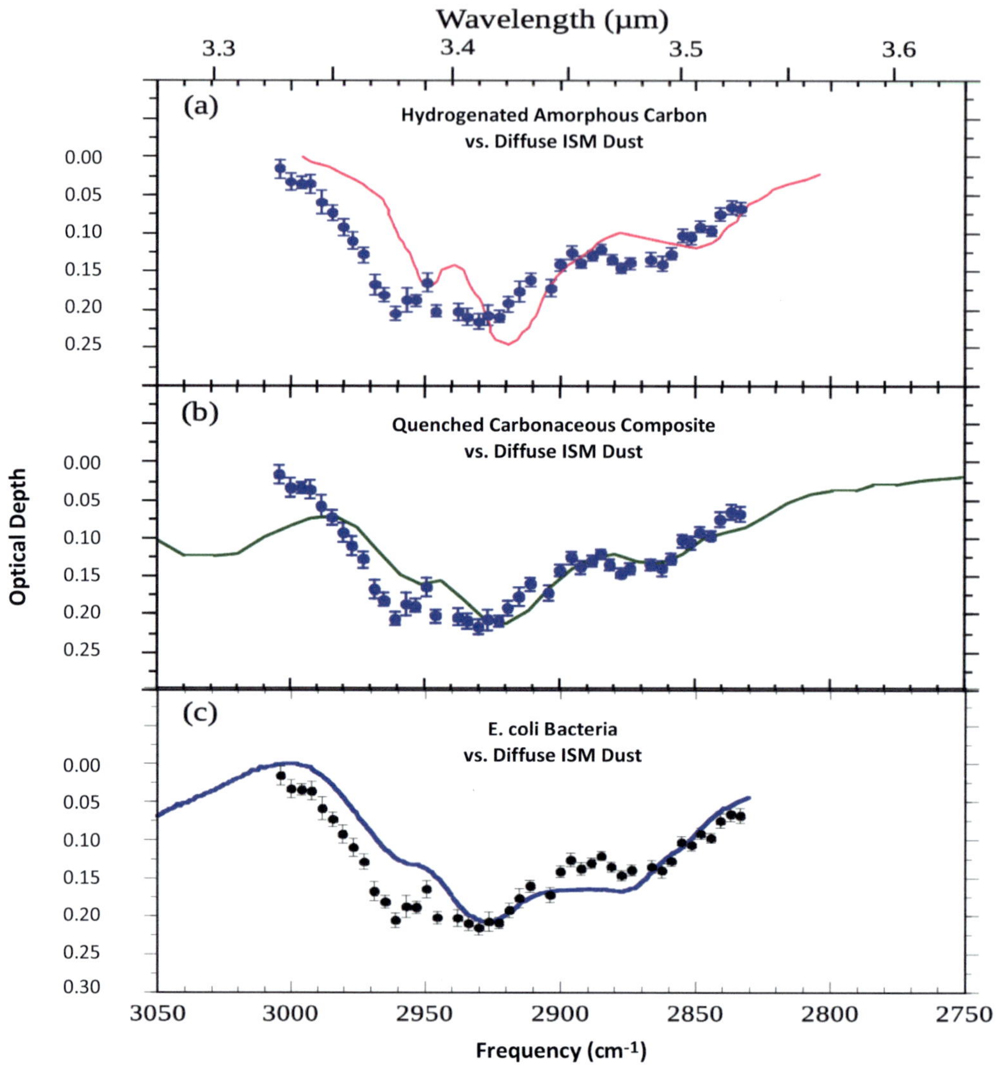

FIG. 1. *A comparison of the optical depth spectrum of galactic center source IRS 6E (solid points) with (a) the optical depth spectrum of a room temperature hydrogenated amorphous carbon (HAC) taken from Borghesi et al. (1987; solid line), (b) the optical depth spectrum of a room temperature filmy quenched carbonaceous composite (QCC) taken from Sakata and Wada (1989; solid line), and (c) the optical depth spectrum of* E. coli *suspended in a KBr pellet (solid line).*

It has previously been shown that the diffuse medium C–H stretching feature is very similar to that produced by residues formed by the UV irradiation and subsequent warming of interstellar ice analogs (Sandford et al. 1991). Infrared spectra of mixed molecular interstellar ice analogs containing H_2O, CH_3OH, CO, and NH_3 reproduce many of the major spectral features attributed to ice in dense clouds (cf. Tielens and Allamandola 1987, Allamandola and Sandford 1988,

Allamandola et al. 1988). Ultraviolet photolysis of ices made up of these molecules produces new, more complex compounds. Warming to 150 K leads to the evaporation of the original volatile components of the ices while leaving the more complex C-rich residues behind (see Allamandola et al. 1988 and Bernstein et al. 1994 and 1995 for more detailed discussions of the properties of the laboratory photolysis residues produced from CH_3OH-containing ices). Such a process (photolysis followed by warm-up) is similar to what is thought to occur to ices in dense molecular clouds that are subsequently dispersed into the diffuse medium by star formation and other dynamic processes.

Figure 2a shows a comparison of the spectrum of GC IRS 6E and the spectrum of a residue produced by the UV photolysis and subsequent warm-up to 200 K of an $H_2O:CH_3OH:NH_3:CO = 10:5:1:1$ ice. The quality of the "fit" provided by this laboratory residue is similar to that of the materials discussed earlier; that is, the overall residue feature is similar to the interstellar feature, but differs from it in detail. As with the earlier materials, we note that the lack of a stronger subfeature near 2955 cm^{-1} indicates that the abundance of $-CH_3$ groups in the laboratory residue is lower than in the carrier of the interstellar feature. This conclusion is supported by Figure 2b, where we show a comparison of the GC IRS 6E spectrum and that of a laboratory residue produced by the UV photolysis and warm-up of an $H_2O:CH_3OH:NH_3:CO:C_3H_8 = 10:5:1:1:1$ ice—an ice enriched in $-CH_3$ groups relative to the one shown in Figure 2a. The resulting laboratory spectrum provides an improved fit to the diffuse medium C–H stretching feature which results from the enhancement in the relative $-CH_3$ abundance.

Some of the hydrocarbon components in primitive meteorites show strong deuterium enrichments and other isotopic anomalies, which indicate a connection to interstellar grains and molecules (Kerridge et al. 1987, Ming and Anders 1988, De Vries et al. 1993, Amari et al. 1994, Lewis et al. 1994). The presence of isotopically anomalous diamonds, SiC, TiC, and graphite in meteorites confirms that some interstellar materials were incorporated into meteorites and survived subsequent processing on the meteorite parent bodies (cf. Anders and Zinner 1993). If the same were true for the organic component of interstellar dust, then the carbonaceous materials observed in the DISM may be one of the "parent" components from which the carbonaceous material in meteorites was derived.

Earlier comparisons of the spectra of the organic extract of the primitive CM meteorite Murchison with the interstellar C–H stretching feature demonstrated a strong similarity (Ehrenfreund et al. 1991). This comparison was made prior to the availability of the high-resolution, high signal-to-noise spectra of the diffuse ISM. Figure 2c shows a comparison of the spectrum of GC IRS 6E and the spectrum of a carbonaceous component of the Murchison chondrite taken by De Vries et al. (1993). It is obvious from Figure 2c that the two spectra are remarkably similar in peak positions, widths, and profiles.

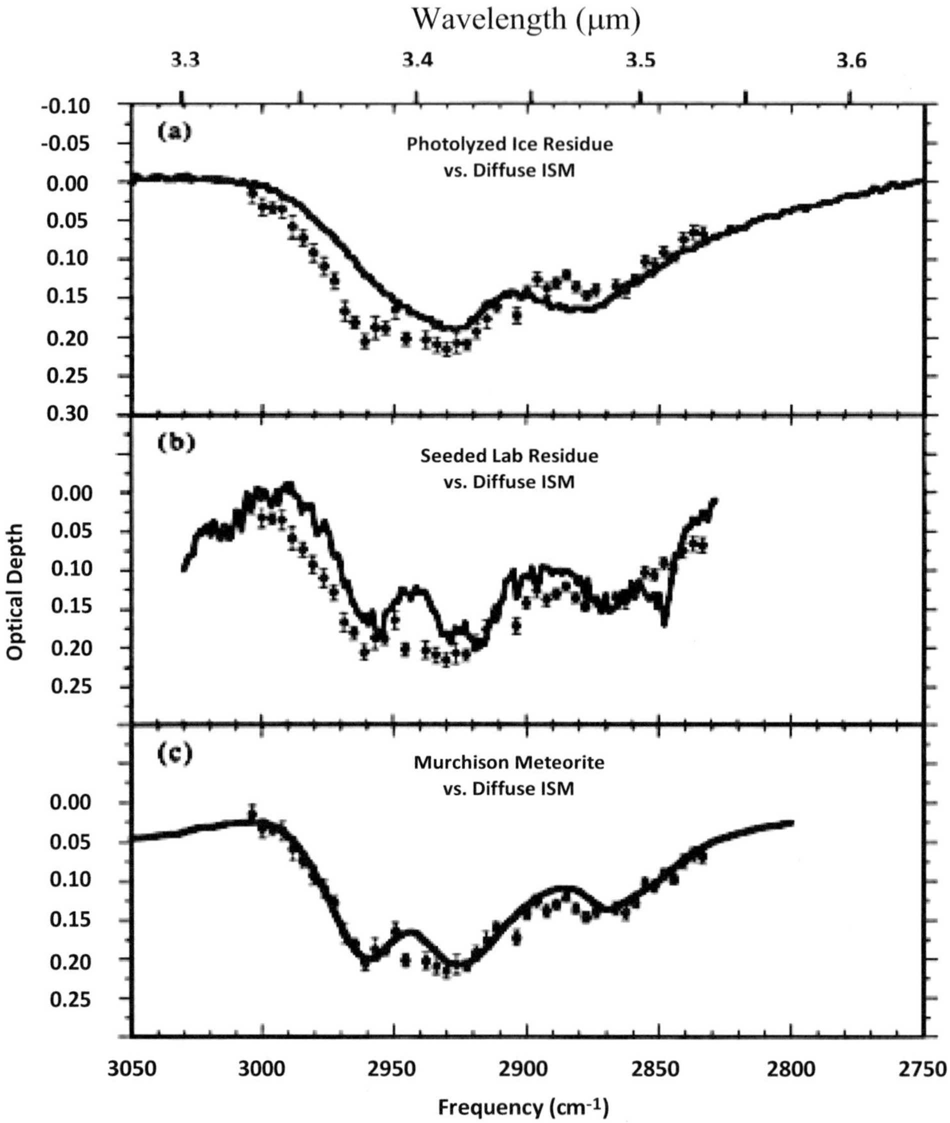

Fig. 2. *A comparison of the optical depth spectrum of galactic center source IRS 6E (solid points) with (a) the spectrum of a laboratory residue produced by the UV irradiation of a 10 K H_2O:CH_3OH:NH_3:CO = 10:5:1:1 interstellar ice analog followed by warm-up to 200 K (solid line), (b) the spectrum of a laboratory residue produced by the UV irradiation of a 10 K H_2O:CH_3OH:NH_3:CO:C_3H_8 = 10:5:1:1:1 interstellar ice analog followed by warm-up to 200 K (solid line), and (c) the spectrum of an organic extract from the Murchison meteorite (solid line). The photolysis residue spectra in (a) and (b) were taken from Allamandola et al. (1988) and the meteoritic spectrum was taken from De Vries et al. (1993) and was kindly provided by W. Golden.*

It should be noted that another interesting comparison with the carbonaceous component of the Murchison meteorite has recently been made by Lee and Wdowiak (1993) who have obtained the spectrum of a residue produced by passing a discharge through a gaseous mixture of hydrogen and the simplest polycyclic aromatic hydrocarbon, naphthalene ($C_{10}H_8$). In general, the spectrum of their residue provides a good overall match to the meteoritic data throughout the midinfrared region but, like all the other laboratory materials discussed above, their residue appears to have a higher ratio of $-CH_2-$ to $-CH_3$ groups than the carrier of the interstellar band. Therefore, while the spectral fit of their residue to the meteoritic data is quite impressive and provides additional support for the idea that the interstellar carrier may contain aromatic materials, it does not provide as compelling a match to the interstellar C–H stretching feature as does the meteoritic material itself.

Several points can be made from the comparisons shown in Figures 1 and 2. First, as has long been known, it is clear that virtually any material that contains substantial amounts of aliphatic $-CH_3$ and $-CH_2-$ groups will produce a feature whose overall position and profile is similar to that of the diffuse interstellar C–H stretching feature. Thus, the general fits provided by all the spectra in Figures 1 and 2 support the idea that dust in the diffuse medium contains abundant aliphatic $-CH_3$ and $-CH_2-$ groups. In this regard, it is unlikely that spectral comparisons of the astronomical and laboratory data will ever yield the "unique" identification of a single interstellar carrier. However, it is also clear that many of the analog materials proposed as the carriers of the interstellar feature on the basis of their fits to older, low resolution astronomical data do not fit the details of the actual profile of the interstellar feature as determined by the high-resolution data. High signal-to-noise, high-resolution ($\Delta\lambda/\lambda \geq 750$) astronomical data therefore represent a means by which the spectral properties of various laboratory materials can be used to further constrain the chemical composition and functionality of the carrier of the diffuse C–H stretching feature. This illustrates the need for both additional high-resolution, full-spectral-coverage, high-quality data like that presented here for GC IRS 6E, and corresponding spectroscopic laboratory data from additional, relevant, well-characterized materials having a range of appropriate compositions.

SUMMARY

In summary, the best match to the overall interstellar C–H stretching feature is provided by the carbonaceous fraction of the Murchison meteorite (Fig. 2c) and, to a somewhat lesser extent, the residues produced by the UV photolysis of some ice mixtures (Fig. 2b). Improved spectral fits might be produced by HACs and QCCs if the relative fractions of their molecular subcomponents were fine-tuned. Ion bombardment of methane and charged particle irradiation of ice tholins provide comparable matches to the interstellar material; however, the UV photolysis process may be more astrophysically appropriate. Thus, even if these materials are not identical to the carrier of the interstellar feature, they provide good potential analogs for the chemical composition of the diffuse dust.

Recent detections of the 3.4 μm absorption feature in other galaxies (Pendleton 1995, 1996; Wright et al. 1996) have revealed the widespread distribution of organic material. The availability of these organic building blocks for incorporation into planetary systems as they formed around stars like the Sun is directly relevant to the origin of life.

ACKNOWLEDGMENTS

This work benefited from collaborations with the following researchers: Louis Allamandola, Sherwood Chang, Dale Cruikshank, Carl Sagan, Scott Sandford, Kris Sellgren, and A.G.G.M. Tielens as well as the excellent services of the support staff at the NASA IRTF (Infrared Telescope Facility). A special word of thanks goes to Rose Grymes and Linda Jahnke in the Life Sciences Division at NASA Ames for providing the laboratory samples of *E. coli*, to Scott Sandford for the spectral analyses of those samples, to G. Strazzulla for data in advance of publication, and to A. Sakata for the QCC data. This work was partially supported by NASA grants 185-52-12-09 and 199-52-12-04 (Exobiology), 188-44-21-04 (Astrophysics), and 452-33-93-03 (Origins of Solar Systems).

REFERENCES

ADAMSON, A.J., D.C.B. WHITTET, AND W.W. DULEY (1990), "The 3.4-μm Interstellar Absorption Feature in Cyg OB2 No. 12," *MNRAS* **243:** 400–404.

ALLAMANDOLA, L.J., AND S.A. SANDFORD (1988), "Laboratory Simulation of Dust Spectra," in *Dust in the Universe*, M.E. Bailey and D.A. Williams, eds. (Cambridge: Cambridge University Press), pp. 229–263.

ALLAMANDOLA, L.J., S.A. SANDFORD, AND G.J. VALERO (1988), "Photochemical and Thermal Evolution of Interstellar/Precometary Ice Analogs," *Icarus* **76:** 225–252.

AMARI, S., R.S. LEWIS, AND E. ANDERS (1994), "Interstellar Grains in Meteorites: I. Isolation of SiC, Graphite, and Diamonds; Size Distribution of SiC and Graphite," *Geochim. Cosmochim. Acta* **58:** 459–470.

ANDERS, E., AND E. ZINNER (1993), "Interstellar Grains in Primitive Meteorites: Diamond, Silicon Carbide, and Graphite," *Meteoritics* **28:** 490–514.

BERNSTEIN, M.P., S.A. SANDFORD, L.J. ALLAMANDOLA, AND S. CHANG (1994), "The Infrared Spectrum of Matrix-Isolated Hexamethylenetetramine in Ar and H_2O at Cryogenic Temperatures," *J. Phys. Chem.* **98:** 12206.

BERNSTEIN, MAX P., SCOTT A. SANDFORD, LOUIS J. ALLAMANDOLA, S. CHANG, AND MAUREEN A. SCHARBERG (1995), "Organic Compounds Produced by Photolysis of Realistic Interstellar and Cometary Ice Analogs Containing Methanol," *ApJ* **454:** 327–344.

BORGHESI, A., E. BUSSOLETTI, AND L. COLANGELI (1987), "Amorphous Carbon and the Unidentified Infrared Bands," *ApJ* **314:** 422–428.

BRIDGER, A., G. WRIGHT, AND T. GEBALLE (1993), in *Conference Abstract Volume of Infrared Astronomy with Arrays: The Next Generation* (Los Angeles: UCLA Press).

BROOKE, T.Y., A.T. TOKUNAGA, AND R.J. KNACKE (1991), "Detection of the 3.4 μm Emission Feature in Comets P/Borosen-Metcalf and Okazaki-Levy-Rudenko (1989R) and an Observational Summary," *AJ* **101:** 268–278.

BUTCHART, I., A.D. MCFADZEAN, D.C.B. WHITTET, T.R. GEBALLE, AND J.M. GREENBERG (1986), "Three Micron Spectroscopy of the Galactic Centre Source IRS 7," *A&A* **154:** L5–L7.

CRONIN, J., AND S. PIZZARELLO (1990), "Aliphatic Hydrocarbons of the Murchison Meteorite," *Geochim. Cosmochim. Acta* **54:** 2859–2868.

CRONIN, J.R., AND S. CHANG (1995), "Organic Matter in Meteorites: Molecular and Isotopic Analyses of the Murchison Meteorite," in *The Chemistry of Life's Origin*, J.M. Greenberg and V. Pirronello, eds. (Dordrecht: Kluwer Academic Publishers), pp. 209–258.

CRUIKSHANK, D.P., AND R.H. BROWN (1987), "Organic Matter on Asteroid 130 Elektra," *Science* **238:** 183–185.

DE VRIES, M.S., K. REIHS, H.R. WENDT, W.G. GOLDEN, AND H.E. HUNZIKER (1993), "A Search for C_{60} in Carbonaceous Chondrites," *Geochim. Cosmochim. Acta* **57:** 933–938.

EHRENFREUND, P., F. ROBERT, L. D'HENDECOURT, AND F. BEHAR (1991), "Comparison of Interstellar and Meteoritic Organic Matter at 3.4 μm," *A&A* **252:** 712–717.

GREENBERG, J.M. (1982), "Dust in Dense Clouds—One Stage in a Cycle," in *Submillimeter Wave Astronomy*, J.E. Beckman and J.P. Phillips, eds. (Cambridge: Cambridge University Press), pp. 261–306.

HOYLE, F., AND N.C. WICKRAMASINGHE (1962), "On Graphite Particles as Interstellar Grains," *MNRAS* **124:** 417–433.

HOYLE, F., N.C. WICKRAMASINGHE, S. AL-MUFTI, A.H. OLAVESEN, AND D.T. WICKRAMASINGHE (1982), "Infrared Spectroscopy Over the 2.9–3.9 μm Waveband in Biochemistry and Astronomy," *A&SS* **83:** 405–409.

JONES, A.P., W.W. DULEY, AND D.A. WILLIAMS (1987), "Interstellar Extinction Correlations," *MNRAS* **229:** 213–221.

KERRIDGE, J.F., S. CHANG, AND R. SHIPP (1987), "Isotopic Characteristics of Kerogen-Like Material in the Murchison Carbonaceous Chondrite," *Geochim. Cosmochim. Acta* **51:** 2527–2540.

KHARE, B.N., W.R. THOMPSON, L. CHEUNG, C. CHYBA, C. SAGAN, E.T. ARAKAWA, C. MEISSE, AND P.S. TUMINELLO (1993), "Production and Optical Constants of Ice Tholin from Charged Particle Irradiation of (1 : 6) C_2H_6/H_2O at 77 K," *Icarus* **103:** 290–300.

LEE, W., AND T.J. WDOWIAK (1993), "Origins of the Hydrocarbon Component of Carbonaceous Chondrites: The Star-Meteorite Connection," *ApJ* **417:** L49–L51.

LEWIS, R.S., S. AMARI, AND E. ANDERS (1994), "Interstellar Grains in Meteorites: II. SiC and Its Noble Gases," *Geochim. Cosmochim. Acta* **58:** 471–494.

MING, T., AND E. ANDERS (1988), "Isotopic Anomalies of Ne, Xe, and C in Meteorites. III. Local and Exotic Noble Gas Components and Their Interrelations," *Geochim. Cosmochim. Acta* **52:** 1245–1254.

MUELLER, B.E., D.J. THOLIN, W.K. HARTMAN, AND D.P. CRUIKSHANK (1992), "Extraordinary Colors of Asteroidal Object (5145) 1992 AD," *Icarus* **97:** 150–154.

OGMEN, M., AND W.W. DULEY (1988), "Laboratory Synthesis of Hydrogenated Amorphous Carbon Films," *ApJ* **334:** L117–120.

PENDLETON, Y.J. (1993), "Organics in the Diffuse Interstellar Medium," in *Astronomical Infrared Spectroscopy: Future Observational Directions*, ASP Conf. Ser., Sun Kwok, ed. (San Francisco: Astronomical Society of the Pacific), **41:** 171–181.

PENDLETON, Y.J. (1994), "Infrared Spectroscopy of Organic Material in Diffuse Interstellar Clouds," in *The First Symposium on the Infrared Cirrus and Diffuse Interstellar Clouds*, ASP Conf. Ser., R. Cutri and W. Latter, eds. (San Francisco: Astronomical Society of the Pacific), **58:** 255–265.

PENDLETON, Y.J. (1995), "Laboratory Comparisons of Organic Materials to Interstellar Dust and the Murchison Meteorite," *Planet. Space Sci.*, **43:** 1359–1364.

PENDLETON, Y.J. (1996), "Organics and Ices in Galactic Dust," in *Proceedings from the Cold Dust Morphology Conference, Johannesburg, South Africa*, D. Block, ed. (Dordrecht: Kluwer Academic Publishers).

PENDLETON, Y.J., AND D.P. CRUIKSHANK (1994), "Life from the Stars?," *Sky & Tel.* **87:** 36–42.

PENDLETON, Y.J., S.A. SANDFORD, L.J. ALLAMANDOLA, A.G.G.M. TIELENS, AND K. SELLGREN (1994), "Near Infrared Absorption Spectroscopy of Interstellar Hydrocarbon Grains," *ApJ* **437:** 683–696.

SAKATA, A., AND S. WADA (1989), "Chemical, Optical and Infrared Properties of Quenched Carbonaceous Chondrites," in *Interstellar Dust*, IAU Symp. 135, L.J. Allamandola and A.G.G.M. Tielens, eds. (Dordrecht: Kluwer Academic Publishers), pp. 191–196.

SAKATA, A., S. WADA, T. ONAKA, AND A.T. TOKUNAGA (1987), "Infrared Spectrum of Quenched Carbonaceous Composite (QCC). II. A New Identification of the 7.7 and 8.6 Micron Unidentified Infrared Emission Bands," *ApJ* **320:** L63–L67.

SANDFORD, S.A. (1984), "Infrared Transmission Spectra from 2.5 to 25 μm of Various Meteorite Classes," *Icarus* **60:** 115–126.

SANDFORD, S.A., L.J. ALLAMANDOLA, A.G.G.M. TIELENS, K. SELLGREN, M. TAPIA, AND Y. PENDLETON (1991), "The Interstellar C–H Stretching Band Near 3.4 Microns: Constraints on the Composition of Organic Material in the Diffuse Interstellar Medium," *ApJ* **371:** 607–620.

SANDFORD, S.A., Y.J. PENDLETON, AND L.J. ALLAMANDOLA (1995), "The Galactic Distribution of Aliphatic Hydrocarbons in the Diffuse Interstellar Medium," *ApJ* **440:** 697–705.

SCHUTTE, W. (1988), *The Evolution of Interstellar Organic Grain Mantles*, PhD thesis, Leiden University, Leiden.

SCHUTTE, W.A., AND J.M. GREENBERG (1988), "The Evolution of Organic Refractory Mantles on Interstellar Grains," in *Dust in the Universe*, M.E. Bailey and D.A. Williams, eds. (New York: Cambridge University Press), pp. 403–406.

STRAZZULLA, G., A.C. CASTORINA, M.E. PALUMBO (1995), "Ion Irradiation of Astrophysical Ices," *Planet. Space Sci.* **43:** 1247–1252.

TIELENS, A.G.G.M., AND L.J. ALLAMANDOLA (1987), "Evolution of Interstellar Dust," in *Physical Processes in Interstellar Clouds*, G.E. Morfill and M. Scholer, eds. (Dordrecht: D. Reidel Publishing Company), pp. 333–376.

TOKUNAGA, A.T., AND T.Y. BROOKE (1990), "Did Comets Form from Unaltered Interstellar Dust and Ices? The Evidence from Infrared Spectroscopy," *Icarus* **86:** 208–219.

WICKRAMASINGHE, D.T., AND D.A. ALLEN (1980), "The 3.4 μm Interstellar Absorption Feature," *Nature* **287:** 518–519.

WRIGHT, G.S., A. BRIDGER, T.A. GEBALLE, AND Y.J. PENDLETON (1996), "Studies of NIR Dust Absorption Features in the Nuclei of Active and IRAS Galaxies," in *Proceedings from the Cold Dust Morphology Conference, Johannesburg, South Africa*, D. Block, ed. (Dordrecht: Kluwer Academic Publishers).

Comet West, March 1976

Photo Credit: P. Stättmayer (ESO)
https://www.eso.org/public/images/c-west-mar1976-ps/ [retrieved 12-12-25]

Comet West was discovered in photographs by Richard M. West of the European Southern Observatory on August 10, 1975. It reached peak brightness in March 1976. During February 25–27, observers reported that it was bright enough to study during full daylight. The comet has an estimated orbital period of 558,000 years. A 2014 reanalysis of photometric observations of the comet during its perihelion in 1976 revealed that dust particles emitted from its coma were a mixture of weakly refractive and highly refractive material. Emission lines of C_2, CN, C_3, CH, and NH_2 were detected. Photometric observations of the comet's fragments about two months after it split apart had revealed an uneven distribution of these compounds, indicating that the original nucleus was heterogeneous with significant composition variations in some areas.

Source: Wikipedia

THE ELECTROWEAK ORIGIN OF THE CHIRAL SIGNATURE OF LIFE

A. J. MacDermott

School of Biological and Molecular Sciences
Oxford Brookes University
Oxford OX3 0BP, England
and
Physical Chemistry Laboratory
Oxford University
Oxford OX1 3QZ, England

G. E. Tranter

Wellcome Research Laboratories
Beckenham, Kent, BR3 3BS, England

The search for extraterrestrial intelligence could be approached by searching for extraterrestrial chirality. Molecules sufficiently complex to support life will also be sufficiently complex to support chirality, that is, the existence of left- and right-handed mirror image pairs, or enantiomers. An important component of SETI, the Search for Extraterrestrial Intelligence, should therefore be SETCHI, the Search for Extraterrestrial CHIrality. The detection of optical activity in extraterrestrial systems could be a sign of life: it has been well-known since Fischer's (1894) "lock and key" hypothesis that an efficient biochemistry must be "homochiral" (i.e., using one enantiomer only; cf. all right-hand screws in engineering), and terrestrial life is of course based on L-amino acids and D-sugars.

It is now increasingly believed that this homochirality is not just a consequence of life but also a precondition for life. This is because polymerization to give the necessary long-chain stereoregular polymers (e.g., all-L polypeptides) will not go in racemic solution—addition of the wrong hand tends to terminate the polymerization (Joyce et al. 1984). An almost homochiral monomer solution is needed for efficient polymerization, and so a homochiral prebiotic chemistry may be a precondition for life. What we require is therefore not merely SETCHI but SETH, the Search for Extraterrestrial Homochirality, which could reveal not just life but also life in the making.

Clearly, a symmetry-breaker is needed to achieve the homochirality necessary to get life started. But once the symmetry is broken, initial selection of one hand in ancestral biomolecules fixes the handedness of the rest of biochemistry through diastereomeric connections, such as that between the L-amino acids and D-sugars (Wolfrom et al. 1949 and Melcher 1974), which seems to preclude L-amino acid/L-sugar or D-amino acid/D-sugar life. But D-amino acid/L-sugar "mirror life" should be just as efficient as L-amino acid/D-sugar terrestrial life, so how was the selection made? Was it a frozen accident or the result of a chiral influence? Ever since Pasteur (1884), who believed (incorrectly) that the chiral influence could be a magnetic field, numerous allegedly chiral influences have been proposed, ranging from the Coriolis force to circularly polarized light, but almost all have now been shown to be either not really chiral at all (Barron 1986) or else even-handed on a time and space average. We believe that the only *universal* chiral influence is the weak force. This is one of the four forces of nature—electromagnetic, weak, strong, and gravitational—and it is the only one that is chiral; that is, it can tell the difference between left and right.

The weak force can feel the handedness of elementary particles themselves. By handedness we mean that fermions, such as electrons and quarks, exist in two states of opposite helicity, interconverted by parity—right helicity corresponds to spin and momentum parallel, and left helicity to spin and momentum antiparallel. The two helicity states participate equally in the electromagnetic, strong, and gravitational interactions, but parity is violated by the weak force; that is, it treats mirror images unequally—left-handed electrons are preferred over right-handed electrons in the weak interaction, and right-handed positrons are preferred over left-handed positrons to the same degree. A "true" mirror is therefore not just P but CP, parity plus charge-conjugation. The weak force thus sees electrons as mainly left-handed and positrons as mainly right-handed. So far as the weak force is concerned, therefore, electrons are left-handed while positrons are right-handed. (However, it must be emphasized that electrons are not really left-handed: there are normally equal numbers of left- and right-handed electrons; it is just that the weak force can feel the left-handed ones better.)

The handedness of elementary particles themselves means that L and D molecules are not really enantiomers but diastereoisomers. The true enantiomer of an L-amino acid is the D-amino acid made of antimatter (Barron 1981). Being diastereoisomers, L and D molecules should therefore differ slightly in all properties, including NMR (nuclear magnetic resonance) chemical shifts (Barra et al. 1987) and especially energy (Hegstrom et al. 1979, 1980). We shall be concerned here with the parity-violating energy difference or PVED, ΔE_{pv}, between enantiomers, which arises from weak neutral current interactions, mediated by the Z° boson, between electrons and neutrons in molecules. This small energy difference between enantiomers produces a slight excess of the more stable enantiomer, which we believe could determinately select the enantiomers used in biology (MacDermott and Tranter 1989). We have calculated the PVED by ab initio methods (MacDermott and Tranter 1989, 1990) for a large number of important biomolecules and obtained values in the range 10^{-20} to 10^{-17} hartree, corresponding to values in the range 10^{-17} kT to 10^{-14} kT at room temperature.

These small energy differences need amplifying, and two main mechanisms have been proposed. The Yamagata cumulative mechanism (Yamagata 1966, Tranter 1985) applies to polymerizations or crystallizations of monomers that are either optically labile or achiral, for instance, the crystallization of quartz, which forms chiral crystals made of achiral silica units. During crystallization, an achiral unit may add on to a growing N-unit crystal of either hand, but owing to the PVED it will have a slight preference for one hand or the other, leading to a slight fractional excess of the preferred hand in the (N+1)-unit crystal. This fractional excess is very small, but operates cumulatively at each of N stages of crystallization, so that the PVED of the N-unit crystal is N times the PVED of a single unit, giving a considerable amplification:

$$\Delta E_{pv}\,(N\text{-unit crystal}) = N\,\Delta E_{pv}\,(\text{one unit})$$

The Kondepudi (1987) catastrophic mechanism, based on nonequilibrium statistical mechanics, is a kinetic scheme whose essential features are autocatalysis and enantiomeric antagonism—that is, the presence of one enantiomer encourages production of itself but inhibits production of its enantiomer. Many biopolymerizations have precisely these characteristics. Kondepudi showed that a PVED of 10^{-17} kT can be amplified to homochirality in just 10^4 years (a very short period on an evolutionary time scale), assuming modest concentrations of 10^{-3} M and realistic reaction rates of 10^{-10} M s^{-1}—although a large volume, 4×10^9 dm^3, is required, corresponding to a small lake. The amplification time and lake size can be reduced dramatically for larger PVEDs; for example, a PVED of 10^{-14} kT (the largest value so far calculated) is amplified in just one year in only 4×10^5 dm^3, corresponding to the proverbial warm little pond. Racemization will of course tend to oppose the amplification, but Kondepudi took this into account and showed that the amplification can withstand a racemization half-life of 10^2 years (cf. 10^5 years for most amino acids).

Some amplification has been achieved by Kondepudi-type processes in the laboratory (Brack and Spach 1980), but starting from initial enantiomeric excesses on the order of a few percent. It is difficult to test amplification from the 10^{-17} level on Earth because of the long times and large volumes required, and also the danger of biochiral contamination on a planet with life. This is where SETH, the Search for Extraterrestrial Homochirality, could be particularly valuable. A planet with no life, undergoing purely chemical evolution, would provide an excellent laboratory in which to test the prebiotic emergence of chiral bias, and an ideal candidate would be Titan, with its interesting photochemistry.

We have calculated the PVEDs of a range of important biomolecules, and in almost all cases it is indeed the "natural" enantiomer that is the more stable. The natural L-amino acids, L-alanine, L-valine, L-serine, and L-aspartate, were all found to be more stable than their "unnatural" D-enantiomers, in their solution conformations and also the α-helix and β-sheet conformations, by 10^{-17} kT (Mason and Tranter 1984, 1985). Turning to the natural D-sugars, it was found that D-glyceraldehyde, the parent of the higher sugars, is indeed PVED-stabilized, by about 10^{-17} kT (Tranter 1986). β-D-deoxyribose, in the C2-endo conformation found in DNA, is similarly PVED-stabilized, but β-D-ribose in the C3-endo form found in RNA is less stable than its enantiomer (Tranter et al. 1992). This latter result appeared discouraging until we realized that the precursor of β-D-nucleotides is not β-D-ribose but α-D-ribose, which we have now shown (MacDermott, 1995) *is* PVED-stabilized. The right-hand DNA backbone is also PVED-stabilized by 10^{-17} kT per nucleotide unit (MacDermott and Tranter 1989, MacDermott 1994).

A PVED of 10^{-17} kT is only just amplifiable by the Kondepudi mechanism; a slightly larger PVED would make amplification very much easier. The PVED is proportional to the sixth power of the atomic number, so we turned next (MacDermott et al. 1992) to molecules in which oxygen is replaced by the heavier sulphur atom. Whereas the normal DNA backbone with $-O-PO_2-O-$ linkages has a PVED of 10^{-17} kT per unit, thiosubstitution of the side oxygens to give $-O-PS_2-O-$ gave a PVED of 10^{-16} kT per unit, and thiosubstitution in the helix itself to give $-S-S-CH_2-$ links gave the enormous PVED of 10^{-14} kT per unit. This is our most exciting result yet—whereas PVEDs of 10^{-17} kT take 10^4 years to amplify in a large lake, PVEDs of 10^{-14} kT take only one year in a small pond. This means that homochirality could be attained much more quickly—and so life could get started much more easily—with a thiosubstituted primitive replicator. This ties in with the view that life may have originated in the highly sulphurous reducing environment of deep-sea thermal vents.

Turning now from polymers to crystals, L-quartz consists of right-hand threefold helices of silica tetrahedra. It represents a model system for Cairns-Smith's (1982) proposed clay replicators, which could be alumino-silicate quartz analogues. A 1.4% excess of L-quartz has been reported in a large collection of over 16,000 crystals from all over the world (Palache et al. 1962), although the statistics have been disputed (Frondel 1978). Our calculations show that L-quartz is PVED-stabilized by 10^{-17} kT per SiO_2 unit (MacDermott and Tranter 1989, MacDermott et al. 1992). This explains the 1% excess by the Yamagata mechanism—to obtain ΔE (crystal) = 10^{-2} kT from $\mathcal{N}\,\Delta E$ (one unit) with ΔE (one unit) = 10^{-17} kT, we would need $\mathcal{N} = 10^{15}$, which corresponds to a small crystal of side 0.1 mm (Tranter 1985).

The weak force thus appears to predict the correct hand, whatever came first in the origin of life. For nucleic acids first, the D-sugars are more stable, as is the right-hand helical backbone of DNA. For proteins first, the L-amino acids are more stable. For clays first, the weak force correctly predicts the reported 1% excess of L-quartz—which preferentially adsorbs L-amino acids (Kavasmaneck and Bonner 1977).

To confirm the enantioselective role of the weak force we need SETH: finding molecules of the same hand in many different extraterrestrial systems would lend support to the weak force as a universal symmetry-breaker. Even finding the "wrong" hand in extraterrestrial systems would be enormously significant, because it would at least be the chiral signature of life! So it is of tremendous importance to look for extraterrestrial homochirality, whether or not one believes in its electroweak origin.

SETH has been neglected so far because the normal instrumentation for circular dichroism (CD) and optical rotatory dispersion (ORD) is too bulky for space missions (1–2 cubic feet). CD and ORD would allow the sign and magnitude of optical rotation to be measured, and the molecules responsible for it to be identified. But just to detect optical rotation, the chiral signature of life, we could use a much more minimal apparatus. It would not tell us the sign, magnitude, or which molecules were responsible, but it would show that homochirality was evolving. Our proposed minimal apparatus places the sample in a cell between two polarized films, polarized at right angles to each other. A miniaturized diode emitter is used as a source of light, and a photodiode or photo-transistor detector is used. With a sample pathlength of 10 cm, the whole apparatus would be about the size of a cigar. The idea is that with crossed polarizers, no light will reach the detector at all unless the plane of polarization is rotated. So for racemic samples, no light will reach the detector. If any light at all does reach the detector, the sample must be optically active; that is, there is an excess of one hand. There is no need for a bulky motor to rotate the polarizers to determine the sign and magnitude of the optical rotation if all we want to know is that there *is* optical rotation. There is no need for a bulky variable wavelength source; one wavelength from a miniature diode will do. With only one wavelength (which would have to be very carefully chosen) it would not be possible to identify molecules, but other analysis techniques on the space probe, such as IR, UV, and so forth, would provide clues. The sample would be prepared and introduced into the cell using instrumentation already present for the other analysis, so nothing extra would have to be carried—just the small cigarlike minimal optical activity detector. False positive readings could arise from stray light, which would have to be carefully eliminated, or from scattering from dust in the sample, which would therefore have to be carefully filtered; scattering could also affect IR and UV, so presumably some filtration apparatus would already be present on the space probe. With careful design several light sources of different wavelengths, accompanied by matching detectors, could be built into the "cigar," thus enhancing the analytical potential.

We propose that the "cigar" should be routinely taken on all missions to other planets to test for the chiral signature of life. If optical rotation were found—and one would probably have to visit several planets before it was found—this would mean it was worth sending a second mission with full CD, ORD, and CD-detected chiral chromatography to identify the chiral molecules and find out if they were the same hand as on Earth.

So, in SETI, we should go for SETH, with a small cigar!

References

Barra, A.L., J.B. Robert, and L. Wiesenfeld (1987), "Parity Non-Conservation: NMR Parameters in Chiral Molecules," *BioSystems* **20:** 57–61.

Barron, L.D. (1981), "Optical Activity and Time Reversal," *Mol. Phys.* **43:** 1395–1406.

Barron, L.D. (1986), "Symmetry and Molecular Chirality," *Chem. Soc. Rev.* **15:** 189–223.

Brack, A., and G. Spach (1980), "β-Structures of Polypeptides with L- and D- Residues," *J. Mol. Evol.* **15:** 231–238.

Cairns-Smith, A.G. (1982), *Genetic Takeover and the Mineral Origins of Life* (Cambridge: Cambridge University Press).

Fischer, E. (1894), "Einfluss der Configuration auf die Wirkung der Enzyme," *Chem. Ber.* **27:** 2985–2993.

Frondel, C. (1978), "Characters of Quartz Fibres," *Am. Mineral.* **63:** 17–27.

Hegstrom, R.A., D.W. Rein, and P.G.H. Sandars (1979), "Parity Non-Conserving Energy Difference Between Mirror-Image Molecules," *Phys. Lett. A* **71:** 499–502.

Hegstrom, R.A., D.W. Rein, and P.G.H. Sandars (1980), "Calculation of the Parity Non-Conserving Energy Difference Between Mirror-Image Molecules," *J. Chem. Phys.* **73:** 2329–2341.

JOYCE, G.F., G.M. VISSER, C.A.A. VAN BOECKEL, J.H. VAN BOOM, L.E. ORGEL, AND J. VAN WESTRESEN (1984), "Chiral Selection in Poly(C)-Directed Synthesis of Oligo (G)," *Nature* **310:** 602–604.

KAVASMANECK, P.R., AND W.A. BONNER (1977), "Adsorption of Amino Acid Derivatives by d- and l-Quartz," *J. Am. Chem. Soc.* **99:** 44–50.

KONDEPUDI, D.K. (1987), "Selection of Molecular Chirality by Extremely Weak Chiral Interactions Under Far-from-Equilibrium Conditions," *BioSystems* **20:** 75–83.

MACDERMOTT, A.J. (1994), "Electroweak Enantioselection and the Origin of Life," in *Origins of Life* **25:** 191.

MACDERMOTT, A.J. (1995), "The Weak Force and the Origin of Life and Self-Organization," *Chemical Evolution: Self-Organization of the Macromolecules of Life*, J. Chela-Flores et al., eds., Deepak Publishing, pp. 237–49.

MACDERMOTT, A.J., AND G.E. TRANTER (1989), "Electroweak Bioenantioselection," *Croatica Chemica Acta* **62:** 165–187.

MACDERMOTT, A.J., AND G.E. TRANTER (1990), "Biomolecular Handedness and the Weak Interaction," in *Symmetries in Science IV*, B. Gruber and J.H. Yopp, eds. (New York: Plenum New York), pp. 67–124.

MACDERMOTT, A.J., G.E. TRANTER, AND S.J. TRAINOR (1992), "The Search for Large Parity-Violating Energy Differences Finds Fruit in Thiosubstituted DNA Analogues," *Chem. Phys. Lett.* **194:** 152–156.

MASON, S.F., AND G.E. TRANTER (1984), "The Parity-Violating Energy Difference Between Enantiomeric Molecules," *Mol. Phys.* **53:** 1091–1111.

MASON, S.F., AND G.E. TRANTER (1985), "The Electroweak Origin of Biomolecular Handedness," *Proc. R. Soc. Lond. A* **397:** 45–65.

MELCHER, G. (1974), "Stereospecificity of the Genetic Code," *J. Mol. Evol.* **3:** 121–140.

PALACHE, C., G. ERMAN, AND C. FRONDEL (1962), *Dana's System of Mineralogy*, 7th ed., vol. 3 (New York: John Wiley & Sons), p. 16.

PASTEUR, L. (1884), "La Dissymetrie Moleculaire," *Rev. Scientifique*, 3eme serie **VII:** 2–6.

TRANTER, G.E. (1985), "Parity Violating Energy Differences of Chiral Minerals and the Origin of Biomolecular Homochirality," *Nature* **318:** 172–173.

TRANTER, G.E. (1986), "Preferential Stabilization of the D-Sugar Series by the Parity-Violating Weak Interaction," *J. Chem. Soc. Chem. Commun.*, **1:** 60–61.

TRANTER, G.E., A.J. MACDERMOTT, R.E. OVERILL, AND P.J. SPEERS (1992), "Computational Studies of the Electroweak Origin of Biomolecular Handedness in Natural Sugars," *Proc. R. Soc. Lond. A* **436:** 603–615.

WOLFROM, J.L., R.U. LUMIEUX, AND S.M. OLIN (1949), "Configurational Correlation of L-(levo)-Glyceraldehyde with Natural (dextro)-Alanine by a Direct Chemical Method," *J. Am. Chem. Soc.* **71:** 2870–2873.

YAMAGATA, Y. (1966), "A Hypothesis for the Asymmetric Appearance of Biomolecules on Earth," *J. Theor. Biol.* **11:** 495–498

Venus (diameter about 12,104 km/7,520 mi)
Photo Credit: NASA/JPL-Caltech
https://science.nasa.gov/venus/venus-facts/
[retrieved 7-23-25]

Earth (diameter about 12,756 km/7,926 mi)
Photo Credit: NASA's Earth Observatory
https://www.nasa.gov/missions/terra/earth-from-space-15-amazing-things-in-15-years [retrieved 10-30-25]
A composite image of the Western Hemisphere [created in 2002]

Mars (diameter about 6,794 km/4,222 mi)
Photo Credit: NASA/JPL/USGS
https://www.livescience.com/facts-about-mars
[retrieved 10-30-25]

Has Life Made Its Global Environment More Habitable?

David J. Des Marais

NASA Ames Research Center
Moffett Field, CA 94035-1000

The concept that life has influenced its global environment in ways that have enhanced its survival has attracted considerable interest. Long-term trends in the evolution of the Sun and Earth certainly could have changed Earth's surface environment, particularly in the absence of any compensating mechanisms. Key long-term trends include the increase in solar luminosity (Sagan and Mullen 1972), a decline in volcanic emanations of CO_2 (Des Marais 1985) and reduced gases (Holland 1984), and the evolution of Earth's continental crust (Lowman 1989). One need only to examine our closest planetary neighbors, Venus and Mars, to appreciate how remarkable it is that a climate favorable for life has been continuously maintained on Earth for more than 3.5 billion years. That the effects of long-term change have somehow been buffered by negative feedback mechanisms seems inescapable. Still, the role played by the biosphere itself remains unclear.

That organisms can interact to optimize their immediate local environ ment and enhance the survival of a diverse community is clearly demonstrated by microbial mats. Benthic microbial mats arose more than 3.5 billion years ago (Schopf et al. 1983) and therefore are probably the oldest biological communities known. These films of bacteria and algae, which inhabit wetted rock and sediment surfaces, create a microenvironment in which wavelengths of light are shared among diverse photosynthetic populations at various depths within the mat (Jørgensen et al. 1987). Nonphotosynthetic bacteria degrade organic matter, releasing key chemical nutrients which are efficiently retained and recycled within the mat (Canfield and Des Marais 1993). Such coordinated, localized environmental control enhances the productivity, growth and survival of the mats.

Biological communities can also influence their external environment (Table 1). Examples of these effects include the net consumption of CO_2 from the ocean and atmosphere, the net burial of organic matter in sediments, and the net release of O_2 to the atmosphere. Biota can also produce trace gases that alter the energy balance of the atmosphere. Microbes can affect rates of weathering. Marine biota can alter the patterns of mineral precipitation and thus modify coastal conditions; tropical reefs are a good example of this capability.

TABLE 1. ***Examples of Life's Capabilities* (C) *and Their Potential Effects* (E) *upon the Environment***

C	Assimilate and utilize molecular hydrogen (H_2) from volcanoes
E	Attenuate the loss of water and reducing power to space
C	Produce organic matter and molecular oxygen (O_2)
E	Over the history of Earth, help to create a net reduction of crustal sediments and a net oxidation of the oceans and atmosphere
C	Produce trace gases that alter the energy balance of the atmosphere
E	Strengthen greenhouse (methane [CH_4] and nitrous oxide [N_2O]) Enhance clouds and hazes which could cause cooling (sulfur species)
C	Enhance weathering rates
E	Lower atmospheric carbon dioxide (CO_2) levels, weakening greenhouse
C	Ability to evolve
E	Survive against externally imposed changes in the global environment Create global environmental change by altering the atmosphere (e.g., adding O_2 or CH_4) and/or through the effects of human technology

One can envision various levels of biological influence on the environment, ranging from simply affecting the environment to actively manipulating it. Organisms certainly have altered global conditions. Examples include their contribution to the atmospheric O_2 inventory (Holland 1984) and their enhancement of weathering (Berner 1993, Schwartzman 1993). Evolutionary innovations have, at times, altered the environment in ways that ultimately enhanced diversity and survival. For example, the development of O_2-producing photosynthesis probably stimulated the rise of diverse populations of oxic respirers (Schopf et al. 1983). The Gaia hypothesis, advanced by Lovelock and others (e.g., see Lovelock 1988), goes a step further and proposes that the biosphere actually regulates properties of the atmosphere, oceans and land surface in ways that maintain conditions favorable for itself. According to Gaia theory, this capacity for control arises both from life's ability to respond exponentially to environmental perturbations and from its ability to adapt and evolve. The theory holds that life must be planetary in scale in order to affect global climate, otherwise conditions would eventually change to the point where the biosphere would be extinguished. Indeed, the Gaia hypothesis proposes that the evolution of life and its environment have been tightly coupled. One way to illustrate this hypothesis is to represent two key environmental parameters—the range of biologically compatible temperatures and the availability of liquid water—as a box in two-dimensional space (Fig. 1). The biosphere began when prebiotic conditions remained within the outermost box for an interval of time sufficient for the origin of life. Life later assumed a globally significant presence due to its growth and evolution and/or to further improvements in conditions. At this point, Gaialike mechanisms gained control of global climate, maintaining conditions within the intermediate-sized box in Figure 1. Perhaps even greater degrees of environmental control were required to sustain more complex biota such as plants and animals. This tighter range of control is depicted by the smallest box in Figure 1. This theory seems plausible, but has it occurred in the real world? Key lingering questions include when and how Gaia gained control, and what specific mechanisms maintained this control.

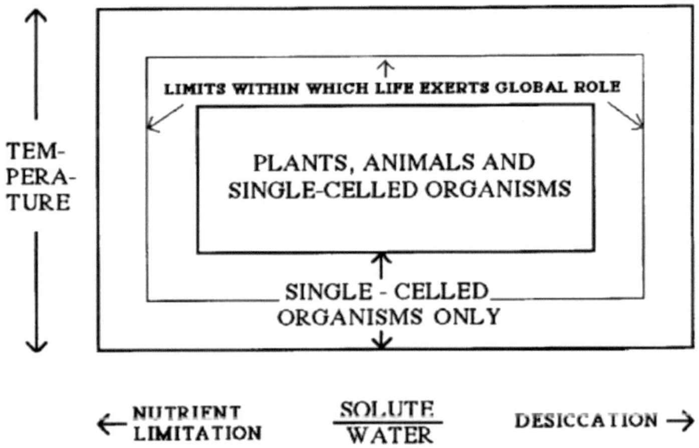

FIG. 1. *Schematic box diagram illustrating the ranges of temperature and water availability in a planetary environment. Outermost box represents conditions necessary to sustain life. Intermediate-sized box depicts conditions perhaps necessary for the biosphere to develop and maintain global-scale control of its environment, as proposed by the Gaia hypothesis. Innermost box represents the narrower range of conditions perhaps necessary to sustain plants and animals.*

One approach to these questions is to assess how Earth's early environment might have been affected by the early biosphere. The ancestor of our modern biosphere arose sometime after the last large bolide impact, which vaporized the oceans and melted the seafloor (Maher and Stevenson 1988). Such an event certainly could have extinguished any preexisting life, and it might have occurred as late as 3.9 billion years ago. In its aftermath, both a relatively unstable crust and substantial volcanic emanations of CO_2 and reduced species (H_2, Fe^{2+}, sulfides, etc.) could have maintained a CO_2-rich, anoxic atmosphere and relatively warm temperatures. Life apparently arose in an aqueous environment under such conditions and utilized reduced volcanic species to reduce CO_2 and produce organic matter. In the prebiotic Earth, volcanic inputs of H_2 to the surface environment would have been balanced by H_2 loss to space (Kasting et al. 1993a). If this loss to space proceeded unabated, both Earth's upper mantle and its surface environment eventually would have become more oxidized (Kasting et al. 1993a). However, practically all bacteria readily assimilate H_2 and produce organic matter (e.g., Fenchel and Blackburn 1979). Perhaps life's first key global role was to incorporate substantial quantities of volcanic H_2, thus reducing atmospheric H_2 levels and the rate of loss of H_2 to space. Some biota with a large appetite for H_2 produce trace gases that can affect climate. For example, methane (CH_4) is produced by a group of H_2-consuming archaebacteria, the methanogens. Because CH_4 is a strong greenhouse gas (Sagan and Mullen 1972), a substantial biological contribution of CH_4 to the atmosphere would have warmed the climate. Thus, conceivably the earliest biosphere might have helped to maintain those warm, anoxic conditions that accompanied its origin.

Environmental changes obviously occurred eventually, as today we live in an O_2-rich, CO_2-poor environment. Trends that could have contributed to such changes include the evolution of oxygen-producing photosynthesis, the increase in solar luminosity (Sagan and Mullen 1972), and the decline in Earth's heat flow (Turcotte 1980), which caused the volcanic emanation of CO_2 (Des Marais 1985) and reduced species (Holland 1984) to decline. The stabilization of continental masses would have promoted the deposition and preservation of organic matter, and therefore the accumulation of these organics, in Earth's crust (Des Marais 1994). This permitted O_2, the other product of oxygenic photosynthesis, to accumulate in the atmosphere. These key processes of vulcanism, continental stabilization and sedimentation are nonbiological; therefore long-term atmospheric changes might have been driven principally by nonbiological processes. Thus, an O_2-rich atmosphere clearly required a biological O_2 source, but the agents that made it possible for O_2 levels to increase—namely, continental stabilization and the long-term decline in vulcanism—are themselves nonbiological processes. One can envision a scenario where life has adapted continually both to changes in the sun and in planet Earth, and to the consequences of its own evolution.

A host of biologically mediated mechanisms for environmental regulation have been proposed (e.g., see Schneider and Boston 1993). Temperature-control mechanisms have received considerable attention (e.g., Lovelock 1993). Lovelock's Daisyworld scenario illustrates the principle of biological temperature control by invoking populations of black and white daisies. Warmer temperatures promote the growth of white daisies, which in turn increase planetary albedo. Greater albedo reflects more solar energy to space and thus mitigates against further deleterious temperature increases. A real-world example of biological albedo control might be plankton that produce sulfate aerosols, which stimulate cloud formation and thus increase planetary albedo (Charlson et al. 1987).

Temperature control can also be achieved by modulating atmospheric concentrations of greenhouse gases such as CO_2 (Fig. 2) or CH_4. For such a feedback mechanism to stabilize temperature, warmer temperatures must decrease the concentration of the greenhouse gas. Rock weathering meets these requirements, because it removes atmospheric CO_2 at rates that increase with temperature (Walker et al. 1981). Soil bacteria can increase weathering rates (Schwartzman 1993). Thus, greater bacterial activity might lower atmospheric CO_2 levels and, thus, decrease global temperatures.

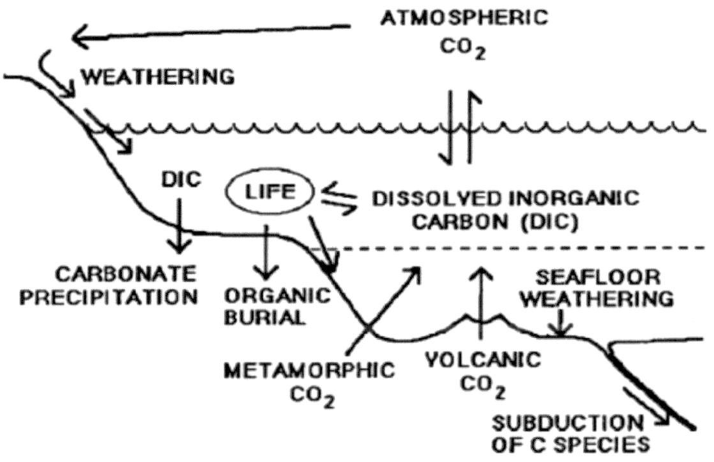

FIG. 2. *Schematic diagram of the carbon cycle, depicting carbon reservoirs and processes affecting their abundance and distribution. Thermal emanations are sources of atmospheric carbon dioxide. Weathering, followed by the burial of carbonates and organic carbon, is a sink. Subduction of carbon species is another sink.*

In summary, the biosphere clearly has altered the environment through its effects upon atmospheric composition and by enhancing key processes such as weathering. However, we cannot yet demonstrate whether a global biological control network, such as that depicted by the Gaia hypothesis, was indeed essential for maintaining a clement environment for more than 3.5 billion years, or whether such a feat required principally the interaction of sunlight, tectonics, vulcanism and the activity of liquid water. This question is central to one of the basic issues put forward in this book, namely, can life actually enlarge or prolong its circumstellar habitable zone? For this to occur, a biosphere must, at some point, "defy" the physical and chemical constraints imposed by its solar system environment. Perhaps our biosphere achieves this defiance because it can alter both atmospheric composition and cloud abundance, thereby modulating the global solar energy budget. This energy budget determines the onset of runaway greenhouses and high-albedo CO_2 clouds, phenomena that might determine, respectively, the inner and outer boundaries of our circumstellar habitable zone (Kasting et al. 1993b). Notable among the possible biological effects in Table 1 is life's ability to adapt, sometimes in response to externally triggered changes, by evolving new capabilities. These new capabilities include human technology. Indeed, because our ultimate evolutionary potential is largely unknown, we cannot yet comprehend the ultimate limits of our own circumstellar habitable zone.

References

Berner, R.A. (1993), "Atmospheric Oxygen, Tectonics and Life," in *Scientists on Gaia*, S.H. Schneider and P.J. Boston, eds. (Cambridge, MA: MIT Press), pp. 161–166.

Canfield, D.E., and D.J. Des Marais (1993), "Biogeochemical Cycles of Carbon, Sulfur, and Free Oxygen in a Microbial Mat," *Geochim. Cosmochim. Acta* **57:** 3971–3984.

Charlson, R.J., J.E. Lovelock, M.O. Andreae, and S.G. Warren (1987), "Oceanic Phytoplankton, Atmospheric Sulphur, Cloud Albedo and Climate," *Nature* **326:** 655–661.

Des Marais, D.J. (1985), "Carbon Exchange Between the Mantle and Crust and Its Effect upon the Atmosphere, Today Compared to Archean Time," in *The Carbon Cycle and Atmospheric CO_2: Natural Variations Archean to Present*, E.T. Sundquist and W.S. Broecker, eds. (Washington, DC: American Geophysical Union), pp. 602–611.

Des Marais, D.J. (1994), "Tectonic Control of the Crustal Organic Carbon Reservoir During the Precambrian," *Chem. Geol.* **114:** 303–314.

Fenchel, T., and T.H. Blackburn (1979), *Bacteria and Mineral Cycling* (New York: Academic Press).

Holland, H.D. (1984), *The Chemical Evolution of the Atmosphere and Oceans* (Princeton, NJ: Princeton University Press).

Jørgensen, B.B., Y. Cohen, and D.J. Des Marais (1987), "Photosynthetic Action Spectra and Adaptation to Spectral Light Distribution in a Benthic Cyanobacterial Mat," *Appl. Env. Microbiol.* **53:** 879–886.

Kasting, J.F., D.H. Eggler, and S.P. Raeburn (1993a), "Mantle Redox Evolution and the Oxidation State of the Archean Atmosphere," *J. Geol.* **101:** 245–257.

KASTING, J.F., D.P. WHITMIRE, AND R.T. REYNOLDS (1993b), "Habitable Zones Around Main Sequence Stars," *Icarus* **101:** 108–128.

LOVELOCK, J.E. (1988), *The Ages of Gaia* (New York: W.W. Norton Co.).

LOVELOCK, J.E. (1993), "Geophysiology—The Science of Gaia," in *Scientists on Gaia*, S.H. Schneider and P.J. Boston, eds. (Cambridge, MA: MIT Press), pp. 3–10.

LOWMAN, P.D.J. (1989), "Comparative Planetology and the Origin of Continental Crust," *Precambrian Res.* **44:** 171–195.

MAHER, K.A., AND D.J. STEVENSON (1988), "Impact Frustration of the Origin of Life," *Nature* **331:** 612–614.

SAGAN, C., AND G. MULLEN (1972), "Earth and Mars: Evolution of Atmospheres and Surface Temperatures," *Science* **177:** 52–56.

SCHNEIDER, S.H., AND P.J. BOSTON, eds. (1993), *Scientists on Gaia* (Cambridge, MA: MIT Press).

SCHOPF, J.W., J.M. HAYES, AND M.R. WALTER (1983), "Evolution of Earth's Earliest Ecosystems: Recent Progress and Unsolved Problems," in *Earth's Earliest Biosphere*, J.W. Schopf, ed. (Princeton, NJ: Princeton University Press), pp. 361–384.

SCHWARTZMAN, D. (1993), "Microbial Weathering and Gaia," in *Scientists on Gaia*, S.H. Schneider and P.J. Boston, eds. (Cambridge, MA: MIT Press), pp. 320–329.

TURCOTTE, D.L. (1980), "On the Thermal Evolution of the Earth," *Earth Planet. Sci. Lett.* **48:** 53–58.

WALKER, J.C.G., P.B. HAYS, AND J.F. KASTING (1981), "A Negative Feedback Mechanism for the Long-Term Stabilization of Earth's Surface Temperature," *JGR* **86:** 9776–9782.

Hubble and Webb Capture Detailed Views of DART Impact

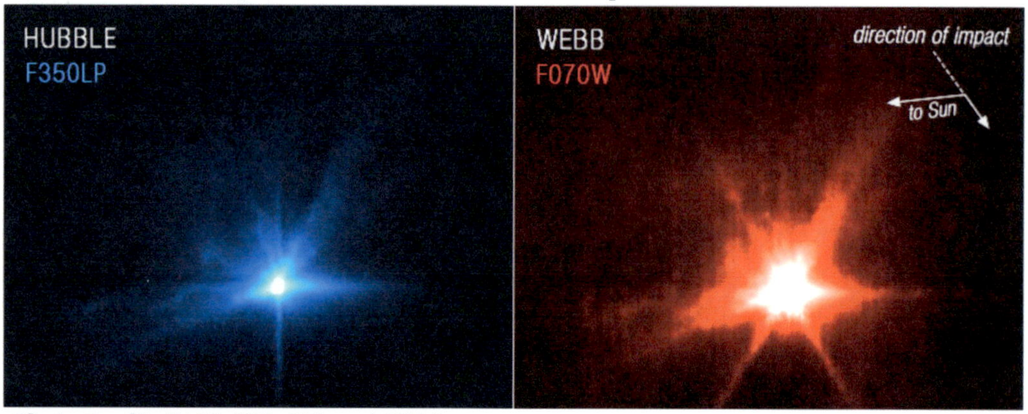

Photo Credit: NASA, ESA, CSA, and STScI
https://esawebb.org/images/weic2215a/ [retrieved 6-20-25]

For the first time, the NASA/ESA/CSA James Webb Space Telescope and the NASA/ESA Hubble Space Telescope have taken simultaneous observations of the same target.

These images, Hubble on the left and Webb on the right, show observations of Dimorphos several hours after NASA's Double Asteroid Redirection Test (DART) intentionally impacted the moonlet asteroid. It was the world's first test of the kinetic impact technique using a spacecraft to deflect an asteroid by modifying its orbit.

Both Webb and Hubble observed the asteroid before and after the collision took place.

Scientists will use the combined observations from Hubble and Webb to gain knowledge about the nature of the surface of Dimorphos, how much material was ejected by the collision, how fast it was ejected, and the distribution of particle sizes in the expanding dust cloud.

In the coming months, scientists will also use Webb's Mid-Infrared Instrument (MIRI) and Near-Infrared Spectrograph (NIRSpec) to observe Dimorphos further. Spectroscopic data will provide researchers with insight into the asteroid's composition. Hubble will monitor Dimorphos ten more times over the next three weeks to learn how the ejecta cloud expands and fades over time.

Hubble observations were conducted in one filter, WFC3/UVIS F350LP (assigned the color blue), while Webb observed at F070W (0.7 micron, assigned the color red).

The Effects of Heavy Meteorite Bombardment on Early Evolution—A New Look at the Molecular Record

Maria Gogarten-Boekels

Department of Molecular and Cell Biology
University of Connecticut
Storrs, CT 06269-3044

J. Peter Gogarten

Department of Molecular and Cell Biology
University of Connecticut
Storrs, CT 06269-3044

Introduction

Two main sources are available to study the early evolution of life on this planet: the fossil record and the molecular, biochemical and anatomical records that survived as a heritage in extant organisms. The early Precambrian fossil record is sparse; however, significant progress has been made in identifying and characterizing microfossils from this era (Schopf 1993). Microfossils of prokaryotes date back to at least 3500 million years before present (BP). The filamentous morphology and size of these ancient microfossils and their occurrence in stromatolite-like structures were interpreted to suggest that these microfossils represent photosynthetic oxygen-producing bacteria (Awramik 1992, Schopf 1993).

The molecular record is stored in biological macromolecules (e.g., Schwartz and Dayhoff 1978) and in biochemical pathways (e.g., Granick 1957, Wächtershäuser 1990) of extant organisms. The comparison of DNA, RNA and protein sequences from different extant species allows one to reconstruct the evolutionary history of these molecules. If their evolution reflects speciation, these molecules can also be used as markers for organismal evolution.

The interpretation of biological records (fossil and molecular) is aided by geological and astrophysical findings and theories. As discussed by Miller (1992), conditions on the early Earth were very different from present day environments. In particular, the analysis of the moon cratering record suggests that large impacts were much more frequent than today (Mahler and Stevenson 1988). Oberbeck and Fogleman (1990) concluded from astrophysical considerations that life that existed at around 3800 million years BP would very likely have been destroyed by giant impacts.

In this paper we discuss the molecular record in light of both the fossil evidence and the cratering record. We evaluate different scenarios for the early evolution of life, taking into account the recent findings of early duplicated genes and the inferred properties of the last common ancestor.

THE MOLECULAR RECORD

The Three Domains and the Progenote

From the study of 16S ribosomal RNA it became apparent that the prokaryotes can be divided into two groups (Woese and Fox 1977b). Originally these were named the eubacterial and the archaebacterial Ur-Kingdoms. More recently these two Ur-Kingdoms were renamed into the domains Bacteria and Archaea (Woese et al. 1990). The division of the prokaryotes is supported by other molecular phylogenies (e.g., ATPases, elongation factors, RNA polymerases; Gogarten et al. 1989, Iwabe et al. 1989, Puhler et al. 1989) and by many other biochemical characters (see Zillig et al. 1992, for a recent summary). The third domain of life is constituted of the eukaryotes or Eucarya.

When the three domains were first recognized it was not obvious which of these groups were more closely related to each other. Woese and Fox (1977a) suggested that the aboriginal separations might not have been a succession of bifurcations, but a trifurcation. They also introduced the term *progenotic state* to denote a primitive stage of development that existed before a defined relation between genotype and phenotype of an organism had been established, that is, before the prokaryotic stage.

Certainly, early life-forms must have existed whose organizational level corresponds to the progenotic stage; however, as will be discussed below, the analysis of duplicated genes and the consideration of the many shared characteristics of extant cellular life suggest that the last common ancestor was not a progenote as originally defined, but a prokaryote not too dissimilar from extant prokaryotes (Gogarten and Taiz 1992, Lazcano 1993a).

The Origins of the Eukaryotes

The endosymbiont theory (cf. Margulis 1993) maintains that some eucaryal cell organelles evolved from Bacteria that functioned as endosymbionts within a host cell. In the case of mitochondria and plastids the bacterial origin has been verified by studies of molecular evolution (e.g., Schwartz and Dayhoff 1978). The information provided by ancient duplicated genes (ATPase subunits, dehydrogenases, elongation factors, ^{met}tRNAs; Gogarten et al. 1989, Iwabe et al. 1989) revealed that a major portion of the nucleocytoplasm of the host cell evolved from an Archaeum-like prokaryote. This finding is further corroborated by the similarities in the transcription machinery in Eucarya and Archaea (Puhler et al. 1989). Other molecular markers appear to contradict this close association between Archaea and the eucaryal nucleocytoplasm (glutamate dehydrogenase, Benachenhou-Lafha et al. 1993; carbamylphosphate synthetase, Lazcano, Puente, Gogarten, unpublished; glutamine synthase, Kumada et al. 1993, Tiboni et al. 1993; heat shock proteins, Gupta and Golding 1993). More detailed analyses are necessary to decide whether these molecular phylogenies represent cases of horizontal gene transfer, additional major contributions to the eukaryotic nucleocytoplasm, or ill-resolved molecular phylogenies (see Hilario and Gogarten 1993 for further discussion).

Tree or Net of Life

The emergence of modern eukaryotes from an endosymbiosis of organisms belonging to different phylogenetic groups demonstrates that the paradigm of a treelike representation of the evolution is questionable. Horizontal gene transfer has also been discussed as an explanation for discrepancies between gene and species trees in the following cases: ATPase subunits of *Thermus thermophilus* (Gogarten et al. 1992), *Methanococcus barkeri* and *Enterococcus hirae* (Hilario and Gogarten 1993), glyceraldehyde phosphate dehydrogenase from *Escherichia coli* (Doolittle et al. 1990), heat shock proteins (Gupta and Golding 1993), glucose-6-phosphate isomerase (Smith and Doolittle 1992), 16S rRNA of plant mitochondria (Gray et al. 1989), P-elements in *Drosophila* (Daniels et al. 1990), and glutamine synthases (Tiboni et al. 1993). For a critical evaluation of some of these cases see Smith et al. (1992). It appears that horizontal transfer of genes between species is not restricted to resistance genes in modern-day bacteria, but occurred throughout evolution.

However, different molecular phylogenies that provide a robust resolution for the deep branches (as measured for example by bootstrap analyses) reveal identical or very similar topologies (e.g., 16S-like rRNAs, Woese 1987; ATPases, Gogarten et al. 1989; RNA polymerases, Puhler et al. 1989; elongation factors, Cammarano et al. 1992). This congruence of several molecular markers indicates that horizontal exchange of genetic information across species boundaries does occur only infrequently. Cases of horizontal transfer can be recognized within the background of the majority consensus of molecular markers. The fusion of separate lineages (net) is revealed by the simultaneous horizontal transfer of several independent genes (e.g., the eucaryal cell organelles that evolved from bacterial endosymbionts).

The Last Common Ancestor

Molecular biology shows the fundamental unity of modern life. All extant organisms are cellular, the genetic information is stored in DNA, transcribed into RNA, and translated into proteins. All organisms use the same (or a very similar) genetic code and they use the same amino acids in their proteins. Although there are differences in the transcription and translation machinery, the process is very similar in all cells. All cells use lipid membranes to separate their protoplasm from the environment or from the cell wall; they use the same energy-rich metabolites; and all living organisms use homologous enzymes to energize their cell membranes.

Furthermore, the study of ancient gene duplications shows that the last common ancestor already possessed a variety of complex enzymatic and regulatory processes. The membrane-energizing, ion-translocating ATPases were already multisubunit enzymes consisting of an ATP hydrolyzing subunit and a paralogous (i.e., derived by a gene duplication from the same ancestral gene) regulatory subunit. Structure function relationships for extant ATPases (Cross and Taiz 1990) suggest that the ATPases present in the last common ancestor already utilized the transmembrane ion gradients for ATP synthesis (Gogarten and Taiz 1992). Furthermore, the last common ancestor already had different paralogous elongation factors (Cammarano et al. 1992), two types of ^{met}tRNA (Iwabe et al. 1989), two glutamine synthases (Kumada et al. 1993, Tiboni et al. 1993), malate and lactate dehydrogenases (Iwabe et al. 1989, Zillig et al. 1992), an internal duplication in the carbamylphosphate synthetase (Puente, Gogarten, Lazcano, unpublished), different heat shock protein homologues (Gupta and Golding 1993), and two glutamate dehydrogenases (Benachenhou-Lafha et al. 1993).

The picture of the last common ancestor emerging from the many shared characters of extant life and from the analysis of ancient duplicated genes is different from that of a primitive progenote. The last common ancestor appears to have been a prokaryotic cell that used DNA, RNA, ribosomes, energy-conserving membranes and a variety of sophisticated, regulated biochemical pathways. The last common ancestor does not seem to have been fundamentally different from present day prokaryotes. There is no a priori reason to assume that this cellular organism would have had difficulties undergoing speciation or adapting to different ecological niches present on the early Earth.

Scenarios That Explain the Absence of Deep Branches

A common characteristic of the above-mentioned molecular phylogenies (i.e., 16S-like rRNAs, ATPases, RNA polymerases, elongation factors) is that the two prokaryotic domains are clearly separated from each other. Although continuously deeper-branching prokaryotes are discovered, so far all of these clearly fall into one of the two domains.

One obvious possibility to explain the absence of deep branches in the rooted tree of life is chance. Different lineages terminate at random; it might be coincidence that so far none of the deep branches have been detected, or they might all have gone extinct. As was pointed out by Zillig et al. (1992), this random death scenario is unlikely. More popular scenarios involve biological reasons as to why there were no or only a few deep branches in the first place. Zillig et al. (1992) proposed a progenote population with rapid exchange of genes among members of the population. In his scenario, a first separation into two subpopulations is brought about by geographic isolation. Within each of the two subpopulations the rapid exchange of genetic information continues and prevents further speciation. The development of properties necessary for speciation (i.e., the step from a progenotic to a prokaryotic stage) occurred independently in both branches and has to be regarded as a case of parallel evolution. One progenote population supposedly evolved into the Bacteria, the other into the Archaea.

A similar proposal was outlined by Kandler (1993). Again, a population with rapid exchange of genetic information is assumed. Within this "precell" population, inventions are passed on among individuals. The ancestors of the three domains are thought to have separated from this precell population sequentially. The differences between the three domains are explained by the following two processes: (1) different subsets of genes from the precell population find their way into the three domain ancestors, (2) some inventions are made in the precell population after one or two of the domain ancestors have already separated from the precell population.

Koch (1994) described a monophyletic phase during which no or only a few side branches were generated. Evolution supposedly occurred slowly and life did not adapt to new ecological niches. As the different life-forms coexisted in the same ecological niche, all organisms were direct competitors, which led to only a few survivors. Only after life adapted to different ecological niches did the different life-forms avoid direct competition, and separate surviving lineages formed.

Alternatively, the absence of independently surviving deep branches in the tree of life can be explained by a catastrophic event that wiped out all but two surviving lineages (Fig. 1). The two lineages surviving the catastrophe evolved into the Bacteria and the Archaea. The latter also contributed most of the eukaryotic nucleo cytoplasm. In contrast to the other scenarios, the assumption of a catastrophic event eradicating most of the lineages from the bottom part of the tree does not assume a progenotic or primitive organization level for the last common ancestor and the organisms that populated the bottom portion of the tree of life. Only the catastrophic extinction scenario is compatible with an already sophisticated prokaryotic last common ancestor.

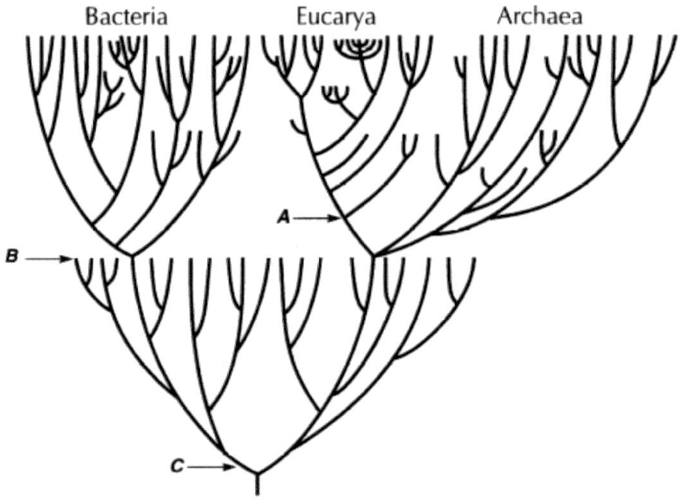

FIG. 1. *The catastrophic extinction scenario explains the absence of deep branching surviving lineages in the molecular record. The branch lengths represent an approximation of the number of substitutions that were calculated from the analyses of ancient duplicated genes (Gogarten et al. 1989, Iwabe et al. 1989). The branches are not scaled with respect to time. Other molecular markers (e.g., 16S rRNA, Olsen and Woese 1993; RNA polymerases, Puhler et al. 1989) provide only unrooted molecular phylogenies; however, also in these cases the three domains are clearly separated from each other by long central branches. A, B, and C denote the origin of the Eucarya, the proposed catastrophic extinction and the last common ancestor, respectively. See text for further discussion.*

The study of recent evolution shows that evolution is not a smooth continuous process, but that it is characterized by major catastrophic events (Wilson 1992). If catastrophic events shaped evolution during the last 800 million years, we have every reason to assume that the same or similar forces also influenced early evolution.

A Catastrophe Theory

The cratering record of the moon suggests that large impacts were likely to frustrate the development of life on Earth between 4000 and 3700 million years BP (Oberbeck and Fogleman 1989). The last impact with an energy sufficient for sterilizing the Earth (i.e., to completely vaporize the oceans) is estimated to have occurred between 4400 and 3800 million years BP; impacts vaporizing the photic zone only were calculated to be highly probable as late as 3800 million years BP (Sleep et al. 1989). A large impact would have drastically altered or obliterated most of the available ecological niches. Such an impact appears as a likely cause for the catastrophe that is suggested by our interpretation of the molecular records. In this context it is of significance that the different proposals for the universal tree of life (Woese 1987, Lake 1988) agree in the deepest branches of the respective trees being occupied by extreme thermophiles. As was also pointed out by Lazcano (1993b), the distribution of thermophiles on the tree of life is not necessarily indicative of life having originated at high temperature (Pace 1991, Holm 1992)—it might be reflective of the selection that took place during the period of heavy bombardment. The idea of life's origin at high temperature has also been criticized because of the instability of biomolecules at elevated temperatures (Miller and Bada 1988), and because an enzyme that seems to be necessary for DNA stability at high temperatures (reverse gyrase) appears to have evolved only later during evolution (Forterre et al. 1995).

The described catastrophic impact scenario reconciles the above-cited criticisms with the distribution of extreme thermophiles on the tree of life. Life could have originated in a mesophilic environment. Prokaryotic life already had diversified and occupied the different ecological niches available on the planet before the occurrence of the late catastrophic impact. Among other niches, prokaryotes also had settled the extremely hot environments of the deep ocean vents. Only those prokaryotes that had adapted to hot environments were able to survive a large impact.

Dating the Molecular Record

There is no justification for assuming a molecular clock that is running at constant speed throughout evolution. To the contrary, the study of more recent molecular evolution, in particular that of vertebrates, showed up to thirteenfold transient increases in substitution rates after a gene duplication event had occurred (Ohta 1991 and refs. therein). The root in the universal tree of life was placed by means of ancient gene duplications; therefore, it seems justified to assume that the bottom part of the tree outlined in Figure 1 is enlarged with respect to the upper portions.

(Note, however, that the deep branches calculated from 16S rRNA are only slightly shorter than the ones depicted in the figure; cf. Olsen and Woese 1993.) On the other hand, the character of the last common ancestor that is inferred from ancient duplicated genes offers no indication for the substitution rate in the bottom portion of the tree being higher by several orders of magnitude. Only few events are available for an attempt to calibrate the molecular record by means of the fossil record.

The Emergence of Eukaryotes

Eukaryotic cells are recognized in the fossil record since about 1750 million years BP. Sediments from about 2000 million years BP contain fossils of probable eukaryotic origin (Schopf 1992). The characterization of microfossils as eukaryotic is based mainly on cell size; therefore, it appears likely that the earliest eukaryotes, even if they are present in the fossil record, will not be classified as such. Our conservative estimate for the separation of the eukaryotic nucleocytoplasmic lineage from the archaebacteria is about 2300 million years BP. However, it is likely that the independent eukaryotic lineage is significantly older than this estimate. The molecular record indicates that the separation between Archaea and Eucarya must have occurred after the postulated impact. Therefore, we can conclude that the late, nearly complete impact frustration of life must have occurred around or well before 2300 million years BP.

Early Archean Microfossils

Microfossils have been described in rocks dating back to around 3500 million years BP. The structurally best-conserved samples are from the Pilbara rocks (Schopf 1993) and the Warrawoona group, both in Western Australia (cf. Awramik 1992). Microfossils were also described in rocks from the Swaziland supergroup in South Africa (Walsh 1992). Stromatolites were described in the Swaziland supergroup (Byerly et al. 1986) and the Warrawoona group (cf. Awramik 1992). Modern stromatolites are the result of interactions of complex microbial communities, including oxygen-producing photosynthetic cyanobacteria. The morphologies of some of the early microfossils are similar to extant *Oscillatoria*. This morphological resemblance and the presence of stromatolites were interpreted as proof (Awramik 1992) or at least strongly suggestive of cyanobacteria being present already about 3500 million years BP. Although some of these findings have been questioned (summarized by Schopf 1993), the resemblance between filamentous microfossils and present day cyanobacteria is impressive. However, other bacterial groups (e.g., the green nonsulfur bacteria *Chloroflexus*, *Heliothrix* and *Oscillochloris* and the nonphotosynthetic

Beggiatoales) contain similar multicellular filamentous forms as well (Pfennig 1989); some of these also resemble *Oscillatoria*. If we assume that the early Archaean cyanobacteria-like microfossils are representatives of lineages that went extinct during the catastrophic impact, then these microfossils cannot be used for calibration. However, if we consider them to be representatives of surviving bacterial groups, then an accelerated substitution rate for the bottom portion of the tree has to be assumed, in order to accommodate the tree of life within the lifetime of this planet.

The properties of the last common ancestor and the study of more recently duplicated genes (see the above discussion) suggest that the substitution rate in the bottom portion of the tree might have been temporarily increased by less than thirteenfold. Assuming a tenfold higher substitution rate throughout the bottom portion of the tree, the last common ancestor is dated to about 3900 million years BP; the point of the catastrophic extinction is estimated to have occurred between 3600 million and 3700 million years BP. If a less accelerated substitution rate is assumed, even earlier times result for the last common ancestor and the nearly complete impact frustration.

Conclusion

The described catastrophic extinction scenario is compatible with the molecular, fossil and astrophysical records. It suggests that life emerged early in Earth's history, even before the time of the heavy bombardment was over. Early life-forms already had colonized extreme habitats that allowed at least two prokaryotic species to survive a late nearly-ocean-boiling impact. The distribution of ecotypes on the rooted universal tree of life cannot necessarily be interpreted to show that life originated in extremely hot environments.

Acknowledgments

We thank Elena Hilario, Henrik Kibak, Scot Magnotta, Lincoln Taiz, Antonio Lazcano and Wolfram Zillig for many stimulating discussions. This work has been supported through NSF grant BSR-9020868.

References

Awramik, S.M. (1992), "The Oldest Records of Photosynthesis," *Photosynth. Res.* **33:** 75–89.

Benachenhou-Lafha, N., P. Forterre, and B. Labedan (1993), "Evolution of Glutamate Dehydrogenase Genes: Evidence for Two Paralogous Protein Families and Unusual Branching Patterns of the Archaebacteria in the Universal Tree of Life," *J. Mol. Evol.* **36:** 335–346.

BYERLY, G.R., D.R. LOWE, AND M.M. WALSH (1986), "Stromatolites from the 3,300–3,500 Myr Swaziland Supergroup, Barberton Mountain Land, South Africa," *Nature* **319:** 489–491.

CAMMARANO, P., P. PALM, R. CRETI, E. CECCARELLI, A.M. SANANGELANTONI, AND O. TIBONI (1992), "Early Evolutionary Relationships Among Known Life Forms Inferred from Elongation Factor ef-2 (ef-g) Sequences. Phylogenetic Coherence and Structure of the Archaeal Domain," *J. Mol. Evol.* **34:** 396–405.

CROSS, R.L., AND L. TAIZ (1990), "Gene Duplication as a Means for Altering H^+/ATP Ratios During the Evolution of FoF1 ATPases and Synthases," *Fed. Eur. Biochem. Soc. Lett.* **259:** 227–229.

DANIELS, S.P., K.R. PETERSON, L.D. STRAUSBAUGH, M.G. KIDWELL, AND A. CHOVNICK (1990), "Evidence for Horizontal Transmission of the P Transposable Element Between *Drosophila* Species," *Genetics* **124:** 339–355.

DOOLITTLE, R.F., D.F. FENG, K.L. ANDERSON, AND M.R. ALBERRO (1990), "A Naturally Occurring Horizontal Gene Transfer from a Eukaryote to a Prokaryote," *J. Mol. Evol.* **31:** 383–388.

FORTERRE, P., F. CONFALONIERI, AND M. DUGUET (1995), "Speculations on the Origin of Life and Thermophily: Review of Available Information on Reverse Gyrase Suggests that Hyperthermophilic Procaryotes Are Not So Primitive," *Origins of Life* **25:** 235–250.

GOGARTEN, J.P., AND L. TAIZ (1992), "Evolution of Proton Pumping ATPases: Rooting the Tree of Life," *Photosynth. Res.* **33:** 137–146.

GOGARTEN, J.P., H. KIBAK, P. DITTRICH, L. TAIZ, E.J. BOWMAN, B.J. BOWMAN, M.F. MANOLSON, R.J. POOLE, T. DATE, T. OSHIMA, J. KONISHI, K. DENDA, AND M. YOSHIDA (1989), "Evolution of the Vacuolar H^+-ATPase: Implications for the Origin of Eukaryotes," *Proc. Natl. Acad. Sci. USA* **86:** 6661–6665.

GOGARTEN, J.P., T. STARKE, H. KIBAK, J. FICHMANN, AND L. TAIZ (1992), "Evolution and Isoforms of V-ATPase Subunits," *J. Exp. Biol.* **172:** 137–147.

GRANICK, S. (1957), "Speculations on the Origin and Evolution of Photosynthesis," *Ann. NY Acad. Sci.* **69:** 292–308.

GRAY, M.W., R. CEDERGREN, Y. ABEL, AND D. SANKOFF (1989), "On the Evolutionary Origin of the Plant Mitochondrion and Its Genome," *Proc. Natl. Acad. Sci. USA* **86:** 2267–2271.

GUPTA, R.S., AND G.B. GOLDING (1993), "Evolution of HSP70 Gene and Its Implications Regarding Relationships Between Archaebacteria, Eubacteria and Eukaryotes," *J. Mol. Evol.* **37:** 573–582.

HILARIO, E.M., AND J.P. GOGARTEN (1993), "Horizontal Transfer of ATPase Genes—The *Tree* of Life Becomes a *Net* of Life," *BioSystems* **31:** 111–119.

HOLM, N.G., ED. (1992), *Marine Hydrothermal Systems and the Origin of Life* (Boston: Kluwer Academic Publishers).

IWABE, N., K.-I. KUMA, M. HASEGAWA, S. OSAWA, AND T. MIYATA (1989), "Evolutionary Relationships of Archaebacteria, Eubacteria and Eukaryotes Inferred from Phylogenetic Trees of Duplicated Genes," *Proc. Natl. Acad. Sci. USA* **86:** 9355–9359.

KANDLER, O. (1993), "The Early Diversification of Life," in *Early Life on Earth*, Nobel Symposium no. 84, S. Bengston, ed. (New York: Columbia University Press), pp. 111–118.

KOCH, A.L. (1994), "Development and Diversification of the Last Common Ancestor," *J. Theor. Biol.* **168:** 269–280.

KUMADA, Y., D.R. BENSON, D. HILLEMANN, T.J. HOSTED, D.A. ROCHFORD, C.J. THOMPSON, W. WOHLLEBEN, AND Y. TATENO (1993), "Evolution of the Glutamine Synthase Gene, One of the Oldest Existing and Functioning Genes," *Proc. Natl. Acad. Sci. USA* **90:** 3009–3013.

LAKE, J.A. (1988), "Origin of the Eukaryotic Nucleus Determined by Rate Invariant Analysis of rRNA Sequences," *Nature* **331:** 184–186.

LAZCANO, A. (1993a), "The RNA World, Its Predecessors and Descendants," in *Early Life on Earth*, Nobel Symposium no. 84, S. Bengston, ed. (New York: Columbia University Press), pp. 69–80.

LAZCANO, A. (1993b), "Biogenesis: Some Like It Hot," *Science* **260:** 1154–1155.

MAHLER, K.A., AND D.J. STEVENSON (1988), "Impact Frustration of the Origin of Life," *Nature* **331:** 612–614.

MARGULIS, L. (1993), *Symbiosis in Cell Evolution* (New York: W.H. Freeman and Company).

MILLER, S.L. (1992), "The Prebiotic Synthesis of Organic Compounds as a Step Toward the Origin of Life," in *Major Events in the History of Life*, J.W. Schopf, ed. (Boston: Jones and Bartlett Publishers), pp. 1–28.

MILLER, S.L., AND J.L. BADA (1988), "Submarine Hot Springs and the Origin of Life," *Nature* **334:** 609–611.

OBERBECK, V.R., AND G. FOGLEMAN (1989), "Impacts and the Origin of Life," *Nature* **339:** 434.

OBERBECK, V.R., AND G. FOGLEMAN (1990), "Impact Constraints on the Environment for Chemical Evolution and the Continuity of Life," *Origins of Life* **20:** 181–195.

OHTA, T. (1991), "Multigene Families and the Evolution of Complexity," *J. Mol. Evol.* **33:** 34–41.

OLSEN, G.J., AND C.R. WOESE (1993), "Ribosomal RNA: A Key to Phylogeny," *FASEB Journal* **7:** 113–123.

PACE, N.R. (1991), "Origin of Life—Facing Up to the Physical Setting," *Cell* **65:** 531–533.

PFENNIG, N. (1989), "Multicellular Filamentous Green Bacteria," in *Bergey's Manual of Systematic Bacteriology*, vol. 3, J.T. Staley, M.P. Bryant, N. Pfennig, and J.G. Holt, eds. (Baltimore: Williams and Wilkins), pp. 1697–1707.

PUHLER, G., H. LEFFERS, F. GROPP, P. PALM, H.P. KLENK, F. LOTTSPEICH, R.A. GARRETT, AND W. ZILLIG (1989), "Archaebacterial DNA-Dependent RNA Polymerases Testify to the Evolution of the Eukaryotic Nuclear Genome," *Proc. Natl. Acad. Sci. USA* **86:** 4569–4573.

SCHOPF, J.W. (1992), "The Oldest Fossils and What They Mean," in *Major Events in the History of Life*, J.W. Schopf, ed. (Boston: Jones and Bartlett Publishers), pp. 54–63.

SCHOPF, J.W. (1993), "Microfossils of the Early Archean Apex Chert: New Evidence of the Antiquity of Life," *Science* **260:** 640–646.

SCHWARTZ, R.M., AND M.O. DAYHOFF (1978), "Origins of Prokaryotes, Mitochondria and Chloroplasts," *Science* **199:** 395–403.

SLEEP, N.H., K.J. ZAHNLE, J.F. KASTING, AND H.J. MOROWITZ (1989), "Annihilation of Ecosystems by Large Asteroid Impacts on the Early Earth," *Nature* **342:** 139–142.

SMITH, M.W., AND R.F. DOOLITTLE (1992), "Anomalous Phylogeny Involving the Enzyme Glucose-6-Phosphate Isomerase," *J. Mol. Evol.* **34:** 544–545.

SMITH, M.W., D.F. FENG, AND R.F. DOOLITTLE (1992), "Evolution by Acquisition: The Case for Horizontal Gene Transfer," *Trends Biochem. Sci.* **17:** 489–493.

TIBONI, O., P. CAMMARANO, AND A.M. SANANGELANTONI (1993), "Cloning and Sequencing of the Gene Encoding Glutamine Synthase I from the Archaeum *Pyrococcus woesei*: Anomalous Phylogenies Inferred from Analysis of Archaeal and Bacterial Glutamine Synthase I Sequences," *J. Bacteriol.* **175:** 2961–2969.

WÄCHTERSHÄUSER, G. (1990), "Evolution of the First Metabolic Cycles," *Proc. Natl. Acad. Sci. USA* **87:** 200–204.

WALSH, M.M. (1992), "Microfossils and Possible Microfossils from the Early Archean Onverwacht Group, Barberton Mountain Land, South Africa," *Precambrian Res.* **54:** 271–293.

WILSON, E.O. (1992), *The Diversity of Life* (Cambridge, MA: Harvard University Press).

WOESE, C.R. (1987), "Bacterial Evolution," *Microbiol. Rev.* **51:** 221–271.

WOESE, C.R., AND G.E. FOX (1977a), "The Concept of Cellular Evolution," *J. Mol. Evol.* **10:** 1–6.

WOESE, C.R., AND G.E. FOX (1977b), "Phylogenetic Structure of the Prokaryotic Domain: The Primary Kingdoms," *Proc. Natl. Acad. Sci. USA* **74:** 5088–5090.

WOESE, C.R., O. KANDLER, AND M.I. WHEELIS (1990), "Towards a Natural System of Organisms: Proposal for the Domains Archaea, Bacteria and Eucarya," *Proc. Natl. Acad. Sci. USA* **87:** 4576–4579.

ZILLIG, W., P. PALM, AND H-P. KLENK (1992), "A Model of the Early Evolution of Organisms: The Arisal of the Three Domains of Life from the Common Ancestor," in *The Origin and Evolution of the Cell*, H. Hartman and K. Matsuno, eds. (Singapore, New Jersey, London, Hong Kong: World Scientific), pp. 163–182.

From the Primitive Soup to Cyanobacteria: It May Have Taken Less Than 10 Million Years

Stanley L. Miller

Department of Chemistry
University of California at San Diego
La Jolla, CA 92093-0317

Antonio Lazcano

Facultad de Ciencias, UNAM
Cd. Universitaria, México, D.F., México

Introduction

Most scientific discussions on the likelihood of extraterrestrial life have been constrained by the characteristics of life on our planet and the environmental conditions under which it may have emerged. Although it has been generally assumed that this process must have been extremely slow, involving hundreds of millions or even billions of years (Oparin 1938, Urey 1952, Wald 1954, Huang 1959a, Simpson 1964, Cloud 1968, Dickerson 1978, Schidlowski 1992), a number of recent discoveries have led to a considerable compression of the time believed necessary for life to appear. It is now recognized that during its early history the Earth and other bodies of the inner Solar System went through a stage of intense collisions. Some of these impacts by large asteroids or comets may have raised the terrestrial surface to sterilizing temperatures and may have evaporated the oceans and killed off life as late as 3.8×10^9 years ago (Sagan 1974, Maher and Stevenson 1988, Sleep et al. 1989, Chyba et al. 1990, Oberbeck and Fogleman 1990). However, there is also ample paleontological evidence derived from the 3.5×10^9 year old Warrawoona sediments showing that only 300 million years after the period of intense impacts ended, our planet was populated by phototactic, stromatolite-forming microorganisms (Schopf 1983, 1993). Although these discoveries are now generally interpreted to imply that the origin and early evolution of life were rapid, no attempts have been made to estimate the actual time required for these processes to occur. As discussed here, all the available evidence derived from (a) the rates of plausible prebiotic chemical reactions, (b) the half-lives of organic compounds in the primitive Earth, and (c) the observed rates of gene amplification in contemporary prokaryotic populations, suggests that the transition from the prebiotic soup to the RNA world, from there to the DNA/protein world, and then to cyanobacteria, may have taken place in less than 10 million years.

The Rates of Syntheses and Destruction of Prebiotic Organic Compounds

The chemistry of prebiotic reactions is fast and robust. The rates of the prebiotic synthesis of amino acids, almost all of which are the result of the Strecker synthesis, have been worked out in some detail. The rates depend on temperature, pH, and HCN, NH_3 and aldehyde concentrations. At pH 8 and 0°C the slow step in amino acid synthesis, the hydrolysis of the corresponding aminonitrile to the amide, is approximately forty years (Miller and Van Trump 1981). The Murchison meteorite amino acids provide an example of a rapid prebiotic synthesis, since the whole process apparently took place in the meteorite parental body in less than 10^5 years (Peltzer et al. 1984). Adenine synthesis is also rapid from concentrated solutions of NH_4CN, requiring hours at 100°C (Oró 1960), a few weeks at 0°C, and months in dilute aqueous solution frozen at −20°C (Sánchez et al. 1966). Even though it is unlikely that sugars were components of the earliest genetic material (Joyce et al. 1987, Shapiro 1988), their synthesis from formaldehyde is also rapid. This is not to imply that all the monomers were formed in a few years. The formation and buildup of the prebiotic soup took place in very few millions of years, but the individual reactions have short half-lives, and there are no known slow steps.

The accumulation of organic compounds is limited by destructive processes. All organic compounds in aqueous solutions undergo decomposition reactions that are significant even at 0°C (Miller and Orgel 1974). As shown in Table 1, although some compounds such as fatty acids (e.g., acetic acid) and aliphatic amino acids (e.g., alanine) are very stable, sugars and many amino acids decompose in hundreds or tens of thousands of years, as do peptide bonds. Furthermore, the destruction rate that applies to all organic compounds on the Earth is controlled by the pyrolysis of organics in the submarine vents at temperatures of 350°C (Miller and Bada 1988). The approximate passage time of the whole ocean through the vents is inversely proportional to the heat flow and to the ocean size, and currently requires about 10 million years (Edmonds et al. 1982), but may have been less than 5 million years on the early Earth because the heat flow was about ten times greater and the oceans may have been smaller. Not all organic compounds on the prebiotic Earth would have been destroyed by the vents, since some of them could survive in inland sediments, salt deposits, evaporites, and in the nonmixed parts of the oceans, but even in these protected environments they would eventually undergo thermal decomposition (Table 1).

TABLE 1. ***Thermal Decomposition of Organic Compounds at 0°C***

Compound	*Mean Life*
Alanine	10^9 years
Fatty Acids	10^9 years
Serine	10^5 years
Sugars	10^4 years
Peptides	10^4 years
RNA and DNA	10^3 years

FROM THE RNA WORLD TO THE DNA/PROTEIN WORLD

Perhaps the major uncertainty in the understanding of the origin of life currently lies in our inability to explain the emergence of a self-replicating system capable of undergoing Darwinian selection. Although it is not possible to estimate the time period for this system to emerge, there are some constraints that may be recognized. An informational polymer must have a lifetime comparable to that of the organism (Westheimer 1987) or, at least, to the time required for its replication. If a slow addition of monomers to polymers is assumed, the rate of polymer formation must nonetheless be rapid compared with hydrolysis rates, especially if a considerable amount of genetic information is to be contained in the polymer. Thus, a 100-base-long RNA molecule needs to be synthesized 100 times faster than the hydrolysis rate of a single phosphodiester bond.

The second major bottleneck in this process is represented by the origin of protein biosynthesis. How this process actually began is unknown, but it is generally assumed that it resulted from the interaction of RNA molecules (or their precursors in a pre-RNA world) with amino acids of prebiotic origin that may have been available. Thus, protein biosynthesis must have begun to take place when suitable concentrations of amino acids were still available, and had not yet diminished by decomposition or destruction by the vents.

We assume that life arose in a prebiotic soup containing most, if not all, of the necessary small molecules, and that the evolution of the basic biosynthetic pathways occurred in a stepwise fashion (Horowitz 1945) or through a related mechanism. It is clear that this provides both for the growth and the energy supply of a large number of organisms, but it would quickly result in the depletion of the available nutrients. There was a potential considerable energy supply from different fermentations of the prebiotic synthesized organic compounds. The usual example cited is the fermentation of glucose, but it is unlikely that large quantities of this sugar were available because of its instability. A more likely early fermentation reaction is that suggested by Clarke and Elsden (1980):

$$\text{glycine} + \text{NADH} + \text{ADP} + \text{P}_\text{i} = \text{acetate} + \text{NH}_4^+ + \text{NAD}^+ + \text{ATP}$$

An ocean with a volume $V_{ocean} = 1.5 \times 10^{21}$ liters with a 1×10^{-4} M concentration of glycine would have contained 1.5×10^{17} moles of glycine. At one mole ATP per mole of glycine, this would be equivalent to 8.4×10^{28} cells, which corresponds to 7×10^4 cells/cm^3 of ocean. A glycine concentration of 10^{-4} M is a rich prebiotic soup, but producible under reducing conditions (Stribling and Miller 1987). The concentration may have been much lower—10^{-8} M of glycine—if the organic compounds were only of extraterrestrial origin (Anders 1989, Chyba et al. 1990), but it would still correspond to 10^{25} cells. As the primitive populations of heterotrophs thrived in the soup, there would have been an exponential decrease in the concentration of the available fermentable organic compounds of prebiotic origin. Such a decrease would have rapidly led to a metabolic energy crisis; that is, there would be strong selection pressure for the development of photosynthesis, as the only really abundant energy source was visible light.

From the Primitive Heterotrophs to Filamentous Cyanobacteria

It is generally accepted that the first oxygen-releasing photosynthetic prokaryotes were the ancestors of contemporary cyanobacteria. The sheathed filamentous cyanobacteria *Oscillatoria* spp. and *Nostoc* spp., which appear to be morphologically similar to the Warrawoona microfossils, have genomes of approximately 7×10^9 base pairs (Herdman 1985). Thus, if we assume that the first DNA/protein organism had about 100 enzymes with ~ 1000 base pairs/gene, we need to estimate the time needed to go from a primordial organism with 100 genes to an oscillatorian-like cyanobacteria with approximately 7000 enzymes.

There are several sources that may increase the cellular DNA dosage (Fig. 1), but it is likely that in early Archean times the most significant one was gene duplication. It has been observed in some directed evolution experiments that populations of haploid microorganisms undergo duplications of large segments covering several genes to provide more enzymes to overcome the lowered concentrations of nutrients (Sonti and Roth 1989) and, at least in some cases, the duplicated copies became fixed in a few weeks (Hartley 1974, Hansche 1975).

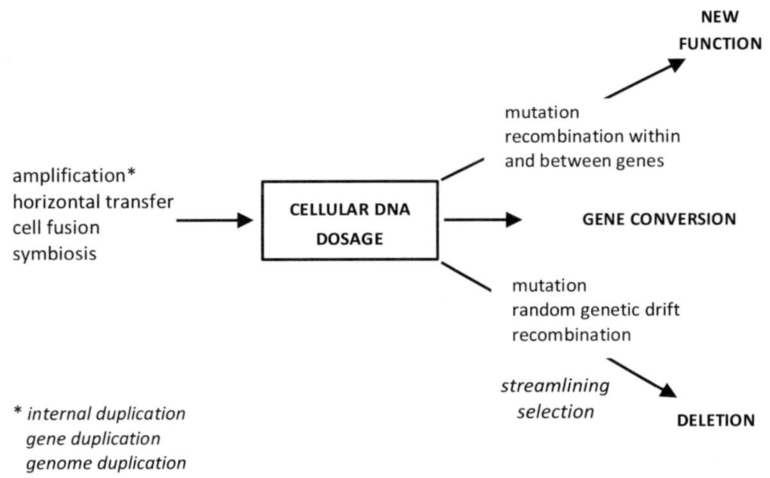

FIG. 1. *Mechanisms that may increase cellular DNA content and their possible evolutionary fate.*

We can estimate the rate of increase of genomes by gene duplication by assuming that the latter are neutral events (Kimura 1968, Li 1982). Accordingly, at any given time in a population there should be numerous duplicates on their way to fixation or loss, and the number of neutral duplicate genes that will become established per generation in a given population will be equal to their rate of occurrence (Doolittle 1979, Li 1982). The observed rates of spontaneous duplication of bacterial genes have values of 10^{-5} to 10^{-3} gene duplications per gene per cell generation (Anderson and Roth 1977, Tlsty et al. 1984). If we use the most conservative value of 10^{-5}, and for the sake of simplicity assume that during the early Archean there were ten generations of cells per year, the rate of accumulation of duplicons in a primordial genome encoding 100 enzymes would be 100 genes × 10 cell generations per year × 10^{-5} gene duplications per gene per cell generation, which equals 0.01 gene duplications per year. If we assume that only 10% of the duplications are neutral (Li 1982), the rate of accretion of duplicons would be 0.001 gene per year, which implies that the genomes would be increasing their DNA content at a rate of one nucleotide pair per year. Thus, under the assumptions discussed here, the time required to go from the 100-gene DNA/protein organisms to a 7000-gene filamentous cyanobacteria would be 7 × 106 years, but it may have been much less.

The situation under strong positive selection should be much faster. If the fixation of new duplicates increases the overall fitness of the population, then the rate of accretion of new genes may be expected to increase significantly. If one of the duplicate gene copies undergoes favorable mutations or internal recombinations (Fig. 1), then it will spread rapidly throughout the population. The rapid divergence of a fixed duplicate is exactly what we envision in the primitive Earth, as the prebiotic soup became exhausted compound by compound. The heterotrophs that developed abilities to metabolize the remaining compounds would have been strongly selected over others, and those bacteria that had developed an enzymatic apparatus allowing the use of photochemical energy sources would be rapidly favored.

The calculations presented here would seem to ignore not only the possibility that some duplicons may be lost by streamlining selection or by chance events (Fig. 1), but also the complexity of some basic metabolic processes, and of photosynthesis in particular. However complex, the enzymes involved in these processes are mostly the result of multiple gene duplications. As shown in Table 2, examples of gene duplication events that took place before the divergence of eubacteria, archaebacteria, and eukaryotes from their last common ancestor include those encoding enzymes involved in protein biosynthesis, DNA replication, ATP production, and basic metabolic pathways. Evidence of duplication and double-duplication events that paved the way to oxygen-releasing cyanobacterial photosynthesis has been preserved in the sequences of ferredoxins (Otaka and Ooi 1989), F-type ATPases (Gogarten et al. 1989), the reductases involved in chlorophyll and bacteriochlorophyll biosynthesis (Burke et al. 1993), the bacterial photosynthetic reaction center (Feher et al. 1989, Blankenship 1992), the two sets of light-harvesting antennae (Youvan and Ismail 1985), and photosystems I and II (Youvan and Marrs 1984, Blankenship 1992). As noted by Chapman and Schopf (1983), a gene duplication of photosystem I and the subsequent development of oxygen-producing photosynthesis still requires the linkage via electron flow of the two photosystems, but this could involve a surprisingly small number of modifications. There is evidence that the photosystems I and II developed in different organisms and combined by horizontal transfer or a fusion event (Büttner et al. 1992, Blankenship 1992), but this would only have accelerated the sequence of evolutionary events.

The rates of microbial evolution may have been accelerated during early Archean times by additional mechanisms, such as intragenic recombination, the modular assembly of new proteins, horizontal transfer, and duplication of the entire primitive genome, as well as by higher mutation rates. Recent measurements have shown that the mutation rate per base pair is inversely proportional to the genome size for DNA viruses, bacteria, and ascomycetes (Drake 1991). Based on this combined evidence, it is reasonable to assume that the rate of evolution during the early Archean would have been considerably faster than that of present day microorganisms, and that evolution to cyanobacteria may have been much faster than the 7×10^6 years calculated here based on genome size growth under neutral conditions.

TABLE 2. ***Products of Gene Duplication Events That Are Known to Have Taken Place Before the Divergence of Eubacteria, Archaebacteria and Eukaryotes****

Elongator and Initiator ^{met}tRNA	DNA topoisomerases
Ribosomal Proteins	A and B DNA polymerases
Amino Acyl tRNA Synthetases	Ferredoxins
Elongation Factors	Glutamine synthases
α and β ATPase Subunits	Heat shock proteins
Carbamoyl Phosphate Synthetase	

* *Based on Forterre et al. 1993 and García-Meza et al. 1994.*

CONCLUSIONS

We would like to emphasize the uncertainties of some of the figures discussed here. At the present time it is not possible to reconstruct in detail either the sequence of the evolutionary events described here, or the pace at which they took place. In spite of the many uncertainties in the current descriptions of the origin and early evolution of life, the data summarized here suggest that the most important bottlenecks in the process leading from the prebiotic soup to the RNA world, and then to DNA/protein organisms and eventually to cyanobacteria, may correspond to (1) the origin of replicating systems, (2) the emergence of protein biosynthesis, and (3) the evolutionary development of the starter types from which later proteins evolved through gene duplication and divergence. It is possible, of course, to imagine a process in which periods of rapid evolution such as those described here alternated with stages of slower evolution. In fact, such episodic schemes are known to have taken place during the evolution of some metazoan lineages. However, because of constraints placed by the rapid exhaustion of prebiotically synthesized compounds, it is extremely unlikely that intermediate stages of slow evolution could have persisted for millions and millions of years. We thus see no compelling reason to assume that the origin and early evolution of life took more than 10 million years, and it may have taken considerably less.

The time scales discussed here also suggest that if the early environmental conditions on Mars were comparable to those of the primitive Earth, life may also have started there. If significant amounts of water and organic compounds were present in the Martian environment at the end of the early bombardment, then life could have arisen if the period of time required was as small as estimated here, but perhaps it could not have survived, as the subsequent geological conditions became very unfavorable. In other words, the development of life on Mars would have been constrained not by time, but rather by geological conditions.

Finally, O, B, and A stars, which are usually omitted in calculations of the abundance of life (Huang 1959a,b) because of their short main sequence lifetimes of approximately 1.5 million, 15 million, and 500 million years, respectively (Limber 1960), may have life on planets surrounding them. Double stars, which had previously been omitted from SETI calculations because they were not believed to have long-term stable orbits (Huang 1960), may also have microbial life on their planets.

Acknowledgments

We thank C. Gómez Eichelman for several useful references, as well as M. Gogarten-Boekels and J.P. Gogarten for providing us with their results prior to publication (Gogarten-Boekels and Gogarten 1995). We thank the NSCORT (NASA Specialized Center for Research and Training) in Exobiology at the University of California, San Diego, for a Visiting Professor fellowship (AL) and grant support (SLM).

References

Anders, E. (1989), "Pre-biotic Organic Matter from Comets and Asteroids," *Nature* **342:** 255–257.

Anderson, R.P., and J.R. Roth (1977), "Tandem Genetic Duplications in Phage and Bacteria," *Ann. Rev. Microbiol.* **31:** 473–505.

Blankenship, R.E. (1992), "Origin and Early Evolution of Photosynthesis," *Photosynth. Res.* **33:** 91–111.

Burke, D.H., J.E. Hearst, and A. Sidow (1993), "Early Evolution of Photosynthesis: Clues from Nitrogenase and Chlorophyll Iron Proteins," *Proc. Natl. Acad. Sci. USA* **90:** 7134–7138.

Büttner, M., D.-L. Xie, H. Nelson, W. Pinther, G. Hauska, and N. Nelson (1992), "Photosynthetic Reaction Centers Genes in Green Sulfur Bacteria and in Photosystem I Are Related," *Proc. Natl. Acad. Sci. USA* **89:** 8135–8139.

Chapman, D.J., and J.W. Schopf (1983), "Biological and Biochemical Effects of the Development of an Aerobic Environment," in *Earth's Earliest Biosphere: Its Origin and Evolution*, J.W. Schopf, ed. (Princeton, NJ: Princeton University Press), pp. 302–320.

CHYBA, C.F., P.J. THOMAS, L. BROOKSHAW, AND C. SAGAN (1990), "Cometary Delivery of Organic Molecules to the Early Earth," *Science* **249:** 366–373.

CLARKE, P.H., AND S.R. ELSDEN (1980), "The Earliest Catabolic Pathways," *J. Mol. Evol.* **15:** 333–338.

CLOUD, P.E., JR. (1968), "Atmospheric and Hydrospheric Evolution of the Primitive Earth," *Science* **160:** 729–736.

DICKERSON, R.E. (1978), "Chemical Evolution and the Origin of Life," *Sci. Am.* **239 (9):** 70–86.

DOOLITTLE, R.F. (1979), "Protein Evolution," in *The Proteins*, H. Neurath, ed. (New York: Academic Press), pp. 1–118.

DRAKE, J.W. (1991), "A Constant Rate of Spontaneous Mutations in DNA-Based Microbes," *Proc. Natl. Acad. Sci. USA* **88:** 7160–7164.

EDMONDS, J.M., K.L. VON DAMM, R.E. MCDUFF, AND C.I. MEASURES (1982), "Chemistry of Hot Springs on the East Pacific Rise and Their Effluent Dispersal," *Nature* **297:** 187–191.

FEHER, G., J.P. ALLEN, M.Y. OKAMURA, AND D.C. REES (1989), "Structure and Function of Bacterial Photosynthetic Reaction Centers," *Nature* **239:** 111–116.

FORTERRE, P., N. BENACHENHOU-LAHFA, F. CONFALONIERI, M. DUGUET, CH. ELIE, AND B. LABEDAN (1993), "The Nature of the Last Universal Ancestor and the Root of the Tree of Life, Still Open Questions," *BioSystems* **28:** 15–32.

GARCÍA-MEZA, V., V. GONZÁLEZ-RODRÍGUEZ, AND A. LAZCANO (1994), "Ancient Paralogous Duplications and the Search for Archean Cells," in *Self Production of Supramolecular Structures: From Synthetic Structures to Models of Minimal Living Systems*, G.R. Fleischaker, S. Colonna, and P.-L. Luisi, eds. (Dordrecht: Kluwer Academic Publishers), pp. 145-153.

GOGARTEN, J.P., H. KIBAK, P. DITTRICH, L. TAIZ, E.J. BOWMAN, B.J. BOWMAN, M.F. MANOLSON, R.J. POOLE, T. DATE, T. OSHIMA, J. KONISHI, K. DENDA, AND M. YOSHIDA (1989), "Evolution of the Vacuolar H^+-ATPase: Implications for the Origin of Eukaryotes," *Proc. Natl. Acad. Sci. USA* **86:** 6661–6665.

GOGARTEN-BOEKELS, M., AND J.P. GOGARTEN (1995), "The Effects of Heavy Meteorite Bombardment on the Early Evolution of Cellular Life—A New Look at the Molecular Record," *Origins of Life* **25:** 251–264.

HANSCHE, P.E. (1975), "Gene Duplication as a Mechanism of Genetic Adaptation in *Saccharomyces cerevisiae*," *Genetics* **79:** 661–674.

HARTLEY, B.S. (1974), "Enzymes Families," in *Evolution in the Microbial World: 24th Symposium of the Society for General Microbiology*, M.J. Carlile and J.J. Skehel, eds. (Cambridge: Cambridge University Press), pp. 151–182.

HERDMAN, M. (1985), "The Evolution of Bacterial Genomes," in *The Evolution of Genome Size*, T. Cavalier-Smith, ed. (London: John Wiley & Sons), pp. 37–68.

HOROWITZ, N.H. (1945), "On the Evolution of Biochemical Synthesis," *Proc. Natl. Acad. Sci. USA* **31:** 153–157.

HUANG, S.-S. (1959a), "Occurrence of Life in the Universe," *Amer. Sci.* **47:** 397–402.

HUANG, S.-S. (1959b), "The Problem of Life in the Universe and the Mode of Star Formation," *PASP* **71:** 421–424.

HUANG, S.-S. (1960), "Life Supporting Regions in the Vicinity of Binary Systems," *PASP* **72:** 106–114.

JOYCE, G.F., A.W. SCHWARTZ, S.L. MILLER, AND L.E. ORGEL (1987), "The Case for an Ancestral Genetic System Involving Simple Analogues of the Nucleotides," *Proc. Natl. Acad. Sci. USA* **84:** 4398–4402.

KIMURA, M. (1968), "Evolutionary Rate at the Molecular Level," *Nature* **217:** 624–626.

LI, W.-H. (1982), "Evolutionary Change of Duplicate Genes," in *Isozymes: Current Topics in Biological and Medical Research*, M.C. Rattazzi, J.G. Scandalios, and G.S. Whitt, eds. (New York: Alan R. Liss), pp. 55–92.

LIMBER, D.N. (1960), "The Universality of the Initial Luminosity Function," *ApJ* **131:** 168–201.

MAHER, K.A., AND D.J. STEVENSON (1988), "Impact Frustration of the Origin of Life," *Nature* **331:** 612–614.

MILLER, S.L., AND L.E. ORGEL (1974), *The Origins of Life on Earth* (Englewood Cliffs, NJ: Prentice-Hall).

MILLER, S.L., AND J.E. VAN TRUMP (1981), "The Strecker Synthesis in the Primitive Ocean," in *Origin of Life*, Y. Wolman, ed. (Dordrecht: D. Reidel Publishing Company), pp. 135–141.

MILLER, S.L., AND J.L. BADA (1988), "Submarine Hot Springs and the Origin of Life," *Nature* **334:** 609–611.

OBERBECK, V.R., AND G. FOGLEMAN (1990), "Impact Constraints on the Environment for Chemical Evolution and the Continuity of Life," *Origins of Life* **20:** 181–195.

OPARIN, A.I. (1938), *The Origin of Life* (New York: The MacMillan Company).

ORÓ, J. (1960), "Synthesis of Adenine from Ammonium Cyanide," *Biochem. Biophys. Res. Comm.* **2:** 407–412.

OTAKA, E., AND T. OOI (1989), "Examination of Protein Sequence Homologies: V. New Perspectives on Evolution Between Bacterial and Chloroplast-Type Ferrodoxins Inferred from Sequence Evidence," *J. Mol. Evol.* **29:** 246–254.

PELTZER, E.T., J.L. BADA, G. SCHLESINGER, AND S.L. MILLER (1984), "The Chemical Conditions on the Parent Body of the Murchison Meteorite: Some Conclusions Based on Amino-, Hydroxy-, and Dicarboxylic Acids," *Adv. Space Res.* **(4) 12:** 61–74.

SAGAN, C. (1974), "The Origin of Life in a Cosmic Context," *Origins of Life* **5:** 497–505.

SÁNCHEZ, R.A., J.P. FERRIS, AND L.E. ORGEL (1966), "Cyanoacetylene in Prebiotic Synthesis," *Science* **154:** 784–786.

SCHIDLOWSKI, M. (1992), "The Initiation of Biological Processes on the Earth: Summary of Empirical Evidence," *Adv. Space Res.* **12 (4):** 143–156.

SCHOPF, J.W., ED. (1983), *The Earth's Earliest Biosphere: Its Origin and Evolution* (Princeton, NJ: Princeton University Press).

SCHOPF, J.W. (1993), "Microfossils of the Early Archean Apex Chert: New Evidence of the Antiquity of Life," *Science* **260:** 640–646.

SHAPIRO, R. (1988), "Prebiotic Ribose Synthesis: A Critical Analysis," *Origins of Life* **18:** 71–85.

SIMPSON, G.G. (1964), "The Nonprevalence of Humanoids," *Science* **143:** 769–775.

SLEEP, N.H., K.J. ZAHNLE, J.F. KASTING, AND H.J. MOROWITZ (1989), "Annihilation of Ecosystems by Large Asteroid Impacts on the Early Earth," *Nature* **342:** 139–142.

SONTI, R.V., AND J.R. ROTH (1989), "Role of Gene Duplications in the Adaptation of *Salmonella typhimurium* to Growth on Limiting Carbon Sources," *Genetics* **123:** 19–28.

STRIBLING, R., AND S.L. MILLER (1987), "Energy Yields for Hydrogen Cyanide and Formaldehyde Syntheses: The HCN and Amino Acid Concentrations in the Primitive Ocean," *Origins of Life* **17:** 261–273.

TLSTY, T.D., A.M. ALBERTINI, AND J.H. MILLER (1984), "Gene Amplification in the *lac* Region of *E. coli*," *Cell* **37:** 217–224.

UREY, H.C. (1952), "On the Early Chemical History of the Earth and the Origin of Life," *Proc. Natl. Acad. Sci. USA* **38:** 351–363.

WALD, G. (1954), "The Origin of Life," *Sci. Am.* **191 (8):** 44–53.

WESTHEIMER, F.H. (1987), "Why Nature Chose Phosphates," *Science* **235:** 1173–1178.

YOUVAN, D.C., AND B.L. MARRS (1984), "Molecular Genetics and the Light Reactions of Photosynthesis," *Cell* **39:** 1–3.

YOUVAN, D.C., AND S. ISMAIL (1985), "Light-Harvesting II (B800–850-Complex) Structural Genes from *Rhodopseudomonas capsulata*," *Proc. Natl. Acad. Sci. USA* **82:** 58–62.

Next Page

The Very Large Array (VLA) in New Mexico consists of 28 antennas (including one that is a spare), each of which is an 82-foot dish with eight receivers tucked inside. The dish moves on an altitude-azimuth mount–what you've probably seen as a classic tripod mount. It tilts up and down and spins around.

The iconic Y shape of the VLA is functional. The wider an array is, the bigger its eye, and the more detail it can see out in space. The VLA's unique shape provides three long arms when we need to zoom in for more detail.

These telescopes are on rails. Three times a year, a specially designed rail truck, called a "transporter," picks up telescopes and hauls them one at a time farther down their track. Over the course of sixteen months, the VLA lengthens each of its legs from 2/3 of a mile to 23 miles long.

For more information, see https://public.nrao.edu/telescopes/vla/

Very Large Array at Blue Hour

Photo Credit: US NSF/AUI/NSF NRAO, Bettymaya Foott [taken at nautical twilight]
https://public.nrao.edu/gallery/very-large-array-at-blue-hour/

TIME FOR INTELLIGENCE ON OTHER PLANETS

Christopher P. McKay

Space Science Division
NASA Ames Research Center
Moffett Field, CA 94035

Abstract

The search for extraterrestrial intelligence (SETI) is usually patterned on the history of life on Earth. It is thus assumed that intelligence—operationally defined as the building of radio telescopes—is a very likely consequence in the evolution of life on any planet, and that the time required from the first appearance of life on Earth to the first construction of a radio telescope on Earth, 3.5 Gyr, is typical for other planets as well. If the intelligence to build radio telescopes is part of a continuum that includes animal intelligence then it is likely to evolve readily, on Earth and elsewhere. Conversely if human intelligence is a qualitative departure from animal intelligence, it may be a trait that has a low probability of evolving in life-forms on other planets. Using the past evolution of vertebrates as a test, the paleontological record suggests that intelligence comparable to humans could have evolved several times on Earth—most notably with the dinosaurs—but did not. Life may be a common planetary phenomenon but intelligence may be rare, perhaps even singular. Considering a world with optimal geology and a direct evolutionary trajectory, the time required for the evolution of intelligence could be as little as 100 million years. The fact that the time to develop intelligence on Earth is within a factor of two of the lifetime of the Sun may indeed be coincidental—the anthropic principle notwithstanding. The main implication of a possible short time scale to intelligence is that the search for extraterrestrial intelligence should be broadened to include stars that maintain habitable zones for much shorter periods than do solar-type stars. The search space is expanded but the chances of a signal are diminished.

INTRODUCTION

One of the profound questions in science is that of life in the Universe. In general this question proceeds in two stages. First there is the question of life of any type beyond the Earth. Logically following this is the question of intelligent life elsewhere. (In this paper we will take the unmodified word *intelligence* to be defined as the ability to construct radio telescopes. The distinction will be made with animal intelligence which will also be considered. All animal intelligence on Earth fails our operational criteria.) Clearly a positive answer to the first is a prerequisite to a positive answer to the second. One might imagine therefore that the search for life outside Earth would begin first with a search for any type of life-form. Indeed, in our own Solar System, where signs of other intelligence are absent, and travel to the other planets possible, this has been and continues to be the focus of the search for life (see, e.g., National Research Council 1990). The study of the earliest organisms on the Earth suggests life appeared rapidly on this planet, and current theories of life posit that on any planet with an environment similar to that of the early Earth, life would also appear on short time scales compared with the lifetime of a planet or star. This scenario may receive its first test with the search for fossil life on Mars (see, e.g., McKay and Stoker 1989)

However, the planets of the Solar System are not a very favorable set of objects on which to base an understanding of life in the Galaxy. Only one example of life—Earth—is currently available for study. Other planets orbiting other stars could harbor life but due to the great distances, searching for life on these other planets is currently not feasible. In fact the mere detection of Earth-sized planets around other main sequence stars is currently just beyond the state of the art.

The difficulty of searching for simple life-forms beyond the Solar System, as well as the fascination with the concept of communication with other intelligent life-forms, drives the search for life toward a search for extra–Solar System intelligence (SETI). The main reason for relying on intelligence as the basis of a life search about other stars is that intelligent life-forms can literally broadcast their presence in a way that simple life-forms cannot—via radiowaves.

Before a rational basis for a search for extra–Solar System intelligence can be formulated we must understand how probable the evolution of intelligence is and how long it might typically take. Based on the principle that the Earth is "typical"—otherwise called the principle of mediocrity or the Copernican principle (Gott 1993)—it has been concluded that intelligence has a very high probability of emerging but only after 3.5 Gyr of evolution.

Dole (1964) in his seminal book *Habitable Planets for Man* suggests that it would require "about 3 billion years for the appearance of complex life." Hart (1979), in an early study of continuously habitable zones (CHZ) about main sequence stars, required that moderate surface temperatures exist continuously through 3 or 4 billion years. More recently, Kasting et al. 1993 define the CHZ as existing for 4.6 billion years although they point out that the width of the CHZ around a star does depend on the time that a planet is required to remain habitable. The current strategy for SETI target stars is that, like the Sun, they would be on the main sequence for over 4 billion years.

The conclusion that intelligence requires 3.5 Gyr to emerge is based on a simple reading of the geological record that presupposes that the evolution of life has been a steady progressive development of ever more complex and sophisticated forms leading ultimately to human intelligence. In this view, 3.5 Gyr is simply the time it takes for evolution to progress to the level of complexity that is manifest in radio telescopes. If this view were supported by the paleontological record it could stand as the basis for SETI. However, evolution has not progressed this way. The history of life on Earth shows clear indications that complexity has both increased and decreased (e.g., Gould 1994). Furthermore, catastrophic external events have clearly played a determining role in evolution. It is even possible that the success of mammals, and hence intelligence in humans, is related to the mass extinction at the end of the Cretaceous—now thought to have been caused by a large impact. Evolution has followed a wandering and random path. Thus, a more detailed look at the particular path that evolution has followed over time and the implications for SETI of the many twists and turns in this path is warranted.

In this paper we approach the paleontological record in two complementary ways. First we look at the record of the evolution of advanced life on Earth and the emergence of sophisticated features such as flight and vision. Then we consider intelligence and ask how many possible opportunities there have been for the evolution of intelligence. The essential question here is: Do we see any suggestion that intelligence is a trait that can be readily evolved, in the way that flight and vision were? As an evolutionary event are radio telescopes likely or unlikely?

We then alter our approach and consider the shortest possible path to intelligence. In this approach we optimize the geological and environmental factors that have played a role in the evolution of life on Earth. Furthermore, we allow evolution to monotonically progress toward human intelligence. Thus, we eliminate any periods of time during which evolution was not moving directly toward the construction of radio telescopes. This would, in some sense, provide one estimate for the minimum time for the evolution of intelligence on another world.

The Evolution of Intelligence

The modern literature on the question of intelligent life in other solar systems suffers from a lack of firm data. So we first review the published opinions. Many paleontologists writing on the subject have expressed pessimism at the prospects of intelligence on other worlds. George Gaylord Simpson, writing at the dawn of the space age, opined that "There probably are forms of life on other planetary systems . . . but . . . it is extremely improbable that such forms of life include humanoids" (Simpson 1964). His definition of a humanoid is essentially equivalent to the SETI definition of radio telescope builders. Stephen Jay Gould has also expressed the opinion that intelligence may be an improbable event (Gould 1989, p. 321). While these authors are knowledgeable in the evolutionary history of life on Earth, they do not state any justifications or rationale for the assertions they present *ex cathedra*. More recently, Ernst Mayr, writing expressly about upcoming SETI plans, stated that while the existence of life somewhere in the Universe is quite probable, intelligence was the singular chance of one evolutionary line out of billions (Mayr 1992). This calculation of the improbability of intelligence (one in a billion) implicitly assumes that all species have equal a priori probability of constructing radio telescopes. This may be unrealistic but suggests that the record might provide a basis for estimating the probability of intelligence. Unusual among his colleagues in the study of evolution, Dawkins (1986, p. 146) holds, without supporting arguments, that "Cumulative selection [is] powerful enough to make the evolution of intelligence probable, if not inevitable." From the apparent lack of intelligence in the nearby stars he then concludes that life is cosmically rare (Dawkins 1986, p. 165). For the most part then, paleontologists feel that intelligence is a qualitative evolutionary development beyond animal intelligence and not one that would have necessarily emerged under even slightly different circumstances.

A major proponent of the view that intelligence is within the natural spectrum of animal intelligence is Carl Sagan. He concludes that human intelligence must be rooted in one or more of four physical properties—the mass of the brain, the mass to body ratio, the presence of large frontal and temporal lobes, or the number of neural connections (Sagan 1977). In any population there will be a natural variation in these parameters across species that rely on animal intelligence for their evolutionary success. The evolution of humanlike intelligence is then the inevitable result of the combination of chance and necessity operating on these species. For example, larger body mass provides the ability to generate the oxygen and glucose needed to operate a larger brain. This factor explains the trend in brain size to body mass in mammals with the noted exception that primates use a larger portion of their total energy reserves for their brain and hence have relatively larger brains (Armstrong 1983). The allocation of additional energy reserves to brain function is an adaptive strategy among the primates as it could well have been for other lineages at other times.

These two views of intelligence—an unusual evolutionary property versus a form of animal intelligence—obviously predict very different probabilities for the occurrence of intelligence on other worlds. In this section we try to assess the probability of intelligence by comparing animal and human intelligence to other complex traits such as vision and flight and by considering the past opportunities for humanlike intelligence to evolve. Although the paleontological record is limited at best, it is the only data available.

It is generally accepted that complex traits such as vision, flight, animal intelligence, and presumably human intelligence, evolve in stages. Intermediate stages must provide evolutionary advantage to ensure selection. It has not been possible to adequately document the intermediate stages for either vision or flight. There are, however, plausibility arguments. For example, it has been suggested that insect flight evolved from protowings used for skimming over the surface of water (Marden and Kramer 1994). The image-forming eye has evolved independently in the vertebrates and the cephalopods, and light-sensitive organs, often resulting from independent evolutionary paths, are found in numerous organisms (Salvini-Plawen and Mayr 1977). Both vision and flight exist over a range of lineages and capabilities; there are examples of convergence and specialization.

Animal intelligence follows a similar pattern; it is found in a range of organisms and at a variety of levels. It is widely believed that animal intelligence at a certain level is an adaptive and readily evolvable feature. In nature we see the independent evolution of this kind of intelligence in mammals, birds, and, arguably, in social insects. The chimpanzee, the fox, the crow, the dolphin, and the white ant amaze us with their clever feats. It is probable that many of the dinosaurs enjoyed similar intellectual capabilities and that animal life on other worlds does as well. This level of animal intelligence—like vision and flight—is readily evolvable.

If the ability to construct radio telescopes is considered a part of, albeit at the extreme of, the animal intelligence continuum then it is no more remarkable than the superlative eyes of the raptors or the hovering flight of hummingbirds. This would suggest that humanlike intelligence could naturally emerge within a distribution of animal intelligences. Viewed in this way, human intelligence is not an extraordinary thing and would be expected to readily evolve in groups of organisms with a suitable range of animal intelligence and morphological capabilities.

It is interesting to note here that Gould, who has extensively studied evolutionary patterns, believes that if the *Homo sapiens* line had been terminated at an early phase, then intelligence may not have developed in any alternate lineages of *Homo*; "*Homo sapiens* is an entity not a tendency" (Gould 1989, p. 320). Earth would then be silent and deaf at radio wavelengths.

The suggestion that humanlike intelligence should be a readily evolvable trait suggests we look for evidence of its independent development in extant and extinct species, similar to the independent development of vision and flight. There is no evidence, however, that there has been another independent emergence of intelligence. It might be argued that among mammals, humans developed intelligence first and are thereby effectively precluding the development of intelligence in any other species. It follows from this argument that intelligence evolves once and only once on a planet because once evolved, it changes the rules of the interaction among species and effectively dominates the planet from then on. This may provide an adequate explanation for the lack of other mammalian intelligence in the present world. However, it does not explain why intelligence did not evolve previously—with the dinosaurs.

Perhaps the best past opportunity for the emergence of intelligence capable of building radio telescopes was during the dinosaur era. It is now considered probable that the dinosaurs were not the lumbering clods of urban myth but that they were biochemically and behaviorally as sophisticated as present mammals (Desmond 1975, Bakker 1986). Evidence continues to point to parenting and social behavior that is on a par with small mammals and birds (Horner and Makela 1979, Norell et al. 1994). Against competition from the cynodonts, the precursors of mammals, the dinosaurs dominated all niches currently occupied by mammals for over 150 million years from the late Triassic to the end of the Cretaceous. This is more than twice the time between the end of the Cretaceous and the construction of the first radio telescope. Ostrom (1992), commenting on this "interruption" in mammalian evolution, states "the archosaurs—the dinosaurs, in particular—disrupted what otherwise would have been a smooth, continuous, uninterrupted evolutionary progression."

If the dinosaurs were so smart, why didn't they build radio telescopes? At least one species of dinosaur may have been capable of eventually doing so. Russell and Séguin (1982) discuss the case of the small carnivorous dinosaur *Stenonychosaurus*, which stood about 120 cm, weighed about 40 kg, had an encephalization quotient about equal to that of a possum or an octopus, and lived over 12 million years before the end of the dinosaurs (65 Myr ago). Russell and Séguin speculate about the evolutionary path these organisms might have followed if dinosaurs continued to survive to present times. They suggest that humanlike intelligence (encephalization quotient near 7) could have been the result. In the context of the present discussion, the question is Why did these creatures not achieve this level of development in the 12 Myr near the end of the Cretaceous if intelligence at the level of human capabilities is a readily evolvable trait?

As an aside one might speculate that perhaps *Stenonychosaurus* or her progeny did build radio telescopes, but their civilization was destroyed by some internal or external catastrophe. Perhaps the lifetime of their civilization was so short compared with the resolution of the geologic record (typically millions of years) that it is simply lost without trace in the depths of time. It is difficult to say what evidence would survive of human civilization—if it was terminated now—after 65 million years of tectonic activity, erosion, and sea level change. It is interesting to note that there is one place where the record of human technology will be preserved for times much longer than 100 million years. If civilization were to totally collapse today, long after all other traces of human civilization had turned to dust, the Apollo landing sites on the Moon would bear mute testimony to technological humans. Did the dinosaurs go to the Moon? Our present resolution of the geological record could not tell us so. Perhaps someday we will know.

It is admittedly difficult to generalize about the evolution of intelligence from the meager data available in the geological record. From the discussion above we suggest that there were at least two independent opportunities for intelligence to evolve on Earth—one lasted about 150 Myr and evidently failed to produce dinosaur intelligence. The other produced mammalian intelligence in 65 Myr. It is not easy to convincingly extend the discussion to include birds and reptiles. In these same groups many other sophisticated features, notably flight, evolved repeatedly. The conservative conclusion from this could be that the odds for developing intelligence are less than one in three (65/215). The lower limit is much more difficult to obtain and one can argue that the set of opportunities was much larger than three. Considering all the possible habitats for dinosaurs, birds, and mammals on continents isolated from hominoid interference until only 20,000 years ago, the opportunities for the independent evolution of intelligence could be numbered in the hundreds. In specific, the New World was isolated from the *Homo* lineage until recently. South America in turn was isolated from placental mammals until 25 Myr ago and saw the development of unusual predatory birds (Simpson 1980, Marshall 1994). Yet there is no evidence that during these millions of years intelligence developed. Thus, the likelihood of evolving intelligence could be more correctly set at one in an uncertain, but possibly large, number.

If the evolution of human intelligence is the result of a unique set of coincidental events, beginning with the K-T impact, running through the formation of the Great Rift Valley in Africa, and then to the timing and extent of the ice ages, one might be forced to conclude that the probability of intelligence evolving is even smaller than suggested above. Perhaps vanishingly small.

Minimum Time to Intelligence

The presence of intelligence on Earth—as evidenced by the presence of radio telescopes—is firm evidence that it is not impossible to evolve intelligent life-forms. How can this one data point be best used to estimate the typical and minimum times that might be required for such evolution?

As discussed above, the view of evolution as a smooth progressive development of complexity leading ultimately to intelligence has been replaced by a view that is much more stochastic. The fossil record suggests that evolution from the earliest common ancestor to technological humans was punctuated by intense periods of rapid change followed by long periods of stasis—it was not smooth. Furthermore it was not progressive; there were long periods of time when the Earth was dominated by organisms not in the direct line leading to radio telescopes—most notably the dinosaurs. Gould (1994) has likened evolution to a staircase with sharp rises separated by periods of level. If we imagine the bottom of the staircase as the first organism and define the top of the staircase as the ability to construct radio telescopes, we might expand Gould's metaphor to suggest that there were also periods during which the staircase was being built in the wrong direction. A minimum time to intelligence could be constructed by removing all those spurious (from this specific perspective) time periods. Within the staircase metaphor we ask, What is the vertical distance between the origin of life and technology? Nothing in our understanding of evolution suggests that these periods of stasis are required. We believe they represent mere historical happenstance.

One approach to the question of time scale for intelligence was developed by Carter (1983) in a discussion of the anthropic principle. He developed a simple mathematical model of evolutionary change and concludes that "the time taken by biological evolution on Earth up to our own present stage of advancement (radio telescopes) has been many tens of thousands of times longer than need have been the case if strong Darwinian selection pressure has at all stages been steadily directed toward the present outcome." Carter (1983), as summarized nicely by Maddox (1984), considers the evolution of sentient beings as consisting of n discrete steps. Only the few steps with lowest probability need to be considered since they set the time scale. Thus, one might expect n to be a small number. Carter (1983) uses the observation that $t_e \cong T_\circ$ where t_e is the observed time to reach intelligence on Earth (3.5 Gyr) and $T_\circ$ is the time frame available for such evolution taken as the lifetime of the Sun (~ 10 Gyr). He considers it a remarkable coincidence that intelligence evolved with just a factor of 2 or so to spare ($t_e \cong T_\circ/2$) and from this finds that $1 \leq n \cong 2$ and concludes therefore that intelligence is astronomically unlikely on other worlds. The coincidence may be even more astounding if one considers recent arguments for the end of the biosphere as being only an additional 1 Gyr away (Caldeira and Kasting 1992). Thus intelligence emerged only after the biosphere had passed 77% of its history. Carter has used one observation, $t_e \cong T_\circ$, to constrain his model. One could take the complementary approach of listing the critical steps in the actual evolution of intelligence on Earth and determining which are the slow, rate-determining events; such a list is shown in Table 1.

The list of critical steps leading to intelligence begins with the origin of life. The duration of this is uncertain. Conditions suitable for life commence at the end of the heavy bombardment about 3.8 Gyr ago. If the Isua sediments (also 3.8 Gyr old) are taken as evidence for phototrophic life (Schidlowski 1988), then the origin of life was a geologically instantaneous event. Setting the Isua issue aside, strong evidence for life is found at 3.5 Gyr ago (Schopf and Packer 1987). Miller and Lazcano (1996) suggest that the origin of life could have taken less than 10 Myr.

TABLE 1. *Critical Events in the Evolution of Intelligence*

	Event	Date (million years ago)	Duration on Earth (million years)	Possible Minimum (million years)
1.	Origin of Life	3800–3500	< 500	10
2.	Oxygenic Photosynthesis	< 3500	< 500	negl.
3a.	Oxygen Environment	2500	1000	100
3b.	Tissue Multicellularity	550	2000	negl.
4.	Development of Animals	510	5	5
5.	Land Ecosystems	400	100	5
6.	Animal Intelligence	250	150	5
7.	Human Intelligence	3	3	3

The second critical step is the development of oxygenic photosynthesis, since this process is critical to the development of an oxygen-rich atmosphere. There is evidence that oxygen production occurred in local basins and lakes as early as 2.7 Gyr ago (Buick 1992), and there are plausible scenarios for the rapid evolution of oxygenic photosynthesis in a globally reducing environment (McKay and Hartman 1991). Schopf (1993) suggests the possibility of oxygenic photosynthesis at 3.5 Gyr ago based on the microfossil evidence.

But oxygen production did not result in an oxygen-rich atmosphere owing to the presence of reductants in the ocean as well as to the continuing emanation of reductants from volcanic sources. Current evidence suggests that it was not until about 2.5 Gyr ago that oxygen levels began to increase on the Earth. Indeed, looking ahead at the remaining steps listed in Table 1 it is clear that the buildup of an oxygen-rich atmosphere and the associated development of tissue multicellularity were the limiting steps in the development of intelligence. All other steps were rapid by comparison.

How rapidly oxygen increased with time after 2.5 Gyr ago is unknown but the development of multicellular life at 550 Myr ago could suggest that oxygen did not reach modern levels until just before that time. The interrelation between multicellularity and the presence of oxygen is usually based on the observation that aerobic metabolism is required to power large multicellular organisms. Thus the development of animals has to await the titration of the reducing atmosphere and the buildup of oxygen. Because of the close tie between the buildup of oxygen and the development of multicellular life, and the uncertainty as to how oxygen increased over time, these two events are listed together in Table 1.

Oxygen may also have been required to provide an ozone shield against UV light, although the levels of oxygen required to fulfill this function are well below the levels needed for respiration. Interestingly, it has also been suggested that metazoans may have contributed to the rise of oxygen by increasing the rate of burial of organic carbon (Runnegar 1992).

Once multicellular organisms developed, evolution proceeded rapidly. Indeed, all major architectural development in the organization of multicellular life occurred some 600 million years ago, beginning in the Ediacara. This was followed by the Cambrian explosion during which, in a span of only 5 million years (Gould 1994), all modern phyla developed except possibly for the Bryozoa—sessile marine organisms not germane to the evolution of human intelligence. Gould describes the Cambrian explosion as "the most remarkable and puzzling event in the history of life." In our approach to determine the minimal time for intelligence to evolve we invoke a similar rapid diversification on other worlds as well. The Cambrian explosion also provides a yardstick by which to measure the time required for subsequent evolutionary stages.

It is interesting to note here that many strange lineages developed during this period that do not have modern counterparts. Runnegar (1992), in discussing this situation, comments that "it is no longer possible to rely on assumed homologies with living animals when attempting to interpret these enigmatic fossils (the Ediacara fossils of 550 million years ago). Instead, the paleobiologist is forced to approach them from 'first principles' using only logic and experiment. It is the kind of analysis that would need to be undertaken if we were to find fossils of extinct multicellular life on Mars, and it is sobering to realize how difficult a task it would be."

The development of surface ecosystems composed of animals and plants is the next major development. This is achieved in 100 Myr but the key steps could have been much faster—if compared with the Cambrian explosion—possibly within 5 Myr.

Land animals for which intelligence is a key adaptive parameter appear with the small carnivores that are to be the precursors of mammals, dinosaurs, and birds. Given that animal intelligence in mammals evolved quite rapidly at the start of the Tertiary it is not impossible that primatelike intelligence could evolve within several million years after the development of land ecosystems, *in an optimal scenario*. The time for the development of radio telescopes from primate precursors that possessed mere animal intelligence is uncertain but it is certainly less than 3 Myr.

Knoll (1985) recognized that the buildup of oxygen was the key variable in the history of life on Earth: "On Earth, surface environments became oxygenic (aerobic) about 1500 to 2000 Ma after the origin of life. All other factors being equal, under different tectonic conditions, the same event might have happened 1000 Ma earlier, 2000 Ma later, or not at all." Knoll related this to the probability analysis of Carter (1983), arguing that the physical evolution of the Solar System determined both the lifetime of the Sun ($T_\odot$) as well as the time required for the emergence of intelligent life (t_e) and that t_e was not necessarily an independent biological parameter.

Following his approach we identify the rise of oxygen and concomitant development of tissue multicellularity as the rate-determining critical step in the evolution of intelligence. All other steps are rapid by comparison. We conclude that

$n = 1$ consistent with Carter's (1983) model. Carter, however, had suggested that $n = 2$ and that these two critical events were the evolution of the genetic code and the organization of the nervous system. From our analysis we find that the two steps considered by Carter were quite rapid compared with the formation of an oxygen-rich atmosphere.

The critical steps leading to intelligence listed in Table 1 could in principle have occurred over a time scale of less than 100 million years on a planet with geological conditions such that oxygen buildup could occur rapidly. In principle then, habitable zones around early-type stars could harbor life, and possibly intelligent life. Although these stars are short lived, and constitute only about 5% of the stars in the solar neighborhood, they have huge habitable zones, since the size of the habitable zone increases as the stellar luminosity increases.

What is not known is whether there is some aspect of the biogeochemical processes on a habitable planet—for example, those dealing with the burial of organic material, the maintenance of habitable temperatures as the stellar luminosity increases gradually over its main sequence lifetime, or global recycling by tectonics—that mandates the long and protracted development of the oxygen-rich biosphere that occurred on Earth. Other important unknowns include the effect of solar system structure on the origin of life and its subsequent evolution to advanced forms. It is possible that cometary impacts early in the history of a planet are necessary to provide volatile elements, and they certainly could play a role in the evolution of advanced life-forms as well.

Setting aside these uncertainties, it is interesting to consider a specific application of this approach to the evolution of intelligence on other worlds to Mars. Unlike the Earth, Mars lacks plate tectonics, and thus deep sedimentary layers on that planet would be more stable than on the Earth. Furthermore, volcanic activity on Mars appears significantly diminished compared with the Earth both in magnitude and duration (e.g., Greeley and Spudis 1981). If life originated on Mars at a time and in a manner similar to Earth, then the biotic production of oxygen would have been much more efficient at creating an oxygen-rich atmosphere on that planet compared with the Earth (Hartman and McKay 1995). An implication of this is that while the transition to an oxidizing atmosphere on Earth required billions of years, the transition may have been much more rapid on Mars. How much faster is hard to estimate, but the relative column mass of the initial water inventory (3000 m for Earth, less than 500 m for Mars; McKay and Stoker 1989), as well as the ratio of masses (10 : 1), suggests that a factor of 10 to 100 is not unlikely. The Precambrian on Mars may have lasted only for 100 million years. Thus Mars may have experienced an oxygen-rich atmosphere, with multicellular life, complex ecosystems and animal intelligence, all over a period during which Earth was still struggling to overcome its load of reductants. Unfortunately, if this was the case the very attributes that allowed Mars to live in the fast lane of evolution—especially the lack of plate tectonics—would have doomed its biosphere to an untimely death due to the lack of global recycling mechanisms (McKay and Stoker 1989). On other worlds other possibilities could exist.

References

Armstrong, E. (1983), "Relative Brain Size and Metabolism in Mammals," *Science* **220:** 1302–1304.

Bakker, R.T. (1986), *The Dinosaur Heresies* (New York: Morrow).

Buick, R. (1992), "The Antiquity of Oxygenic Photosynthesis: Evidence from Stromatolites in Sulphate-Deficient Archaean Lakes," *Science* **255:** 74–77.

Caldeira, K., and J.F. Kasting (1992), "The Life Span of the Biosphere Revisited," *Nature* **360:** 721–723.

Carter, B. (1983), "The Anthropic Principle and Its Implications for Biological Evolution," *Phil. Trans. R. Soc. Lond. A* **310:** 347–363.

Dawkins, R. (1986), *The Blind Watchmaker* (New York: W.W. Norton & Company).

Desmond, A.J. (1975), *The Hot-Blooded Dinosaurs* (New York: Warner Books).

Dole, S.H. (1964), *Habitable Planets for Man* (New York: Blaisdell Publishing Company).

Gott, J.R., III (1993), "Implications of the Copernican Principle for Our Future Prospects," *Nature* **363:** 315–319.

Gould, S.J. (1989), *Wonderful Life* (New York: W.W. Norton & Company).

Gould, S.J. (1994), "The Evolution of Life on the Earth," *Sci. Am.* **271 (4):** 85–91.

Greeley, R., and P.D. Spudis (1981), "Volcanism on Mars," *Rev. Geophys. Space Phys.* **19:** 13–41.

Hart, M.H. (1979), "Habitable Zones Around Main Sequence Stars," *Icarus* **37:** 351–357.

Hartman, H., and C.P. McKay (1995), "Oxygenic Photosynthesis and the Oxidation State of Mars," *Planet. Space Sci.* **43:** 123–128.

Horner, J.R., and R. Makela (1979), "Nest of Juveniles Provides Evidence of Family Structure Among Dinosaurs," *Nature* **282:** 296–298.

Kasting, J.F., D.P. Whitmire, and R.T. Reynolds (1993), "Habitable Zones Around Main Sequence Stars," *Icarus* **101:** 108–128.

Knoll, A.H. (1985), "The Precambrian Evolution of Terrestrial Life," in *The Search for Extraterrestrial Life: Recent Developments*, M.D. Papagiannis, ed. (Dordrecht: D. Reidel Publishing Company) pp. 201–211.

Maddox, J. (1984), "New Twist for the Anthropic Principle," *Nature* **307:** 409.

Marden, J.H., and M.G. Kramer (1994), "Surface-Skimming Stoneflies: A Possible Intermediate Stage in Insect Flight Evolution," *Science* **266:** 427–430.

Marshall, L.G. (1994), "The Terror Birds of South America," *Sci. Am.* **270 (2):** 90.

MAYR, E. (1992), "Is It Worth Looking for Extraterrestrial Intelligence?" (in German), *Naturwissenschaftliche Rundschau* **45:** 264–266.

MCKAY, C.P., AND C.R. STOKER (1989), "The Early Environment and Its Evolution on Mars: Implications for Life," *Rev. Geophys.* **27:** 189–214.

MCKAY, C.P., AND H. HARTMAN (1991), "Hydrogen Peroxide and the Origin of Oxygenic Photosynthesis," *Origins of Life* **21:** 157–163.

MILLER, S.L., AND A. LAZCANO (1996), "From the Primitive Soup to Cyanobacteria: It May Have Taken Less Than 10 Million Years," this volume.

NATIONAL RESEARCH COUNCIL (1990), *The Search for Life's Origins* (Washington, DC: National Academy Press).

NORELL, M.A., J.M. CLARK, D. DEMBERELYIN, B. RHINCHEN, L.M. CHIAPPE, A.R. DAVIDSON, M.C. MCKENNA, P. ALTANGEREL, AND M.J. NOVACEK (1994), "A Theropod Dinosaur Embryo and the Affinities of the Flaming Cliffs Dinosaur Eggs," *Science* **266:** 779–782.

OSTROM, J.H. (1992), "A History of Vertebrate Successes," in *Major Events in the History of Life*, J.W. Schopf, ed. (Boston: Jones and Bartlett Publishers), pp. 119–139.

RUNNEGAR, B. (1992), "Evolution of the Earliest Animals," in *Major Events in the History of Life*, J.W. Schopf, ed. (Boston: Jones and Bartlett Publishers), pp. 65–93.

RUSSELL, D.A., AND R. SÉGUIN (1982), "Reconstructions of the Small Cretaceous Theropod *Stenonychosaurus inequalis* and a Hypothetical Dinosauroid," in *Syllogeus*, **37:** 1–43 (Ottawa: National Museum of Natural Sciences).

SAGAN, C. (1977), *The Dragons of Eden* (New York: Random House).

SALVINI-PLAWEN, L.V., AND E. MAYR (1977), "On the Evolution of Photoreceptors and Eyes," in *Evolutionary Biology*, M.K. Hecht, W.C. Steere, and B. Wallace, eds. (New York: Plenum Press).

SCHIDLOWSKI, M. (1988), "A 3,800-Million-Year Isotopic Record of Life from Carbon in Sedimentary Rocks," *Nature* **333:** 313–318.

SCHOPF, J.W. (1993), "Microfossils of the Early Archean Apex Chert: New Evidence of the Antiquity of Life," *Science* **260:** 640–646.

SCHOPF, J.W., AND B.M. PACKER (1987), "Early Archean (3.3-Billion to 3.5-Billion-Year-Old) Microfossils from Warrawoona Group, Australia," *Science* **237:** 70–73.

SIMPSON, G.G. (1964), "The Nonprevalence of Humanoids," *Science* **143:** 769–775.

SIMPSON, G.G. (1980), *Splendid Isolation: The Curious History Of South American Mammals* (New Haven, CT: Yale University Press).

The Andromeda Galaxy (M31) (a composite image)

Photo Credit: *X-ray:* NASA/CXO/UMass/Z. Li & Q. D. Wang, ESA/XMM-Newton
Infrared: NASA/JPL-Caltech/WISE, Spitzer, NASA/JPL-Caltech/K. Gordon (U. Ariz.), ESA/Herschel, ESA/Planck, NASA/IRAS, NASA/COBE
Radio: NSF/GBT/WSRT/IRAM/C. Clark (STScI)
Ultraviolet: NASA/JPL-Caltech/GALEX
Optical: Andromeda, Unexpected, © Marcel Drechsler, Xavier Strottner, Yann Sainty & J. Sahner, T. Kottary
Composite Image Processing: L. Frattare, K. Arcand, J. Major
https://www.nasa.gov/image-article/nasas-chandra-shares-a-new-view-of-our-galactic-neighbor/ [retrieved 12-12-25]

The Andromeda Galaxy, or M31, is the closest spiral galaxy to the Milky Way, at a distance of about 2.5 million light-years. Studying Andromeda helps astronomers understand the structure and evolution of the Milky Way. M31 has played an important role in many aspects of astrophysics, but particularly in the discovery of dark matter. In the 1960s, astronomer Vera Rubin and her colleagues studied M31 and determined that there was some unseen matter in the galaxy that was affecting how the galaxy and its spiral arms rotated. This unknown material was named "dark matter." Its nature remains one of the biggest open questions in astrophysics today, one that NASA's Nancy Grace Roman Space Telescope is designed to help answer.

BIOTICALLY MEDIATED SURFACE COOLING AND HABITABILITY FOR COMPLEX LIFE

David W. Schwartzman

Department of Biology
Howard University
Washington, DC 20059

Steven N. Shore

Department of Physics and Astronomy
Indiana University at South Bend
South Bend, IN 46634

Abstract

The width of the habitable zone for complex life may be substantially smaller than that for the appearance of biota, constrained by the presence of liquid water. The consequences of this largely ignored fact will be discussed in light of their implications to SETI. Biotic colonization of continents and resultant enhancement of weathering, first by thermophilic bacteria, led to lower steady-state atmospheric carbon dioxide levels and surface cooling on Earth. Evolution of prokaryotes and complex life on terrestrial planets is expected to have similar geochemical and climatic consequences. Biotically mediated cooling increases the width of the habitable zone for the possible occurrence and evolutionary time frame of complex life. For Earthlike planets within the CHZ of stars less massive than the Sun, we predict the earlier emergence of complex life, all other factors being the same.

Introduction

While Dole (1964) focused on planetary habitability for human life in his pioneering study, most discussions since have defined the habitable zone (HZ; CHZ is the continuously habitable zone) by the presence of liquid water alone (e.g., Kasting et al. 1993a, Kasting 1996). A liquid water HZ is compatible with terrestrial life-forms ranging from hyperthermophilic microbes with upper temperature limits for growth up to about 120°C (water is liquid under pressure) to algae growing in water films near 0°C in snow fields. Eukaryotes with mitochrondria, henceforth to be called "complex" life, with apologies to Lynn Margulis who has emphasized the complexity of prokaryotes, have an upper temperature limit for viable growth of about 60°C, a limit apparently determined by the thermolability of the mitochondrial membrane (Brock and Madigan 1991). Hence the width of the HZ for complex life on Earthlike planets could be significantly smaller than that for prokaryotes (since thermophilic prokaryotes do well above that temperature) assuming that ambient temperatures remain above 60°C for the lifetime of the biosphere. The relevance of such a conjecture is of course dependent on the cosmic replication of the general pattern of evolution of life found here on Earth. The consequence of this largely ignored fact will be discussed here in the context of the evolution of the Earth's biosphere, the integrated system comprising the biota and the space-time it inhabits and influences, namely the atmosphere, oceans and solid crust down several kilometers as Gold (1992) has recently emphasized. In particular, we argue that biotically mediated cooling of the surface of Earthlike planets is necessary for the origin and continued evolution of complex life. Temperatures on an abiotic planet would remain above the 60°C limit for complex life.

In this study, we argue that the statement "no life, no complex life" has a nontautological interpretation. We assume that the emergence of relatively simple life-forms is almost inevitable on Earthlike planets. Such organisms are remarkably robust. Complex life, however, requires a more restrictive set of physical conditions—in particular, lower temperatures. While any nonpanspermic origin of complex life plausibly requires the prior existence of "primitive" life, the former may additionally require the latter in a biospheric sense, that is, via biotically mediated cooling along the lines we are proposing. We will show that biotically mediated cooling may substantially alter the size of the habitable zone around stars for the emergence of complex life.

Biotically Mediated Cooling on Earth

Chemical weathering of CaMg silicates on land supplies bicarbonate, calcium and magnesium ions to the ocean where calcium/magnesium carbonate precipitation can occur. These combined reactions constitute a carbon "sink" with respect to the atmosphere/hydrosphere, with volcanic/metamorphic emissions being the source.

The net burial of organic carbon is also a sink, likely always less than the silicate weathering sink. The carbonate-silicate climatic stabilizer operates because of a negative feedback relation between surface temperature changes and atmospheric carbon dioxide levels via the positive relation between surface temperature and weathering rate. This stabilizer was first proposed by Walker et al. (1981). As a result, the surface temperature has been stabilized since the early Archean within the habitable limits.

In Walker et al.'s version of the climatic stabilizer the biota is incidental to its operation, with little effect on temperature (see Kasting et al. 1988). However, if the enhancement of chemical weathering by life is substantial, then the Earth's surface temperature has been controlled not by the inorganic geochemical cycle alone, but rather by the biosphere, within the boundary conditions imposed by plate tectonics and solar evolution (Schwartzman and Volk 1989, 1991a).

The enhancement of weathering by land biota has occurred throughout the history of life by the stabilization of soil under the protective cover of the first microbes (so-called cryptogamic soil) and later higher plants and their living and dead associations (rhizosphere, organic debris), as well as other biotic effects including soil carbonic and organic acid elevation (Schwartzman and Volk 1989, 1991a, 1992) (see Fig. 1). Soil creates a reservoir of high surface area mineral grains for a given land area, accelerating the removal of atmospheric carbon dioxide via its reaction as carbonic acid with CaMg silicates. On an abiotic land surface, water and wind erosion are likely to preempt significant accumulation of soil. In contrast, under biotic conditions, rapid erosion on high slopes leads to acceleration of chemical weathering since soils reform rapidly with biotic colonization and stabilization.

We have defined the biotic-enhancement-of-weathering factor (B) in a quantitative way as follows: B is how much faster the silicate weathering carbon sink is under biotic conditions than abiotic, at the same atmospheric carbon dioxide level (pCO_2) and surface temperature (Schwartzman and Volk 1989). There is a consensus that $B > 1$, if only for biotic elevation of pCO_2 in soils. Just how much bigger is still under contention. We have argued that B is probably greater than 100. Estimates to date have come mainly from field and laboratory studies of recent weathering. The classic study of Jackson and Keller (1970) indicated $B = > 10$ to > 100 for a lichen species growing on Hawaiian basalt, although this study has recently generated some controversy (Berner 1992, Jackson 1993, Schwartzman 1993, Cochran and Berner 1993). Cochran and Berner (1992) reported initial enhancements by vascular plants of one or more orders of magnitude. Berner's modeling (1993, 1994) of the Phanerozoic atmospheric carbon dioxide levels, partly confirmed by paleoatmospheric pCO_2 determinations, gives an effective enhancement of about seven times for vascular plants relative to prevascular plants, similar to a field estimate from a comparison of vegetated to unvegetated (lichenized?) portions of the same watershed in the Swiss Alps (Drever and Zobrist 1992). It is important to note here that B is defined relative to an abiotic rate. Since no purely abiotic weathering at present occurs on the Earth's surface, indirect means are necessary to set limits on this parameter.

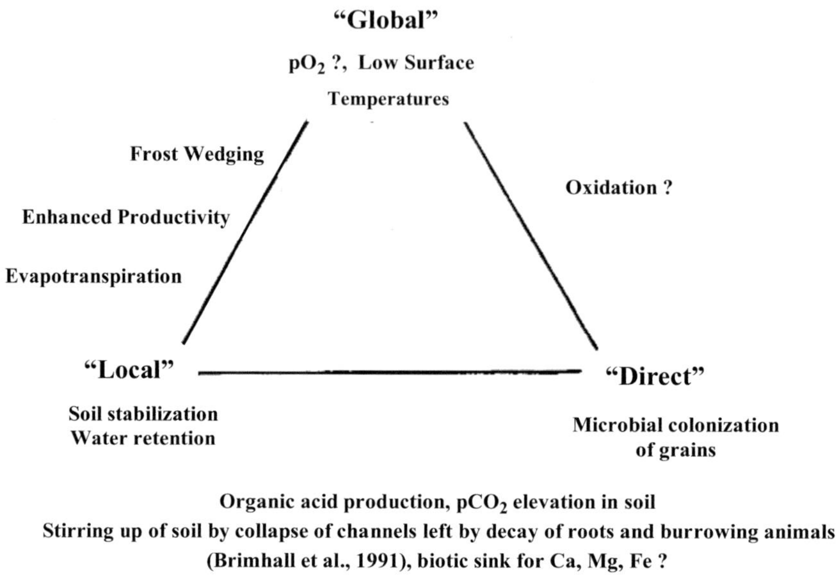

FIG. 1. *Biotic enhancement of weathering: summary of factors.*

One such estimate of B comes from the ratio (about 1000) of the rate of chemical weathering in a tropical soil to that of a computed abiotic two-dimensional bare rate at similar temperatures and runoff conditions (Schwartzman and Volk 1989), though several caveats are in order (e.g., the likely extent of soil cover under abiotic conditions). Another approach is the computation of B from the inversion of the surface temperature history of the Earth, a procedure followed for the prevascular plant surface in the Phanerozoic by Berner (1993, 1994).

We have attempted the same approach for the Precambrian (Schwartzman and Volk 1991a, Schwartzman and McMenamin 1993, Schwartzman et al. 1993, Schwartzman 1995). Our first attempt (Schwartzman and Volk 1991a) assumed the conventional scenario (C) for the temperature history, while the more recent attempt assumed a much warmer Precambrian than heretofore assumed (H) (see Fig. 2). Scenario H is consistent with the hypothesis that surface temperatures have been a critical constraint on microbial evolution, determining the timing of major innovations (see next section). This scenario is supported by a wide range of geochemical and geological evidence (see Schwartzman and McMenamin 1993, Schwartzman et al. 1993, Schwartzman 1995). In particular, the paleotemperature isotopic record of cherts (see review by Knauth 1992) coincides with scenario H (there are of course perturbations in the Phanerozoic; scenario H is a first approximation, ignoring these deviations from the trend). A necessary assumption for these temperature calculations, that the Precambrian oxygen isotopic composition of seawater was close to the present, received recent support from Holmden and Muehlenbacks (1993).

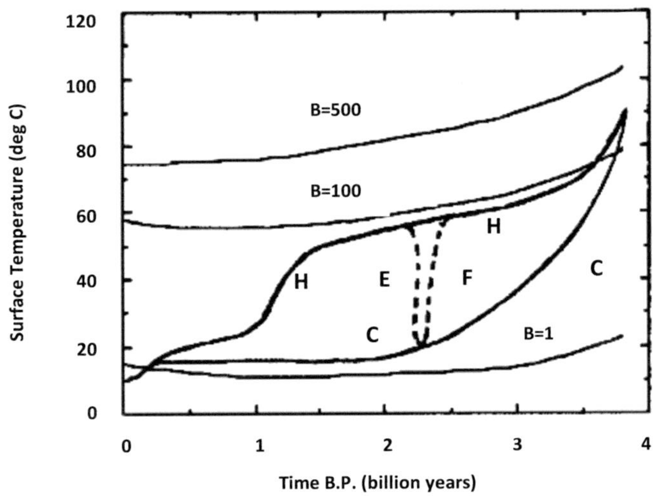

FIG. 2. *Earth surface temperature versus time BP (billion years). Two alternative scenarios are noted (H and C, with possible excursion E-F). Temperature trajectories for* B = *500, 100 and 1 are also indicated (for preferred model b; see Appendix at the end of this paper).*

Also shown in Figure 2 is a large excursion of temperature (E-F), required by a Huronian glaciation at about 2.3 Gyr. Such an excursion was suggested by Knauth and Epstein (1976). It is not necessary if the Huronian "tillites" are really the product of impacts, a recently suggested possibility (Oberbeck and Marshall 1992, Rampino 1992, Oberbeck et al. 1993). Unfortunately there is no isotopic paleotemperature data from these deposits. The plausibility of such a large excursion is subject to climatic modeling. We show the scenario H probably requires $B \geq 80$, just using the inversion of inferred temperatures prior to the Huronian event, provided carbon dioxide/water vapor were the sole greenhouse gases (Schwartzman and McMenamin 1993 and Schwartzman et al. 1993 give a limit of about 50, using a different greenhouse function than here). If post-Huronian temperatures in the early Proterozoic follow scenario H, then the lower limit on *B* is even stronger since a reduced-gas (e.g., methane) greenhouse is unlikely with the rise of atmospheric oxygen during this time. This limit on *B* is a minimum for biotic enhancement over abiotic conditions since some biotic acceleration of weathering is plausible even in the Archean, with partial thermophilic colonization of the land (Schwartzman and Volk 1989, 1991a); under such conditions *B* is equivalent to the ratio of the present biotic-enhancement-of-weathering factor to the factor at time t (see Appendix at the end of this paper).

A methane and/or ammonia greenhouse (Sagan 1977, Kasting et al. 1983, Zahnle 1986, Kasting and Grinspoon 1990, Sagan and Chyba 1996, abstract, this volume) has been proposed for the Archean, implying that atmospheric carbon dioxide achieved steady state by high surface temperature alone, or partially supplemented by pCO_2. The plausibility of a methane greenhouse during the Archean may be diminished if an additional sink with respect to the atmosphere aside from photochemical dissociation of methane (Kasting et al. 1983) were present. Perhaps both aerobic and anaerobic (utilizing sulfate reduction) methane-oxidizing microbes provided such a sink (Large 1983, Strauss et al. 1992). Aerobic microenvironments probably existed in the Archean, at least by 3.5 Gyr if cyanobacteria did indeed emerge then, as recent evidence supports. The case for an ammonia greenhouse may be weakened if the pH of the Archean ocean was lower than the required minimum for the required production rate (pH = 7.3; see Summers and Chang 1993). High pCO_2 during the Archean is consistent with just such low pH model oceans (Grotzinger and Kasting 1993). Finally, a high-pCO_2 pressure-cooker atmosphere at 3.8 Gyr, inherited from the late bombardment, at steady state with modest continents, requires B to be on the order of 100 or more (Schwartzman and Volk 1991a). A steady-state CO_2 greenhouse in the Archean, with the proposed high temperatures, provides a solution to the faint young Sun paradox that is consistent with our best estimates of continental area and volcanic outgassing rates, without the need for other greenhouse gases. In our view, other arguments for requiring a reducing atmosphere (i.e., necessary condition for origin of life, mechanism for liquid water on early Mars) have plausible alternatives.

Several abiotic factors, namely impact-derived soil (regolith) and greater volcanic activity (volcanics weather faster than other silicate rocks), would have favored a higher weathering intensity, relative to the present, and lower temperatures, strengthening the requirement for a high present biotic enhancement of weathering to keep temperatures high in the Archean.

Also plotted in Figure 2 are the temperature curves for B values of 500, 100 and 1. These trajectories were computed from our model of the steady-state atmospheric pCO_2 determined by the balancing of the volcanic source and weathering sink (see Appendix for details). For $B > 160$, the surface temperature remains above 60°C on an abiotic Earth; that is, such an Earth is habitable for thermophiles but not "complex" life (similar to conclusions we reached in Schwartzman and Volk 1989). We think B could very well be above this lower limit. Further, if B is close to one, a possibility we think unlikely, then biotic cooling is not necessary for the emergence and evolution of complex life; the original abiotic silicate-carbonate climatic stabilizer will do the job.

THE EVOLUTION OF THE BIOSPHERE AND ITS BIOTA

The upper temperature limit for viable growth has been determined for the main organismal groups (Table 1). This limit is apparently determined by the thermolability of biomolecules (nucleic acids, proteins), organellar membranes, and enzyme systems (Brock and Madigan 1991). For example, the mitochondrial membrane is particularly thermolabile, apparently resulting in an upper temperature limit of 60°C for aerobic eukaryotes. Could the upper temperature limit of 50°C for Metazoa be linked to the thermolability of proteins essential to blastula formation?

TABLE 1. ***Upper Temperature Limits for Growth of Living Organisms, Times of Their Emergence***

Group	*Approximate Upper Temperature (°C)*	*Time of Emergence (Gyr)*
Plants	45–50	~ 0.5
Animals (Metazoa)	50	1–1.5
Eukaryotic Microbes	60	2.1–2.8
Prokaryotic Microbes		
Cyanobacteria	70–73	3.5
Methanogens	> 100	~ 3.8
Extreme Thermophiles	> 100	~ 3.8

Temperatures from Brock and Madigan 1991.

We have proposed that surface temperatures have been the critical constraint on microbial evolution (Schwartzman et al. 1993), determining the timing of major innovation (e.g., the emergence of cyanobacteria at 70°C, its upper temperature limit for growth, at 3.5 Gyr). A high carbon dioxide "pressure-cooker" atmosphere, inherited from the vigorous degassing and cometary impacts on the early Earth, is assumed, with surface temperatures close to 100°C (Kasting and Ackerman 1986, Chyba et al. 1990, Delsemme 1992). Soon after the origin of life, extreme thermophiles colonized the land, enhancing weathering rates, sequestering carbon dioxide into limestone deposits and thereby cooling the Earth's surface (Schwartzman and Volk 1989, 1991a). Temperature is constrained by the upper temperature limits for the assumed times of emergence of cyanobacteria (70°C, at 3.5 Gyr), eukaryotes (60°C, at about 2.6 Gyr), and Metazoa (50°C, at 1.0–1.5 Gyr). Surface cooling is determined by the evolution of the biosphere, constrained by abiotic boundary conditions (i.e., luminosity of the Sun, continental area, outgassing rate).

Each new innovation in the microbial soil community resulted in greater biotic enhancement of weathering, culminating in the rhizosphere of higher plants (Schwartzman et al. 1993). This scenario for surface temperature as a function of geologic age is shown in Figure 3. Second-order perturbations arising from such factors as continental drift, episodic burial of organic carbon, and pulses of intense vulcanism, are ignored (particularly relevant in the Phanerozoic, when atmospheric pCO_2 levels drop). The temperature curve in Figure 3 is slightly modified from our previous papers (Schwartzman and McMenamin 1993, Schwartzman et al. 1993). An earlier cooling in the mid-Proterozoic is indicated, consistent with the apparent emergence of Metazoa at 1.0 to 1.5 Gyr (Chapman 1992, Morris 1993) and paleotemperatures of 25–43°C on 1.1 to 1.2 Gyr cherts (Kenny and Knauth 1992). This cooling may have been the result of the rise of atmospheric oxygen and more extensive colonization of the land surface by algal mats (Beeunas and Knauth 1985, Horodyski and Knauth 1994). While the first definitive fossil evidence for Metazoa dates at 0.65 Gyr, an older problematic record does exist (e.g., Robbins et al. 1985). The story of the earliest Metazoa may be analogous to that of Eucarya (Sogin et al. 1989), in the unlikely preservation of microscopic soft-bodied organisms as fossils.

A geophysiological model for the evolution of the biosphere, summarizing the above discussion, is shown in Figure 4. The evolution of the biosphere does not optimize conditions for existing biota, unlike the original Gaia hypothesis. At least two catastrophes for existing life occurred: the well-known oxygen catastrophe and an earlier temperature catastrophe for thermophilic bacteria, once the likely colonizers of the ocean and land, now restricted to hot springs, hydrothermal vents on the ocean floor, porosity in the first few kilometers of the crust, and hot water heaters.

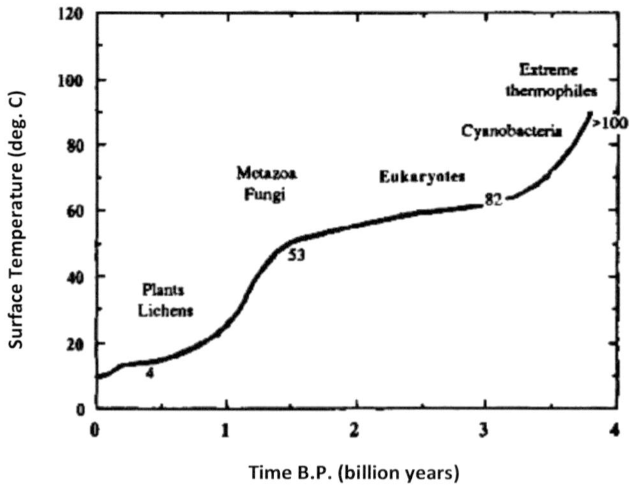

FIG. 3. *Proposed surface temperature history of Earth, with key evolutionary developments noted. Numbers near curves are* B *values for preferred model b.*

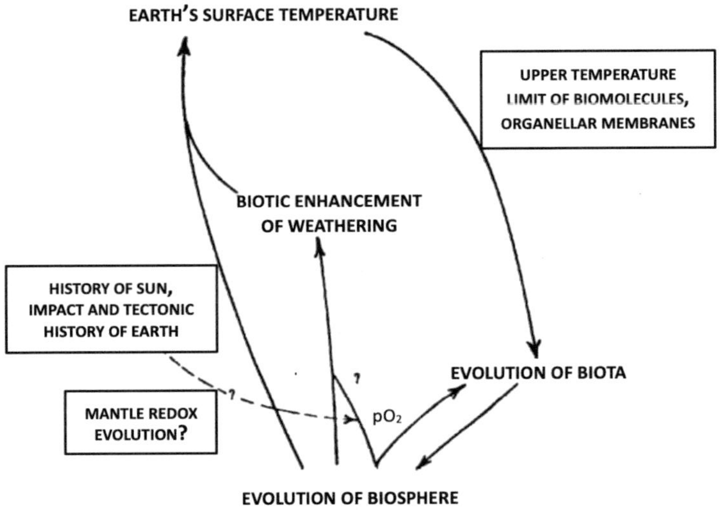

FIG. 4. *Geophysiological model of evolution of biosphere. Constraints on feedback are in boxes. A mantle buffering of atmospheric pCO_2 is suggested by Kasting et al. 1993b.*

The possible variables affecting the timing of emergence and evolution of complex life include surface temperatures and the oxygen level in the surface environment (pO_2). If pO_2 in surface microenvironments is limiting to the emergence of aerobic eukaryotes and Metazoa, then the critical level may be reached earlier than the upper temperature limit (60°C and 50°C, respectively). Thus, aerobic microenvironments in the Archean may well have preceded the emergence of aerobic eukaryotes, which in turn preceded their first likely presence in the fossil record.

Des Marais et al. (1992) and Des Marais (1996) have proposed that the rise in atmospheric pO_2 near the time of the Huronian event was triggered by organic carbon burial, a consequence of rifting and orogeny. This scenario combined with our own could plausibly account for the timing of the emergence of megascopic eukaryotes (Han and Runnegar 1992) followed by Metazoa some 1–1.5 Gyr ago. Both Des Marais et al. (1992) and Kasting et al. (1993b) have proposed that the tectonic history of the Earth (plate tectonics) has codetermined with oxygenic photosynthesis the pO_2 at the surface (shown as feedbacks in Fig. 4). We agree with Des Marais' conclusion (1996) that the biota's most prominent role has been to affect the oxidation state and the temperature of the Earth's surface.

FURTHER IMPLICATIONS TO THE CHZ FOR COMPLEX LIFE

The inner boundary of the CHZ for complex life is determined by a solar flux relative to the present, S_{eff}, equal to 1.13, corresponding to a surface temperature (T_s) of 60°C at pCO_2 = 0 (Caldeira and Kasting 1992), since this is the limit to biotically mediated cooling (except perhaps cooling arising from a Lovelockian biotic effect on the planetary albedo). Thus the inner boundary of the CHZ is close to that for water loss (S_{eff} = 1.10; Kasting et al. 1993a).

We show the results of an Earth/Sun system in Figure 5a, computed assuming our preferred model b for volcanic outgassing and land area as a function of time (see Appendix). This CHZ is limited by the boundaries for water loss and first CO_2 condensation (Kasting et al. 1993a, Kasting 1996). Note that t_c, the time (Gyr) since the emergence of complex life ($T_s \leq 60°C$), varies inversely with the assumed *B* for a given distance of the planet from the Sun (*D* in AU) (see Fig. 5b). Further, as *D* increases it becomes easier to cool the planet (t_c increases) for a given *B*. *B* is here defined as the biotic enhancement factor operating on the present Earth.

Figure 6 illustrates the relation between *D* and t_c for given assumed *B* values. For $t_c \geq 2$ Gyr, that is, eukaryotes emerged at least 2 Gyr ago, and $B > 160$, $D > 1.15$, the conservative outer limit of the CHZ; that is, on an abiotic Earthlike planet around a Sunlike star $T_s > 60°C$ over the whole CHZ.

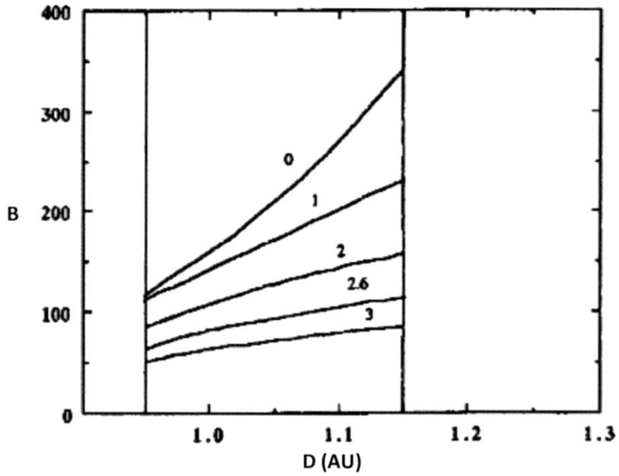

FIG. 5a. *Biotic enhancement factor versus* D, *for an Earth-Sun model, with the distance of the Earth from the Sun in AU (preferred model b results). Numbers on curves are* t_c, *the time (Gyr) since the emergence of complex life* ($T_S \leq 60°C$). *The inner* (D = *0.95) and conservative outer* (D = *1.15) boundaries of the liquid water CHZ are shown.*

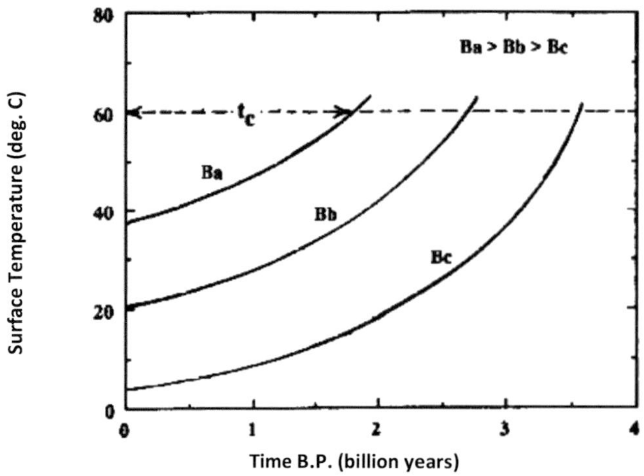

FIG. 5b. *Surface temperature versus time for a given* D. *The Bs are biotic enhancement factors;* t_c *is indicated.*

FIG. 6. D *versus* t_c *for an Earth-Sun model; numbers on curves are* B *values (preferred model b results). The inner* (D = *0.95) and conservative outer* (D = *1.15) boundaries of the liquid water CHZ are shown.*

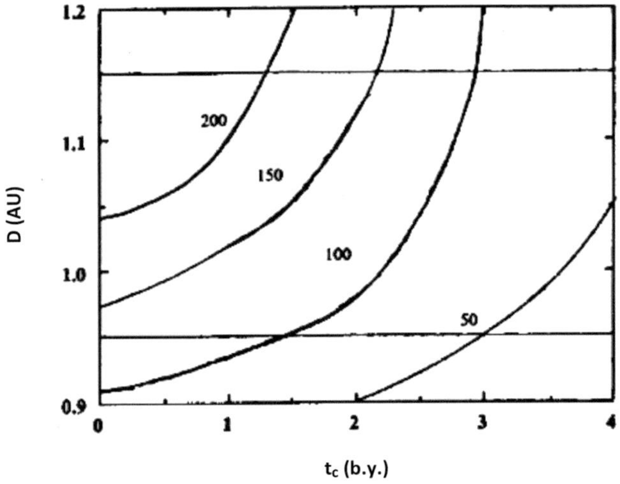

We now consider the application of our model of biotically mediated cooling to other stars. We have suggested that biotically mediated surface cooling may have significantly enhanced the possibility of the evolution of complex life and intelligence on Earthlike planets around Sunlike stars (Schwartzman and Volk 1991b, Schwartzman 1995). If surface temperature is indeed a critical constraint on the emergence of complex life, then the time needed for its appearance dated from the origin of life (Δt_c) may be shorter on biospheres of terrestrial planets within the habitable zone that cool more quickly than the Earth (conversely for slower cooling). In fact, the assumption of mediocrity (Shklovskii and Sagan 1966) may be misleading as a guide to the evolution of alien biospheres. Complex life and intelligence may be more or less common in the Galaxy than deduced by simply extrapolating the time frame of evolutionary change on the Earth (McKay 1996). However, a better understanding of the evolution of our own biosphere could inform a predictive theory of alien biospheres. For terrestrial planets, Δt_c and Δt_i (time needed for intelligence to emerge) may prove to be an explicit function of planetary mass, distance from its star, and stellar mass.

In particular, for an Earthlike planet within a liquid water CHZ scaled to our Sun (Kasting et al. 1993a) we would predict the following relations, assuming all other variables are held constant:

$$M_{\text{star}} > M_{\text{Sun}};\ \Delta t_{c,i\ \text{star}} > \Delta t_{c,i\ \text{Sun}}$$

$$M_{\text{star}} < M_{\text{Sun}};\ \Delta t_{c,i\ \text{star}} < \Delta t_{c,i\ \text{Sun}}$$

(M is the mass of the star.)

These results follow from the rate of luminosity increase with time, dL/dt, relative to that of the Sun; biotically mediated cooling must "fight" dL/dt. For an

Earthlike planet in the CHZ around an F0 star, there is "double trouble" because of the star's short lifetime on the main sequence (about 3 Gyr). These results suggest that complex life should emerge and evolve faster around stars less massive than and of course more numerous than Sunlike stars in the Galaxy. This news will be welcomed by the SETI optimist camp. However, differences in other variables from Earth's unique history may make Earthlike planets around, say, K or M stars less likely to harbor complex life (see Table 2).

TABLE 2. ***Some of the Possible Variables Affecting the Time Required for the Emergence of Complex Life, Intelligence ($\Delta t_{c,i}$) on Earthlike Planets Around Other Stars****

Variable (change from past to present time, relative to Earth-Sun)	$\Delta t_{c,i}$
Luminosity ↑	↑ $M_{star} > M_{Sun}$
Volcanic Outgassing Rate ↓	↓ No plate tectonics, lower K, U
Continental Land Area Growth Rate ↑	↓ Smaller initial oceans
Biotic Enhancement of Weathering ↓	↑ Different evolutionary pattern
Impact Size, Frequency ↓	↓ Supply of greenhouse gases drops

* *Scaled to the liquid water CHZ of the Earth-Sun system.*

See Boston and Thompson (1991) for an interesting discussion of possible scenarios for biospheric evolution.

IMPLICATIONS TO EVOLUTION ON MARS

In a final note of wild speculation, we predict it will be found that complex life emerged earlier on Mars than on Earth, since the Martian surface cooled faster than that of Earth. This prediction does not require that biotically mediated cooling did occur on Mars, or even that Mars was in the habitable zone defined by persistent liquid water at its surface with temperatures necessarily above its freezing point (Kasting et al.'s 1993a HZ) in its early history, an issue still under debate. We assume only that the origin of life did occur on Mars, at the same time as on Earth, that habitable environments did exist (e.g., in ice-covered lakes [McKay and Davis 1991] or in hot springs [Walter and Des Marais 1993]), and that life was not aborted in, say, the first billion years by, for example, impacts. An extant biostratigraphic record in the Martian crust will test this admittedly outrageous hypothesis.

ACKNOWLEDGMENTS

Ken Caldeira provided his and Jim Kasting's updated version of the Kasting and Ackerman greenhouse function, as well as suggested the significance of the S = 1.13 limit for biotically mediated cooling. As always, Jim Kasting patiently explained his papers and clarified many issues for us. We thank Laurance Doyle for inviting our participation in this exciting conference.

APPENDIX

Model Description

TABLE A1. ***Parameters for Assumed Variation of Land Area (A) and Carbon Dioxide Outgassing Rate (V) as Function of Age (t)***

Model	$(V/V_0)(A_0/A)$ *at* t = *3.8 Gyr*	cV_0	ω
a ("constant")	1 × 1 = 1	–	0
b ("preferred")	3 × 4 = 12	0.375	0.289
c ("upper limit")	8 × 10 = 80	0.129	0.547

A_0/A and V_0/V are the present and past continental land areas and volcanic outgassing rates (of carbon dioxide), respectively (see Figs. A1 and A2). We assume the following functions for volcanic outgassing, V, and land area, A, variation:

$$V = V_o\, e^{(\omega t)} \tag{A1}$$

V parallels the decrease in radioactive heat generation in the Earth to present where t is time (BP) in billions of years. We couple the rates of continental land area growth and volcanic outgassing:

$$(dA/dt)_t = -c(dV/dt)_t \tag{A2}$$

Our rationale for assuming linkage of land area and outgassing rate is that both outgassing of carbon dioxide and continental crust generation are presumably a function of subduction and juvenile outgassing rates. Further, intracontinental growth via underplating results in both outgassing and land area increases through uplift. This parameterization also simplifies model calculations. In any case, alternative parameterizations of land area as a function of time consistent with the isotopic evolution of the crust/mantle system give similar model results for computed surface temperatures as a function of time.

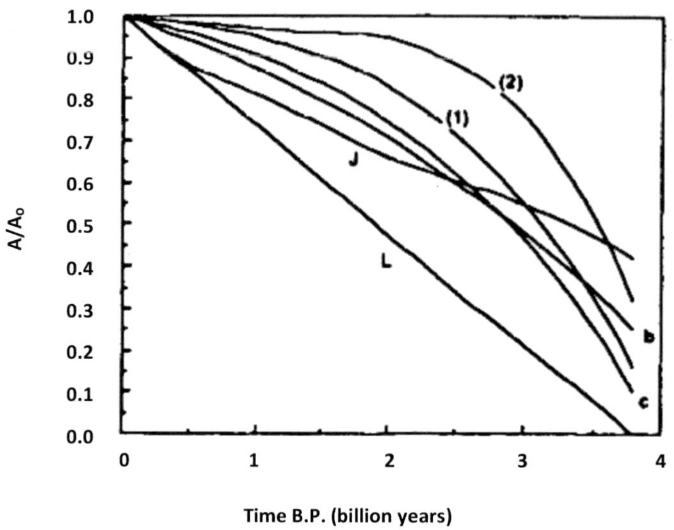

FIG. A1. *Growth of continental land area (mass) versus time;* A/A_o *is the ratio of land area at time* t *to present. Curves b and c correspond to our models b and c, respectively, described in text. Curves (1) and (2) are from limiting tectonic/geochemical models of mean age of sediments and continents of Allegre and Jaupart (1985), curve J is from Jacobsen's (1988) inversion of Sm-Nd mass balance for the depleted mantle–continental crust system. Curve L, the linear growth of continental area through time, with* A = 0 *at 3.8 Gyr, is taken as a lower limit to* A/A_o *as a function of age. Substantial very early growth of the continents has received recent support from Bowring et al. (1993).*

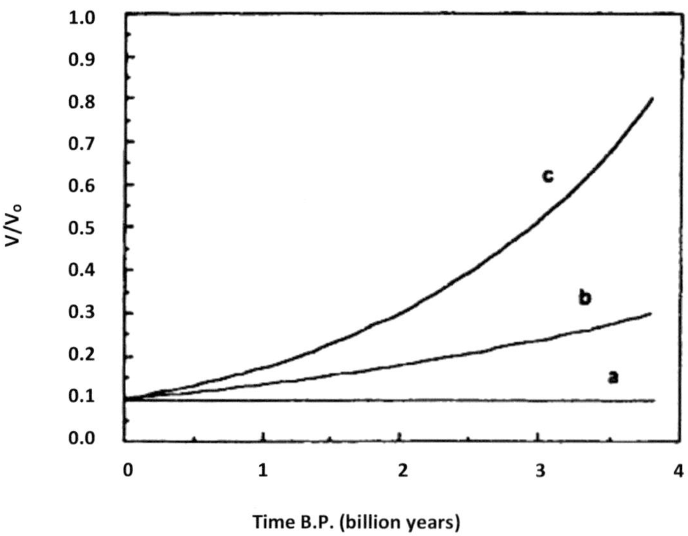

FIG. A2. *Carbon outgassing rate relative to present rate* (V/V_o) *versus time (BP) for models a, b, and c.*

Let $B_o = W_o/W_{\text{abiotic}}$ where W_o is the present biotically enhanced chemical weathering rate and W_{abiotic} is the hypothetical abiotic rate at the same atmospheric carbon dioxide level and surface temperature. Then

$$B_o/\{(A/A_o)(V_o/V)\} = (P_{ab}/P_o)^{\alpha} e^{(\beta\Delta T)} e^{(\gamma\Delta T)}. \qquad \text{(A3)}$$

This formulation adds the biotic enhancement term B_o to the carbonate-silicate model developed by Walker et al. (1981) and assumes a steady state between carbon sinks and sources with respect to the atmosphere (Schwartzman and Volk 1989). P_o is the present atmospheric partial pressure of CO_2, P_{ab} the partial pressure of CO_2 required under abiotic conditions for steady state, ΔT is the temperature elevation required for an abiotic condition over the present global mean, T_o, taken as 288°K ($\Delta T = T_{ab} - T_o$; T_{ab} is the abiotic surface temperature). The factors α, β, and γ express the dependence of silicate weathering rate on pCO_2, temperature, and run-off, respectively; assumed $\beta = 0.056$ and $\gamma = 0.017$. Note that $e^{(\gamma\Delta T)} \leq 2$ based on energy considerations (Pollack et al. 1987). The following expression can be derived for a biotic Earth surface:

$$B = (A/A_o)(V_o/V)(P_t / P_o)^{\alpha} e^{\beta(T_t - T_o)} e^{\gamma(T_t - T_o)} \tag{A4}$$

where $B = B_o / B_t$. Recall from the definition of B that $B_o = W_o/W_{\text{abiotic}}$, $B_t = W_t/W_{\text{abiotic}}$. The variation of land area and outgassing rate as a function of time follows from models a, b, and c given above. The relation between surface temperature T_t and P_t, pCO_2 at any time t, was computed from two different greenhouse functions.

(1) For $P_t > 0.03$ bar, an updated version of Kasting and Ackerman's (1986) function, given in equation form by Caldeira (pers. comm.), with $T_t = f(P_t, S_t)$, where S_t is the relative solar flux at time t:

$$\begin{aligned} T_t = {} & 138.114 - 73.179(p) - 73.960(p^2) + 56.048(p^3) + 405.836(S_t) \\ & + 595.774(p)(S_t) + 385.004(p^2)(S_t) - 296.420(p^3)(S_t) - 316.907(S_t^2) \\ & - 839.205(p)(S_t^2) - 548.962(p^2)(S_t^2) + 461.125(p^3)(S_t^2) \\ & + 129.545(S_t^3) + 345.867(p)(S_t^3) + 251.4629(p^2)(S_t^3) \\ & - 216.438(p^3)(S_t^3), \end{aligned} \tag{A5}$$

where $p = \log_{10} (P_t/1 \text{ bar})$

The maximum error is 1.95 K, the rms error is 0.65 K.

$$S_t = (1 + 0.0835t)^{-1} D^{-2}, \tag{A6}$$

where D is the distance of the Earth to the Sun (AU).

(2) For $0.03 > P_t \geq 0.0003$ bar, a slightly modified function from Walker et al. (1981) ($T_o = 288$ K rather than 285 K, present $T_e = 255$ K rather than 253 K):

$$T = 2\,T_e + 4.6\,(P_t/P_o)^{0.364} - 226.4, \tag{A7}$$

where T_e is the effective radiating temperature of the Earth (K) (no greenhouse effect) at time t.

We used the following relation for T_e from Kasting (1987):

$$T_e = 255/(1 + 0.087\,t\,)^{0.25} \tag{A8}$$

Sensitivity of Scenario H to Model Parameters

TABLE A2

T°C	Atmos pCO_2 (bars)	Time BP (Gyr)	Required B			
			Model	a	b	c
70	2.56	3.5		658	82	25
60	1.22	2.6		301	82	43
50	0.31	1.5		103	53	38

We assume $\alpha = 0.3$, corresponding to abiotic weathering of CaMg silicates in an open system (see Schwartzman and Volk 1989, Berner 1992, 1993); in a closed system $\alpha > 0.3$. The choice of α, β (probable minimum), greenhouse function (pCO_2 is probable minimum because cloud feedbacks are assumed absent) all make the computed B values minima. Probable higher-impact-derived regolith, volcanic contribution to weathering and the postulated seafloor weathering carbon sink (François and Walker 1992) relative to present have the same effect on the limit to B. Two possible factors would make the computed B values maxima—reducing gas greenhouses in the Archean (see previous discussion) and greater organic carbon burial now relative to the past (Des Marais et al. 1992). The latter would reduce the required B values by no less than a factor of 0.6, assuming a present ratio of the silicate weathering to organic carbon sink of 1.3 (Berner 1994) and no Archean organic carbon sink. If assumed model ratios of (V_o/V) and (A/A_o) are too low, as a result of the neglect of degassing from subduction of Cenozoic deposits of pelagic $CaCO_3$ (Caldeira 1992) and early continental crust formation (Bowring et al. 1993), respectively, then computed B values are again minima. Subduction of exogenic carbon into the mantle and its reappearance in volcanic outgassing (Hauri et al. 1993) has probably decreased the variation of (V_o/V) ratios over time.

REFERENCES

ALLEGRE, C.J., AND C. JAUPART (1985), "Continental Tectonics and Continental Kinetics," *Earth Planet. Sci. Lett.* **74:** 171–186.

BEEUNAS, M.A., AND L.P. KNAUTH (1985), "Preserved Stable Isotopic Signature of Subaerial Diagenesis in the 1.2-b.y. Mescal Limestone, Central Arizona: Implications for the Timing and Development of a Terrestrial Plant Cover," *Geol. Soc. Am. Bull.* **96:** 737–745.

BERNER, R.A. (1992), "Weathering, Plants, and the Long-Term Carbon Cycle," *Geochim. Cosmochim. Acta* **56:** 3225–3232.

BERNER, R.A. (1993), "Paleozoic Atmospheric CO_2: Importance of Solar Radiation and Plant Evolution," *Science* **261:** 68–70.

BERNER, R.A. (1994), "3GEOCARB II: A Revised Model of Atmospheric CO_2 over Phanerozoic Time," *Amer. J. Sci.* **294:** 56–91.

BOSTON, P.J., AND S.L. THOMPSON (1991), "Theoretical Microbial and Vegetation Control of Planetary Environments," in *Scientists on Gaia*, S.H. Schneider and P.J. Boston, eds. (Cambridge: MIT Press), pp. 99–117.

BOWRING, S.A., T. HOUSH, AND C. ISACHSEN (1993), "Nd Isotope Systematics of the Early Archean Acasta Gneisses: Implications for Models of the Early Earth," abstract, Annual Meeting of the Geological Society of America, p. A73.

BRIMHALL, G.H., O.A. CHADWICK, C.J. LEWIS, W. COMPSTON, I.S. WILLIAMS, K.J. DANTI, W.E. DIETRICH, M.E. POWER, D. HENDRICKS, AND J. BRATT (1991), "Deformational Mass Transport and Invasive Process in Soil Evolution," *Science* **255:** 695–702.

BROCK, T.D., AND M.T. MADIGAN (1991), *Biology of Microorganisms*, 6th ed. (Englewood Cliffs, NJ: Prentice-Hall).

CALDEIRA, K. (1992), "Enhanced Cenozoic Chemical Weathering and the Subduction of Pelagic Carbonate," *Nature* **357:** 578–581.

CALDEIRA, K., AND J.F. KASTING (1992), "The Life Span of the Biosphere Revisited," *Nature* **360:** 721–723.

CHAPMAN, D.J. (1992), "Origin and Divergence of Protists," in *The Proterozoic Biosphere*, J.W. Schopf and C. Klein, eds. (Cambridge: Cambridge University Press), pp. 477–483.

CHYBA, C.F., P.J. THOMAS, L. BROOKSHAW, AND C. SAGAN (1990), "Cometary Delivery of Organic Molecules to the Early Earth," *Science* **249:** 366–373.

COCHRAN, M.F., AND R.A. BERNER (1992), "Initial Effects of Vegetation on Hawaiian Basalt Weathering Rates," abstract, Annual Meeting of the Geological Society of America, p. A169.

COCHRAN, M.F., AND R.A. BERNER (1993), reply to the comments on "Weathering, Plants, and the Long-Term Carbon Cycle," *Geochim. Cosmochim. Acta* **57:** 2147–2148.

DELSEMME, A.H. (1992), "Cometary Origin of Carbon, Nitrogen and Water on the Earth," *Origins of Life* **21:** 279–298.

DES MARAIS, D.J. (1996), "Has Life Made Its Global Environment More Habitable?," this volume.

DES MARAIS, D.J., H. STRAUSS, R.E. SUMMONS, AND J.M. HAYES (1992), "Carbon Isotope Evidence for the Stepwise Oxidation of the Proterozoic Environment," *Nature* **359:** 605–609.

DOLE, S.H. (1964), *Habitable Planets for Man* (New York: Blaisdell Publishing Company).

DREVER, J.I., AND J. ZOBRIST (1992), "Chemical Weathering of Silicate Rocks as a Function of Elevation in the Southern Swiss Alps," *Geochim. Cosmochim. Acta* **56:** 3209–3216.

FRANÇOIS, L.M., AND J.C.G. WALKER (1992), "Modeling the Phanerozoic Carbon Cycle and Climate: Constraints from the $^{87}Sr/^{86}Sr$ Isotopic Ratio of Seawater," *Amer. J. Sci.* **292:** 81–135.

GOLD, T. (1992), "The Deep, Hot Biosphere," *Proc. Natl. Acad. Sci. USA* **89:** 6045–6049.

GROTZINGER, J.P., AND J.F. KASTING (1993), "New Constraints on Precambrian Ocean Composition," *J. Geol.* **101:** 235–243.

HAN, T.-M., AND B. RUNNEGAR (1992), "Megascopic Eukaryotic Algae from the 2.1-Billion-Year-Old Negaunee Iron-Formation, Michigan," *Science* **257:** 232–235.

HAURI, E.H., N. SHIMIZU, J.J. DIEU, AND S.R. HART (1993), "Evidence for Hotspot-Related Carbonatite Metasomatism in the Oceanic Upper Mantle," *Nature* **365:** 221–227.

HOLMDEN, C., AND K. MUEHLENBACKS (1993), "The $^{18}O/^{16}O$ Ratio of 2-Billion-Year-Old Seawater Inferred from Ancient Oceanic Crust," *Science* **259:** 1733–1736.

HORODYSKI, R.J., AND L.P. KNAUTH (1994), "Life on Land in the Precambrian," *Science* **263:** 494–498.

JACKSON, T.A. (1993), comment on "Weathering, Plants, and the Long-Term Carbon Cycle," by Robert A. Berner, *Geochim. Cosmochim. Acta* **57:** 2141–2144.

JACKSON, T.A., AND W.D. KELLER (1970), "A Comparative Study of the Role of Lichens and 'Inorganic' Processes in the Chemical Weathering of Recent Hawaiian Lava Flows," *Amer. J. Sci.* **269:** 446–466.

JACOBSEN, S.B. (1988), "Isotopic and Chemical Constraints on Mantle-Crust Evolution," *Geochim. Cosmochim. Acta* **52:** 1341–1350.

KASTING, J.F. (1987), "Theoretical Constraints on Oxygen and Carbon Dioxide Concentrations in the Precambrian Atmosphere," *Precambrian Res.* **34:** 205–229.

KASTING, J.F. (1996), "Habitable Zones Around Stars: An Update," this volume.

KASTING, J.F., AND T.P. ACKERMAN (1986), "Climatic Consequences of Very High Carbon Dioxide Levels in the Earth's Early Atmosphere," *Science* **234:** 1383–1385.

KASTING, J.F., AND D.H. GRINSPOON (1990), "The Faint Young Sun Problem," in *The Sun in Time*, C.P. Sonnett, M.S. Giampapa, and M.S. Matthews, eds. (Tucson: University of Arizona Press), pp. 447–462.

KASTING, J.F., K.J. ZAHNLE, AND J.C.G. WALKER (1983), "Photochemistry of Methane in the Earth's Early Atmosphere," *Precambrian Res.* **20:** 121–148.

KASTING, J.F., O.B. TOON, AND J.B. POLLACK (1988), "How Climate Evolved on the Terrestrial Planets," *Sci. Am.* **258 (2):** 90–97.

KASTING, J.F., D.P. WHITMIRE, AND R.T. REYNOLDS (1993a), "Habitable Zones Around Main Sequence Stars," *Icarus* **101:** 108–128.

KASTING, J.F., D.H. EGGLER, AND S.P. RAEBURN (1993b), "Mantle Redox Evolution and the Oxidation State of the Archean Atmosphere," *J. Geol.* **101:** 245–257.

KENNY, R., AND L.P. KNAUTH (1992), "Continental Paleoclimates from δD and $\delta^{18}O$ of Secondary Silica in Paleokarst Chert Lags," *Geology* **20:** 219–222.

KNAUTH, L.P. (1992), "Origin and Diagenesis of Cherts: An Isotopic Perspective," in *Isotopic Signatures and Sedimentary Records, Lecture Notes in Earth Sciences #43*, N. Clauer and S. Chaudhuri, eds. (Berlin: Springer-Verlag), pp. 123–152.

KNAUTH, L.P., AND S. EPSTEIN (1976), "Hydrogen and Oxygen Isotope Ratios in Nodular and Bedded Cherts," *Geochim. Cosmochim. Acta* **40:** 1095–1108.

LARGE, P.J. (1983), *Methylotrophy and Methanogenesis* (Washington, DC: American Society for Microbiology).

MCKAY, C.P. (1996), "Time for Intelligence on Other Planets," this volume.

MCKAY, C.P., AND W.L. DAVIS (1991), "Duration of Liquid Water Habitats on Early Mars," *Icarus* **90:** 214–221.

MORRIS, S.C. (1993), "The Fossil Record and the Early Evolution of the Metazoa," *Nature* **361:** 219–225.

OBERBECK, V.R., AND J.R. MARSHALL (1992), "Impacts, Flood Basalts, and Continental Breakup," abstract, *Lunar and Planetary Science Proceedings, March 23, 1992*, p. 113.

OBERBECK, V.R., J.R. MARSHALL, AND H. AGGARWAL (1993), "Impacts, Tillites, and the Breakup of Gondwanaland," *J. Geol.* **101:** 1–19.

POLLACK, J.B., J.F. KASTING, S.M. RICHARDSON, AND K. POLIAKOFF (1987), "The Case for a Wet, Warm Climate on Early Mars," *Icarus* **71:** 203–224.

RAMPINO, M.R. (1992), "Ancient 'Glacial' Deposits Are Ejecta of Large Impacts: The Ice Age Paradox Explained," abstract, *EOS Transactions*, American Geophysical Union, vol. 73, no. 43, October 27, 1992, Supplement A32C-1, p. 99.

ROBBINS, E.I., K.G. PORTER, AND K.A. HABERYAN (1985), "Pellet Microfossils: Possible Evidence for Metazoan Life in Early Proterozoic Time," *Proc. Natl. Acad. Sci. USA* **82:** 5809–5813.

SAGAN, C. (1977), "Reducing Greenhouses and the Temperature History of Earth and Mars," *Nature* **269:** 224–226.

SAGAN, C., AND C. CHYBA (1996), "Organic Shielding of UV-Labile Greenhouse Gases on Early Earth and Mars," abstract, this volume.

SCHWARTZMAN, D. (1993), comment on "Weathering, Plants, and the Long-Term Carbon Cycle," by Robert A. Berner, *Geochim. Cosmochim. Acta* **57:** 2145–2146.

SCHWARTZMAN, D. (1995), "Temperature and the Evolution of the Earth's Biosphere," in *Progress in the Search for Extraterrestrial Life*, ASP Conf. Ser., S. Shostak, ed. (San Francisco: Astronomical Society of the Pacific), **74:** 152–161.

SCHWARTZMAN, D.W., AND T. VOLK (1989), "Biotic Enhancement of Weathering and the Habitability of Earth," *Nature* **340:** 457–460.

SCHWARTZMAN, D., AND T. VOLK (1991a), "Biotic Enhancement of Weathering and Surface Temperatures on Earth Since the Origin of Life," *Palaeogeogr., Palaeoclimatol., Palaeoecol. (Global Planet. Change Sect.)* **90:** 357–371.

SCHWARTZMAN, D., AND T. VOLK (1991b), "Geophysiology and Habitable Zones Around Sun-Like Stars," in *Bioastronomy, Proceedings, Val Cenis, France, 1990, Lecture Notes in Physics 390*, J. Heidmann and M.J. Klein, eds. (Berlin: Springer-Verlag), pp. 155–162.

SCHWARTZMAN, D., AND T. VOLK (1992), "Biotic Enhancement of Earth Habitability," in *Encyclopedia of Earth System Science*, vol. 1, W.A. Nierenberg, ed. (New York: Academic Press), pp. 387–394.

SCHWARTZMAN, D., AND M. MCMENAMIN (1993), "A Much Warmer Earth Surface for Most of Geologic Time: Implications to Biotic Weathering," *Chem. Geol.* **107:** 221–223.

SCHWARTZMAN, D., M. MCMENAMIN, AND T. VOLK (1993), "Did Surface Temperatures Constrain Microbial Evolution?," *BioScience* **43:** 390–393.

SHKLOVSKII, I.S., AND C. SAGAN (1966), *Intelligent Life in the Universe* (San Francisco: Holden-Day).

SOGIN, M.L., J.H. GUNDERSON, H.J. ELWOOD, D.A. ALONSO, AND D.A. PEATTIE (1989), "Phylogenetic Meaning of the Kingdom Concept: An Unusual Ribosomal RNA from *Giardia lambia*," *Science* **243:** 75–77.

STRAUSS, H., D.J. DES MARAIS, J.M. HAYES, AND R.E. SUMMONS (1992), "The Carbon-Isotopic Record," in *The Proterozoic Biosphere*, J.W. Schopf and C. Klein, eds. (Cambridge: Cambridge University Press), pp. 117–127.

SUMMERS, D.P., AND S. CHANG (1993), "Prebiotic Ammonia from Reduction of Nitrite by Iron (II) on the Early Earth," *Nature* **365:** 630–633.

WALKER, J.C.G., P.B. HAYS, AND J.F. KASTING (1981), "A Negative Feedback Mechanism for the Long-Term Stabilization of Earth's Surface Temperature," *JGR* **86:** 9776–9782.

WALTER, M.R., AND D.J. DES MARAIS (1993), "Preservation of Biological Information in Thermal Spring Deposits: Developing a Strategy for the Search for Fossil Life on Mars," *Icarus* **101:** 129–143.

ZAHNLE, K.J. (1986), "Photochemistry of Methane and the Formation of Hydrocyanic Acid (HCN) in the Earth's Early Atmosphere," *JGR* **91:** 2819–2834.

The Forest-Habitability of Earthlike Planets

Martin Heath

Greenwich Community College
London, England

The Concept of Forest-Habitability

Recognizing that forest biomes are not only one of the most distinctive features of the Earth's surface, but also that they contain some 90% of the carbon bound up in living things (Bolin et al. 1979), it is surprising that they have not been the specific focus of previous discussions of planetary habitability. The width of the most favourable zone about the Sun for Earth's present types of forest is a question that begs to be explored through detailed models for tree species ecophysiology and forest community dynamics. For the purposes of this crude preliminary treatment, however, it is instructive to ask questions about biological responses to three key parameters, namely light, temperature, and partial pressure of CO_2. For each of these, and for any given species of plant, under given conditions, there may be determined compensation points at which photosynthetic fixation of carbon from atmospheric CO_2 matches exactly losses through photorespiration and aerobic respiration. The latter is required for the metabolism of growth in higher plants. Experimental work (Crawford 1989 provides a recent discussion) confirms that growth can fall off sharply when the partial pressure of oxygen is reduced to half present atmospheric level.

It must be remembered, of course, that a healthy forest-dominated biosphere will not thrive at the very limits of environmental tolerance, where net productivity is zero, but rather, conditions must remain sufficiently within these limits if a substantial biomass is to be maintained. It should be stressed also that what is discussed below is not the absolute width of the forest habitable zone, but rather a tentative estimation of the width of the most favourable zone for existing forest biomes. It cannot be assumed that all other biospheres need be closely similar to ours. However, investigations of the generalized tolerances of ours do suggest the sort of biological adaptations that might be necessary if biospheres as massive as ours were to exist under some of the different global environmental regimes suggested by current habitability models.

Temperature and partial pressure of CO_2 (pCO_2) have been investigated in the context of the carbonate-silicate-biogeochemical cycle by J. F. Kasting and various coworkers, and it is their models that provide much of the framework in which conjecture about higher life habitability must proceed. A detailed explanation of this cycle and the way in which it can function as part of a planetary thermostat is provided elsewhere in this volume (see Kasting 1996). It is sufficient here to note an outline. The greenhouse gas CO_2 is outgassed by volcanoes, coming from juvenile mantle sources and from the metamorphism of carbonates at convergent plate boundaries. It raises the surface temperature of the Earth, and this encourages evaporation of water from standing bodies (water vapour is a much more efficient greenhouse gas than CO_2). Atmospheric CO_2 dissolves in rainwater to form carbonic acid, which promotes rock weathering. Streams carry resulting bicarbonate ions to the sea, and the carbon is incorporated in marine carbonates. These can suffer metamorphism and degassing at a future date. With greater insolation, higher temperatures would encourage precipitation and weathering. This would reduce the pCO_2, hence greenhouse warming, and the Earth's temperature rise would also be reduced. With less insolation, weathering rates would decrease, and given the same rate of outgassing, pCO_2 would rise until equilibrium conditions were achieved. This would constrain any drop in temperature. The time scale for adjusting equilibrium pCO_2 is on the order of $> 10^5$ years (Walker et al. 1981).

The Outer Margin of the Forest-Habitable Zone

On Earth, a useful guide to the location of the tree line is Köppen's Rule, which states that full-sized trees (> 5 m tall) can survive where temperatures exceed 10°C for thirty days a year. There are, however, several alternative ways of defining the vegetational changes at the boundaries of the northern forests. Payette (1983) places the "tree line" north of the "continuous forest limit," but south of the "tree species limit." There have also been several attempts to more closely identify the environmental factors that control these changes (see review in Sirois 1992). It has been suggested, for example (Bryson 1966, Krebs and Barry 1970), that the tundra-taiga boundary corresponds closely with the mean summer location of the Arctic frontal zone, a meteorological feature identifiable from available data with some subjective bias. The extent to which the front determines the tree line, or the tree line influences the development of frontal systems, deserves further research.

In general terms it would not be unreasonable to use the ability of the carbonate-silicate-cycle planetary thermostat to keep much of a planet's surface above 10°C for much of the year as a crude guide to the forest-habitability of a planet receiving less insolation than Earth. Unfortunately, it is not certain exactly how the temperature of an Earthlike planet would decline with reduction in insolation, although Marshall et al. (1988) did publish results of simulations for different continental configurations. They appear to have been too optimistic in concluding that an ice line could stabilize at low latitudes. Caldeira and Kasting (1992a) considered that if ice advanced closer to the equator than about latitude 30°, runaway glaciation would follow regardless of how high pCO_2 might be. Kasting et al. (1993) offered 1.37 AU as an explicitly pessimistic estimate of how far a planet might lie from the Sun and still maintain liquid water on its surface. At this distance, the formation of highly reflective clouds of CO_2 ice high in the atmosphere would start to counteract the CO_2 greenhouse. This problem need not be acute at 1.37 AU, though the mean global temperature of 0°C assumed by Kasting et al. is not encouraging in the present context.

Under conditions of abundant water and CO_2, the type of photosynthesis that would be favoured is C3, that used by most forest trees. In this, carbon is fixed by the enzyme RuBP (ribulose bisphosphate) carboxylase, and forms the three-carbon molecule phosphoglyceric acid. Temperature optimums are typically in the range 20–25°C, and light saturation is not unusual at 0.25 to 0.30 full sunlight. CO_2 compensation points of 30–70 ppm (PAL ≈ 350 ppm) have been determined (Black 1973). A disadvantage of C3 plants is that O_2 can combine with RuBP carboxylase instead of CO_2, with formation of phosphoglycolic acid. Seventy-five percent of the carbon involved can be rescued by the glycolate pathway, but CO_2 is produced due to metabolism in the peroxisomes and mitochondria, a process termed *photorespiration*—so reducing net carbon fixation.

It is true that photosynthesis can be continued down to very low light levels, and some trees, such as *A. saccharum* Marsh. (sugar maple), can survive in the understory with around 1–3% of full daylight (Spurr and Barnes 1980). We cannot dismiss the possibility that given otherwise adequate environmental parameters, some organisms could sustain net photosynthetic production at around, say, 10.00 AU from the Sun, but optimism must be tempered with the recognition that ecosystems like ours could not survive at this distance. It must be recognized that forests create their own shaded environments, and juvenile trees may have to survive under these conditions for some years while waiting for a gap to appear in the tree canopy. Spurr and Barnes (1980) offer a table of understory tolerances. "Very intolerant" trees, including certain pioneer species such as Jack pine (*Pinus banksiana* Lamb.), may need in excess of 60% full sunlight. Stands of the "very tolerant" *A. saccharum* can actually produce such a deep shade as to ensure that its own slow-growing seedlings can outcompete other species. Shading then, can be decisive, but survival on the forest floor must be understood also in terms of a range of factors, including competition for soil moisture and nutrients. It should be observed that by 1.3 AU from the Sun no tree could receive as much as 0.6 Earth insolation. Understory species and saplings surviving near their light compensation points would be disadvantaged under tree canopies as dense as those in Earth's forests, and flecks of sunlight would contribute significantly less energy to the forest floor.

On Earth, plants have a number of strategies for coping with increasing shade (Krebs 1985), and these might be of value also on planets receiving less insolation than ours. One of the more obvious morphological adaptations is that leaves can be arranged so as to reduce mutual shading. Metabolic functions too can be adjusted for lower light. Achieving a net carbon balance for the year may require a reduction in both respiration and growth rates. If temperatures were kept high by atmospheric greenhouse gases, but insolation remained consistently below Earth values, it would be advantageous for plants to slow down metabolically, so that they minimized the possibility of depletion of carbohydrate reserves. Also, shade plants tend to produce a smaller quantity of the enzyme RuBP carboxylase. (Arctic plants do the opposite.) Perhaps the most common protein in nature, this enzyme, with a molecular weight of around 500,000, is metabolically costly to synthesise. On the other hand, the quantity of the photosynthetic pigment chlorophyll (and the proportion of the light-harvesting chlorophyll b) can be usefully increased in plants that live in deep shade. Increased resistance to fungal attack is another useful shade adaptation. Seeds, it may be conjectured, might need larger carbohydrate reserves to feed seedlings that must contend with low-productivity conditions, and this would make them heavier and less easily dispersed.

It is uncertain how far greatly elevated equilibrium pCO_2 would permit higher plants to compensate for reduced light, though the possibility that such plants could evolve cannot be dismissed. Experiments by Rothschild and Mancinelli (1990) demonstrated that the productivity of the cyanobacterium *Lyngbya* rose sharply with increased pCO_2. Existing higher plants do not necessarily benefit from higher pCO_2. Norby et al. (1992) grew saplings of yellow poplar (*Liriodendron tulipifera* L.) for nearly three growing seasons with pCO_2 up to almost twice PAL, with no fertilizer or supplemental water. They observed an increase in photosynthetic efficiency and reduction in foliar respiration, but no increase in whole plant carbon storage.

The Inner Margin of the Forest-Habitable Zone

Inward of 1.00 AU, key problems for plants would be posed by increasing temperature and extreme CO_2 drawdown through accelerated weathering.

The maximum temperatures that plants can endure for prolonged periods vary substantially, and a major cause of thermal damage appears to be phase changes in chloroplast membranes which disrupt photosystem II electron transport (Raison et al. 1980). A generalized upper limit for plant growth in the range 45–55°C is often quoted, but though many plants can tolerate such temperatures for about half an hour, long-term thermal stress can prove lethal beyond about 35–40°C (Fitter and Hay 1987). The highest tissue temperature recorded, 65°C, is from a species of the cactus genus *Opuntia*, and it is evident that cacti can tolerate such temperatures for many hours a day. Cacti seedlings, growing near the soil, may have to endure over 70°C (Nobel 1984).

Under conditions of elevated illumination and temperature C3 photosynthesis is superseded by C4 photosynthesis and crassulacean acid metabolism (CAM). In C4 photosynthesis (summarized in many texts; for example, in Grierson and Covey 1984) CO_2 is fixed after conversion to HCO_3^-, by a very efficient enzyme—PEP (phosphoenol pyruvate) carboxylase—that does not bind with O_2. The four-carbon molecule so formed, oxaloacetate, is converted to malate and/or aspartate, which is transported to a ring of cells, which form a sheath around vascular bundles—a characteristic termed *Kranz anatomy*. Here, CO_2 is regenerated by decarboxylation of aspartic acid and malic acid, and refixed by RuBP carboxylase under favourable internal conditions of elevated pCO_2, with little opportunity for O_2 to compete to bind with this enzyme. C4 plants can have CO_2 compensation points of just a few parts per million, and light saturation for photosynthesis may occur at full sunlight, or even higher levels. Optimum temperatures typically lie in the range 30–35°C. In CAM, frequent among desert succulents, stomata are opened at night to admit CO_2, which is fixed by PEP carboxylase, and stored in cell vacuoles as malic acid. Decarboxylation follows in the daytime, when stomata are closed to prevent water loss, and CO_2 is then refixed by RuBP carboxylase. Under very extreme conditions, when plants are obliged to keep their stomata shut for long periods, CO_2 produced by respiration may be photosynthetically refixed in the process of CAM-idling. CO_2 compensation points are typically very low, but light saturation can be well below full sunlight (Black 1973). It should be borne in mind that such adaptations can work only where nocturnal conditions are sufficiently cool.

Several morphological adaptations to high irradiance are seen in existing plants (Fitter and Hay 1987). Large amounts of stored water bestow a modicum of thermal inertia. Small leaves can reduce transpirational losses of water per unit ground area.

Moreover, the smaller a leaf, the more effective is convectional heat dissipation. Since the saturation content of water in air is temperature-sensitive, the gradient of water vapour concentration, therefore rate of water loss through transpiration, will be reduced for a cooler leaf. Where there is abundant water to transpire, large leaves may remain significantly cooler than the surrounding air, and develop a considerable boundary layer resistance to sensible heat flux (Smith 1978). Leaf reflectivity may be increased (Ehleringer 1980) and nongrowth responses can adjust leaf angle with respect to sunlight (Ehleringer and Forseth 1980).

The inner edge of the habitable zone adopted by Kasting et al. (1993) was 0.95 AU. Here, the oceans could gradually be lost to space under a moist greenhouse regime. This was a conservative estimate, ignoring possible cooling of the Earth by cloud climate feedbacks. A similarly conservative estimate for the inner margin of the most favourable zone for forest biomes must be more pessimistic. Lovelock and Whitfield (1982) argued that by 0.1 Gyr from now, increases in solar luminosity required by astrophysical models could be compensated by reduction of pCO_2 below 150 ppm, which they took to be the lower limit for photosynthesis. The equivalent present day distance from the Sun is near 0.99 AU. Worsely and Nance (1989) pointed out that C4 plants often have very low CO_2 compensation points, and so did Caldeira and Kasting (1992b), who modeled CO_2 drawdown, mean global temperature, and biological productivity as the Sun increases in luminosity by 15% over the next 1.6 Gyr.

Reapplying their model here in terms of decreasing heliocentric distance, instead of future enhanced solar luminosity, the prospects for forests appear to decline rapidly inward of 1.00 AU. For 0.70 Gyr from now, when solar output is 1.06 its current value, the modeled quantity of atmospheric CO_2, and biological productivity, was an order of magnitude smaller than now, though the mean global temperature was a not unhealthy 20°C (compared with about 15°C today). The fate of standing biomass is uncertain, but an obvious possibility is that it will dwindle with atmospheric CO_2. If so, major environmental deterioration may follow as the present amelioration of surface regime by evapotranspiration is undermined. Computer simulations by Shukla and Mintz (1982) and Shukla et al. (1990) suggested that for the present Earth system, large-scale clearance of forests could allow rainfall to be much reduced, and summer temperatures to rise to around 50°C in some areas. Another possibility is that plants might somehow become more efficient at fixing CO_2 from the atmosphere, and so manage to fix the same amount even if the partial pressure of CO_2 is much reduced. Today, the biosphere takes up on the order of 60×10^9 tonnes of carbon per year (Bolin et al. 1979), and incorporates it into biomass. Once the total amount of carbon in the atmosphere has fallen below this level, however, it will not be possible to do this. A substantial plant biomass might then survive if trees had much slower growth rates and greater longevity, thus increasing the turnover time for carbon through the biosphere, and reducing the rate of uptake. Unfortunately, biological productivity falls off to two orders of magnitude below its present value by 0.96 AU. It is not unreasonable to doubt whether it is likely that a large biomass could survive once CO_2 has become a scarce resource.

Planetary History

The background of the Earth's habitability may be summarized in conservative terms as follows. Axial obliquity is stabilized by the existence of a large satellite—the Moon (Laskar and Robutel 1993, Laskar et al. 1993). The planet is sufficiently geologically active to have a carbonate-silicate cycle (Walker et al. 1981), and for the horizontal motions associated with plate tectonics to rejuvenate freeboard (Wise 1974). The primordial bombardment has tailed off, and a substantial continental mass has differentiated. With the Archaean/Proterozoic transition, near 2.5 Gyr ago, outgassing of CO_2 has declined, and the area of exposed rock available to promote CO_2 drawdown through weathering has increased (Walker 1990). The overall fall in CO_2, continuing through the Phanerozoic, may have posed a substantial problem for the earliest of the higher plants, since by increasing opportunities for dwindling CO_2 to enter stomatal pores, plants also increase opportunities for water loss. Robinson (1991) postulated that the need to invest in a greater area for gas exchange, while improving water-use efficiency, may have been a key factor in driving higher plant evolution. It was only after around 0.37 Gyr ago that plants evolved leaves, roots, secondary tissues, such as wood, and the arborescent habit.

Whether Robinson's hypothesis is correct or not, the fossil record does not, strictly speaking, require us to assume that the Earth, while basically habitable, has offered conditions suitable for the evolution of the trees for a total interval of more than 0.40 Gyr. Such a conclusion might be too severe, but there are grounds to argue that the Late Proterozoic/Phanerozoic Earth system has been and remains capable of undergoing sufficiently severe perturbations to terminate the present window of enhanced favourability for forests.

A thought-provoking example of environmental change proceeding faster than evolution is provided by the fate of one Gondwanaland fragment—Antarctica. Antarctica separated from Australia about 50 Myr ago. Until then, the continents shared a similar flora of southern conifers and austral evergreen dicotyledons. As Antarctica drifted toward the South Pole, forests became patchy, finally giving way to scrub, cushion plants and finally ice fields (Axelrod 1979).

The Late Proterozoic glaciations of about 1.00 to 0.60 Gyr ago appear to have affected low latitudes (review in Windley 1984), and the end of these might have made an important contribution to opening the window for forest biomes.

Climatic models differ, but some (Marshall et al. 1988) would permit low global temperatures if continents moved into an equatorial configuration, with accelerated weathering promoting CO_2 drawdown. Caldeira and Kasting (1992a) lead to even more pessimistic conclusions. These workers point out that the time scale on which ice sheets could grow and possibly lead to global glaciation is much shorter than the response time of the carbonate-silicate cycle, so runaway glaciation remains a possibility, despite the existence of a crude planetary thermostat. Above 0.8 present Earth insolation, deglaciation would be possible, but the sequence of events postulated would see the elimination of forests. An ice-covered Earth would offer little opportunity for CO_2 drawdown through weathering, and within 30 Myr, CO_2 outgassed by volcanoes would build up to about 0.12 bar. The Earth's mean global temperature could rise to > 70°C, and ice cover would become unstable. Below 0.8 present insolation, global glaciation if it occurred would be associated with the formation of high-albedo clouds of CO_2 ice, and glaciation would be irreversible. The model was intended to explore climatic problems on an early Earth receiving reduced insolation from a fainter Sun, but it may have interesting implications for the width of the most favourable zone about the Sun for forest biomes.

On the other hand, while Caldeira and Kasting (1992b) argue that CO_2 drawdown in response to rising solar output could have reduced pCO_2 by an order of magnitude by 0.70 Gyr from now, Caldeira (1991) has argued that within 0.1 Gyr, subduction of pelagic carbonates of planktonic origin may cause an order of magnitude increase in the rate of outgassing of metamorphic CO_2. The time scale of biological response to such a possible rise in pCO_2 and associated rise in temperature remains to be investigated.

For slowly rotating planets with no large satellite, chaotic changes in obliquity are possible on a relatively short time scale (Laskar and Robutel 1993, Laskar et al. 1993). Studies by D. M. Williams (pers. comm.) imply that even with an obliquity of 85° the present Earth system need not experience midsummer north polar temperatures above roughly 40°C, and winters could be warm. Antarctic temperatures might range between about 50°C and –50°C. This need not pose an insurmountable problem for higher plants, despite the continuous insolation experienced in summer. In the event that the continents formed into a polar supercontinent, however, predicted midsummer temperatures might exceed 100°C over the entire landmass. This would constitute closure of the forest biome window.

Aside from the most extreme conditions, temperature tolerances of existing plants encourage hopes that wide seasonal swings in temperature could be endured. Notably, one genus, *Salix* (willows), which occurs from the tropics to the Arctic, possesses the genetic potential to survive an extraordinary range of temperature. Sakai (1970) found that twigs of species from hot climates could, if permitted a preparatory period of frost hardening, survive freezing at –50°C for many hours, without suffering damage, and twigs subjected to –90°C could still root and produce new shoots. Twigs of a willow from southwestern Japan survived immersion in liquid nitrogen at −196°C. The giant rosette plants of tropical mountains in Africa and South America grow to about 3 m, and survive a diurnal temperature cycle similar to the yearly cycle of other environments, living tissues, including water-storing pith, being protected by a thick insulating layer of dead leaves (Hedburg and Hedburg 1979).

Some workers have even conjectured that major changes in global biomass may occur with far less drastic environmental disturbances than those discussed above. Adams et al. (1990) have quoted palynological, pedological and sedimentological evidence to argue that during the Last Glacial Maximum (LGM), 18,000 years ago, total biomass was near 40% that of today. This is consistent with the results of experimental work by Polley et al. (1993), who observed that both water-use efficiency and above-ground biomass of C3 species increased linearly from typical LGM values (160–200 ppm) to present levels of CO_2 (350 ppm).

Conclusions

The definition of the most favourable zone about the Sun for forest habitability is somewhat arbitrary and based, moreover, on model-dependent arguments. The foregoing discussion suggests that it might be located within the limits 0.97 AU (where biological productivity will be reduced by an order of magnitude) and 1.12 AU (beyond which irreversible glaciation is possible). Future research may modify these limits, or suggest others based on different considerations. Work in progress includes investigation of the implications for Earth-type forest biomes of planets with parent stars with a range of masses and luminosity histories, and of the prospects for extreme ecosystems.

Acknowledgments

My thanks to Laurance Doyle and Neil Heather for their encouragement, and to Darren Williams for a helpful discussion of his work.

References

Adams, J.M., H. Faure, L. Faure-Denard, J.M. McGlade, and F.I. Woodward (1990), "Increases in Terrestrial Carbon Storage from the Last Glacial Maximum to the Present," *Nature* **348:** 711–714.

Axelrod, D.I. (1979), "The Roles of Plate Tectonics in Angiosperm History," in *Historical Biogeography, Plate Tectonics, and the Changing Environment. Proceedings of the Thirty-Seventh Annual*

Biology Colloquium and Selected Papers, J. Gray and A.J. Boucot, eds. (Corvallis: Oregon State University Press), pp. 435–447.

BLACK, C.C. (1973), "Photosynthetic Carbon Fixation in Relation to Net CO_2 Uptake," *Ann. Rev. Plant Physiol.* **24:** 253–286.

BOLIN, B., E.T. DEGENS, P. DUVIGNEAUD, AND S. KEMPE (1979), "The Global Biogeochemical Carbon Cycle" (Table 1.2, P. Duvigneaud), in *The Global Carbon Cycle*, SCOPE 13, B. Bolin, E.T. Degens, S. Kempe, and P. Ketner, eds. (Chichester: John Wiley & Sons), pp. 1–56.

BRYSON, R.A. (1966), "Air Masses, Streamlines, and the Boreal Forest," *Geogr. Bull.* **8:** 228–269.

CALDEIRA, K. (1991), "Continental-Pelagic Carbonate Partitioning and the Global Carbonate-Silicate Cycle," *Geology* **19:** 204–206.

CALDEIRA, K., AND J.F. KASTING (1992a), "Susceptibility of the Early Earth to Irreversible Glaciation Caused by Carbon Dioxide Clouds," *Nature* **359:** 226–228.

CALDEIRA, K., AND J.F. KASTING (1992b), "The Life Span of the Biosphere Revisited," *Nature* **360:** 721–723.

CRAWFORD, R.M.M. (1989), *Studies in Plant Survival: Ecological Case Histories of Plant Adaptation* (Oxford: Blackwell Scientific Publications).

EHLERINGER, J.R. (1980), "Leaf Morphology and Reflectance in Relation to Water and Temperature Stress," in *Adaptations of Plants to Water and High Temperature Stress*, N.C. Turner and P.J. Kramer, eds. (New York: Wiley Interscience), pp. 295–308.

EHLERINGER, J.R., AND I. FORSETH (1980), "Solar Tracking by Plants," *Science* **210:** 1094–1098.

FITTER, A.H., AND R.K.M. HAY (1987), *Environmental Physiology of Plants*, 2d ed. (London: Academic Press).

GRIERSON, D., AND S. COVEY (1984), *Plant Molecular Biology* (London: Blackie & Son).

HEDBURG, I., AND O. HEDBURG (1979), "Tropical-Alpine Life Forms of Vascular Plants," *Oikos* **33:** 297–307.

KASTING, J.F. (1996), "Habitable Zones Around Stars: An Update," this volume.

KASTING, J.F., D.P. WHITMIRE, AND R.T. REYNOLDS (1993), "Habitable Zones Around Main Sequence Stars," *Icarus* **101:** 108–128.

KREBS, C.J. (1985), *Ecology—The Experimental Analysis of Distribution and Abundance*, 3d ed. (New York: Harper & Row).

KREBS, J.S., AND R.G. BARRY (1970), "The Arctic Front and the Tundra-Taiga Boundary in Eurasia," *Geogr. Rev.* **60:** 548–554.

LASKAR, J., AND P. ROBUTEL (1993), "The Chaotic Obliquity of the Planets," *Nature* **361:** 608–612.

LASKAR, J., F. JOUTEL, AND P. ROBUTEL (1993), "Stabilization of the Earth's Obliquity by the Moon," *Nature* **361:** 615–617.

LOVELOCK, J.E., AND M. WHITFIELD (1982), "Life Span of the Biosphere," *Nature* **296:** 516–563.

MARSHALL, H.G., J.C.G. WALKER, AND W.R. KUHN (1988), "Long-Term Climate Change and the Geochemical Cycle of Carbon," *JGR* **93:** 791–801.

NOBEL, P.S. (1984), "Extreme Temperatures and Thermal Tolerances for Seedlings of Desert Succulents," *Oecologia* **62:** 310–317.

NORBY, R.J., C.A. GUNDERSON, S.D. WULLSCHLEGER, E.G. O'NEILL, AND M.K. MCCRACKEN (1992), "Productivity and Compensatory Responses of Yellow-Poplar Trees in Elevated CO_2," *Nature* **357:** 322–324.

PAYETTE, S. (1983), "The Forest-Tundra and Present Tree-Lines of the Northern Quebec–Labrador Peninsula," in *Tree-Line Ecology: Proceedings of the Northern Quebec Tree-Line Conference*, Collection Nordicana 47, P. Morisset and S. Payette, eds. (Quebec: Universite Laval), pp. 3–23.

POLLEY, H.W., H.B. JOHNSON, B.D. MARINO, AND S.M. HERMAN (1993), "Increase in C3 Plant Water-Use Efficiency and Biomass over Glacial to Present CO_2 Concentrations," *Nature* **361:** 61–64.

RAISON, J.K., J.A. BERRY, P.A. ARMOND, AND C.S. PIKE (1980), "Membrane Properties in Relation to the Adaptation of Plants to High and Low Temperature Stress," in *Adaptation of Plants to Water and High Temperature Stress*, N.C. Turner and P.J. Kramer, eds. (New York: Wiley Interscience), pp. 261–273.

ROBINSON, J.M. (1991), "Phanerozoic Atmospheric Reconstructions: A Terrestrial Perspective," *Palaeogeogr., Palaeoclimatol., Palaeoecol. (Global Planet. Change Sect.)* **97:** 51–62.

ROTHSCHILD, L.J., AND R.L. MANCINELLI (1990), "Model of Carbon Fixation in Microbial Mats from 3,500 Myr Ago to the Present," *Nature* **345:** 710–712.

SAKAI, A. (1970), "Freezing Resistance in Willows from Different Climates," *Ecology* **51:** 485–491.

SHUKLA, J., AND Y. MINTZ (1982), "Influence of Land-Surface Evapotranspiration on the Earth's Climate," *Science* **215:** 1498–1501.

SHUKLA, J., C. NOBRE, AND P. SELLERS (1990), "Amazon Deforestation and Climate Change," *Science* **247:** 1322–1325.

SIROIS, L. (1992), "The Transition Between Boreal Forest and Tundra," in *A Systems Analysis of the Global Boreal Forest*, H.H. Shugart, R. Leemans, and G.B. Bonan, eds. (Cambridge: Cambridge University Press).

SMITH, W.K. (1978), "Temperatures of Desert Plants: Another Perspective on the Adaptability of Leaf Size," *Science* **201:** 614–616.

SPURR, S.H., AND B.V. BARNES (1980), *Forest Ecology*, 3d ed. (Chichester: John Wiley & Sons).

WALKER, J.C.G. (1990), "Precambrian Evolution of the Climate System," *Palaeogeogr., Palaeoclimatol., Palaeoecol. (Global Planet. Change Sect.)* **82:** 261–289.

WALKER, J.C.G., P.B. HAYS, AND J.F. KASTING (1981), "A Negative Feedback Mechanism for the Long-Term Stabilization of Earth's Surface Temperature," *JGR* **86:** 9776–9782.

WINDLEY, B.F. (1984), *The Evolving Continents*, 2d ed. (Chichester: John Wiley & Sons).

WISE, D.U. (1974), "Continental Margins, Freeboard and the Volumes of Continents and Oceans Through Time," in *The Geology of Continental Margins*, C.A. Burke and C.L. Drake, eds. (Berlin: Springer-Verlag), pp. 45–58.

WORSELY, T.R., AND R.D. NANCE (1989), "Carbon Redox and Climate Control Through Earth History, a Speculative Reconstruction," *Palaeogeogr., Palaeoclimatol., Palaeoecol. (Global Planet. Change Sect.)* **75:** 259–282.

The California Nebula (NGC 1499)

Photo Credit: Stephen Rahn, Macon, Georgia, United States
https://commons.wikimedia.org/wiki/File:The_California_Nebula_%2848824328181%29.jpg/
[retrieved 12-18-25]

The California Nebula is an emission nebula in the constellation Perseus. This cosmic cloud is drifting through the Orion Arm of the spiral Milky Way Galaxy. Our Sun, also in the Orion Arm, is only about 1,500 light-years from the California Nebula. By coincidence, the California Nebula transits in the zenith in Central California, as the latitude matches the declination of the object. The most prominent glow of the nebula is the red light characteristic of hydrogen atoms recombining with electrons stripped away (ionized) by energetic starlight. The fluorescence is due to excitation of the Hβ line in the nebula by the nearby prodigiously energetic O7 star, Xi Persei (also known as Menkhib), which was formed in the California Nebula only a few million years ago. Visually, the California Nebula and the Pleiades are not too far from each other in the night sky and are a popular target for astrophotographers, especially given the redness of the nebula and the blue of the Pleiades. The California Nebula was discovered by E. E. Barnard in 1884.

EVOLUTIONARY BEHAVIOR OF A SELF-COPYING COMPUTER PROGRAM

Rosemarie Swanson and Jun-Jie Liu

Biochemistry-Biophysics
Texas A&M University
College Station, TX 77843-2128

Abstract

A nonbiological alternative "world" of self-reproducing computer programs offers an opportunity to consider questions about evolution from a broadened perspective. Must particular features exist in any system capable of evolution, regardless of the medium in which it is expressed? Is it necessary or beneficial to have a genetic code? The observed natural genetic code has an error-resistant form (it is almost a Gray code). What would be the effects on evolutionary diversity and progress of a more error-sensitive or more error-tolerant code? Some discussion of the benefits and liabilities of a genetic code and some intermediate results on both the diversity and the features in common of "creatures" produced in independent evolutionary runs will be described and evaluated. The original human-written ancestral program will be compared with some of the programs that evolved to illustrate the nature and limits of creativity generated by mutation and selection.

INTRODUCTION

The rightful focus of this conference is on defining conditions suitable for the origin of water-dependent life; but broader possibilities for identifying life are also open to consideration. For example, a significant feature of life is that it evolves. A system capable of evolution seems to need three characteristics: (1) it copies itself (2) with errors (3) under selective pressure.

The features of an evolvable system can clearly exist in the form of a computer program. This opens a way to study events very early in evolution. For example, the relationship between the twenty genetically occurring amino acids and their codons gives the natural genetic code a certain level of tolerance to error. Although it is not feasible to repeat early life evolution to explore the consequences of a radically different correspondence between amino acids and codons, it is possible to investigate such a question within a computer-based evolutionary system.

We found that in a computer-based evolutionary system, an error-tolerant genetic code, compared with an error-sensitive one, led to consistently greater genetic diversity, suggesting that a system based on an error-tolerant code is likely to be more robust than others.

The background for asking the question we addressed and the way in which we approached an answer is the subject of this paper.

Background

A simple relationship emerged in a study (Swanson 1984) of the properties of the amino acids and their patterns of substitution (Dayhoff et al. 1978) for each other in related proteins. The twenty genetically occurring amino acids could be ordered so that neighboring entries in a list were most similar. The resulting list written around a circle showed that the amino acids fell into four groups: hydrophobic, hydrophilic, large, and small. Furthermore, when the codons for these amino acids were arranged in the same order, a pattern also emerged in the codons: hydrophobic amino acids tended to have pyrimidines in the *middle* codon position, while hydrophilics tended to have purines; large amino acids tended to have pyrimidines in the *first* codon position, while small ones tended to have purines.

Such a codon–amino acid relationship corresponds to a very simple Gray code (Flores 1976), a form of error-tolerant code (see Fig. 1). Although the biological apparatus that reads codons now is too accurate to need the level of error tolerance that this kind of system can provide, it might have offered an advantage very early in evolution. Obviously it is not possible to investigate this question in a biological experiment. However, recently it became feasible to explore this question in a computer system, the Tierra system conceived and developed by Tom Ray at the University of Delaware.

The Tierra system (Ray 1992) is a computer program that simulates both the hardware and the operating system of a computer designed to run "creature" programs—programs that copy themselves and are subject to mutation. That is, the Tierra system provides an environment for a population of programs to reproduce themselves and compete for resources and be displaced by newly developed genomes.*

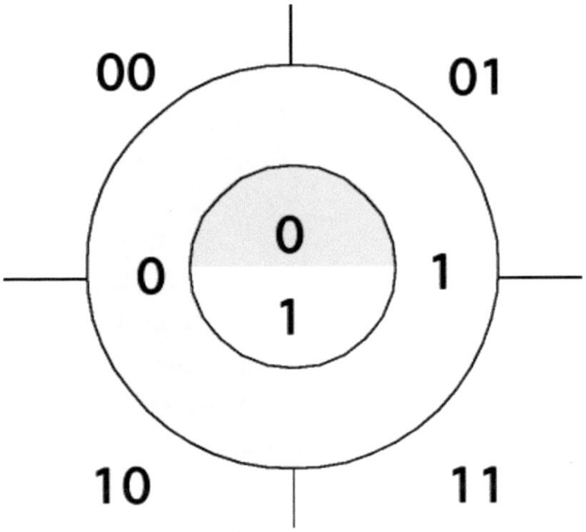

FIG. 1. *Simple Gray code. The inner circle represents middle base purine and hydrophilic amino acids (0) or middle base pyrimidine and hydrophobic amino acids (1). The outer circle represents first base purine and small amino acids (0) or first base pyrimidine and large amino acids (1). Note that a single bit change can shift amino acid properties only to an adjacent quadrant, not to the diametrically opposite quadrant. Mutations are conservative in this sense: the most radical amino acid property change in this scheme cannot occur with a single mutation.*

A Tierra run normally begins with a single copy of a creature program, the "ancestor," written by Tom Ray, in a world that consists of 60,000 bytes of computer memory and a CPU to execute the code of the creature. The code of the creature does nothing but examine itself to find out its length, request that much memory from the operating system, and proceed to make a copy of itself into the newly requested memory location. It then starts the copy running. The copy does the same thing its parent did, except now the parent and descendent programs must timeshare the CPU cycles. Soon there are 3, 5, 6, 8, and more copies of the program running, distributed in the 60,000 bytes of memory. However, the operating system randomly makes occasional errors, so that some of the copies are not identical to their parent, and the difference affects how the code runs. Normally the changed code does still run, but it behaves a little differently. The 60,000 bytes of memory fill up rapidly with creatures, and the reaper queue kicks in. If there is not enough space to fill a request of a creature to make a new copy, the creature at the top of the reaper queue is killed off and its space given to the requesting creature. The operating system keeps statistics on how accurate each creature is in copying itself and how long it has existed. Creatures that make accurate copies of themselves move back in the reaper queue, while creatures that make a lot of errors get moved forward in the queue. Thus, there is competition among the creatures for memory space and accuracy of copying. There is also competition for CPU cycles, because creatures that are able to copy themselves faster (by becoming shorter) have more opportunities to move back from the head of the reaper queue, and they also can

* Tierra source code, documentation, discussion, and references available by FTP can be found at https://tomray.me/tierra/source/

contribute more individuals to the total population. Thus, the Tierra system has the important features of an evolutionary system: self-replicating entities, a source of variation in the entities (errors in self-copying—mutations), and competition for resources (selective pressure). Even though the population starts from a single "organism," it quickly develops variability, and it continues to change its composition with time. No "species" lives forever.

THEORY

The Tierra system was designed with the natural genetic code as a kind of paradigm: the creatures are based on a set of 32 possible instructions; these correspond to the twenty amino acids of the natural world. The instructions are encoded with 5 bits; the bits correspond to the nucleotide-triplet codons (6 bits) of the natural world. Mutation in the Tierra system consists of changing one of the 5 bits encoding an instruction, thereby changing the instruction. Now, an error-tolerant encoding scheme, such as a Gray code (Fig. 1), has similar instructions encoded by similar bit patterns, so that changing a bit has a higher probability of yielding an instruction that might have a related function than of yielding an instruction that has a very different function. We hypothesize that a mutation that is based on a small change has a better chance of yielding a working creature than a mutation that makes a large change. However, the distinct possibility exists that large changes are beneficial because insufficient variety in the material that selection has available will hamper evolutionary progress. We asked the question: Will an error-tolerant Tierran genetic code result in any significant difference in the evolutionary behavior of the system compared with its behavior using an error-sensitive genetic code?

If selection pressure is strong, and few mutants survive, an error-tolerant code should be better, whereas if selection pressure is weak and most mutants survive, an error-sensitive code will generate a more radically varied set of creatures, enabling a wider search range in evolutionary space. We imagine a "cloud" of acceptable variants around each creature, with closer locations corresponding to more similar variants. If selection is strong, the cloud will be small. If selection is weak, the cloud will be broad. The best strategy for a creature is to spread its mutants around as far as its cloud reaches, but not much farther, so that it doesn't waste reproduction on variants that have too small a chance of survival.

But if mutation is a random event, how can a creature have a "strategy" for controlling the range of mutations? One answer lies in using a genetic code. Consider the following idea.

If an organism "encoded" its proteins directly as amino acids and mutation is random, each amino acid would have equal chances of being mutated to any other amino acid, similar or not. If an organism encoded its proteins using more than one "coding unit" per amino acid and mutation is random on the coding units, then each amino acid would have unequal probabilities of being mutated to other amino acids, and mutation could be weighted with respect to the amino acids even though it is random on the codons. An example will illustrate this. Use the numbers 0 to 15 as amino acids and represent them with the digits 0–9 and letters A-F. If mutation is random on the set 0 . . . F, 0 will change with equal probability to any other number in the set. In contrast, consider the set 0 . . . F in terms of their binary "codon" representation.

0=0000 1=0001 2=0010 3=0011 4=0100 5=0101 6=0110 7=0111
8=1000 9=1001 A=1010 B=1011 C=1100 D=1101 E=1110 F=1111

Choosing 0=0000 as an example, it is clear that 1=0001, 2=0010, 4=0100, and 8=1000 are the closest mutants to 0=0000, in terms of a single change to a *codon*. 3=0011, 5=0101, 6=0110, 9=1001, A=1010, and C=1100 all require *two* changes to a 0000 codon to be reached by mutation, while 7=0111, B=1011, D=1101 and E=1110 require *three* changes to a 0000 codon, and F=1111 requires *four* changes to be reached by mutation from 0=0000. Thus, the result of uniform or random mutation probabilities depends on what set of items the mutation is random on. Making mutation random on a set of codons instead of on the unencoded amino acids makes it possible to bias the probabilities of mutation toward favorable changes, even though the process of mutation is random.

Experimental Procedure

We designed two different Tierran genetic codes, in addition to the original one created by Tom Ray, which appears to be neither maximally error tolerant nor maximally error sensitive. First we carried out a series of Tierran evolutionary runs and, using standard sequence comparison algorithms (program PILEUP; Genetics Computer Group 1991), determined what kind of instruction substitutions occurred in closely related, successful Tierran creatures. The results are not as simple and understandable as is the case with the natural amino acids and natural genetic code, but we gained enough insight to see how to organize the instruction set into a Gray code form, and also into an "anti"–Gray code form. We organized the instructions according to whether they modified the CPU registers, the stack, and/or the instruction pointer, or none of the above. Then we assigned bit patterns that mapped these characteristics as well as possible onto a Gray code. As in the natural genetic code, the correspondence is not perfect, but the trend can be expressed. To make an error-sensitive version of the Tierran genetic code, we deliberately assigned groups of instructions (four to six instructions) with the most similar characteristics to bit encodings that all differed in at least two positions.

We then carried out a series of paired evolutionary runs, pairs differing only in the genetic code assigned. A random number generator is used to determine the sites of mutation during a run. There is considerable variation in detail among runs with different seeds, so paired genetic code comparison runs used the same seed. There are many choices with respect to details of a run—length of run, rate of mutation, criteria for preserving "fossils" of creatures for later analysis, and so forth. Paired runs matched in all these details, but they still differed in their detailed progression and in their states after equal numbers of generations.

RESULTS

Different instances of Tierran evolution are always similar in broad outline, but always different in detail, like the parallels and differences that exist among Australian marsupials and European mammals that fill similar ecological niches.

The most interesting quantitative difference (Table 1, col. c) between the paired runs is that the runs with the error-tolerant genetic code resulted in a greater number of distinct genomes, a result consistent with our hypothesis that conservative mutations should have a greater probability of survival than radical mutations.

TABLE 1. ***Comparison of Several Pairs of Evolutionary Runs Using Error-Tolerant vs. Error-Sensitive Genetic Codes***

		c		*d*			*f*	
a	*b*	*Tolerant*	*Sensitive*	*Tolerant*	*Sensitive*	*e*	*Tolerant*	*Sensitive*
125	C10	1010	836	32	34	4	1.85	1.92
590	C5	2460	2095	596	578	2	2.15	2.06
600	D7	2770	2052	191	167	4	1.61	1.45
660	C8	2934	1828	346	278	3	1.73	1.26
1000	B6	4602	4432	314	438	3	1.73	1.41
2000	B9	7279	6533	518	650	3	1.15	1.41

Diversity (last two columns) is an information entropy calculated for creature sizes.
a *Generations in run.*
b *Id for pair of runs.*
c *Total number of genomes arising.*
d *Number of species exceeding a minimum percent of the total population.*
e *Minimum percent required in column d.*
f *Average diversity of sizes.*

A striking qualitative difference is that the distribution of mutants with the error-sensitive and error-tolerant genetic codes was different from all the runs with the original genetic code.

In our investigations (dozens of runs) with the original Tierran genetic code, no two runs have ever yielded identical creatures, but a typical "phenotypic" community has always developed, with analogous species developing in different runs. For example, those runs always yielded, as an early successful mutant, a length 45 parasite that lacks the copy subroutine. The runs with the modified genetic codes did *not* show this typical behavior. Parasites developed, but they had a different length. Runs with different genetic codes diverge from each other, and runs with *different* genetic codes and the *same* random number seed differ more in details than runs with the *same* genetic code and *different* random number seeds.

The creatures produced by evolution are more disorganized in their code than human-written code for the creatures would be, but this characteristic affects mainly the ease of understanding the code, not its execution. Some creatures were quite inefficient in their copying, but still managed to reproduce many times.

Conclusions

From the work described here, we conclude that evolution in the Tierra system (and likely also in the primeval natural world) is a chaotic process, in which many different "species," but broadly similar sets of self-replicating forms, can develop. However, these "communities" are also systematically constrained in their development by the kinds of mutations likely to occur. One simple form of constraint is the use of a genetic code. The particular form of that genetic code can play a stronger role than chance in making an ecology richer or sparser and thus presumably more or less resilient to disturbance. The broadly error-tolerant form of the natural genetic code could be the result of early selective advantage over alternative primitive forms of genetic code.

In brief, in the Tierra system, and probably elsewhere, (1) evolution is fast (also rich and creative), (2) evolution is systematically constrained by the spectrum of mutation available to it, and (3) a genetic code is a means to favorably *bias* the results of a *random* process.

Finally, continuing work on this system using tournament competitions between the most successful creatures from error-sensitive and error-tolerant runs showed that creatures from error-sensitive runs usually out-competed creatures from error-tolerant runs. This unanticipated outcome leads us to conclude that Tierran selective pressure (which comes primarily from other creatures) must be weak overall, so that exploring a larger mutation space and undergoing apparently stronger selection in the error-sensitive runs eventually results in better self-copiers.

REFERENCES

DAYHOFF, M.O., R.M. SCHWARTZ, AND B.C. ORCUTT (1978), "A Model of Evolutionary Change in Proteins," in *Atlas of Protein Sequence and Structure*, vol. 5 (Silver Spring, MD: National Biomedical Research Foundation), pp. 345–352.

FLORES, I. (1976), "Codes," in *Encyclopedia of Computer Science*, A. Ralston and C.L. Meek, eds. (New York: Petrocelli/Charter), pp. 210–215.

GENETICS COMPUTER GROUP (1991), *Program Manual for GCG Package*, vers. 7, April 1991.

RAY, T.S. (1991), "An Approach to the Synthesis of Life," in *Artificial Life II*, Santa Fe Institute Studies in the Sciences of Complexity, vol. XI, J.D. Farmer, C. Langton, S. Rasmussen, and C. Taylor, eds. (Redwood City, CA: Addison-Wesley), pp. 371–408.

RAY, T.S. (1992), "Evolution, Ecology, and Optimization of Digital Organisms," Working Paper 92-08-042 (Santa Fe: Santa Fe Institute).

SWANSON, R. (1984), "A Unifying Concept for the Amino Acid Code," *Bull. Math. Biol.* **46:** 187–203.

On Production of the Chirality of Biogenetic Material on the Early Earth

Liaofu Luo

Laboratory of Theoretical Physics and Biology
Department of Physics
Inner Mongolia University
Huhehote, China

Introduction

Francis Crick said that the chirality of biological macromolecules is the first principle of biochemistry. In fact, the universal chirality of amino acid and ribose molecules is the most mysterious and striking event at the molecular level in the evolution of life. One may say with no exaggeration that chirality is one of the most important marks of life. The origin of biological chirality is closely related to the evolution of life on the Earth, which provides a clue in exploring the origin of life. How can one explain the origin of the chirality of biogenetic material? There are two possible answers. The first is that it is delivered by comets, meteorites, or other loads of interstellar organics. The second is that it is produced on the Earth. The analysis of the chirality of amino acids in the 1969 Murchison meteorite seems to support the latter since all amino acids with asymmetrical carbon on it are racemic. If there is no evidence on the L-R asymmetry of amino acids in meteorites in further observations, we must accept the point that life is not nonnative. In this note, we shall investigate the possibility of the production of the chirality of biogenetic material on the early Earth. We will point out that the weak interaction as an asymmetrical force can drive the amino acids into one of the two probable configurations opposite to each other. The left-right asymmetry of chemical reaction rates induced by inelastic scattering of beta electrons within macromolecules is the dominant factor for producing biogenetic chirality. We will demonstrate that the rate asymmetry is on the order of 10^{-6}. The asymmetry introduces a term of asymmetrical force to the reaction-diffusion equation and gains amplification in the solution of the equation. So the concentration difference between L- and D- type molecules may reach a considerable figure. The next step is the polymerization process of chiral monomers. We will give a theoretical proof that the selectivity of the polymerization, namely, the relative difference of probabilities of adding an L- or D- isomer to the chain, would cause the occurrence of homochiral polymers from a medium with small polarization. So, the process of the chemical evolution of life on Earth eventually pronounces the victory of one kind of chirality.

Weak Interaction as the Origin of Biochirality

The proposal of parity-nonconservation of the weak interaction as the origin of biochirality is very attractive because of its simplicity and universality (see also MacDermott and Tranter 1996). Since the first attempt of Vester et al. (1959) and Ulbricht et al. (1962) to link the violation of parity at the subatomic level with that observed at the biomolecular level, a lot of work—both experimental and theoretical—has been done (Bonner 1991). Recently, Salam (1992) proposed a theory of phase transition of amino acids induced by a weak neutral current. However, the parity-nonconservative neutral current effect is extremely small ($< 10^{-17}$) and is easily dominated by any fluctuation factors. In fact, the maximal effect of the weak force does not come from the neutral current but from the irradiation of β-electrons on the chiral molecules. Hegstrom (1985) indicated that the chiral polarization resulting from β-radiolysis can be up to five or six orders of magnitude larger than the value resulting from weak neutral currents. We demonstrated that the irradiation of β-electrons on chiral molecules can induce a left-right asymmetrical effect on the order of 10^{-6} (Zeldovitch and Saakyan 1980, Wang and Luo 1985). Experimental investigations of β-radiolysis of amino acids have been done by several authors and the results were quite uncertain (see Garay 1968, Bonner et al. 1975, Darge et al. 1976, Hodge et al. 1979, and references cited in Bonner 1991). It seems that a thorough analysis has not yet been done and more theoretical investigations should be proceeding.

As is well known, the electrons emitted from β-decay are all left-polarized with polarization $\xi = -v/c$ (v is the speed of the electron and c is the speed of light).

Let the momentum of incident and scattered electrons be denoted by $\vec{k}_o$ and $\vec{k}$, respectively, and the momentum transfer $\vec{q} = \vec{k} - \vec{k}_o$. After the inelastic collision, the quantum state of the chiral molecule changes from a ground state O to an excited state n. The inelastic cross section $\sigma(O \rightarrow n)$ is dependent on the dipole strength D_n and rotatory strength R_n of the molecule (Wang and Luo 1985), according to

$$\sigma(O \rightarrow n) \approx D_n + (2/3)\, \hbar\, \xi\, (\vec{q} \cdot \vec{k}_o / mc\vec{k}_o)\, R_n, \tag{1}$$

where h is Planck's constant ($\hbar = h/2\pi$), and m is the electron's mass.

For enantiomers R_n values are equal in absolute value and opposite in sign. Denote the rotatory strength of L- (D-) type molecule by R_n^+ (R_n^-), then $R_n^- = -R_n^+$. The asymmetry of the cross section is

$$F = \{\sigma_L(O \rightarrow n) - \sigma_D(O \rightarrow n)\}/\{\sigma_L(O \rightarrow n) + \sigma_D(O \rightarrow n)\}$$

$$= (2/3)\, \hbar\, \xi\, (\vec{q} \cdot \vec{k}_o / mck_o)\, R_n^+ / D_n \tag{2}$$

Setting $R_n/D_n = 10^{-2}$ gives $F = 10^{-6}$. The transition of a molecule from the ground state to an excited state would change its activity and reaction rate. So the asymmetry of cross sections would cause an asymmetry of the reaction rates of chiral molecules under the irradiation of β-electrons,

$$(k_L - k_D)/(k_L + k_D) = 10^{-6} \tag{3}$$

The left-right asymmetry is on the order of one-millionth, which is much greater than other weak interaction processes, such as the asymmetry of cross section and energy shift due to neutral current, the asymmetrical absorption of *bremsstrahlung* radiation of β-electrons by chiral molecules, and so forth.

Furthermore, the asymmetry of reaction rate constants will introduce an asymmetrical force (g) in the reaction-diffusion equation and gain amplification in the solution of the equation. The asymmetry in concentration of L- and D- type molecules will be on the order of $g^{1/3}$ (Kondepudi and Nelson 1983, Kondepudi and Nelson 1985, Ding and Xu 1985 and references therein). Therefore, under β-electron irradiation the concentration difference of enantiomers of amino acids may reach to the order of 10^{-2}. Of course, the amplification mechanism can work only in the vicinity of the bifurcation point with a scale of 10^{-4}. Perhaps the asymmetrical effect (enantiomeric excess) may be experimentally observed more easily in the process far from equilibrium.

What about the sign of the handedness? Equation 2 gives the asymmetry of the decomposition rate of enantiomers of amino acids under irradiation of β-electrons. When $R_n^+ > 0$, $\sigma_D > \sigma_L$, favoring the L-amino acid surviving. When $R_n^+ < 0$, $\sigma_L > \sigma_D$, the survival of the D-amino acid is favored. In fact, among twenty amino acids, glycine (Gly), alanine (Ala), serine (Ser), aspartic acid (Asp), glutamic acid (Glu) and valine (Val) are early members of the primitive Earth's environment and others can be biosynthetically produced from these early amino acids (Wong 1988). Gly has no handedness. Ala, Val, Asp and Glu all have positive rotatory strengths for their L-isomers; L-Ser has positive rotatory strength in hydrochloric acid (HCl). For these amino acids—from optical to ultraviolet wavelengths, in a wide range of the spectrum—there is no absorption band and anomalous dispersion of optical activity. So the theoretical prediction of the handedness seems to agree with observational data.

About the origin of biochirality, if we don't attribute the solution of the problem to extraterrestrial origin, then the asymmetrical reaction of chiral molecules under the irradiation of β-electrons may be the first step of the process. We know that on the primitive Earth, especially in seawater, there were many radioactive elements (e.g., the decay series of uranium, thorium and ^{40}K [potassium] nuclei). Among these, ^{40}K is the most important radioactive source. Its half-life is 1.26×10^9 years and its average β-electron kinetic energy is 0.44ev. Four billion years ago, the number of decay events per kilogram per second provided by ^{40}K was 3000 in the Earth's crust, and 1200 in seawater. Taking the above data for seawater as an example and taking an average free path of 0.3 gm/cm^2 for 1 Mev electrons, gives a β-electron flux as large as 0.4×10^{-11}/sec. Thus, after β-electron irradiating for 10^4 years on the average, the transition of chiral molecules to excited states would take place with 100% probability and the enantiomeric excess of amino acids occurs. This is the first stage of the formation of biochirality. The amplification of the asymmetry through chemical reactions and polymerization would complete the process of the production of biochirality in the stage of chemical evolution.

Selective Polymerization of Chiral Molecules

In the last section we showed that the concentration difference of enantiomers of amino acids may reach to the order of 10^{-2} under β-electron irradiation and other appropriate conditions on the primitive Earth. In order to explain the homochirality of biomolecules one should answer how the polymerization process amplifies the left-right asymmetry—that is, how the polymerization of monomers results in homochiral polymers rather than heterochiral ones. Now we will give a general discussion of the polymerization process of chiral monomers (Luo 1995).

Let the energy of L- type chiral monomers be denoted by E_L and that of D- type by E_D. Define the chiral energy $\mathcal{J}$ as

$$J = (E_D - E_L)/2k_BT, \tag{4}$$

where k_B is Boltzmann's constant and T is the temperature.

Evidently, the ratio concentration of chiral monomers and its polarization are

$$n_L/n_D = e^{2J} \tag{5}$$

and

$$\eta = (n_L - n_D)/(n_L + n_D) = \tanh J, \tag{6}$$

respectively. To discuss the polymerization of monomers one should consider the interaction energy of neighboring monomers. Let the interaction energy (averaged over all possible orientations of monomers relative to each other) be $E_{LL} = E_{DD}$, $E_{LD} = E_{DL}$ from symmetry considerations. Defining the chiral interaction energy as

$$U = (E_{LD} - E_{LL})/2k_BT, \tag{7}$$

one has $U > 0$, in general. The stereoselectivity γ describes the difference in relative probabilities of adding the L- isomer and D- isomer to the chain. It relates to U as follows:

$$\gamma = \{\exp(-E_{LL}/k_BT) - \exp(-E_{LD}/k_BT)\}/\{\exp(-E_{LL}/k_BT) + \exp(-E_{LD}/k_BT)\} = \tanh U \tag{8}$$

Consider a one-dimensional system of N chiral monomers. The chirality of the i-th monomer is described by spin σ_i. $\sigma_i = +1$ means L and $\sigma_i = -1$ means D. The Hamiltonian of the system is described by

$$H = -k_BT(J \sum_i \sigma_i + U \sum_i \sigma_i \sigma_{i+1}) \tag{9}$$

Though the molecules in L- and D- units of a polymer cannot interconvert freely we may introduce a statistical ensemble that is defined by the Hamiltonian (Eqn. 9) and calculate the equilibrium distribution by use of Boltzmann's law. The calculation method is taken from the Ising model (Huang 1963). We obtain the partition function Z

$$Z = \lambda_1^N + \lambda_2^N$$
$$= \lambda_1^N \text{ (as } N \text{ becomes large)} \tag{10}$$

$$\lambda_{1.2} = e^{U}\{\cosh J + [\sinh^2 J + e^{-4U}]^{1/2}\}, \tag{11}$$

where λ is a parameter denoting the nonequilibrium constraints on the system, typically the concentration of one of the reactants that is maintained constant.

Defining the chiral polarization of polymer f_L through the fraction of $\sigma = +1$ monomers in the chain, we deduce

$$\begin{aligned} f_L &= 1/2\,(1 + N^{-1}\,\partial\, lnZ/\partial J) \\ &= 1/2\,\{1 + \sinh J/[\sinh^2 J + e^{-4U}]^{1/2}\} \end{aligned} \tag{12}$$

If $f_L = 1$ the polymer is of chirally pure L- type, if $f_L = 0$ the polymer is of chirally pure D- type. Equation 12 gives a curve of f_L versus $\mathcal{J}$ which has a sharp transition from 0 to 1 at $\mathcal{J} = 0$. The sharpness of the transition depends on the largeness of the interaction energy U. So the equation shows that the polymer is chirally pure when $e^{-4U} << \sinh^2\mathcal{J}$ and it is L- or D- type depending on the sign of $\mathcal{J}$. In fact, for a given $\mathcal{J}$, from

$$\exp(-4U_{\min}) = \sinh^2 J \tag{13}$$

we get $U_{\min}$. When $U > U_{\min}$ the polymer can be viewed as chirally pure, approximately. It is called the "cooperativity" in physics. Thus we have succeeded in deducing the homochiral polymer by use of the above-mentioned model. The basic assumptions implied in our theory are (1) a definite nonvanishing chiral energy $\mathcal{J}$ which gives the advantageous direction of "up" and "down" of the "spin" statistically, and (2) a definite stereoselectivity (or the difference between the interaction energies E_{LD} and E_{LL}) which guarantees a sharp cooperative transition of the chiral polarization f_L versus chiral energy $\mathcal{J}$. The two assumptions correspond to the two energy terms in Equation 9. Table 1 gives the results for different chiral parameters. For simplicity, only the $\mathcal{J} > 0$ case (corresponding to L- type) is listed; the similar relation for the case of $\mathcal{J} < 0$ (D- type) can easily be found.

From Table 1 we find that there are two possible scenarios for the formation of chiral polymers (Avetisov et al. 1991). The first scenario corresponds to the case of $\mathcal{J} > 0.1$. It requires a small stereoselectivity, $\gamma < 0.8$. If the temperature is lowered, then γ will be smaller than 0.5, which could be satisfied by usual chemical reactions. However, it seems difficult to obtain the chiral monomers with $\eta > 0.1$ through the asymmetrical weak force. The only way of explaining the origin of biochirality in this scenario is to introduce the spontaneous breaking of mirror symmetry of chemical reactions (Avetisov et al. 1991, Avetisov 1992).

TABLE 1. *The Relations of Chiral Parameters*

J	n_L/n_D	η	$sinh^2J$	U_{min}	γ_{min}	T_c
10^{-6}	~ 1	10^{-6}	10^{-12}	6.9	~ 1	25 K
10^{-2}	1.02	10^{-2}	10^{-4}	2.3	0.99	76 K
0.1	1.22	0.1	10^{-2}	1.15	0.82	151 K
0.32	1.90	0.31	0.11	0.56	0.51	311 K
0.87	5.70	0.70	0.968	0.008	0.008	
0.88	5.81	0.71	0.996	10^{-3}	10^{-3}	
1	7.39	0.76	1.38	~ 0	~ 0	
2	54.6	0.96	13.1	~ 0	~ 0	

The second scenario corresponds to the case of $J < 0.1$, which is more attractive for us. In this case a large stereoselectivity is required for production of the chiral polymer. From Equation 7 the large chiral interaction energy U can be obtained by lowering the temperature. If $\Delta E = E_{LD} - E_{LL}$ takes 1% of the C–N bond energy, that is, 0.7 Kcal/mol, which means $\gamma \sim 0.5$ at $T \sim 300$ K, then the maximal temperatures T_c necessary for forming homochiral polymers are listed in the last column of Table 1. Of course, since ΔE is difficult to determine accurately, the listed values of T_c in Table 1 are only for reference. In order to obtain a large chiral energy U or large stereoselectivity γ, another approach is to introduce a catalytic mechanism to increase the specificity of the polymerization of monomers. In fact, the bioselection of enantiomers with participation of enzymes has reached $\gamma = 0.99999999$. In summary, in the second scenario of the formation of chiral polymers, the parameters J and η are permitted to take a relatively small value which could be created by weak interaction. So the assumption of production of biochirality through the weak force does not seem to conflict with any physical principle. The possibility is still open to investigation.

REFERENCES

AVETISOV, V.A. (1992), Lecture in College on Methods and Exp. Techniques in Biophysics (Trieste).

AVETISOV, V.A., V.I. GOLDANSKII, AND V.V. KUZMIN (1991), "Handedness, Origin of Life and Evolution,"*Physics Today* **44 (7):** 33–41.

BONNER, W.A. (1991), "The Origins and Amplification of Biomolecular Chirality," *Origins of Life* **21:** 60–111.

BONNER, W.A., M A. VAN DORT, AND M.R. YEARIAN (1975), "Asymmetric Degradation of DL-Leucine with Longitudinally Polarised Electrons," *Nature* **258:** 419–421.

DARGE, W., I. LACZKO, AND W. THIEMANN (1976), "Stereoselectivity of β-Irradiation of D,L-Tryptophan in Aqueous Solution," *Nature* **261:** 522–524.

DING, D.F., AND J.H. XU (1985), "The Symmetry-Breaking in Living Systems—The Selection Rules, Sensitivity and Chiral Matching," *Scientia Sinica B* **28:** 250–263.

GARAY, A.S. (1968), "Origin and Role of Optical Isomery in Life," *Nature* **219:** 338–340.

HEGSTROM, R. (1985), "Weak Neutral Current and β-Radiolysis Effects on the Origin of Biomolecular Chirality," *Nature* **315:** 749.

HODGE, L.A., F.B. DUNNING, G.K. WALTERS, R.H. WHITE, AND G.J. SCHROEPFER, JR. (1979), "Degradation of DL-Leucine with Longitudinally Polarised Electrons," *Nature* **280:** 250–252.

HUANG, K.S. (1963), *Statistical Mechanics* (New York: John Wiley & Sons), p. 346.

KONDEPUDI, D.K., AND G.W. NELSON (1983), "Chiral Symmetry Breaking in Nonequilibrium Systems," *Phys. Rev. Lett.* **50:** 1023–1026.

KONDEPUDI, D.K., AND G.W. NELSON (1985), "Weak Neutral Currents and the Origin of Biomolecular Chirality," *Nature* **314:** 438–441.

LUO, L.F. (1995?), *Acta Scientiarum Naturalium Universitatis Intramongolicae* (Neimenggu) **25**.

MACDERMOTT, A.J., AND G.E. TRANTER (1996), "The Electroweak Origin of the Chiral Signature of Life," this volume.

SALAM, A. (1992), "The Role of Chirality in the Origin of Life," *J. Mol. Evol.* **33:** 105–113.

ULBRICHT, T.L.V., ET AL. (1962), "Attempts to Induce Optical Activity with Polarized β-Radiation," *Tetrahedron* **18:** 629–637.

VESTER, F., ET AL. (1959), "Optical Activity and Parity Violation in β-Decay," *Naturwissenschaften* **46:** 68.

WANG, J.Y., AND L.F. LUO (1985), "The Origin of Biological Chirality and Weak Interaction," *Scientia Sinica B* **28:** 1265–1277.

WONG, JT-F. (1988), "Evolution of the Genetic Code," *Microbiol. Sci.* **5:** 174–181.

ZELDOVITCH, YA.B. AND D.B. SAAKYAN (1980), "Asymmetry in the Separation of Optically Active Molecules by Longitudinally Polarized Relativistic Electrons," *J. Exp. Theor. Phys.* **78:** 2233–2237.

APPENDIXES

—Appendix I—

List of Conference Participants

Sara Acevedo	*SETI Institute*
Dale Andersen	*SETI Institute*
Peter Backus	*SETI Institute*
Emma Bakes	*Princeton University*
Akiva Bar-Nun	*Tel Aviv University*
Max Bernstein	*SETI Institute*
John Billingham	*NASA Ames Research Center*
J. Ellen Blue	*SRI International*
William Borucki	*NASA Ames Research Center*
Stefan Brandmeier	*Polytechnic University of Konstanz, Germany*
Theodore Bunch	*NASA Ames Research Center*
Ken Caldeira	*Lawrence Livermore National Laboratory*
John (Chris) Campbell	*University of Alabama at Birmingham*
Michael Carr	*United States Geological Survey*
Sherwood Chang	*NASA Ames Research Center*
Christopher Chyba	*National Security Council*
Benton C. Clark	*Martin Marietta*
Lynnette Cook	*Morrison Planetarium*
Dale Cruikshank	*NASA Ames Research Center*
D. Kent Cullers	*NASA Ames Research Center*
David W. Deamer	*University of California at Davis*
David J. Des Marais	*NASA Ames Research Center*
Donald DeVincenzi	*NASA Ames Research Center*
Anthony Dobrovolskis	*NASA Ames Research Center*
Stephen H. Dole	*Laguna Hills, CA*
Luke Dones	*UC Santa Cruz & NASA Ames Research Center*
Joanne Donovan	*Hierophont Corporation*
Laurance Doyle	*SETI Institute*
Frank Drake	*UC Santa Cruz & SETI Institute*
John Dreher	*SETI Institute*
Kerry Edwards	*Hierophont Corporation*

APPENDIX I—LIST OF CONFERENCE PARTICIPANTS (cont.)

Marina Fomenkova	*NASA Ames Research Center*
Friedemann Freund	*San Jose State University & SETI Institute*
J. Peter Gogarten	*University of Connecticut*
John Grotzinger	*Massachusetts Institute of Technology*
Robert Haberle	*NASA Ames Research Center*
Alan Hale	*Southwest Institute for Space Research*
Martin Heath	*Greenwich Community College, England*
Neil Heather	*Synernet*
Icko Iben	*University of Illinois*
Miriam G. Iben	*University of Illinois*
Linda L. Jahnke	*NASA Ames Research Center*
Jon M. Jenkins	*SETI Institute*
Harry Jones	*NASA Ames Research Center*
James Kasting	*Pennsylvania State University*
Harold P. Klein	*Santa Clara University*
David Koch	*NASA Ames Research Center*
Robert Krekorian	*SETI Institute*
Antonio Lazcano	*Universidad Nacional Autónoma de México*
Alain Léger	*Institut d'Astrophysique Spatiale—Orsay*
Jack Lissauer	*State University of New York at Stony Brook*
Donald Lowe	*Stanford University*
Keith Lynar	*SRI International*
Alexandra J. MacDermott	*Oxford University*
Clifford Matthews	*University of Illinois*
Gene McDonald	*UC San Diego*
Christopher McKay	*NASA Ames Research Center*
Michael Meyer	*NASA Headquarters*
David Morrison	*NASA Ames Research Center*
Michelle N. Murray	*SETI Institute*
Hakobu Nakamura	*Biological Institute, Konan University*
Koji Ohnishi	*Niigata University*
Bernard M. Oliver	*NASA Ames Research Center*
John Oró	*University of Houston*
Sandra Owen	*NASA Ames Research Center*
Julie Paque	*SETI Institute*
Yvonne Pendleton	*NASA Ames Research Center*
Thomas Pierson	*SETI Institute*

APPENDIX I—LIST OF CONFERENCE PARTICIPANTS (cont.)

Ray Reynolds	*NASA Ames Research Center*
Jan Rosenqvist	*Observatoire de Paris—Meudon*
Carl Sagan	*Cornell University*
Jean Schneider	*Observatoire de Paris—Meudon*
David Schwartzman	*Howard University*
Charles Seeger	*SETI Institute*
Thomas Shen	*SETI Institute*
John Stauffer	*Harvard-Smithsonian Center for Astrophysics*
Carol Stoker	*NASA Ames Research Center*
David Summers	*SETI Institute*
Michael Summers	*SETI Science Education Group*
Rosemarie Swanson	*Texas A&M University*
Jill Tarter	*SETI Institute*
O. Brian Toon	*NASA Ames Research Center*
Alexey F. Topunov	*A.N. Bakh Institute of Biochemistry, Russian Academy of Science*
Dominique Toublanc	*Observatoire de Bordeaux*
C.T. Vanajakshi	*NASA Ames Research Center*
Rajendra Vikramsingh	*Palo Alto*
George Wetherill	*The Carnegie Institution of Washington*
Robert A. Wharton, Jr.	*Desert Research Institute*
Daniel P. Whitmire	*University of Southwestern Louisiana*
Darren M. Williams	*Pennsylvania State University*

—APPENDIX II—

COMBINED LIST OF REFERENCES

Planetary Aspects

AHRENS, T.J., J.D. O'KEEFE, AND M.A. LANGE (1989), "Formation of Atmospheres During Accretion of the Terrestrial Planets," in *Origin and Evolution of Planetary and Satellite Atmospheres*, S.K. Atreya, J.B. Pollack, and M.S. Mathews, eds. (Tucson: University of Arizona Press), pp. 328–385.

AINES, R.D., AND G.R. ROSSMAN (1984), "Water in Minerals—A Peak in the Infrared?," *JGR* **89:** 4059–4071.

ALLEN, C.W. (1976), *Astrophysical Quantities*, 3d ed. (London: The Athlone Press), p. 249 (Table 1).

ANDERS, A. (1989), "Pre-biotic Organic Matter from Comets and Asteroids," *Nature* **342:** 255–257.

ANTHONY, C. (1982), *The Biochemistry of Methylotrophs* (London: Academic Press).

BARNES, J.R. (1980), "Time Spectral Analysis of Midlatitude Disturbances in the Martian Atmosphere," *J. Atmos. Sci.* **37:** 2002–2015.

BARSHAY, S.S., AND J.S. LEWIS (1975), "Chemistry of Solar Material," in *The Dusty Universe*, G.B. Field and A.G.W. Cameron, eds. (New York: Neale Watson Academic Publ.), pp. 33–40.

BARSHAY, S.S., AND J.S. LEWIS (1976), "Chemistry of Primitive Solar Material," *Ann. Rev. Astron. Astrophys.* **14:** 81–90.

BARSHAY, S.S., AND J.S. LEWIS (1978), "Chemical Structure of the Deep Atmosphere of Jupiter," *Icarus* **33:** 593–611.

BELL, D.R., AND G.R. ROSSMAN (1992), "Water in Earth's Mantle: The Role of Nominally Anhydrous Minerals," *Science* **255:** 1391–1397.

BERGER, A. (1988), "Milankovitch Theory and Climate," *Rev. Geophys.* **26:** 624–657.

BIRD, C.W., J.M. LYNCH, F.J. PIRT, W.W. REID, C.J.W. BROOKS, AND B.S. MIDDLEDITCH (1971), "Steroids and Squalene in *Methylococcus capsulatus* Grown on Methane," *Nature* **230:** 473–474.

BLOCH, K. (1983), "Sterol Structure and Membrane Function," *CRC Crit. Rev. Biochem.* **14:** 47–92.

BOON, J.J. (1984), "Tracing the Origin of Chemical Fossils in Microbial Mats: Biogeochemical Investigations of Solar Lake Cyanobacterial Mats Using Analytical Pyrolysis Methods," in *MBL Lectures in Biology: Microbial Mats: Stromatolites*, vol. 3, Y. Cohen, R.W. Castenholz, and H.O. Halvorson, eds. (New York: Alan R. Liss, Inc.), pp. 313–342.

APPENDIX II—PLANETARY ASPECTS REFERENCES (cont.)

BOUVIER, P., M. ROHMER, P. BENVENISTE, AND G. OURISSON (1976), "$\Delta^{8(14)}$-Steroids in the Bacterium *Methylococcus capsulatus*," *Biochem. J.* **159:** 267–271.

BROWN, G.E. (1980), *Olivine and Silicate Spinels, Reviews in Mineralogy*, R.H. Ribbe, ed. (Washington, DC: American Mineralogical Society), pp. 275–381.

BULLOCK, M.A., D.H. GRINSPOON, AND J.W. HEAD (1992), "Monte Carlo Modeling of the Resurfacing of Venus," paper presented at the International Colloquium on Venus, August 10–12, Pasadena, California.

CALDEIRA, K., AND J.F. KASTING (1992), "Susceptibility of Early Earth to Irreversible Glaciation Caused by Carbon Dioxide Clouds," *Nature* **359:** 226–228.

CHAMEIDES, W.L., AND J.C.G. WALKER (1981), "Rates of Fixation by Lightning of Carbon and Nitrogen in Possible Primitive Atmospheres," *Origins of Life* **11:** 291–302.

CHANG, S. (1988), "Planetary Environments and the Conditions of Life," *Phil. Trans. R. Soc. Lond. A* **325:** 601–610.

CHANG, S. (1993), "Prebiotic Synthesis in Planetary Environments," in *The Chemistry of Life's Origins*, J.M. Greenberg, ed. (Amsterdam: Kluwer Academic Publishers), pp. 259–299.

CHANG, S., D. DES MARAIS, R. MACK, S.L. MILLER, AND G.E. STRATHEARN (1983), "Prebiotic Organic Synthesis and the Origin of Life," in *Earth's Earliest Biosphere*, J.W. Schopf, ed. (Princeton, NJ: Princeton University Press), pp. 53–92.

CHITTENDEN, G.J.F., AND A.W. SCHWARTZ (1981), "Prebiotic Photosynthetic Reactions," *BioSystems* **14:** 15–32.

CHYBA, C.F. (1990), "Impact Delivery and Erosion of Planetary Oceans in the Early Inner Solar System," *Nature* **343:** 129–133.

CHYBA, C., AND C. SAGAN (1992), "Endogenous Production, Exogenous Delivery and Impact-Shock Synthesis of Organic Molecules: An Inventory for the Origins of Life," *Nature* **355:** 125–132.

CHYBA, C.F., P.J. THOMAS, L. BROOKSHAW, AND C. SAGAN (1990), "Cometary Delivery of Organic Molecules to the Early Earth," *Science* **249:** 366–373.

COLBY, J., D.I. STIRLING, AND H. DALTON (1977), "The Soluble Methane Monooxygenase of *Methylococcus capsulatus* (Bath): Its Ability to Oxygenate *n*-Alkanes, *n*-Alkenes, Ether, and Alicyclic, Aromatic and Heterocyclic Compounds," *Biochem. J.* **165:** 395–402.

DALTON, H. (1992), "Methane Oxidation by Methanotrophs," in *Biotechnology Handbooks: Methane and Methanol Utilizers*, J.C. Murrell and H. Dalton, eds. (New York: Plenum Press), pp. 85–114.

DERRY, L.A., AND S.B. JACOBSEN (1990), "The Chemical Evolution of Precambrian Seawater: Evidence from REEs in Banded Iron Formations," *Geochim. Cosmochim. Acta* **54:** 2965–2975.

DICKINSON, J.T., L.C. JENSEN, AND M.R. MCKAY (1987), "Neutral Molecule Emission from the Fracture of Crystalline Magnesium Oxide," *J. Vac. Sci. Technol. A* **1987.5(4, Pt. 2):** 1162–1168.

APPENDIX II—PLANETARY ASPECTS REFERENCES (cont.)

DICKINSON, J.T., L.C. JENSEN, M.R. MCKAY, AND F. FREUND (1986), "The Emission of Atoms and Molecules Accompanying Fracture of Single Crystal Magnesium Oxide," *J. Vac. Sci. Technol. A* **1986.4(3, Pt. 2):** 1648–1652.

DICKINSON, J.T., M. DRESSER, AND L.C. JENSEN (1985), "Time Correlation of Ion and Electron Emission from Surfaces Following Fracture," Springer Series Surface Science, *Desorption Induced Electron Transitions (DIET)* (Heidelberg: Springer-Verlag), pp. 281–289.

DOLE, S.H. (1964), *Habitable Planets for Man* (New York: Blaisdell Publishing Company).

DONAHUE, T.M., AND J.B. POLLACK (1983), "Origin and Evolution of the Atmosphere of Venus," in *Venus*, D.M. Hunten, L. Colin, T.M. Donahue, and V.I. Moroz, eds. (Tucson: University of Arizona Press), pp. 1003–1036.

DONAHUE, T.M., AND R.E. HARTLE (1992), "Solar Cycle Variations in H^+ and D^+ Densities in the Venus Ionosphere: Implications for Escape," *Geophys. Res. Lett.* **19:** 2449–2452.

DONAHUE, T.M., J.H. HOFFMAN, R.R. HODGES, AND A.J. WATSON (1982), "Venus Was Wet: A Measurement of the Ratio of Deuterium to Hydrogen," *Science* **216:** 630–633.

DOYLE, M.P., AND S.N. MAHAPATRO (1984), "Nitric Oxide Dissociation from Trioxodinitrate (II) in Aqueous Solution," *J. Am. Chem. Soc.* **106:** 3678–3679.

DREVER, J.I., Y.H. LI, AND J.B. MAYNARD (1988), "Geochemical Cycles: The Continental Crust and the Oceans," in *Chemical Cycles in the Evolution of the Earth*, C.B. Gregor, R.M. Garrels, F.T. Mackenzie, and J.B. Maynard, eds. (New York: John Wiley & Sons), pp. 17–51.

FEGLEY, B., JR., R.G. PRINN, H. HARTMAN, AND G.H. WATKINS (1986), "Chemical Effects of Large Impacts on the Earth's Primitive Atmosphere," *Nature* **319:** 305–308.

FEGLEY, M.B., JR., AND J.S. LEWIS (1979), "Thermodynamics of Selected Trace Elements in the Jovian Atmosphere," *Icarus* **38:** 166–179.

FEGLEY, M.B., JR., AND J.S. LEWIS (1980), "Volatile Element Chemistry in the Solar Nebula: Na, K, F, Cl, Br, and P," *Icarus* **41:** 439–455.

FREEMAN, K.H., J.M. HAYES, J.-M. TRENDEL, AND P. ALBRECHT (1990), "Evidence from Carbon Isotope Measurements for Diverse Origins of Sedimentary Hydrocarbons," *Nature* **343:** 254–256.

FREIER, R.K., ED. (1976), *Aqueous Solutions*, vol. 1 (Berlin: Walter de Gruyter).

FREUND, F. (1984), "Abiotic Synthesis of Organic Molecules from Minerals Containing Traces of Dissolved H_2O, CO_2 and N_2. Part I: Basic Principles," in *Proc. 7th Int. Conf. Origins of Life*, K. Dose, A.W. Schwarz, and W.H.P. Tieman, eds. (Dordrecht: D. Reidel Publishing Company), pp. 185–196.

FREUND, F. (1985), "Conversion of Dissolved 'Water' into Molecular Hydrogen and Peroxy Linkages," *J. Non-Cryst. Solids* **71:** 195–202.

FREUND, F. (1986), "Solute Carbon and Carbon Segregation in Magnesium Oxide Single Crystals—Secondary Ion Mass Spectrometer Study," *Phys. Chem. Minerals* **13:** 262–276.

FREUND, F. (1987), "Hydrogen and Carbon in Solid Solution in Oxides and Silicates," *Phys. Chem. Minerals* **15:** 1–18.

APPENDIX II—PLANETARY ASPECTS REFERENCES (cont.)

FREUND, F., AND G. OBERHEUSER (1986), "Water Dissolved in Olivine: A Single Crystal Infrared Study," *JGR* **91:** 745–761.

FREUND, F., AND H. WENGELER (1982), "The Infrared Spectrum of OH-Compensated Defect Sites in C-Doped MgO and CaO Single Crystals," *J. Phys. Chem. Solids* **43:** 129–145.

FREUND, F., H. KATHREIN, H. WENGELER, R. KNOBEL, AND H.J. HEINEN (1980), "Carbon in Solid Solution in Forsterite—A Key to the Untractable Nature of Reduced Carbon in Terrestrial and Cosmogenic Rocks," *Geochim. Cosmochim. Acta* **44:** 1319–1333.

FREUND, F., M.M. FREUND, AND F. BATLLO (1993), "Critical Review of Electrical Conductivity Measurements and Charge Distribution Analysis of Magnesium Oxide," *JGR* **98:** 22,209–22,229.

FREUND, F., M.M. MASUDA, AND M.M. FREUND (1991), "Highly Mobile Oxygen Hole-type Charge Carriers in Fused Silica," *J. Mater. Sci.* **8:** 1619–1622.

GIERASCH, P.J., AND O.B. TOON (1973), "Atmospheric Pressure Variation and the Climate of Mars," *J. Atmos. Sci.* **30:** 1502–1508.

GOUGH, D.O. (1981), "Solar Interior Structure and Luminosity Variations," *Solar Phys.* **74:** 21–34.

GREEN, H.W., AND Y. GUEGEN (1983), "Deformation of Peridotite in the Mantle and Extraction by Kimberlite: A Case History Documented by Fluid and Solid Precipitates in Olivine," *Tectonophys.* **92:** 71–92.

GREGOR, C.B., R.M. GARRELS, F.T. MACKENZIE, AND J.B. MAYNARD, EDS. (1988), *Chemical Cycles in the Evolution of the Earth* (New York: John Wiley & Sons).

GRINSPOON, D.H. (1993), "Implications of the High D/H Ratio for the Sources of Water in Venus' Atmosphere," *Nature* **363:** 428–431.

GURWELL, M.A., AND Y.L. YUNG (1993), "Fractionation of Hydrogen and Deuterium on Venus Due to Collisional Ejection," *Planet. Space Sci.* **41:** 91–104.

HABERLE, R.M., D. TYLER, C.P. MCKAY, AND W.L. DAVIS (1994), "A Model for the Evolution of CO_2 on Mars," *Icarus* **109:** 102–120.

HANSON, R.S. (1980), "Ecology and Diversity of Methylotrophic Organisms," *Adv. Appl. Microbiol.* **26:** 3–39.

HART, M.H. (1978), "The Evolution of the Atmosphere of the Earth," *Icarus* **33:** 23–39.

HART, M.H. (1979), "Habitable Zones Around Main Sequence Stars," *Icarus* **37:** 351–357.

HARTMANN, W.K., R.J. PHILLIPS, AND G.J. TAYLOR (1986), *Origin of the Moon* (Houston: Lunar and Planetary Science Institute).

HAYES, J.M. (1983), "Geochemical Evidence Bearing on the Origin of Aerobiosis, a Speculative Hypothesis," in *Earth's Earliest Biosphere: Its Origin and Evolution*, J.W. Schopf, ed. (Princeton, NJ: Princeton University Press), pp. 291–301.

HAYES, J.M. (1994), "Global Methanotrophy at the Archean-Proterozoic Transition," in *Early Life on Earth*, Proceedings of Nobel Symposium 84, S. Bengtson, ed. (New York: Columbia University Press).

APPENDIX II—PLANETARY ASPECTS REFERENCES (cont.)

HAYES, J.M., K.H. FREEMAN, B.N. POPP, AND C.H. HOHAM (1990), "Compound-Specific Isotopic Analyses: A Novel Tool for Reconstruction of Ancient Biogeochemical Processes," *Org. Geochem.* **16:** 1115–1128.

HENNET, J.C., N.G. HOLM, AND M.H. ENGEL (1992), "Abiotic Synthesis of Amino Acids Under Hydrothermal Conditions and the Origin of Life: A Perpetual Phenomenon?," *Naturwiss* **79:** 361–365.

HOLLAND, H.D. (1973), "The Oceans: A Possible Source of Iron in Iron Formations," *Econ. Geol.* **68:** 1169–1172.

HOLLAND, H.D. (1984), *The Chemical Evolution of the Atmosphere and Oceans* (Princeton, NJ: Princeton University Press), ch. 4.

HOLLOWAY, J.R. (1981), "Volatile Interactions in Magmas," *Adv. Phys. Geochem.* **1:** 273–293.

HUANG, S.-S. (1959), "Occurrence of Life in the Universe," *Amer. Sci.* **47:** 397–402.

HUANG, S.-S. (1960), "Life Outside the Solar System," *Sci. Am.* **202 (4):** 55–63.

HUNTEN, D.M. (1973), "The Escape of Light Gases from Planctary Atmospheres," *J. Atmos. Sci.* **30:** 1481–1494.

HUNTEN, D.M., T.M. DONAHUE, J.C.G. WALKER, AND J.F. KASTING (1989), "Escape of Atmospheres and Loss of Water," in *Origin and Evolution of Planetary and Satellite Atmospheres*, S.K. Atreya, J.B. Pollack, and M.S. Mathews, eds. (Tucson: University of Arizona Press), pp. 386–422.

IBEN, I. (1967a), "Stellar Evolution Within and Off the Main Sequence," *Ann. Rev. Astron. Astrophys.* **5:** 571–626.

IBEN, I. (1967b), "Stellar Evolution. VI. Evolution from the Main Sequence to the Red Giant Branch for Stars of Mass 1 $M_\odot$, 1.25 $M_\odot$, and 1.5 $M_\odot$," *ApJ* **147:** 624–649.

INGERSOLL, A.P. (1969), "The Runaway Greenhouse on Venus: A History of Water on Venus," *J. Atmos. Sci.* **26:** 1191–1198.

INGMANSON, D.E., AND M.J. DOWLER (1977), "Chemical Evolution and the Evolution of the Earth's Crust," *Origins of Life* **8:** 221–224.

JACKSON, M.J., T.G. POWELL, R.E. SUMMONS, AND I.P. SWEET (1986), "Hydrocarbon Show and Petroleum Source Rocks in Sediments as Old as 1.7 × 10^9 Years," *Nature* **322:** 727–729.

JAHNKE, L.L., AND H.P. KLEIN (1979), "Oxygen as a Factor in Eukaryote Evolution: Some Effects of Low Levels of Oxygen on *Saccharomyces cerevisiae*," *Origins of Life* **9:** 329–334.

JAHNKE, L.L., AND H.P. KLEIN (1983), "Oxygen Requirements for Formation and Activity of the Squalene Epoxidase in *Saccharomyces cerevisiae*," *J. Bacteriol.* **155:** 488–492.

JAHNKE, L.L., AND P.D. NICHOLS (1986), "Methyl Sterol and Cyclopropane Fatty Acid Composition of *Methylococcus capsulatus* Grown at Low Oxygen Tensions," *J. Bacteriol.* **167:** 238–242.

JAHNKE, L.L., H. STAN-LOTTER, K. KATO, AND L.I. HOCHSTEIN (1992), "Presence of Methyl Sterol and Bacteriohopanepolyol in an Outer-Membrane Preparation from *Methylococcus capsulatus* (Bath)," *J. Gen. Microbiol.* **138:** 1759–1766.

APPENDIX II—PLANETARY ASPECTS REFERENCES (cont.)

JAOUL, O. (1990), "Multicomponent Diffusion and Creep in Olivine," *JGR* **95:** 17,631–17,642.

KASTING, J.F. (1987), "Theoretical Constraints on Oxygen and Carbon Dioxide Concentrations in the Precambrian Atmosphere," *Precambrian Res.* **34:** 205–229.

KASTING, J.F. (1988), "Runaway and Moist Greenhouse Atmospheres and the Evolution of Earth and Venus," *Icarus* **74:** 472–494.

KASTING, J.F. (1990), "Bolide Impacts and the Oxidation State of Carbon in the Earth's Early Atmosphere," *Origins of Life* **20:** 199–231.

KASTING, J.F. (1991), "CO_2 Condensation and the Climate of Early Mars," *Icarus* **94:** 1–13.

KASTING, J.F., AND J.B. POLLACK (1983), "Loss of Water from Venus. I. Hydrodynamic Escape of Hydrogen," *Icarus* **53:** 479–508.

KASTING, J.F., AND T.P. ACKERMAN (1986), "Climatic Consequences of Very High Carbon Dioxide Levels in the Earth's Early Atmosphere," *Science* **234:** 1383–1385.

KASTING, J.F., D.P. Whitmire, AND R.T. REYNOLDS (1993), "Habitable Zones Around Main Sequence Stars," *Icarus* **101:** 108–128.

KASTING, J.F., J.B. POLLACK, AND T.P. ACKERMAN (1984), "Response of Earth's Surface Temperature to Increases in Solar Flux and Implications for Loss of Water from Venus," *Icarus* **57:** 335–355.

KNOBEL, R. (1983), *Gaseous H–C–N–O Compounds Released from Magnesium Oxide and Silicate Surfaces*, Ph.D. dissertation, Cologne (in German).

KNOBEL, R., H. BREUER, AND F. FREUND (1984), "Abiotic Synthesis of Organic Molecules from Minerals Containing Traces of Dissolved H_2O, CO_2 and N_2. Part II: Experimental Data," in *Proc. 7th Int. Conf. Origins of Life*, K. Dose, A.W. Schwarz, and W.H.P. Tieman, eds. (Dordrecht: D. Reidel Publishing Company), pp. 197–202.

KRÖNER, A., AND P.W. LAYER (1992), "Crust Formation and Plate Motion in the Early Archean," *Science* **256:** 1405–1411.

LASKAR, J., AND P. ROBUTEL (1993), "The Chaotic Obliquity of the Planets," *Nature* **361:** 608–614.

LASKAR, J., F. JOUTEL, AND P. ROBUTEL (1993), "Stabilization of the Earth's Obliquity by the Moon," *Nature* **361:** 615–617.

LAWLER, M.E., AND D.E. BROWNLEE (1992), "CHON as a Component of Dust from Comet Halley," *Nature* **359:** 810–812.

LEWIS, J.S. (1972), "Low-Temperature Condensation from the Solar Nebula," *Icarus* **16:** 241–252.

LEWIS, J.S. (1973), "Origin and Composition of the Terrestrial Planets and Satellites of the Outer Planets," in *The Origin of the Solar System*, H. Reeves, ed. (Paris: CNRS), pp. 202–205.

LISSAUER, J.J. (1996), "Spacing of Planetary Orbits About Stars of Differing Masses," this volume.

LOVELOCK, J.E. (1979), *Gaia: A New Look at Life on Earth* (Oxford: Oxford University Press).

APPENDIX II—PLANETARY ASPECTS REFERENCES (cont.)

MACKWELL, S.J., Q. BAI, AND D.L. KOHLSTEDT (1990), "Rheology of Olivine and the Strength of the Lithosphere," *Geophys. Res. Lett.* **17:** 9–12.

MANCINELLI, R.L., AND C.P. MCKAY (1988), "The Evolution of Nitrogen Cycling," *Origins of Life* **18:** 311–325.

MATHEZ, E.A., J.D. BLACIC, J. BERRY, C. MAGGIORE, AND M. HOLLANDER (1987), "Carbon in Olivine: Results from Nuclear Reaction Analysis," *JGR* **92:** 3500–3506.

MATTIOLI, G.S., AND B.J. WOOD (1986), "Upper Mantle Oxygen Fugacity Recorded by Spinel Iherzolites," *Nature* **322:** 626–628.

MIDGLEY, M. (1987), "An Efflux System for Cationic Dyes and Related Compounds in *Escherichia coli*," *Microbiol. Sci.* **4:** 125–127.

MILLER, G.H., G.R. ROSSMAN, AND G.E. HARLOW (1987), "The Natural Occurrence of Hydroxide in Olivine," *Phys. Chem. Minerals* **14:** 461–472.

MILLER, S.L. (1953), "A Production of Amino Acids Under Possible Primitive Earth Conditions," *Science* **117:** 527–528.

MUKHIN, L.M., M.V. GERASIMOV, AND E.N. SAFONOVA (1989), "Origin of Precursors of Organic Molecules During Evaporation of Meteorites and Mafic Terrestrial Rocks," *Nature* **340:** 46–48.

NORTH, G.R., AND J.A. COAKLEY (1979), "Differences Between Seasonal and Mean Annual Energy Balance Model Calculations of Climate and Climate Sensitivity," *J. Atmos. Sci.* **36:** 1189–1203.

NORTH, G.R., J.G. MENGEL, AND D.A. SHORT (1983), "Simple Energy Balance Model Resolving the Seasons and the Continents: Application to the Astronomical Theory of the Ice Ages," *JGR* **88:** 6576–6586.

NORTH, G.R., R.F. CALAHAN, AND J.A. COAKLEY (1981), "Energy Balance Climate Models," *Rev. Geophys. Space Phys.* **19:** 91–121.

ORÓ, J. (1961), "Comets and the Formation of Biochemical Compounds on the Primitive Earth," *Nature* **190:** 389–390.

OURISSON, G., M. ROHMER, AND K. PORALLA (1987), "Procaryotic Hopanoids and Other Polyterpenoid Sterol Surrogates," *Ann. Rev. Microbiol.* **41:** 301–333.

PANGANIBAN, A.T., JR., T.E. PATT, W. HART, AND R.S. HANSON (1979), "Oxidation of Methane in the Absence of Oxygen in Lake Water Samples," *Appl. Env. Microbiol.* **37:** 303–309.

PARMENTIER, E.M., AND P.C. HESS (1992), "Chemical Differentiation of a Convecting Planetary Interior: Consequences for a One-Plate Planet Such as Venus," paper presented at the International Colloquium on Venus, August 10–12, Pasadena, California.

PINTO, J.P., G.R. GLADSTONE, AND Y.L. YUNG (1980), "Photochemical Reduction of Carbon Dioxide to Formaldehyde in the Earth's Primitive Atmosphere," *Science* **210:** 183–185.

POLLACK, J.B., J.F. KASTING, S.M. RICHARDSON, AND K. POLIAKOFF (1987), "The Case for a Wet, Warm Climate on Early Mars," *Icarus* **71:** 203–224.

APPENDIX II—PLANETARY ASPECTS REFERENCES (cont.)

RASOOL, S.I., AND C. DE BERGH (1970), "The Runaway Greenhouse and Accumulation of CO_2 in the Venus Atmosphere," *Nature* **226:** 1037–1039.

REIBACH, P.H., AND C.R. BENEDICT (1977), "Fractionation of Stable Carbon Isotopes by Phosphoenolpyruvate Carboxylase from C_4 Plants," *Plant Physiol.* **59:** 564–568.

ROHMER, M., P. BOUVIER-NAVE, AND G. OURISSON (1984), "Distribution of Hopanoid Triterpenes in Prokaryotes," *J. Gen. Microbiol.* **130:** 1137–1150.

ROHMER, M., P. BOUVIER, AND G. OURISSON (1980), "Non-Specific Lanosterol and Hopanoid Biosynthesis by a Cell-Free System from the Bacterium *Methylococcus capsulatus*," *Eur. J. Biochem.* **112:** 557–560.

RUDDIMAN, W.F., M.E. RAYMO, D.G. MARTINSON, B.M. CLEMENT, AND J. BACKMAN (1989), "Pleistocene Evolution of Northern Hemisphere Climate," *Paleoceanography* **4:** 353–412.

SAGAN, C., AND C. CHYBA (1996), "Organic Shielding of UV-Labile Greenhouse Gases on Early Earth and Mars," abstract, this volume.

SCHABER, G.G., R.G. STROM, H.J. MOORE, L.A. SODERBLOM, R.L. KIRK, D.J. CHADWICK, D.D. DAWSON, L.R. GADDIS, J.M. BOYCE, AND J. RUSSELL (1992), "Geology and Distribution of Impact Craters on Venus: What Are They Telling Us?," *JGR* **97:** 13257–13301.

SCHIDLOWSKI, M., AND P. AHARON (1992), "Carbon Cycle and Carbon Isotope Record: Geochemical Impact of Life over 3.8 Ga of Earth History," in *Early Organic Evolution: Implications for Mineral and Energy Resources*, M. Schidlowski, S. Golubic, M.M. Kimberley, D.M. McKirdy, and P.A. Trudinger, eds. (Berlin: Springer-Verlag), pp. 147–175.

SCHOPF, J.W., ED. (1983), *The Earth's Earliest Biosphere: Its Origin and Evolution* (Princeton, NJ: Princeton University Press), pp. 53–92, and references therein.

SELLERS, W.D. (1965), *Physical Climatology* (London: University of Chicago Press).

SHKLOVSKII, I.S., AND C. SAGAN (1966), *Intelligent Life in the Universe* (San Francisco: Holden-Day, Inc.).

SMITH, L.L., AND S.H. GROSS (1972), "The Evolution of Water Vapor in the Atmosphere of Venus," *J. Atmos. Sci.* **29:** 173–178.

SOLOMON, S.C., AND J.W. HEAD (1991), "Fundamental Issues in the Geology and Geophysics of Venus," *Science* **252:** 252–260.

STOLPER, E.M. (1982), "Water in Silicate Glasses: An Infrared Spectroscopic Study," *Contrib. Mineral. Petrol.* **81:** 451–462.

STRIBLING, R., AND S.L. MILLER (1987), "Energy Yields for Hydrogen Cyanide and Formaldehyde Syntheses: The HCN and Amino Acid Concentrations in the Primitive Ocean," *Origins of Life* **17:** 261–273.

SUMMONS, R.E., AND M.R. WALTER (1990), "Molecular Fossils and Microfossils of Prokaryotes and Protists from Proterozoic Sediments," *Amer. J. Sci.* **290A:** 212–244.

SUMMONS, R.E., L.L. JAHNKE, AND Z. ROKSANDIC (1994), "Carbon Isotopic Fractionation in Lipids from Methanotrophic Bacteria: Relevance for Interpretation of the Geochemical Record of Biomarkers," *Geochim. Cosmochim. Acta* **58:** 2853–2863.

APPENDIX II—PLANETARY ASPECTS REFERENCES (cont.)

SUMMONS, R.E., T.G. POWELL, AND C.J. BOREHAM (1988), "Petroleum Geology and Geochemistry of the Middle Proterozoic McArthur Basin, Northern Australia. III. Composition of Extractable Hydrocarbons," *Geochim. Cosmochim. Acta* **52:** 1747–1763.

THOMPSON, A.B. (1992), "Water in the Earth's Upper Mantle," *Nature* **358:** 295–302.

THOMPSON, L.S., AND E.J. BARRON (1981), "Comparison of Cretaceous and Present Earth Albedos: Implications for the Causes of Paleoclimates," *J. Geol.* **89:** 143–165.

TINGLE, T.N., E.A. MATHEZ, AND M.F. HOCHELLA (1991), "Carbonaceous Matter in Peridotites and Basalts Studied by XPS, SALI, and LEED," *Geochim. Cosmochim. Acta* **55:** 1345–1352.

TINGLE, T.N., M.F. HOCHELLA, C.H. BECKER, AND R. MALHOTRA (1990), "Organic Compounds on Crack Surfaces in Olivine from San Carlos, Arizona, and Hualalai Volcano, Hawaii," *Geochim. Cosmochim. Acta* **54:** 477–485.

TOPS (1992), *Toward Other Planetary Systems*, NASA.

TOUMA, J., AND J. WISDOM (1993), "The Chaotic Obliquity of Mars," *Science* **259:** 1294–1297.

TURCOTTE, D. (1992), "Episodic Plate Tectonics on Venus," paper presented at the International Colloquium on Venus, August 10–12, Pasadena, California.

TURCOTTE, D.L. (1980), "On the Thermal Evolution of the Earth," *Earth Planet. Sci. Lett.* **48:** 53–58.

WALKER, J.C.G. (1975), "Evolution of the Atmosphere of Venus," *J. Atmos. Sci.* **32:** 1248–1256.

WALKER, J.C.G. (1977), *Evolution of the Atmosphere* (London: Macmillan Publishing Company), p. 205.

WALKER, J.C.G. (1985), "Carbon Dioxide on the Early Earth," *Origins of Life* **16:** 117–127.

WALKER, J.C.G., AND P. BRIMBLECOMBE (1985), "Iron and Sulfur in the Pre-Biologic Ocean," *Precambrian Res.* **28:** 205–222.

WALKER, J.C.G., K.K. TUREKIAN, AND D.M. HUNTEN (1970), "An Estimate of the Present-Day Deep-Mantle Degassing Rate from Data on the Atmosphere of Venus," *JGR* **75:** 3558–3561.

WALKER, J.C.G., P.B. HAYS, AND J.F. KASTING (1981), "A Negative Feedback Mechanism for the Long-Term Stabilization of Earth's Surface Temperature," *JGR* **86:** 9776–9782.

WARD, W.R. (1973), "Large Scale Variations in the Obliquity of Mars," *Science* **181:** 260–262.

WARD, W.R. (1974), "Climatic Variations on Mars. I. Astronomical Theory of Insolation," *JGR* **79:** 3375–3386.

WARD, W.R. (1979), "Present Obliquity Oscillations of Mars: Fourth-Order Accuracy in Orbital *e* and *I*," *JGR* **84:** 237–241.

WARD, W.R. (1992), "Long-Term Orbital and Spin Dynamics of Mars," in *Mars*, H. Kieffer, B. Jakosky, and C. Snyder, eds. (Tucson: University of Arizona Press), pp. 298–320.

WATSON, A.J., T.M. DONAHUE, AND J.C.G. WALKER (1981), "The Dynamics of a Rapidly Escaping Atmosphere: Applications to the Evolution of Earth and Venus," *Icarus* **48:** 150–166.

APPENDIX II—PLANETARY ASPECTS REFERENCES (cont.)

WETHERILL, G.W. (1985), "Occurrence of Giant Impacts During the Growth of the Terrestrial Planets," *Science* **228:** 877–879.

WETHERILL, G.W. (1994), "The Plurality of Habitable Worlds," in *Fifth Exobiology Symposium and Mars Workshop*, NASA Ames Research Center, April.

WHITICAR, M.J., F. FARBER, AND M. SCHOELL (1986), "Biogenic Methane Formation in Marine and Freshwater Environments: CO_2 Reduction vs. Acetate Fermentation—Isotope Evidence," *Geochim. Cosmochim. Acta* **50:** 693–709.

WHITMIRE, D.P., R.T. REYNOLDS, AND J.F. KASTING (1991), "Habitable Zones for Earth-like Planets Around Main Sequence Stars," in *Bioastronomy: The Search for Extraterrestrial Life*, J. Heidmann and M.J. Klein, eds. (Berlin: Springer-Verlag), pp. 173–178.

WILLIAMS, D.M., J.F. KASTING, AND K. CALDEIRA (1996), "Chaotic Obliquity Variations and Planetary Habitability," this volume.

WINDLEY, B.F., ED. (1976), *The Early History of the Earth* (New York: John Wiley & Sons).

YUNG, Y.L., AND M.B. MCELROY (1979), "Fixation of Nitrogen in the Prebiotic Atmosphere," *Science* **203:** 1002–1004.

Astrophysical Aspects

ABT, H., AND S. LEVY (1976), "Multiplicity Among Solar-Type Stars," *ApJSupp* **30:** 273–306.

ALLEN, C.W. (1976), *Astrophysical Quantities*, 3d ed. (London: The Athlone Press).

ANGEL, J.R., A.Y. CHENG, AND N.J. WOOLF (1984), "A Space Telescope for IR Spectroscopy of Earth-Like Planets," *Nature* **322:** 341.

ANGEL, R. (1989), "Use of a 16 m Telescope to Detect Earth-like Planets," in *The Next Generation Space Telescope*, P. Bély, C. Burrows, and G. Illingworth, eds. (Baltimore, MD: Hubble Space Telescope Science Institute), p. 81.

ANGEL, R. (1994), "Ground-Based Imaging of Extrasolar Planets Using Adaptive Optics," *Nature* **368:** 203.

ARGYLE, E. (1974), "On the Observability of Extrasolar Planetary Systems," *Icarus* **21:** 199.

ARTYMOWICZ, P., C.J. CLARKE, S.H. LUBOW, AND J.E. PRINGLE (1991), "The Effect of an External Disk on the Orbital Elements of a Central Binary," *ApJ* **370:** L35–L38.

AUMANN, H.H., F.C. GILLETT, C.A. BEICHMAN, T. DE JONG, J.R. HOUCK, F.J. LOW, G. NEUGEBAUER, R.G. WALKER, AND P.R. WESSELIUS (1984), "Discovery of a Shell Around Alpha Lyrae," *ApJ* **278:** L23–L27.

BACKMAN, D.E., AND F. PARESCE (1993), "Main-Sequence Stars with Circumstellar Solid Material: The Vega Phenomenon," in *Protostars and Planets III*, E.H. Levy and J.I. Lunine, eds. (Tucson: University of Arizona Press), pp. 1253–1304.

BADALYAN, O.G., AND M.A. LIVSHITS (1992), "Mass Loss by Active Late-Type Stars," *Sov. Astron.* **36:** 70–74.

APPENDIX II—ASTROPHYSICAL ASPECTS REFERENCES (cont.)

BARROW, J.D., AND F.J. TIPLER (1986), *The Anthropic Cosmological Principle* (Oxford: Oxford University Press).

BECKWITH, S.V.W., A.I. SARGENT, R.S. CHINI, AND R. GÜSTEN (1990), "A Survey for Circumstellar Disks Around Young Stellar Objects," *AJ* **99:** 924–945.

BELL, J.F., III, AND W.J. BORUCKI (1995), "Characteristics of Transits by Earth-Sized Planets in Binary Star Systems," in *Progress in the Search for Extraterrestrial Life*, ASP Conf. Ser., S. Shostak, ed., **74:** 165–170.

BENEST, D. (1988), "Planetary Orbits in the Elliptic Restricted Problem. I. The α Centauri System," *A&A* **206:** 143–146.

BENEST, D. (1991), "Habitable Planetary Orbits Around α Centauri and Other Binaries," in *Bioastronomy: The Search for Extraterrestrial Life*, J. Heidmann and M.J. Klein, eds. (Berlin: Springer-Verlag), pp. 44–47.

BILLS, B.G. (1990), "The Rigid Body Obliquity History of Mars," *JGR* **95:** 14137–14153.

BLACK, D.C. (1980), "In Search of Other Planetary Systems," *Space Sci. Rev.* **25:** 35–81.

BODENHEIMER, P. (1989), "The Impact of Stellar Evolution on Planetary System Development," in *The Formation and Evolution of Planetary Systems*, H.A. Weaver and L. Danly, eds. (Cambridge: Cambridge University Press), pp. 243–273.

BOESGAARD, A.M. (1990), "Lithium and Beryllium in Stellar Atmospheres," in *Sixth Cambridge Workshop on Cool Stars, Stellar Systems, and the Sun*, ASP Conf. Ser., M.S. Giampapa and J.A. Bookbinder, eds., **9:** 317–327.

BOOTHROYD, A.I., I. SACKMANN, AND F.A. FOWLER (1991), "Our Sun II. Early Mass Loss of $0.1M_\odot$ and the Case of the Missing Lithium," *ApJ* **377:** 318–329.

BORUCKI, W., AND D. KOCH (1994), "Multiplex Approach to the Photometric Detection of Planets," *A&SS* **212:** 293.

BORUCKI, W., AND D. KOCH (1995), "Comparing the Expectations for Different Planetary Search Methods," in *Progress in the Search for Extraterrestrial Life*, ASP Conf. Ser., S. Shostak, ed., **74:** 173–180.

BORUCKI, W.J., AND A.L. SUMMERS (1984), "The Photometric Method of Detecting Other Planetary Systems," *Icarus* **58:** 121–134.

BORUCKI, W.J., J.D. SCARGLE, AND H.S. HUDSON (1985), "Detectability of Extrasolar Planetary Transits," *ApJ* **291:** 852–854.

BOSS, A.P. (1995), "Proximity of Jupiter-like Planets to Low-Mass Stars," *Science* **267:** 360–362.

BRACEWELL, R.N. (1978), "Detecting Non-Solar Planets by Spinning IR Interferometer," *Nature* **274:** 780.

BRANCEWICZ, H.K., AND T.Z. DWORAK (1980), "A Catalogue of Parameters for Eclipsing Binaries," *Acta Astronomica* **30:** 501–524.

BRANDMEIER, S. (1994), *Simulation of Planetary Transits in Eclipsing Binary Star Systems,* Master's Thesis in Computer Science at Polytechnic University of Konstanz, Germany.

APPENDIX II—ASTROPHYSICAL ASPECTS REFERENCES (cont.)

BRANDMEIER, S., AND L.R. DOYLE (1996), "Simulation of Planetary Transits in Eclipsing Binary Star Systems," this volume.

BROECKER, W.S., AND G.H. DENTON (1990), "What Drives Glacial Cycles?," *Sci. Am.* 262 (1): 49–56.

BROMLEY, B.C. (1992), "Detecting Faint Echoes in Stellar-Flare Light Curves," *PASP* **104:** 1049.

BROWN, A., A. VEALÉ, P. JUDGE, J. BOOKBINDER, AND I. HUBENY (1990), "Mass Loss Upper Limits for A and F Dwarfs," in *Sixth Cambridge Workshop on Cool Stars, Stellar Systems, and the Sun*, ASP Conf. Ser., M.S. Giampapa and J.A. Bookbinder, eds., **9:** 183–185.

BURKE, B.F. (1986), "Detection of Planetary Systems and the Search for Evidence of Life," *Nature* **322:** 340.

BURNS, J.A. (1986), "The Evolution of Satellite Orbits," in *Satellites*, J.A. Burns and M.S. Matthews, eds. (Tucson: University of Arizona Press), pp. 117–158.

CAFFEE, M.W., C.M. HOHENBERG, R.H. NICHOLS, JR., C.T. OLINGER, R. WIELER, A. PEDRONI, P. SIGNER, T.D. SWINDLE, AND J.N. GOSWAMI (1991), "Do Meteorites Contain Irradiation Records from Exposure to an Enhanced-Activity Sun?," in *The Sun in Time*, C.P. Sonett, M.S. Giampapa, and M.S. Matthews, eds. (Tucson: University of Arizona Press).

CAFFEE, M.W., C.M. HOHENBERG, T.D. SWINDLE, AND J.N. GOSWAMI (1987), "Evidence in Meteorites for an Active Early Sun," *ApJ* **313:** L31–L35.

CALDEIRA, K., AND J.F. KASTING (1992), "Susceptibility of the Early Earth to Irreversible Glaciation Caused by Carbon Dioxide Clouds," *Nature* **359:** 226–228.

CAMERON, A.C., AND C. CAMPBELL (1993), "Rotational Evolution of Magnetic T Tauri Stars with Accretion Disks," *A&A* **274:** 309–318.

CAMERON, A.G.W. (1963), "Stellar Life Zones," in *Interstellar Communication*, A.G.W. Cameron, ed. (New York: W.A. Benjamin), pp. 107–114.

CAMERON, A.G.W. (1988), "Origin of the Solar System," *Ann. Rev. Astron. Astrophys.* **26:** 441–472.

CAMERON, A.G.W., AND W. BENZ (1991), "The Origin of the Moon and the Single Impact Hypothesis IV," *Icarus* **92:** 204–216.

CARR, M. (1987), "Water on Mars," *Nature* **326:** 30–35.

CARTER, B. (1983), "The Anthropic Principle and Its Implications for Biological Evolution," *Phil. Trans. R. Soc. Lond. A* **310:** 347–363.

CAYREL DE STROBEL, G., R. CAYREL, E. FRIEL, J.P. ZAHN, AND C. BENTOLILA (1994), "A Case of the Quadruple System ξUMa: Its Activity and Lithium Depletion," *A&A* **291:** 505–516.

CLOUD, P. (1983), "The Biosphere," *Sci. Am.* 249 (3): 176–189, 202.

CONNES, P. (1994), "Development of Absolute Accelerometry," *A&SS* **212:** 357.

COUNSELMAN, C. (1973), "Outcomes of Tidal Evolution," *ApJ* **180:** 307–314.

CRAINE, E., E. ENGEL, AND R. TUCKER (1992), "A Cool Star Sample and Extra-Solar System Planetary Search Program," in *Seventh Cambridge Workshop on Cool Stars, Stellar Systems and the Sun*, ASP Conf. Ser., M.S. Giampapa and J.A. Bookbinder, eds., **26:** 588.

APPENDIX II—ASTROPHYSICAL ASPECTS REFERENCES (cont.)

CUZZI, J.N., A.R. DOBROVOLSKIS, AND R.C. HOGAN (1994), "What Initiated Planetesimal Formation?," *Lunar and Planetary Science XXV*, 307–308.

DALE, G.E., AND J.L. LINSKY (1981), "First Detection of Nonflare Microwave Emission from the Coronae of Single Late-Type Stars," *ApJ* **250:** 284–292.

DELSEMME, A.H. (1991), "Origin of the Biosphere of the Earth," in *Bioastronomy: The Search for Extraterrestrial Life*, J. Heidmann and M.J. Klein, eds. (Berlin: Springer-Verlag), pp. 117–123.

DOBROVOLSKIS, A. (1980), "Atmospheric Tides and the Rotation of Venus. II. Spin Evolution," *Icarus* **41:** 18–35.

DOBROVOLSKIS, A., C. LEVIT, J. SCARGLE, J. CUZZI, L. DONES, M. SHOWALTER, AND K. YOUNG (1996), "Orbital Stability of Planets of Binary Stars," abstract, this volume.

DOLE, S.H. (1964), *Habitable Planets for Man* (New York: Blaisdell Publishing Company).

DOLE, S.H. (1970), *Habitable Planets for Man*, 2d ed. (New York: American Elsevier).

DONAHUE, T.M., AND R.E. HARTLE (1992), "Solar Cycle Variations in H^+ and D^+ Densities in the Venus Ionosphere: Implications for Escape," *Geophys. Res. Lett.* **19:** 2449–2452.

DONES, L., AND S. TREMAINE (1993), "Why Does the Earth Spin Forward?," *Science* **259:** 350–354.

DONNISON, J.R., AND D.F. MIKULSKIS (1992), "Three-Body Orbital Stability Criteria for Circular Orbits," *MNRAS* **254:** 21–26.

DOYLE, L. (1988), "Progress in Determining the Space Orientation of Stars," in *Bioastronomy—The Next Step* (Dordrecht: Kluwer Academic Publishers), p. 101.

DOYLE, L.R. (1993), "Binary Star Ecoshells," unpublished proposal.

DOYLE, L.R., AND C.P. MCKAY (1992), "Exobiological Habitats: An Overview," in *Bioastronomy*, J. Heidmann and M.J. Klein, eds. (New York: Springer-Verlag), pp. 163–172.

DOYLE, L.R., N.C. HEATHER, R. VIKRAMSINGH, AND D.P. WHITMIRE (1996), "Stellar Mass Loss and the Young Solar System Habitable Zone," this volume.

DOYLE, L.R., R. VIKRAMSINGH, D.P. WHITMIRE, AND N.C. HEATHER (1995), "Circumstellar Habitable Zones and Mass Loss from Young Solar-type Stars. II. Observational Considerations," in *Progress in the Search for Extraterrestrial Life*, ASP Conf. Ser., S. Shostak, ed., **74:** 195–204.

DOYLE, L.R., T.J. WILCOX, AND J.J. LORRE (1984), "The Space Orientation of Stars," *ApJ* **287:** 307–314.

DUNCAN, M., T. QUINN, AND S. TREMAINE (1989), "The Long-Term Evolution of Orbits in the Solar System: A Mapping Approach," *Icarus* **82:** 402–418.

DUQUENNOY, A., AND M. MAYOR (1991), "Multiplicity Among Solar-type Stars in the Solar Neighborhood. II. Distribution of the Orbital Elements in an Unbiased Sample," *A&A* **248:** 485–524.

DVORAK, R. (1986), "Critical Orbits in the Elliptic Three-Body Problem," *A&A* **167:** 379–386.

DVORAK, R., CH. FROESCHLÉ, AND CL. FROESCHLÉ (1989), "Stability of Outer Planetary Orbits (P-Types) in Binaries," *A&A* **226:** 335–342.

APPENDIX II—ASTROPHYSICAL ASPECTS REFERENCES (cont.)

EDWARDS, S., S. STROM, W. HERBST, J. ATTRIDGE, K. MERRILL, R. PROBST, AND I. GATLEY (1993), "Angular Momentum Regulation in Low-Mass Young Stars Surrounded by Accretion Disks," *AJ* **106:** 372–382.

EMBLETON, B.J.J., AND G.E. WILLIAMS (1986), "Low Palaeolatitude of Deposition for Late Precambrian Periglacial Varvites in South Australia: Implications for Palaeoclimatology," *Earth Planet. Sci. Lett.* **79:** 419–430.

FEKEL, F.C., JR. (1981), "The Properties of Close Multiple Stars," *ApJ* **246:** 879–898.

FISCHER, D., AND G. MARCY (1992), "Multiplicity Among M Dwarfs," *ApJ* **396:** 178–194.

FOGG, M.J. (1992), "An Estimate of the Prevalence of Biocompatible and Habitable Planets," *J. Brit. Interplanet. Soc.* **45:** 3–12.

FOMALONT, E.B., K.I. KELLERMANN, J.V. WALL, AND D. WEISTROP (1984), "A Deep 6-Centimeter Radio Source Survey," *Science* **225:** 23–28.

FOMALONT, E.B., K.I. KELLERMANN, M.C. ANDERSON, D. WEISTROP, J.V. WALL, R.A. WINDHOUST, AND J.A. KRISTIAN (1988), "New Limits to Fluctuations in the Cosmic Background Radiation at 4.86 GHz Between 12 and 60 Arcsecond Resolution," *AJ* **96:** 1187–1191.

GHEZ, A. (1993), *The Multiplicity of T Tauri Stars in the Star Forming Regions Taurus-Auriga and Ophiucus-Scorpius: A 2.2 μm Speckle Imaging Survey*, PhD Thesis (Pasadena: California Institute of Technology).

GLADMAN, B. (1993), "Dynamics of Systems of Two Close Planets," *Icarus* **106:** 247–263.

GOLDIN, D. (1994), "Address to the 75th Anniversary Meeting of the AGU, Washington D.C.," *Nature* **369:** 349.

GOLDREICH, P. (1966), "History of the Lunar Orbit," *Rev. Geophys.* **4:** 411–439.

GOLDREICH, P. (1972), "Tides and the Earth-Moon System," *Sci. Am.* 226 (4): 43–52, 118.

GOLDREICH, P., AND S.J. PEALE (1968), "The Dynamics of Planetary Rotations," *Ann. Rev. Astron. Astrophys.* **6:** 287–320.

GOLDREICH, P., AND S.J. PEALE (1970), "The Obliquity of Venus," *AJ* **75:** 273–283.

GOLDREICH, P., AND W.R. WARD (1973), "The Formation of Planetesimals," *ApJ* **183:** 1051–1061.

GÓMEZ, M., L. HARTMANN, S. KENYON, AND R. HEWETT (1993), "On the Spatial Distribution of Pre–Main Sequence Stars in Taurus," *AJ* **105:** 1927–1937.

GOUGH, D.O. (1981), "Solar Interior Structure and Luminosity Variations," *Solar Phys.* **74:** 21–34.

GOULD, S.J. (1994), "The Evolution of Life on the Earth," *Sci. Am.* 271 (4): 85–91.

GRAEDEL, T., I. SACKMANN, AND A. BOOTHROYD (1991), "Early Solar Mass Loss: A Potential Solution to the Weak Sun Paradox," *Geophys. Res. Lett.* **18:** 1881–1884.

GRAZIANI, F., AND D.C. BLACK (1981), "Orbital Stability Constraints on the Nature of Planetary Systems," *ApJ* **251:** 337–341.

APPENDIX II—ASTROPHYSICAL ASPECTS REFERENCES (cont.)

GÜDEL, M., J.H.M.M. SCHMITT, AND A.O. BENZ (1994), "Discovery of Microwave Emission from Four Nearby Solar-Type Stars," *Science* **265:** 933–935.

GUILLOT, T., G. CHABRIER, P. MOREL, AND D. GAUTIER (1994), "Nonadiabatic Models of Jupiter and Saturn," *Icarus* **112:** 354.

HABERLE, R.M., AND J. SCHAEFFER (1995), "Obliquity Experiments with a Mars General Circulation Model," *Bull. Amer. Astron. Soc.* **27:** 1066.

HABERLE, R.M., C.P. MCKAY, D. TYLER, AND R.T. REYNOLDS (1996), "Can Synchronously Rotating Planets Support an Atmosphere?," this volume.

HALE, A. (1994), "Orbital Coplanarity in Solar-type Binary Systems: Implications for Planetary System Formation and Detection," *AJ* **107:** 306–332.

HALE, A. (1996), "Determining Habitable Planetary Environments Within Binary and Multiple Systems," this volume.

HARLAND, W.B., AND M.J.S. RUDWICK (1964), "The Great Infra-Cambrian Ice Age," *Sci. Am.* 211 (2): 28–36.

HARRINGTON, R.S. (1972), "Stability Criteria for Triple Stars," *Celest. Mech.* **6:** 322–327.

HARRINGTON, R.S. (1977), "Planetary Orbits in Binary Stars," *AJ* **82:** 753–756.

HARRIS, A.W., AND W.R. WARD (1982), "Dynamical Constraints on the Formation and Evolution of Planetary Bodies," *Ann. Rev. Earth Planet. Sci.* **10:** 61–108.

HART, M.H. (1978), "The Evolution of the Atmosphere of the Earth," *Icarus* **33:** 23–39.

HART, M.H. (1979), "Habitable Zones Around Main Sequence Stars," *Icarus* **37:** 351–357.

HARTMANN, L., AND J. STAUFFER (1989), "Additional Measurements of Pre-Main-Sequence Stellar Rotation," *AJ* **97:** 873–880.

HEPPENHEIMER, T.A. (1974), "Outline of a Theory of Planet Formation in Binary Systems," *Icarus* **22:** 436–447.

HEPPENHEIMER, T.A. (1978), "On the Formation of Planets in Binary Star Systems," *A&A* **65:** 421–426.

HERBIG, G., AND D. TERNDRUP (1986), "The Trapezium Cluster of the Orion Nebula," *ApJ* **307:** 609–618.

HIDE, R., AND J.O. DICKEY (1991), "Earth's Variable Rotation," *Science* **253:** 629–637.

HOFFLEIT, D., AND C. JASCHEK (1982), *The Bright Star Catalogue* (New Haven, CT: Yale University Observatory).

HOLMAN, M.J., AND J. WISDOM (1993), "Dynamical Stability in the Outer Solar System and the Delivery of Short Period Comets," *AJ* **105:** 1987–1999.

HOURDIN, F. (1994), personal communication.

HUANG, S.-S. (1960), "Life Outside the Solar System," *Sci. Am.* **202 (4):** 55–63.

HUANG, S.-S. (1960), "The Sizes of Habitable Planets," *PASP* **72:** 489.

APPENDIX II—ASTROPHYSICAL ASPECTS REFERENCES (cont.)

HUANG, T.-Y., AND K.A. INNANEN (1983), "The Gravitational Escape/Capture of Planetary Satellites," *AJ* **88:** 1537–1548.

HUYGENS, C. (1698), *The Celestial Worlds Discoverd*, Fac-simile: F. Cass and Co., 1968.

IBEN, I. (1967), "Stellar Evolution. VI. Evolution from the Main Sequence to the Red-Giant Branch for Stars of Mass $1_{\odot}$," *ApJ* **147:** 624–663.

IBEN, I. (1967a), "Stellar Evolution Within and Off the Main Sequence," *Ann. Rev. Astron. Astrophys.* **5:** 571–626.

IBEN, I. (1967b), "Stellar Evolution. VI. Evolution from the Main Sequence to the Red Giant Branch for Stars of Mass 1 $M_{\odot}$, 1.25 $M_{\odot}$ and 1.5 $M_{\odot}$," *ApJ* **147:** 624–649.

IBEN, I. (1974), "Post Main Sequence Evolution of Single Stars," *Ann. Rev. Astron. Astrophys.* **12:** 215–256.

IBEN, I., AND A. RENZINI (1983), "Asymptotic Giant Branch Evolution and Beyond," *Ann. Rev. Astron. Astrophys.* **21:** 271–342.

IDA, S. (1990), "Stirring and Dynamic Friction Rates of Planetesimals in the Solar Gravitational Field," *Icarus* **88:** 129–145.

IMBRIE, J., AND J.Z. IMBRIE (1980), "Modeling the Climatic Response to Orbital Variations," *Science* **207:** 943–953.

ISAACMAN, R., AND C. SAGAN (1977), "Computer Simulation of Planetary Accretion Dynamics: Sensitivity to Initial Conditions," *Icarus* **31:** 510–533.

JENKINS, G.S., H.G. MARSHALL, AND W.R. KUHN (1993), "Precambrian Climate: The Effects of Land Area and the Earth's Rotation Rate," *JGR* **98:** 8785–8791.

JENKINS, J.M., AND L.R. DOYLE (1994), "Detection of Extrasolar Planets in Eclipsing Binary Systems: A New Cross-Correlation Technique," abstract in *Bull. Amer. Astron. Soc.* **26:** 1089.

JONES, B.F., AND M. WALKER (1988), "Proper Motions and Variabilities of Stars near the Orion Nebula," *AJ* **95:** 1755–1782.

KANT, I. (1755), *Allegmeine Naturgeschichte und Theorie des Himmels* (Konigsburg und Leipzig: Johann Friederich Petersen). English translation: W. Hastie, "Universal Natural History and Theories of the Heavens," in *Kant's Cosmology* (New York: Greenwood Publishing), 1968.

KASTING, J.F. (1988), "Runaway and Moist Greenhouse Atmospheres and the Evolution of Earth and Venus," *Icarus* **74:** 472–494.

KASTING, J.F. (1991), "CO_2 Condensation and the Climate of Early Mars," *Icarus* **94:** 1–13.

KASTING, J.F. (1996), "Habitable Zones Around Stars: An Update," this volume.

KASTING, J.F., D.P. WHITMIRE, AND R.T. REYNOLDS (1993), "Habitable Zones Around Main Sequence Stars," *Icarus* **101:** 108–128.

KASTING, J.F., O.B. TOON, AND J.B. POLLACK (1988), "How Climate Evolved on the Terrestrial Planets," *Sci. Am.* **258 (2):** 90–97.

KERRIDGE, J.F. (1993), "What Can Meteorites Tell Us About Nebular Conditions and Processes During Planetesimal Accretion?," *Icarus* **106:** 135–150.

APPENDIX II—ASTROPHYSICAL ASPECTS REFERENCES (cont.)

KOCH, D., AND W. BORUCKI (1996), "A Search for Earth-Sized Planets in Habitable Zones Using Photometry," this volume.

Kolvoord, R.A., and R. Greenberg (1992), "A Critical Reanalysis of Planetary Accretion Models," *Icarus* **98:** 2–19.

KONIGL, A. (1991), "Disk Accretion onto Magnetic T Tauri Stars," *ApJ* **370:** L39–L43.

KUBALA, A., D. BLACK, AND V. SZEBEHELY (1993), "Stability of Outer Planetary Orbits Around Binary Stars: A Comparison of Hill's and Laplace's Stability Criteria," *Celest. Mech. Dyn. Astron.* **56:** 51–68.

LACY, C.H. (1977), "Absolute Dimensions and Masses of the Remarkable Spotted dM4e Eclipsing Binary Flare Star CM Draconis," *ApJ* **218:** 444–460.

LADA, E., D. DEPOTY, N. EVANS, AND I. GATLEY (1991), "A 2.2 Micron Survey in the L1630 Molecular Cloud," *ApJ* **371:** 171–182.

LANG, K.R. (1980), *Astrophysical Formulae*, 2d ed. (Berlin: Springer-Verlag).

LAPLACE, P.S. (1796), *Exposition du Système du Monde* (Paris: Cercle-Social). English translation: J. Pond, *The System of the World* (London: Richard Phillips), 1809.

LASKAR, J. (1994), "Large Scale Chaos in the Solar System," *A&A* **287:** L9–L12.

LASKAR, J., AND P. ROBUTEL (1993), "The Chaotic Obliquity of the Planets," *Nature* **361:** 608–612.

LASKAR, J., F. JOUTEL, AND F. BOUDIN (1993a), "Orbital, Precessional, and Insolation Quantities for the Earth from −20 Myr to +10 Myr," *A&A* **270:** 522–533.

LASKAR, J., F. JOUTEL, AND P. ROBUTEL (1993b), "Stabilization of the Earth's Obliquity by the Moon," *Nature* **361:** 615–617.

LAZZARO, D., B. SICARDY, F. ROQUES, AND R. GREENBERG (1994), "Is There a Planet Around β Pictoris? Perturbations of a Planet on a Circumstellar Dust Disk. 2. The Analytical Model," *Icarus* **108:** 59–80.

LECACHEUX, A. (1990), "On the Feasibility of Extra-Solar Planetary Detection at Very Low Radio Frequencies," in *Bioastronomy—The Search for Extraterrestrial Life*, J. Heidmann and M.J. Klein, eds. (Berlin: Springer-Verlag), p. 21.

LECACHEUX, A. (1994), "Solar System, Low Frequency Radio Astronomy from the Moon," *Adv. Space Res.* **14:** 193.

LECACHEUX, A. (1994), personal communication.

LECAR, M., J.C. WHEELER, AND C.F. MCKEE (1976), "Tidal Circulation of the Binary X-Ray Sources Hercules X-1 and Centaurus X-3," *ApJ* **205:** 556–562.

LÉGER, A., J.M. MARIOTTI, J.L. PUGET, D. ROUAN, AND J. SCHNEIDER (1996), "How to Evidence Primitive Life on an Exoplanet?—The DARWIN Project," this volume.

LÉGER, A., M. PIRRE, AND F.J. MARCEAU (1993), "Search for Primitive Life on a Distant Planet: Relevance of O_2 and O_3 Detections," *A&A* **227:** 309.

APPENDIX II—ASTROPHYSICAL ASPECTS REFERENCES (cont.)

LEITHERER, C., AND C. ROBERT (1991), "Observations of Stellar Winds from Hot Stars at 1.3 Millimeters," *ApJ* **377:** 629–638.

LISSAUER, J.J. (1987), "Time Scales for Planetary Accretion and Structure of the Protoplanetary Disk," *Icarus* **69:** 245–265.

LISSAUER, J.J. (1989), "Which Stars Have Planets?," in *The Formation and Evolution of Planetary Systems*, H.A. Weaver and L. Danly, eds. (Cambridge: Cambridge University Press), pp. 304–308.

LISSAUER, J.J. (1993), "Planet Formation," *Ann. Rev. Astron. Astrophys*. **31:** 129–174.

LISSAUER, J.J. (1995), "Urey Prize Lecture: On the Diversity of Plausible Planetary Systems," *Icarus* **114:** 217–236.

LISSAUER, J.J., AND D.M. KARY (1991), "The Origin of the Systematic Component of Planetary Rotation. I. Planet on a Circular Orbit," *Icarus* **94:** 126–159.

LISSAUER, J.J., AND G.R. STEWART (1993), "Growth of Planets from Planetesimals," in *Protostars and Planets III*, E. Levy, J.I. Lunine, and M.S. Matthews, eds. (Tucson: University of Arizona Press), pp. 1061–1088.

LISSAUER, J.J., AND V.S. SAFRONOV (1991), "The Random Component of Planetary Rotation," *Icarus* **93:** 288–297.

LISSAUER, J.J., J.B. POLLACK, G.W. WETHERILL, AND D.J. STEVENSON (1995), "Formation of the Neptune System," in *Neptune and Triton*, D. Cruikshank and M.S. Matthews, eds. (Tucson: University of Arizona Press), pp. 37–108.

LOVELOCK, J.E. (1975), "Thermodynamics and the Recognition of Alien Biospheres," *Proc. R. Soc. Lond. B* **189:** 167.

LOVELOCK, J.E., AND A.J. WATSON (1982), "The Regulation of Carbon Dioxide and Climate," *Planet. Space Sci.* **30:** 795–802.

MAO, S., AND B. PACZYNSKI (1991), "Gravitational Microlensing by Double Stars and Planetary Systems," *ApJ* **374:** L37.

MASON, D., H.A. MCALISTER, W.I. HARTKOPF, AND M.M. SHARA (1995), "Binary Star Orbits from Speckle Interferometry. VII. The Multiple System Xi Ursae Majoris," *AJ* **109:** 332–340.

MAYOR, M., AND J.-C. MERMILLIOD (1991), "Rotational Velocities of Stars in Open Clusters: The Time Dependency Revisited," in *Angular Momentum Evolution of Young Stars*, S. Catalano and J.R. Stauffer, eds. (Dordrecht: Kluwer Academic Publishers), pp. 143–150.

MAZEH, T., D. GOLDBERG, A. DUQUENNOY, AND M. MAYOR (1992), "On the Mass-Ratio Distribution of Spectroscopic Binaries with Solar-Type Primaries," *ApJ* **401:** 265–268.

MCCAUGHREAN, M., AND J. STAUFFER (1994), "High Resolution Near-Infrared Imaging of the Trapezium: A Stellar Census," *AJ* **108:** 1382–1397.

MCKAY, C.P., AND C.R. STOKER (1989), "The Early Environment and Its Evolution on Mars: Implications for Life," *Rev. Geophys. Space Phys.* **27:** 189–214.

APPENDIX II—ASTROPHYSICAL ASPECTS REFERENCES (cont.)

McKAY, C.P., AND W.L. DAVIS (1991), "Duration of Liquid Water Habitats on Early Mars," *Icarus* **90:** 214–221.

MULLAN, D.J., E.M. SION, F.C. BRUHWEILER, AND K.G. CARPENTER (1989), "Evidence for a Cool Wind from the K2 Dwarf in the Detached Binary V471 Tauri," *ApJ* **339:** L33–L36.

MULLAN, D.J., J.G. DOYLE, R.O. REDMAN, AND M. MATHIOUDAKIS (1992), "Limits on Detectability of Mass Loss from Cool Dwarfs," *ApJ* **397:** 225–231.

NAEYE, R. (1994), "Moon of Our Delight," *Discover* 15 (1): 72–74.

NERON DE SURGY, O., AND J. LASKAR (1995), "On the Long Term Evolution of the Spin of the Earth," *Bull. Amer. Astron. Soc.* **27:** 1172.

OATES, W.J., ED. (1940), "Letter to Herodotus," in *The Stoic and Epicurean Philosophers: The Complete Extant Writings of Epicurus, Epictetus, Lucretius [and] Marcus Aurelius* (New York: Random House), pp. 3–15:5.

O'DELL, C., Z. WEN, AND X. HU (1993), "Discovery of New Objects in the Orion Nebula on HST Images—Shocks, Compact Sources, and Protoplanetary Disks," *ApJ* **410:** 696–700.

O'DELL, R., AND Z. WEN (1994), As reported in *Sci. News* **145:** 391.

ÖPIK, E.J. (1951), "Collision Probabilities with the Planets and the Distribution of Interplanetary Matter," *Proc. R. Irish Acad.* **A54:** 165–199.

OWEN, T. (1980), "The Search for Early Forms of Life in Other Planetary Systems: Future Possibilities Afforded by Spectroscopic Techniques," in *Strategies for the Search for Life in the Universe*, M.D. Papagiannis, ed. (Dordrecht: D. Reidel Publishing Company).

PARRY, I.R., B. ESPERANZA CARRASCO, AND F.G. WATSON (1994), "Multi-Fibre Photometry—A Technique for Detecting Extra-Solar Planets," in *SPIE Astronomy Instrumentation Conference*, Hawaii.

PASQUINI, L., Q. LUI, AND R. PALLAVICIMI (1994), "Lithium Abundances of Nearby Solar-like Stars," *A&A* **287:** 191–205.

PEALE, S.J. (1977), "Rotation Histories of the Natural Satellites," in *Planetary Satellites*, J.A. Burns, ed. (Tucson: University of Arizona Press).

PEALE, S.J. (1986), "Consequences of Tidal Evolution," in *The Solar System: Observations and Interpretations*, M.G. Kivelson, ed. (Englewood Cliffs, NJ: Prentice-Hall), pp. 275–288.

PEALE, S.J. (1989), "Some Unsolved Problems in the Evolutionary Dynamics of the Solar System," *Celest. Mech. Dyn. Astron.* **46:** 253–275.

PENDLETON, Y.J., AND D.C. BLACK (1983), "Further Studies on Criteria for the Onset of Dynamical Instability in General Three-Body Systems," *AJ* **88:** 1415–1419.

PHILLIPS, J.A., S.E. THARSETT, AND S.R. KULKARNI, EDS. (1993), *Planets Around Pulsars*, ASP Conf. Ser. **36**.

POLLACK, J.B. (1979), "Climatic Change on the Terrestrial Planets," *Icarus* **37:** 479–553.

POLLACK, J.B., AND R.T. REYNOLDS (1974), "Implications of Jupiter's Early Contraction History for the Composition of the Galilean Satellites," *Icarus* **21:** 248–253.

APPENDIX II—ASTROPHYSICAL ASPECTS REFERENCES (cont.)

POLLACK, J.B., J.F. KASTING, S.M. RICHARDSON, AND K. POLIAKOFF (1987), "The Case for a Wet, Warm Climate on Early Mars," *Icarus* **71:** 203–224.

POLLARD, W.G. (1979), "The Prevalence of Earthlike Planets," *Amer. Sci.* **67:** 653–659.

PROSSER, C. (1992), "Membership of Low-Mass Stars in the Open Cluster Alpha Persei," *AJ* **103:** 488–513.

PROSSER, C., J. STAUFFER, L. HARTMANN, D. SODERBLOM, B. JONES, M. WERNER, AND M. MCCAUGHREAN (1994), "HST Photometry of the Trapezium Cluster," *ApJ* **421:** 517–541.

RABL, G., AND R. DVORAK (1988), "Satellite-Type Planetary Orbits in Double Stars: A Numerical Approach," *A&A* **191:** 385–391.

RAMSEY, L.W. (1990), "Optical and UV Spectra of RS CVn Stars," in *Sixth Cambridge Workshop on Cool Stars, Stellar Systems, and the Sun*, ASP Conf. Ser., M.S. Giampapa and J.A. Bookbinder, eds., **9:** 195–205.

REIPURTH, B., AND H. ZINNECKER (1994), "Visual Binaries Among Pre–Main Sequence Stars," *A&A* **278:** 81–108.

REYNOLDS, R.T., C.P. MCKAY, AND J.F. KASTING (1987), "Europa, Tidally Heated Oceans, and Habitable Zones Around Giant Planets," *Adv. Space Res.* **7:** 125–132.

REYNOLDS, R.T., S.W. SQUYRES, D.S. COLBURN, AND C.P. MCKAY (1983), "On the Habitability of Europa," *Icarus* **56:** 246–254.

ROBERTS, M. (1957), "The Numbers of Early-Type Stars in the Galaxy and Their Relation to Galactic Clusters and Associations," *PASP* **99:** 141.

ROOD, R.T., AND J.S. TREFIL (1981), "Are We Alone?," in *The Possibility of Extraterrestrial Civilizations* (New York: Scribner Publishing Company).

ROQUES, F., H. SCHOLL, B. SICARDY, AND B.A. SMITH (1994), "Is There a Planet Around β Pictoris? Perturbations of a Planet on a Circumstellar Dust Disk. 1. The Numerical Model," *Icarus* **108:** 37–58.

ROSENBLATT, F. (1971), "A Two-Color Photometric Method for Detection of Extra-Solar Planetary Systems," *Icarus* **14:** 71–93.

SACKMANN, I.-J., A.I. BOOTHROYD, AND K.E. KRAEMER (1993), "Our Sun. III. Present and Future," *ApJ* **418:** 457–468.

SAFRONOV, V.S. (1969), *Evolution of the Protoplanetary Cloud and Formation of the Earth and Planets* (Moscow: Nauka Press). English translation: NASA TTF-677, 1972.

SAGAN, C., AND C. CHYBA (1996), "Organic Shielding of UV-Labile Greenhouse Gases on Early Earth and Mars," abstract, this volume.

SAGAN, C., AND G. MULLEN (1972), "Earth and Mars: Evolution of Atmospheres and Surface Temperatures," *Science* **177:** 52–56.

SCHECHTER, M. (1984), "Planets in Binary Star Systems," *Sky & Tel.* **68:** 394–395.

SCHNEIDER, J. (1994), "On the Search for O_2 in Extrasolar Planets," *A&SS* **212:** 321.

APPENDIX II—ASTROPHYSICAL ASPECTS REFERENCES (cont.)

SCHNEIDER, J. (1995), "On the Occultations of a Binary Star by a Circum-Orbiting Dark Companion," *Planet. Space Sci.* **42:** 539.

SCHNEIDER, J. (1996), "The Search for Habitable Extrasolar Planets: Current Programs and Long-Term Plans," this volume.

SCHNEIDER, J., AND L.R. DOYLE (1995), "Ground-Based Detection of Terrestrial Extrasolar Planets by Photometry: The Case for CM Draconis," *Earth, Moon, and Planets* **71:** 153–173.

SCHNEIDER, J., AND M. CHEVRETON (1990), "The Photometric Search for Earth-Sized Extrasolar Planets by Occultation in Binary Systems," *A&A* **232:** 251–257.

SCHNEIDER, J., M. CHEVRETON, AND E. MARTIN (1990), "New Efforts in the Search for Extrasolar Planets," in *24th ESALB Symposium*, ESA SP-315, C. Battrick, ed. (Noordwijk: European Space Agency), p. 67.

SCHWARTZMAN, D.W., AND T. VOLK (1989), "Biotic Enhancement of Weathering and the Habitability of Earth," *Nature* **340:** 457–460.

SHU, S., J. NAJITA, D. GALLI, E. OSTRIKER, AND S. LIZANO (1993), "The Collapse of Clouds and the Formation of Stars and Disks," in *Protostars and Planets III*, E. Levy, J.I. Lunine, and M.S. Matthews, eds. (Tucson: University of Arizona Press), pp. 3–45.

SIMON, T. (1992), "Bonas + 11: The Activity-Rotation-Age Relations for Solar-Type Stars," in *Seventh Cambridge Workshop on Cool Stars, Stellar Systems, and the Sun*, ASP Conf. Ser., M.S. Giampapa and J.A. Bookbinder, eds., **26:** 3–14.

SKRUTSKIE, M., D. DUTKEVITCH, S. STROM, S. EDWARDS, AND K. STROM (1990), "A Sensitive 10-micron Search for Emission Arising from Circumstellar Dust Associated with Solar-Type Pre–Main Sequence Stars," *AJ* **99:** 1187–1195.

SKUMANICH, A. (1972), "Time Scales for CaII Emission Decay, Rotational Braking and Lithium Depletion," *ApJ* **171:** 565–567.

SMITH, B.A., AND R.J. TERRILE (1984), "A Circumstellar Disk Around Beta Pictoris," *Science* **226:** 1421–1424.

SODERBLOM, D.R. (1986), "The Temperatures of Alpha Centauri A and B," *A&A* **158:** 273–274.

SODERBLOM, D.R. (1991), "Main Sequence Angular Momentum Loss in Low-Mass Stars," in *Angular Momentum Evolution of Young Stars*, S. Catalano and J.R. Stauffer, eds. (Dordrecht: Kluwer Academic Publishers), pp. 151–165.

SODERBLOM, D.R., D.K. DUNCAN, AND D.R.H. JOHNSON (1991), "The Chromospheric Emission-Age Relation for Stars of the Lower Main Sequence and Its Implications for the Star Formation Rate," *ApJ* **375:** 722–739.

SODERBLOM, D.R., M.S. OEY, D.R.H. JOHNSON, AND R.P.S. STONE (1990), "The Evolution of the Lithium Abundances of Solar-Type Stars. I. The Hyades and Coma Berenices Clusters," *AJ* **99:** 595–607.

SOLOMON, S.C., AND J.W. HEAD (1991), "Fundamental Issues in the Geology and Geophysics of Venus," *Science* **252:** 252–260.

APPENDIX II—ASTROPHYSICAL ASPECTS REFERENCES (cont.)

SQUYRES, S.W., AND J.F. KASTING (1994), "Early Mars: How Warm and How Wet?," *Science* **265:** 744–749.

STAUFFER, J., AND L. HARTMANN (1987), "The Distribution of Rotational Velocities for Low-Mass Stars in the Pleiades," *ApJ* **318:** 337–355.

STAUFFER, J., C. PROSSER, L. HARTMANN, AND M. MCCAUGHREAN (1994), "Additional Constraints on Circumstellar Disks in the Trapezium Cluster," *AJ* **108:** 1375–1381.

STAUFFER, J., L. HARTMANN, AND B. JONES (1989), "Rotational Velocities of Newly Discovered, Low-Mass Members of the Alpha Persei Cluster," *ApJ* **346:** 160–167.

STAUFFER, J., L. HARTMANN, AND D. LATHAM (1987), "Rotational Velocities of Low-Mass Stars in the Hyades," *ApJ* **320:** L51–L55.

STERN, R.A., B.M. HAISCH, Y. UCHIDA, F. WALTER, A. VEALE, A. BROWN, O. VILHU, AND D. HANNIKAINEN (1990), "Co-Ordinated *Ginga*, IUE and VLA Observations of Flaring Activity in σ^2 CR B," in *Sixth Cambridge Workshop on Cool Stars, Stellar Systems, and the Sun*, ASP Conf. Ser., M.S. Giampapa and J.A. Bookbinder, eds., **9:** 227–229.

STEWART, G.R., AND G.W. WETHERILL (1988), "Evolution of Planetesimal Velocities," *Icarus* **74:** 542–553.

STEWART, G.R., AND W.M. KAULA (1980), "Gravitational Kinetic Theory for Planetesimals," *Icarus* **44:** 154–171.

STROM, K., S. STROM, AND M. MERRILL (1993), "Infrared Luminosity Functions for the Young Stellar Population Associated with the L1641 Molecular Cloud," *ApJ* **412:** 233–253.

STROM, K., S. STROM, S. EDWARDS, S. CABRIT, AND M. SKRUTSKIE (1989), "Circumstellar Material Associated with Solar-Type Pre–Main Sequence Stars—A Possible Constraint on the Timescale for Planet Building," *AJ* **97:** 1451–1470.

STROM, S.E., S. EDWARDS, AND M.F. SKRUTSKIE (1993), "Evolution Time Scales for Circumstellar Disks Associated with Intermediate- and Solar-Type Stars," in *Protostars and Planets III*, E. Levy, J.I. Lunine, and M.S. Matthews, eds. (Tucson: University of Arizona Press), pp. 837–866.

SUSSMAN, G.J., AND J. WISDOM (1992), "Chaotic Evolution of the Solar System," *Science* **257:** 56–59.

SWENSON, F.J., AND J. FAULKNER (1992), "Lithium Dilution Through Main-Sequence Mass Loss," *ApJ* **395:** 654–674.

SZEBEHELY, V. (1980), "Stability of Planetary Orbits in Binary Systems," *Celest. Mech.* **22:** 7–12.

SZEBEHELY, V., AND R. MCKENZIE (1981), "Stability of Outer Planetary Systems," *Celest. Mech.* **23:** 3–7.

TOON, O.B., J.B. POLLACK, W. WARD, J.A. BURNS, AND K. BILSKI (1980), "The Astronomical Theory of Climate Change on Mars," *Icarus* **44:** 552–607.

TOPS (1992), *Toward Other Planetary Systems*, NASA.

TOUMA, J., AND J. WISDOM (1993), "The Chaotic Obliquity of Mars," *Science* **259:** 1294–1297.

APPENDIX II—ASTROPHYSICAL ASPECTS REFERENCES (cont.)

TOUMA, J., AND J. WISDOM (1994), "Evolution of the Earth-Moon System," *AJ* **108:** 1943–1961.

TREMAINE, S., AND L. DONES (1993), "On the Statistical Distribution of Massive Impactors," *Icarus* **106:** 335–341.

TSCHARNUTER, W.M., AND A.P. BOSS (1993), "Formation of the Protosolar Nebula," in *Protostars and Planets III*, E. Levy, J.I. Lunine, and M.S. Matthews, eds. (Tucson: University of Arizona Press), pp. 921–938.

TSESEVICH, V.P. (1973), "Photometric Phases of Eclipses," in *Eclipsing Variable Stars*, V.P. Tsesevich, ed., translated by R. Hardin (New York: Halsted-Wiley).

VIDAL-MADJAR, A., R. FERLET, A. LECAVELIER DES ETANGS, G. PERRIN, F. SÈVRE, F. COLAS, F. DELADERRIÈRE, AND A.-M. LAGRANGE-HENRI (1993), "68 Ophiuchi," *IAU Circ. 5795*.

WALKER, G. (1995?), "Spectroscopic Techniques for the Search for Other Planetary Systems," in *Proceedings of the Second TOPS Conference on Planetary Systems: Formation, Evolution and Detection*.

WALKER, G.A., A.R. WALKER, A.W. IRWIN, S.L.S. YANG, A.M. LARSON, AND D.C. RICHARDSON (1995), "A Search for Jupiter-Mass Companions to Near-by Stars," *Icarus* **116:** 359–375.

WALKER, J.C.G., AND K.J. ZAHNLE (1986), "Lunar Nodal Tide and Distance to the Moon During the Precambrian," *Nature* **320:** 600–602.

WALKER, J.C.G., P.B. HAYS, AND J.F. KASTING (1981), "A Negative Feedback Mechanism for the Long-Term Stabilization of Earth's Surface Temperature," *JGR* **86:** 9776–9782.

WARD, W.R. (1973), "Large-Scale Variations in the Obliquity of Mars," *Science* **181:** 260–262.

WARD, W.R. (1974), "Climatic Variations on Mars. I. Astronomical Theory of Insolation," *JGR* **79:** 3375–3386.

WARD, W.R. (1982), "Comments on the Long-Term Stability of the Earth's Obliquity," *Icarus* **50:** 444–448.

WARD, W.R. (1992), "Long-Term Orbital and Spin Dynamics of Mars," in *Mars*, H. Kieffer, B. Jakosky, and C. Snyder, eds. (Tucson: University of Arizona Press), pp. 298–320.

WARD, W.R., J.A. BURNS, AND O.B. TOON (1979), "Past Obliquity Oscillations of Mars: The Role of the Tharsis Uplift," *JGR* **84:** 243–259.

WEIDENSCHILLING, S.J., AND J.N. CUZZI (1993), "Formation of Planetesimals in the Solar Nebula," in *Protostars and Planets III*, E. Levy, J.I. Lunine, and M.S. Matthews, eds. (Tucson: University of Arizona Press), pp. 1031–1060.

WETHERILL, G.W. (1990), "Formation of the Earth," *Ann. Rev. Earth Planet. Sci.* **18:** 205–256.

WETHERILL, G.W. (1991), "Occurrence of Earth-like Bodies in Planetary Systems," *Science* **253:** 535–538.

WETHERILL, G.W. (1992), "An Alternative Model for the Formation of the Asteroid Belt," *Icarus* **100:** 307–325.

WETHERILL, G.W. (1994), "Possible Consequences of Absence of 'Jupiters' in Planetary Systems," *A&SS* **212:** 23–32.

APPENDIX II—ASTROPHYSICAL ASPECTS REFERENCES (cont.)

WETHERILL, G.W. (1994a), "Possible Consequences of Absence of 'Jupiters' in Planetary Systems," in *Planetary Systems: Formation, Evolution, and Detection*, B.F. Burke, J.H. Rahe and E. Roettger, eds. (Dordrecht: Kluwer Academic Publishers), pp. 23–32.

WETHERILL, G.W. (1994b), "Provenance of the Terrestrial Planets," *Geochim. Cosmochim. Acta* **58:** 4513–4520.

WETHERILL, G.W. (1996), "The Formation and Habitability of Extra-Solar Planets," *Icarus* **119:** 219–238.

WETHERILL, G.W. (1996), "The Formation of Habitable Planetary Systems," this volume.

WETHERILL, G.W., AND G.R. STEWART (1989), "Accumulation of a Swarm of Small Planetesimals," *Icarus* **77:** 350–357.

WETHERILL, G.W., AND G.R. STEWART (1993), "Formation of Planetary Embryos: Effects of Fragmentation, Low Relative Velocity, and Independent Variation of Eccentricity and Inclination," *Icarus* **106:** 190–209.

WHITE, S.M., M.R. KUNDU, Y. UCHIDA, AND N. NITTA (1990), "The Radio Spectrum of RS CVN Stars," in *Sixth Cambridge Workshop on Cool Stars, Stellar Systems, and the Sun*, ASP Conf. Ser., M.S. Giampapa and J.A. Bookbinder, eds., **9:** 239–242.

WHITMIRE, D.P., AND L.R. DOYLE (1995), "A Slightly More Massive Sun as an Explanation of Warm Temperatures on Early Mars," *JGR Planets* **100:** 5457–5464.

WHITMIRE, D.P., AND R.T. REYNOLDS (1996), "Circumstellar Habitable Zones: Astronomical Considerations," this volume.

WHITMIRE, D.P., J.J. MATESE, AND L.J. TOMLEY (1988), "A Brown Dwarf Companion as an Explanation of the Asymmetry in the Beta Pictoris Disk," *A&A* **203:** L13–L15.

WHITMIRE, D.P., L.R. DOYLE, R.T. REYNOLDS, AND J.J. MATESE (1995a), "Circumstellar Habitable Zones and Mass Loss from Young Solar-type Stars. I. Theory," in *Progress in the Search for Extraterrestrial Life*, ASP Conf. Ser., S. Shostak, ed., **74:** 183–193.

WHITMIRE, D.P., L.R. DOYLE, R.T. REYNOLDS, AND J.J. MATESE (1995b), "A Slightly More Massive Young Sun as an Explanation for Warm Temperatures on Early Mars," *JGR* **100:** 5457–5464.

WHITMIRE, D.P., R.T. REYNOLDS, AND J.F. KASTING (1991), "Habitable Zones for Earth-like Planets Around Main Sequence Stars," in *Bioastronomy: The Search for Extraterrestrial Life*, J. Heidmann and M.J. Klein, eds. (Berlin: Springer-Verlag), pp. 173–178.

WILLIAMS, D.M., J.F. KASTING, AND K. CALDEIRA (1996), "Chaotic Obliquity Variations and Planetary Habitability," this volume.

WILLIAMS, G.E. (1993), "History of the Earth's Obliquity," *Earth Science Reviews* **34:** 1–45.

WISDOM, J. (1982), "The Origin of the Kirkwood Gaps: A Mapping for Asteroidal Motion Near the 3 : 1 Commensurability," *AJ* **87:** 577–593.

WOLSZCZAN, A. (1994), "Confirmation of Earth-Mass Planets Orbiting the Millisecond Pulsar PSR B1257+12," *Science* **264:** 538–542.

APPENDIX II—ASTROPHYSICAL ASPECTS REFERENCES (cont.)

WORLEY, C.E., AND W.D. HEINTZ (1983), "Fourth Catalog of Orbits of Visual Binary Stars," *Publ. US Naval Obs.*, 2d ser. **24**, Part VII.

WRIGHT, A.E., AND M.F. BARLOW (1975), "The Radio and Infrared Spectrum of Early-type Stars Undergoing Mass Loss," *MNRAS* **170:** 41–51.

YODER, C.F. (1995), "Venus' Free Obliquity," *Icarus* **117:** 250–286.

ZAHN, J.-P. (1977), "Tidal Friction in Close Binary Stars," *A&A* **57:** 383–394.

ZAHN, J.-P. (1994), "Rotation and Lithium Depletion in Late-Type Binaries," *A&A* **288:** 829–841.

ZUCKERMAN, B., AND E.E. BECKLIN (1993), "Infrared Observations of the Remarkable Main-Sequence Star HD 98800," *ApJ* **406:** L25–L28.

Biological Aspects

A'HEARN, M.F. (1992), "Cometary Chemistry," in *Astrochemistry of Cosmic Phenomena*, P.D. Singh, ed. (Netherlands: IAU), pp. 415–420.

ADAMS, J.M., H. FAURE, L. FAURE-DENARD, J.M. MCGLADE, AND F.I. WOODWARD (1990), "Increases in Terrestrial Carbon Storage from the Last Glacial Maximum to the Present," *Nature* **348:** 711–714.

ADAMSON, A.J., D.C.B. WHITTET, AND W.W. DULEY (1990), "The 3.4-μm Interstellar Absorption Feature in Cyg OB2 No. 12," *MNRAS* **243:** 400–404.

AGARWAL, V.K., W. SCHUTTE, J.M. GREENBERG, J.P. FERRIS, R. BRIGGS, S. CONNOR, C.P.E.M. VAN DE BULT, AND F. BAAS (1985), "Photochemical Reactions in Interstellar Grains. Photolysis of CO, NH_3 and H_2O," *Origins of Life* **16:** 21–40.

AIHARA, J. (1992), "Why Aromatic Compounds Are Stable," *Sci. Am.* **240 (3):** 62–68.

ALLAMANDOLA, L.J., AND S.A. SANDFORD (1988), "Laboratory Simulation of Dust Spectra," in *Dust in the Universe*, M.E. Bailey and D.A. Williams, eds. (Cambridge: Cambridge University Press), pp. 229–263.

ALLAMANDOLA, L.J., S.A. SANDFORD, A.G.G.M. TIELENS, AND T.M. HERBST (1988a), "Infrared Spectra of Dense Clouds in the C–H Stretch Region: Methanol and Diamonds," *ApJ* **399:** 134–146.

ALLAMANDOLA, L.J., S.A. SANDFORD, A.G.G.M. TIELENS, AND T.M. HERBST (1993), "Diamonds in Dense Molecular Clouds: A Challenge to the Standard Interstellar Paradigm," *Science* **260:** 64–66.

ALLAMANDOLA, L.J., S.A. SANDFORD, AND B. WOPENKA (1987), "Interstellar Polycyclic Aromatic Hydrocarbons and Carbon in Interplanetary Dust Particles and Meteorites," *Science* **237:** 56–59.

ALLAMANDOLA, L.J., S.A. SANDFORD, AND G.J. VALERO (1988b), "Photochemical and Thermal Evolution of Interstellar/Precometary Ice Analogs," *Icarus* **76:** 225–252.

APPENDIX II—BIOLOGICAL ASPECTS REFERENCES (cont.)

ALLEGRE, C.J., AND C. JAUPART (1985), "Continental Tectonics and Continental Kinetics," *Earth Planet. Sci. Lett.* **74:** 171–186.

ALLER, L.H. (1961), *The Abundance of the Elements* (New York: Interscience Publishers, Inc.), pp. 220–231.

ALVAREZ, L.W., W. ALVAREZ, F. ASARO, AND H.V. MICHE (1980), "Extraterrestrial Cause for the Cretaceous-Tertiary Extinction," *Science* **208:** 1095–1108.

AMARI, S., R.S. LEWIS, AND E. ANDERS (1994), "Interstellar Grains in Meteorites: I. Isolation of SiC, Graphite, and Diamonds; Size Distribution of SiC and Graphite," *Geochim. Cosmochim. Acta* **58:** 459–470.

ANDERS, E. (1989), "Pre-biotic Organic Matter from Comets and Asteroids," *Nature* **342:** 255–257.

ANDERS, E., AND E. ZINNER (1993), "Interstellar Grains in Primitive Meteorites: Diamond, Silicon Carbide, and Graphite," *Meteoritics* **28:** 490–514.

ANDERS, E., R. HAYATSU, AND M.H. STUDIER (1974), "Catalytic Reactions in the Solar Nebula: Implications for Interstellar Molecules and Organic Compounds in Meteorites," in *Cosmochemical Evolution and the Origins of Life*, vol. 1, J. Oró, S.L. Miller, C. Ponnamperuma, and R.S. Young, eds. (Dordrecht: D. Reidel Publishing Company), pp. 57–67.

ANDERSON, R.P., AND J.R. ROTH (1977), "Tandem Genetic Duplications in Phage and Bacteria," *Ann. Rev. Microbiol.* **31:** 473–505.

ARMSTRONG, E. (1983), "Relative Brain Size and Metabolism in Mammals," *Science* **220:** 1302–1304.

AVETISOV, V.A. (1992), Lecture in College on Methods and Exp. Techniques in Biophysics (Trieste).

AVETISOV, V.A., V.I. GOLDANSKII, AND V.V. KUZMIN (1991), "Handedness, Origin of Life and Evolution,"*Physics Today* **44 (7):** 33–41.

AWRAMIK, S.M. (1992), "The Oldest Records of Photosynthesis," *Photosynth. Res.* **33:** 75–89.

AXELROD, D.I. (1979), "The Roles of Plate Tectonics in Angiosperm History," in *Historical Biogeography, Plate Tectonics, and the Changing Environment. Proceedings of the Thirty-Seventh Annual Biology Colloquium and Selected Papers*, J. Gray and A.J. Boucot, eds. (Corvallis: Oregon State University Press), pp. 435–447.

BADA, J., K.L.F. BRINTON, G.D. MCDONALD, AND X. WANG (1996), "A Search for Extra-terrestrial Amino Acids in Polar Ice: A Progress Report," this volume.

BAILEY, M.E., S.V.M. CLUBE, AND W.N. NAPIER (1990), *The Origin of Comets* (Oxford: Pergamon Press), pp. 451–454.

BAKKER, R.T. (1986), *The Dinosaur Heresies* (New York: Morrow).

BAR-NUN, A., A. LAZCANO-ARAUJO, AND J. ORÓ (1981), "Could Life Have Originated in Cometary Nuclei?," *Origins of Life* **11:** 387–394.

BAR-NUN, A., N. BAR-NUN, S.H. BAUER, AND C. SAGAN (1970), "Shock Synthesis of Amino Acids in Simulated Primitive Environments," *Science* **168:** 470–473.

APPENDIX II—BIOLOGICAL ASPECTS REFERENCES (cont.)

BARRA, A.L., J.B. ROBERT, AND L. WIESENFELD (1987), "Parity Non-Conservation: NMR Parameters in Chiral Molecules," *BioSystems* **20:** 57–61.

BARRON, L.D. (1981), "Optical Activity and Time Reversal," *Mol. Phys.* **43:** 1395–1406.

BARRON, L.D. (1986), "Symmetry and Molecular Chirality," *Chem. Soc. Rev.* **15:** 189–223.

BECKER, L.A., J. BADA, R.E. WINANS, J.E. HUNT, T.E. BUNCH, AND B.M. FRENCH (1994), "Fullerenes in the 1.8 byr Old Sudbury Impact Structure," *Science* **265:** 642–645.

BEEUNAS, M.A., AND L.P. KNAUTH (1985), "Preserved Stable Isotopic Signature of Subaerial Diagenesis in the 1.2-b.y. Mescal Limestone, Central Arizona: Implications for the Timing and Development of a Terrestrial Plant Cover," *Geol. Soc. Am. Bull.* **96:** 737–745.

BENACHENHOU-LAFHA, N., P. FORTERRE, AND B. LABEDAN (1993), "Evolution of Glutamate Dehydrogenase Genes: Evidence for Two Paralogous Protein Families and Unusual Branching Patterns of the Archaebacteria in the Universal Tree of Life," *J. Mol. Evol.* **36:** 335–346.

BERNER, R.A. (1992), "Weathering, Plants, and the Long-Term Carbon Cycle," *Geochim. Cosmochim. Acta* **56:** 3225–3232.

BERNER, R.A. (1993), "Atmospheric Oxygen, Tectonics and Life," in *Scientists on Gaia*, S.H. Schneider and P.J. Boston, eds. (Cambridge, MA: MIT Press), pp. 161–166.

BERNER, R.A. (1993), "Paleozoic Atmospheric CO_2: Importance of Solar Radiation and Plant Evolution," *Science* **261:** 68–70.

BERNER, R.A. (1994), "3GEOCARB II: A Revised Model of Atmospheric CO_2 over Phanerozoic Time," *Amer. J. Sci.* **294:** 56–91.

BERNHARD, R.P., T.H. SEE, AND F. HÖRZ (1992), "Projectile Compositions and Modal Frequencies on the 'Chemistry of Micrometeoroids' LDEF Experiment," in *Second LDEF Symp.*, NASA CP-3194, pp. 551–573.

BERNSTEIN, M.P., S.A. SANDFORD, L.J. ALLAMANDOLA, AND S. CHANG (1994), "The Infrared Spectrum of Matrix-Isolated Hexamethylenetetramine in Ar and H_2O at Cryogenic Temperatures," *J. Phys. Chem.* **98:** 12206.

BERNSTEIN, MAX P., SCOTT A. SANDFORD, LOUIS J. ALLAMANDOLA, S. CHANG, AND MAUREEN A. SCHARBERG (1995), "Organic Compounds Produced by Photolysis of Realistic Interstellar and Cometary Ice Analogs Containing Methanol," *ApJ* **454:** 327–344.

BEUST, H., A.M. LAGRANGE-HENRI, A. VIDAL-MAJDAR, AND R. FERLET (1990), "The β Pictoris Circumstellar Disk X. Numerical Simulations of Infalling Evaporating Bodies," *A&A* **236:** 202–216.

BLACK, C.C. (1973), "Photosynthetic Carbon Fixation in Relation to Net CO_2 Uptake," *Ann. Rev. Plant Physiol.* **24:** 253–286.

BLANKENSHIP, R.E. (1992), "Origin and Early Evolution of Photosynthesis," *Photosynth. Res.* **33:** 91–111.

BLUMER, M. (1976), "Polycyclic Aromatic Compounds in Nature," *Sci. Am.* **234 (3):** 34–38.

BOLIN, B., E.T. DEGENS, P. DUVIGNEAUD, AND S. KEMPE (1979), "The Global Biogeochemical Carbon Cycle" (Table 1.2, P. Duvigneaud), in *The Global Carbon Cycle*, SCOPE 13, B. Bolin, E.T. Degens, S. Kempe, and P. Ketner, eds. (Chichester: John Wiley & Sons), pp. 1–56.

APPENDIX II—BIOLOGICAL ASPECTS REFERENCES (cont.)

BONNER, W.A. (1991), "The Origins and Amplification of Biomolecular Chirality," *Origins of Life* **21:** 60–111.

BONNER, W.A., M.A. VAN DORT, AND M.R. YEARIAN (1975), "Asymmetric Degradation of DL-Leucine with Longitudinally Polarised Electrons," *Nature* **258:** 419–421.

BORGHESI, A., E. BUSSOLETTI, AND L. COLANGELI (1987), "Amorphous Carbon and the Unidentified Infrared Bands," *ApJ* **314:** 422–428.

BOSTON, P.J., AND S.L. THOMPSON (1991), "Theoretical Microbial and Vegetation Control of Planetary Environments," in *Scientists on Gaia*, S.H. Schneider and P.J. Boston, eds. (Cambridge: MIT Press), pp. 99–117.

BOWRING, S.A., T. HOUSH, AND C. ISACHSEN (1993), "Nd Isotope Systematics of the Early Archean Acasta Gneisses: Implications for Models of the Early Earth," abstract, Annual Meeting of the Geological Society of America, p. A73.

BRACK, A., AND G. SPACH (1980), "β-Structures of Polypeptides with L- and D- Residues," *J. Mol. Evol.* **15:** 231–238.

BRIDGER, A., G. WRIGHT, AND T. GEBALLE (1993), in *Conference Abstract Volume of Infrared Astronomy with Arrays: The Next Generation* (Los Angeles: UCLA Press).

BRIDSON, P.K., AND L.E. ORGEL (1980), "Catalysis of Accurate poly(C)-Directed Synthesis of 3′-5′- Linked Oligoguanylates of Zn^{++}," *J. Mol. Biol.* **144:** 567–577.

BRIGGS, R., G. ERTEM, J.P. FERRIS, J.M. GREENBERG, J.P. MCCAIN, C.X. MENDOZA-GÓMEZ, AND W. SCHUTTE (1992), "Comet Halley as an Aggregate of Interstellar Dust and Further Evidence for the Photochemical Production of Organics in the Interstellar Medium," *Origins of Life* **22:** 287.

BRIMHALL, G.H., O.A. Chadwick, C.J. Lewis, W. Compston, I.S. Williams, K.J. Danti, W.E. Dietrich, M.E. Power, D. Hendricks, and J. Bratt (1991), "Deformational Mass Transport and Invasive Process in Soil Evolution," *Science* **255:** 695–702.

BRINTON, K.L.F., AND J.L. BADA (1994), "The Use of the Ninhydrin Reaction for the Elimination of Protein Amino Acids in Cosmogeochemical Samples," *Origins of Life* **24:** 125–126.

BROCK, T.D., AND M.T. MADIGAN (1991), *Biology of Microorganisms*, 6th ed. (Englewood Cliffs, NJ: Prentice-Hall).

BROOKE, T.Y., A.T. TOKUNAGA, AND R.J. KNACKE (1991), "Detection of the 3.4 μm Emission Feature in Comets P/Borosen-Metcalf and Okazaki-Levy-Rudenko (1989R) and an Observational Summary," *AJ* **101:** 268–278.

BRYSON, R.A. (1966), "Air Masses, Streamlines, and the Boreal Forest," *Geogr. Bull.* **8:** 228–269.

BUICK, R. (1992), "The Antiquity of Oxygenic Photosynthesis: Evidence from Stromatolites in Sulphate-Deficient Archaean Lakes," *Science* **255:** 74–77.

BULTEROV, A. (1859), "Hexamethylenetetramine," *Justus Liebigs Annalen Der Chemie* **11:** 250.

BUNCH, T.E., F. RADICATI DI BROZOLO, R.H. FLEMING, D.W. HARRIS, D.E. BROWNLEE, AND T. REILLY (1992), "LDEF Impact Craters Formed by Carbon-Rich Impactors: A Preliminary Report," in *First LDEF Symp.*, NASA CP-3134, pp. 549–564.

APPENDIX II—BIOLOGICAL ASPECTS REFERENCES (cont.)

BUNCH, T.E., L. BECKER, J. BADA, J. MACKLIN, F. RADICATI DI BROZOLO, R.H. FLEMING, AND J. ERLICHMAN (1993), "Hypervelocity Impact Survivability Experiments for Carbonaceous Impactors," in *Second LDEF Symp.*, NASA CP-3194, pp. 453–477.

BUNCH, T.E., L. BECKER, J. BADA, J.M. PAQUE, AND J. MACKLIN (1996), "The Effects of Hypervelocity Impact on Carbonaceous Compounds: Experimental Insights," this volume.

BURKE, D.H., J.E. HEARST, AND A. SIDOW (1993), "Early Evolution of Photosynthesis: Clues from Nitrogenase and Chlorophyll Iron Proteins," *Proc. Natl. Acad. Sci. USA* **90:** 7134–7138.

BUTCHART, I., A.D. MCFADZEAN, D.C.B. WHITTET, T.R. GEBALLE, AND J.M. GREENBERG (1986), "Three Micron Spectroscopy of the Galactic Centre Source IRS 7," *A&A* **154:** L5–L7.

BUTLEROW, A. (1861), "Formation Sintetique d'une Substance Sucreé," *Compte. Rend. Acad. Sci.* **53:** 145–147.

BÜTTNER, M., D.-L. XIE, H. NELSON, W. PINTHER, G. HAUSKA, AND N. NELSON (1992), "Photosynthetic Reaction Centers Genes in Green Sulfur Bacteria and in Photosystem I Are Related," *Proc. Natl. Acad. Sci. USA* **89:** 8135–8139.

BYERLY, G.R., D.R. LOWE, AND M.M. WALSH (1986), "Stromatolites from the 3,300–3,500 Myr Swaziland Supergroup, Barberton Mountain Land, South Africa," *Nature* **319:** 489–491.

CAIRNS-SMITH, A.G. (1982), *Genetic Takeover and the Mineral Origins of Life* (Cambridge: Cambridge University Press).

CALDEIRA, K. (1991), "Continental-Pelagic Carbonate Partitioning and the Global Carbonate-Silicate Cycle," *Geology* **19:** 204–206.

CALDEIRA, K. (1992), "Enhanced Cenozoic Chemical Weathering and the Subduction of Pelagic Carbonate," *Nature* **357:** 578–581.

CALDEIRA, K., AND J.F. KASTING (1992a), "Susceptibility of the Early Earth to Irreversible Glaciation Caused by Carbon Dioxide Clouds," *Nature* **359:** 226–228.

CALDEIRA, K., AND J.F. KASTING (1992b), "The Life Span of the Biosphere Revisited," *Nature* **360:** 721–723.

CAMERON, A.G.W., AND W. BENZ (1991), "The Origin of the Moon and the Single Impact Hypothesis IV," *Icarus* **92:** 204–216.

CAMMARANO, P., P. PALM, R. CRETI, E. CECCARELLI, A.M. SANANGELANTONI, AND O. TIBONI (1992), "Early Evolutionary Relationships Among Known Life Forms Inferred from Elongation Factor ef-2 (ef-g) Sequences. Phylogenetic Coherence and Structure of the Archaeal Domain," *J. Mol. Evol.* **34:** 396–405.

CANFIELD, D.E., AND D.J. DES MARAIS (1993), "Biogeochemical Cycles of Carbon, Sulfur, and Free Oxygen in a Microbial Mat," *Geochim. Cosmochim. Acta* **57:** 3971–3984.

CARTER, B. (1983), "The Anthropic Principle and Its Implications for Biological Evolution," *Phil. Trans. R. Soc. Lond. A* **310:** 347–363.

CHANG, S. (1979), "Comets, Cosmic Connections with Carbonaceous Meteorites and Interstellar Molecules," in *Space Missions to Comets*, NASA CP-2089, M. Neugebauer, D.K. Yeomans, J.C. Brandt, and R.W. Hobbs, eds., pp. 59–111.

APPENDIX II—BIOLOGICAL ASPECTS REFERENCES (cont.)

CHAPMAN, C.R., AND D. MORRISON (1994), "Impacts on the Earth by Asteroids and Comets: Assessing the Hazard," *Nature* **367:** 33–40.

CHAPMAN, D.J. (1992), "Origin and Divergence of Protists," in *The Proterozoic Biosphere*, J.W. Schopf and C. Klein, eds. (Cambridge: Cambridge University Press), pp. 477–483.

CHAPMAN, D.J., AND J.W. SCHOPF (1983), "Biological and Biochemical Effects of the Development of an Aerobic Environment," in *Earth's Earliest Biosphere: Its Origin and Evolution*, J.W. Schopf, ed. (Princeton, NJ: Princeton University Press), pp. 302–320.

CHARLSON, R.J., J.E. LOVELOCK, M.O. ANDREAE, AND S.G. WARREN (1987), "Oceanic Phytoplankton, Atmospheric Sulphur, Cloud Albedo and Climate," *Nature* **326:** 655–661.

CHYBA, C.F. (1990), "Impact Delivery and Erosion of Planetary Oceans in the Early Inner Solar System," *Nature* **343:** 129–133.

CHYBA, C.F. (1993), "Explosions of Small Spacewatch Objects in the Earth's Atmosphere," *Nature* **363:** 701–703.

CHYBA, C., AND C. SAGAN (1992), "Endogenous Production, Exogenous Delivery and Impact-Shock Synthesis of Organic Molecules: An Inventory for the Origins of Life," *Nature* **355:** 125–132.

CHYBA, C.F., P.J. THOMAS, AND K.J. ZAHNLE (1993), "Atmospheric Disruption of Small Comets and Asteroids and the 1908 Tunguska Explosion," *Nature* **361:** 40–44.

CHYBA, C.F., P.J. THOMAS, L. BROOKSHAW, AND C. SAGAN (1990), "Cometary Delivery of Organic Molecules to the Early Earth," *Science* **249:** 366–373.

CHYBA, C.F., T.C. OWEN, AND W.-H. IP (1994), "Impact Delivery of Volatiles and Organic Molecules to Earth," in *Hazards Due to Comets and Asteroids*, T. Gehrels, ed. (Tucson: University of Arizona Press), pp. 9–58.

CINTALA, M.J. (1993), "Impact-Induced Thermal Effects in the Lunar and Mercurian Regolith," *JGR* **97:** 947–973.

CLARK, B.C. (1988), "Primeval Procreative Comet Pond," *Origins of Life* **18:** 209–238.

CLARKE, P.H., AND S.R. ELSDEN (1980), "The Earliest Catabolic Pathways," *J. Mol. Evol.* **15:** 333–338.

CLOUD, P.E., JR. (1968), "Atmospheric and Hydrospheric Evolution of the Primitive Earth," *Science* **160:** 729–736.

COCHRAN, M.F., AND R.A. BERNER (1992), "Initial Effects of Vegetation on Hawaiian Basalt Weathering Rates," abstract, Annual Meeting of the Geological Society of America, p. A169.

COCHRAN, M.F., AND R.A. BERNER (1993), reply to the comments on "Weathering, Plants, and the Long-Term Carbon Cycle," *Geochim. Cosmochim. Acta* **57:** 2147–2148.

CRAWFORD, R.M.M. (1989), *Studies in Plant Survival: Ecological Case Histories of Plant Adaptation* (Oxford: Blackwell Scientific Publications).

CRONIN, J.R. (1976), "Acid-Labile Amino Acid Precursors in the Murchison Meteorite," *Origins of Life* **7:** 337–348.

APPENDIX II—BIOLOGICAL ASPECTS REFERENCES (cont.)

CRONIN, J.R. (1989), "Origin of Organic Compounds in Carbonaceous Chondrites," *Adv. Space Res.* **9:** 54–64.

CRONIN, J.R., AND S. CHANG (1995), "Organic Matter in Meteorites: Molecular and Isotopic Analyses of the Murchison Meteorite," in *The Chemistry of Life's Origin*, J.M. Greenberg and V. Pirronello, eds. (Dordrecht: Kluwer Academic Publishers), pp. 209–258.

CRONIN, J., AND S. PIZZARELLO (1990), "Aliphatic Hydrocarbons of the Murchison Meteorite," *Geochim. Cosmochim. Acta* **54:** 2859–2868.

CRONIN, J.R., AND S. PIZZARELLO (1983), "Amino Acids in Meteorites," *Adv. Space Res.* **3:** 5–12.

CROSS, R.L., AND L. TAIZ (1990), "Gene Duplication as a Means for Altering H^+/ATP Ratios During the Evolution of FoF1 ATPases and Synthases," *Fed. Eur. Biochem. Soc. Lett.* **259:** 227–229.

CRUIKSHANK, D.P., AND R.H. BROWN (1987), "Organic Matter on Asteroid 130 Elektra," *Science* **238:** 183–185.

CRUIKSHANK, D.P., W.K. HARTMANN, D.J. THOLEN, L.J. ALLAMANDOLA, R.H. BROWN, C.N. MATTHEWS, AND J.F. BELL (1991), "Solid C≡N Bearing Material on Outer Solar System Bodies," *Icarus* **94:** 345–353.

DANIELS, S.P., K.R. PETERSON, L.D. STRAUSBAUGH, M.G. KIDWELL, AND A. CHOVNICK (1990), "Evidence for Horizontal Transmission of the P Transposable Element Between *Drosophila* Species," *Genetics* **124:** 339–355.

DARGE, W., I. LACZKO, AND W. THIEMANN (1976), "Stereoselectivity of β-Irradiation of D,L-Tryptophan in Aqueous Solution," *Nature* **261:** 522–524.

DAWKINS, R. (1986), *The Blind Watchmaker* (New York: W.W. Norton & Company).

DAYHOFF, M.O., R.M. SCHWARTZ, AND B.C. ORCUTT (1978), "A Model of Evolutionary Change in Proteins," in *Atlas of Protein Sequence and Structure*, vol. 5 (Silver Spring, MD: National Biomedical Research Foundation), pp. 345–352.

DE VRIES, M.S., K. REIHS, H.R. WENDT, W.G. GOLDEN, AND H.E. HUNZIKER (1993), "A Search for C_{60} in Carbonaceous Chondrites," *Geochim. Cosmochim. Acta* **57:** 933–938.

DEAMER, D.W., AND G.R. FLEISCHAKER (1994), *Origins of Life: The Central Concepts* (Boston: Jones and Bartlett Publishers), p. 431.

DEAMER, D.W., AND R.M. PASHLEY (1989), "Amphiphilic Components of the Murchison Carbonaceous Chondrite: Surface Properties and Membrane Formation," *Origins of Life* **19:** 21–38.

DELSEMME, A.H. (1991), "Nature and History of the Organic Compounds in Comets: An Astrophysical View," in *Comets in the Post-Halley Era*, vol. 1, R.L. Newburn, Jr., M. Neugebauer, and J. Rahe, eds. (Dordrecht: Kluwer Academic Publishers), pp. 377–428.

DELSEMME, A.H. (1992), "Cometary Origin of Carbon, Nitrogen and Water on the Earth," *Origins of Life* **21:** 279–298.

DES MARAIS, D.J. (1985), "Carbon Exchange Between the Mantle and Crust and Its Effect upon the Atmosphere, Today Compared to Archean Time," in *The Carbon Cycle and Atmospheric CO_2: Natural Variations Archean to Present*, E.T. Sundquist and W.S. Broecker, eds. (Washington, DC: American Geophysical Union), pp. 602–611.

APPENDIX II—BIOLOGICAL ASPECTS REFERENCES (cont.)

DES MARAIS, D.J. (1994), "Tectonic Control of the Crustal Organic Carbon Reservoir During the Precambrian," *Chem. Geol.* **114:** 303–314.

DES MARAIS, D.J. (1996), "Has Life Made Its Global Environment More Habitable?," this volume.

DES MARAIS, D.J., H. STRAUSS, R.E. SUMMONS, AND J.M. HAYES (1992), "Carbon Isotope Evidence for the Stepwise Oxidation of the Proterozoic Environment," *Nature* **359:** 605–609.

DESMOND, A.J. (1975), *The Hot-Blooded Dinosaurs* (New York: Warner Books).

DICKERSON, R.E. (1978), "Chemical Evolution and the Origin of Life," *Sci. Am.* **239 (9):** 70–86.

DICKINSON, R.G., AND A.L. RAYMOND (1923), "The Crystal Structure of Hexamethylene-tetramine," *J. Am. Chem. Soc.* **45:** 22.

DING, D.F., AND J.H. XU (1985), "The Symmetry-Breaking in Living Systems—The Selection Rules, Sensitivity and Chiral Matching," *Scientia Sinica B* **28:** 250–263.

DOLE, S.H. (1964), *Habitable Planets for Man* (New York: Blaisdell Publishing Company).

DOOLITTLE, R.F. (1979), "Protein Evolution," in *The Proteins*, H. Neurath, ed. (New York: Academic Press), pp. 1–118.

DOOLITTLE, R.F., D.F. FENG, K.L. ANDERSON, AND M.R. ALBERRO (1990), "A Naturally Occurring Horizontal Gene Transfer from a Eukaryote to a Prokaryote," *J. Mol. Evol.* **31:** 383–388.

DRAKE, J.W. (1991), "A Constant Rate of Spontaneous Mutations in DNA-Based Microbes," *Proc. Natl. Acad. Sci. USA* **88:** 7160–7164.

DREVER, J.I., AND J. ZOBRIST (1992), "Chemical Weathering of Silicate Rocks as a Function of Elevation in the Southern Swiss Alps," *Geochim. Cosmochim. Acta* **56:** 3209–3216.

EDMONDS, J.M., K.L. VON DAMM, R.E. MCDUFF, AND C.I. MEASURES (1982), "Chemistry of Hot Springs on the East Pacific Rise and Their Effluent Dispersal," *Nature* **297:** 187–191.

EHLERINGER, J.R. (1980), "Leaf Morphology and Reflectance in Relation to Water and Temperature Stress," in *Adaptations of Plants to Water and High Temperature Stress*, N.C. Turner and P.J. Kramer, eds. (New York: Wiley Interscience), pp. 295–308.

EHLERINGER, J.R., AND I. FORSETH (1980), "Solar Tracking by Plants," *Science* **210:** 1094–1098.

EHRENFREUND, P., F. ROBERT, L. D'HENDECOURT, AND F. BEHAR (1991), "Comparison of Interstellar and Meteoritic Organic Matter at 3.4 μm," *A&A* **252:** 712–717.

ENGELKE, F., J.H. HAHN, W. HENKE, AND R.N. ZARE (1991), "Determination of Phenylthio-hydantoin–Amino Acids by Two-Step Laser Desorption/Multiphoton Ionization," *Anal. Chem.* **59:** 909–912.

EVANS, R.A., P. LORENCAK, T.-K. HA, AND C. WENTRUP (1991), "HCN Dimers: Imino-acetonitrile and N-Cyanomethanimine," *J. Am. Chem. Soc.* **113:** 7261–7276.

FEHER, G., J.P. ALLEN, M.Y. OKAMURA, AND D.C. REES (1989), "Structure and Function of Bacterial Photosynthetic Reaction Centers," *Nature* **239:** 111–116.

APPENDIX II—BIOLOGICAL ASPECTS REFERENCES (cont.)

FENCHEL, T., AND T.H. BLACKBURN (1979), *Bacteria and Mineral Cycling* (New York: Academic Press).

FERNÁNDEZ, J.A. (1985), "The Formation and Dynamical Survival of the Comet Cloud," in *Dynamics of Comets: Their Origin and Evolution*, A. Carusi and G.B. Valsecchi, eds. (Dordrecht: D. Reidel Publishing Company), pp. 45–70.

FERRIS, J.P., AND W.J. HAGAN (1984), "HCN and Chemical Evolution: The Possible Role of Cyano Compounds in Prebiotic Synthesis," *Tetrahedron* **40:** 1093–1120.

FERRIS, J.P., P.C. JOSHI, E.H. EDELSON, AND J.G. LAWLESS (1978), "HCN: A Plausible Source of Purines, Pyrimidines and Amino Acids on the Primitive Earth," *J. Mol. Evol.* **11:** 293–311.

FESENKOV, V.G. (1971), "Organic Substances in the Universe," in *Chemical Evolution and the Origin of Life*, R. Buvet and C. Ponnamperuma, eds. (Amsterdam: North-Holland Publishing Company), pp. 495–498.

FISCHER, E. (1894), "Einfluss der Configuration auf die Wirkung der Enzyme," *Chem. Ber.* **27:** 2985–2993.

FITTER, A.H., AND R.K.M. HAY (1987), *Environmental Physiology of Plants*, 2d ed. (London: Academic Press).

FLORES, I. (1976), "Codes," in *Encyclopedia of Computer Science*, A. Ralston and C.L. Meek, eds. (New York: Petrocelli/Charter), pp. 210–215.

FORTERRE, P., F. CONFALONIERI, AND M. DUGUET (1995), "Speculations on the Origin of Life and Thermophily: Review of Available Information on Reverse Gyrase Suggests that Hyperthermophilic Procaryotes Are Not So Primitive," *Origins of Life* **25:** 235–250.

FORTERRE, P., N. BENACHENHOU-LAHFA, F. CONFALONIERI, M. DUGUET, CH. ELIE, AND B. LABEDAN (1993), "The Nature of the Last Universal Ancestor and the Root of the Tree of Life, Still Open Questions," *BioSystems* **28:** 15–32.

FRANÇOIS, L.M., AND J.C.G. WALKER (1992), "Modeling the Phanerozoic Carbon Cycle and Climate: Constraints from the $^{87}Sr/^{86}Sr$ Isotopic Ratio of Seawater," *Amer. J. Sci.* **292:** 81–135.

FRONDEL, C. (1978), "Characters of Quartz Fibres," *Am. Mineral.* **63:** 17–27.

GARAY, A.S. (1968), "Origin and Role of Optical Isomery in Life," *Nature* **219:** 338–340.

GARBOW, J.R., J. SCHAEFER, R.A. LUDICKY, AND C.N. MATTHEWS (1987), "Detection of Secondary Amides in HCN Polymers by Dipolar Rotational Spin-Echo ^{15}N NMR," *Macromolecules* **20:** 305–309.

GARCÍA-MEZA, V., V. GONZÁLEZ-RODRÍGUEZ, AND A. LAZCANO (1994), "Ancient Paralogous Duplications and the Search for Archean Cells," in *Self Production of Supramolecular Structures: From Synthetic Structures to Models of Minimal Living Systems*, G.R. Fleischaker, S. Colonna, and P.-L. Luisi, eds. (Dordrecht: Kluwer Academic Publishers), pp. 145-153.

GENETICS COMPUTER GROUP (1991), *Program Manual for GCG Package*, vers. 7, April 1991.

GOGARTEN-BOEKELS, M., AND J.P. GOGARTEN (1995), "The Effects of Heavy Meteorite Bombardment on the Early Evolution of Cellular Life—A New Look at the Molecular Record," *Origins of Life* **25:** 251–264.

APPENDIX II—BIOLOGICAL ASPECTS REFERENCES (cont.)

GOGARTEN, J.P., AND L. TAIZ (1992), "Evolution of Proton Pumping ATPases: Rooting the Tree of Life," *Photosynth. Res.* **33:** 137–146.

GOGARTEN, J.P., H. KIBAK, P. DITTRICH, L. TAIZ, E.J. BOWMAN, B.J. BOWMAN, M.F. MANOLSON, R.J. POOLE, T. DATE, T. OSHIMA, J. KONISHI, K. DENDA, AND M. YOSHIDA (1989), "Evolution of the Vacuolar H^+-ATPase: Implications for the Origin of Eukaryotes," *Proc. Natl. Acad. Sci. USA* **86:** 6661–6665.

GOGARTEN, J.P., T. STARKE, H. KIBAK, J. FICHMANN, AND L. TAIZ (1992), "Evolution and Isoforms of V-ATPase Subunits," *J. Exp. Biol.* **172:** 137–147.

GOLD, T. (1992), "The Deep, Hot Biosphere," *Proc. Natl. Acad. Sci. USA* **89:** 6045–6049.

GOTT, J.R., III (1993), "Implications of the Copernican Principle for Our Future Prospects," *Nature* **363:** 315–319.

GOULD, S.J. (1989), *Wonderful Life* (New York: W.W. Norton & Company).

GOULD, S.J. (1994), "The Evolution of Life on the Earth," *Sci. Am.* **271 (4):** 85–91.

GRANICK, S. (1957), "Speculations on the Origin and Evolution of Photosynthesis," *Ann. NY Acad. Sci.* **69:** 292–308.

GRAY, M.W., R. CEDERGREN, Y. ABEL, AND D. SANKOFF (1989), "On the Evolutionary Origin of the Plant Mitochondrion and Its Genome," *Proc. Natl. Acad. Sci. USA* **86:** 2267–2271.

GREELEY, R., AND P.D. SPUDIS (1981), "Volcanism on Mars," *Rev. Geophys. Space Phys.* **19:** 13–41.

GREENBERG, J.M. (1977), "From Interstellar Dust to Comets to Dust," in *Comets, Asteroids and Meteorites*, A.H. Delsemme, ed. (Toledo, OH: University of Toledo Press), p. 491.

GREENBERG, J.M. (1982), "Dust in Dense Clouds—One Stage in a Cycle," in *Submillimeter Wave Astronomy*, J.E. Beckman and J.P. Phillips, eds. (Cambridge: Cambridge University Press), pp. 261–306.

GREENBERG, J.M. (1989), "Interstellar Dust: An Overview of Physical and Chemical Evolution," in *Evolution of Interstellar Dust and Related Topics*, A. Bonetti, J.M. Greenberg, and S. Aeillo, eds. (Amsterdam: Elsevier North-Holland), pp. 7–52.

GREENBERG, M. (1982), "What Are Comets Made Of? A Model Based on Interstellar Dust," in *Comets*, L.L. Wilkening, ed. (Tucson: University of Arizona Press), pp. 131–163.

GRIERSON, D., AND S. COVEY (1984), *Plant Molecular Biology* (London: Blackie & Son).

GROTZINGER, J.P., AND J.F. KASTING (1993), "New Constraints on Precambrian Ocean Composition," *J. Geol.* **101:** 235–243.

GUPTA, R.S., AND G.B. GOLDING (1993), "Evolution of HSP70 Gene and Its Implications Regarding Relationships Between Archaebacteria, Eubacteria and Eukaryotes," *J. Mol. Evol.* **37:** 573–582.

HAHN, J.H., R. ZENNOBI, AND R.N. ZARE (1987), "Subfemtomole Quantitation of Molecular Absorbates by Two-Step Laser Mass Spectrometry," *J. Am. Chem. Soc.* **109:** 2842–2843.

HAN, T.-M., AND B. RUNNEGAR (1992), "Megascopic Eukaryotic Algae from the 2.1-Billion-Year-Old Negaunee Iron-Formation, Michigan," *Science* **257:** 232–235.

APPENDIX II—BIOLOGICAL ASPECTS REFERENCES (cont.)

HANSCHE, P.E. (1975), "Gene Duplication as a Mechanism of Genetic Adaptation in *Saccharomyces cerevisiae*," *Genetics* **79:** 661–674.

HART, M.H. (1979), "Habitable Zones Around Main Sequence Stars," *Icarus* **37:** 351–357.

HARTLEY, B.S. (1974), "Enzymes Families," in *Evolution in the Microbial World: 24th Symposium of the Society for General Microbiology*, M.J. Carlile and J.J. Skehel, eds. (Cambridge: Cambridge University Press), pp. 151–182.

HARTMAN, H., AND C.P. MCKAY (1995), "Oxygenic Photosynthesis and the Oxidation State of Mars," *Planet. Space Sci.* **43:** 123–128.

HAURI, E.H., N. SHIMIZU, J.J. DIEU, AND S.R. HART (1993), "Evidence for Hotspot-Related Carbonatite Metasomatism in the Oceanic Upper Mantle," *Nature* **365:** 221–227.

HAWKER, J.R., JR., AND J. ORÓ (1981), "Cyanamide Mediated Syntheses of Leu, Ala, and Phe Peptides Under Plausible Primitive Earth Conditions," in *Origins of Life*, Y. Wolfman, ed. (Dordrecht: D. Reidel Publishing Company), pp. 225–232.

HEDBURG, I., AND O. HEDBURG (1979), "Tropical-Alpine Life Forms of Vascular Plants," *Oikos* **33:** 297–307.

HEGSTROM, R. (1985), "Weak Neutral Current and β-Radiolysis Effects on the Origin of Biomolecular Chirality," *Nature* **315:** 749.

HEGSTROM, R.A., D.W. REIN, AND P.G.H. SANDARS (1979), "Parity Non-Conserving Energy Difference Between Mirror-Image Molecules," *Phys. Lett. A* **71:** 499–502.

HEGSTROM, R.A., D.W. REIN, AND P.G.H. SANDARS (1980), "Calculation of the Parity Non-Conserving Energy Difference Between Mirror-Image Molecules," *J. Chem. Phys.* **73:** 2329–2341.

HERDMAN, M. (1985), "The Evolution of Bacterial Genomes," in *The Evolution of Genome Size*, T. Cavalier-Smith, ed. (London: John Wiley & Sons), pp. 37–68.

HILARIO, E.M., AND J.P. GOGARTEN (1993), "Horizontal Transfer of ATPase Genes—The *Tree* of Life Becomes a *Net* of Life," *BioSystems* **31:** 111–119.

HILLENKAMP, F., M. KARAS, R.C. BEAVIS, AND B.T. CHAIT (1991), "Matrix-Assisted Laser Desorption/Ionization Mass Spectrometry of Biopolymolecules," *Anal. Chem.* **63:** 24–37.

HOBAN, S., D.C. REUTER, M.A. DISANTI, AND M.J. MUMMA (1993), "Comets," *Icarus* **105:** 548.

HOBAN, S., M.J. MUMMA, D.C. REUTER, M.A. DISANTI, R.R. JOYCE, AND A. STORRS (1991), "A Tentative Identification of Methanol as the Progenitor of the 3.52-μm Emission Feature in Several Comets," *Icarus* **93:** 122–134.

HODGE, L.A., F.B. DUNNING, G.K. WALTERS, R.H. WHITE, AND G.J. SCHROEPFER, JR. (1979), "Degradation of DL-Leucine with Longitudinally Polarised Electrons," *Nature* **280:** 250–252.

HOLLAND, H.D. (1984), *The Chemical Evolution of the Atmosphere and Oceans* (Princeton, NJ: Princeton University Press).

HOLM, N.G., ED. (1992), *Marine Hydrothermal Systems and the Origin of Life* (Boston: Kluwer Academic Publishers).

APPENDIX II—BIOLOGICAL ASPECTS REFERENCES (cont.)

HOLMDEN, C., AND K. MUEHLENBACKS (1993), "The $^{18}O/^{16}O$ Ratio of 2-Billion-Year-Old Seawater Inferred from Ancient Oceanic Crust," *Science* **259:** 1733–1736.

HORNER, J.R., AND R. MAKELA (1979), "Nest of Juveniles Provides Evidence of Family Structure Among Dinosaurs," *Nature* **282:** 296–298.

HORODYSKI, R.J., AND L.P. KNAUTH (1994), "Life on Land in the Precambrian," *Science* **263:** 494–498.

HOROWITZ, N.H. (1945), "On the Evolution of Biochemical Synthesis," *Proc. Natl. Acad. Sci. USA* **31:** 153–157.

HÖRZ, F., H. FECHTIG, AND J. JANICKE (1983), "Morphology and Chemistry of Projectile Residues in Small Experimental Craters," *Proc. 14th LPSC, JGR* **88:** B353–363.

HOYLE, F., AND N.C. WICKRAMASINGHE (1962), "On Graphite Particles as Interstellar Grains," *MNRAS* **124:** 417–433.

HOYLE, F., N.C. WICKRAMASINGHE, S. AL-MUFTI, A.H. OLAVESEN, AND D.T. WICKRAMASINGHE (1982), "Infrared Spectroscopy Over the 2.9–3.9 μm Waveband in Biochemistry and Astronomy," *A&SS* **83:** 405–409.

HUANG, K.S. (1963), *Statistical Mechanics* (New York: John Wiley & Sons), p. 346.

HUANG, S.-S. (1959a), "Occurrence of Life in the Universe," *Amer. Sci.* **47:** 397–402.

HUANG, S.-S. (1959b), "The Problem of Life in the Universe and the Mode of Star Formation," *PASP* **71:** 421–424.

HUANG, S.-S. (1960), "Life Supporting Regions in the Vicinity of Binary Systems," *PASP* **72:** 106–114.

HUEBNER, W.F., D.C. BOICE, AND A. KORTH (1989), "Halley Polymeric Organic Molecules," *Adv. Space Res.* **9:** 29–32.

HUSS, G.R. (1990), "Ubiquitous Interstellar Diamond and SiC in Primitive Chondrites: Abundances Reflect Metamorphism," *Nature* **347:** 159–162.

IP, W.-H., AND J.A. FERNÁNDEZ (1988), "Exchange of Condensed Matter Among the Outer and Terrestrial Protoplanets and the Effect on Surface Impact and Atmospheric Accretion," *Icarus* **74:** 47–61.

IRVINE, W.M., P.F. GOLDSMITH, AND A. HJALMARSON (1987), "Chemical Abundances in Molecular Clouds," in *Interstellar Processes*, D.J. Hollenbach and H.A. Thronson, Jr., eds. (Dordrecht: D. Reidel Publishing Company), pp. 561–609.

ISHIWATARI, R., AND K. FUKUSHIMA (1979), "Generation of Saturated and Aromatic Hydrocarbons by Thermal Alteration of Young Kerogen," *Geochim. Cosmochim. Acta* **43:** 1343–1349.

IWABE, N., K.-I. KUMA, M. HASEGAWA, S. OSAWA, AND T. MIYATA (1989), "Evolutionary Relationships of Archaebacteria, Eubacteria and Eukaryotes Inferred from Phylogenetic Trees of Duplicated Genes," *Proc. Natl. Acad. Sci. USA* **86:** 9355–9359.

JACKSON, T.A. (1993), comment on "Weathering, Plants, and the Long-Term Carbon Cycle," by Robert A. Berner, *Geochim. Cosmochim. Acta* **57:** 2141–2144.

APPENDIX II—BIOLOGICAL ASPECTS REFERENCES (cont.)

JACKSON, T.A., AND W.D. KELLER (1970), "A Comparative Study of the Role of Lichens and 'Inorganic' Processes in the Chemical Weathering of Recent Hawaiian Lava Flows," *Amer. J. Sci.* **269:** 446–466.

JACOBSEN, S.B. (1988), "Isotopic and Chemical Constraints on Mantle-Crust Evolution," *Geochim. Cosmochim. Acta* **52:** 1341–1350.

JONES, A.P., W.W. DULEY, AND D.A. WILLIAMS (1987), "Interstellar Extinction Correlations," *MNRAS* **229:** 213–221.

JØRGENSEN, B.B., Y. COHEN, AND D.J. DES MARAIS (1987), "Photosynthetic Action Spectra and Adaptation to Spectral Light Distribution in a Benthic Cyanobacterial Mat," *Appl. Env. Microbiol.* **53:** 879–886.

JOYCE, G.F., A.W. SCHWARTZ, S.L. MILLER, AND L.E. ORGEL (1987), "The Case for an Ancestral Genetic System Involving Simple Analogues of the Nucleotides," *Proc. Natl. Acad. Sci. USA* **84:** 4398–4402.

JOYCE, G.F., G.M. VISSER, C.A.A. VAN BOECKEL, J.H. VAN BOOM, L.E. ORGEL, AND J. VAN WESTRESEN (1984), "Chiral Selection in Poly(C)-Directed Synthesis of Oligo (G)," *Nature* **310:** 602–604.

KANDLER, O. (1993), "The Early Diversification of Life," in *Early Life on Earth*, Nobel Symposium no. 84, S. Bengston, ed. (New York: Columbia University Press), pp. 111–118.

KARGEL, J.S. (1992), "Ammonia-Water Volcanism on Icy Satellites," *Icarus* **100:** 556–574.

KASTING, J.F. (1987), "Theoretical Constraints on Oxygen and Carbon Dioxide Concentrations in the Precambrian Atmosphere," *Precambrian Res.* **34:** 205–229.

KASTING, J.F. (1993), "Earth's Early Atmosphere," *Science* **259:** 920–926.

KASTING, J.F. (1996), "Habitable Zones Around Stars: An Update," this volume.

KASTING, J.F., AND D.H. GRINSPOON (1990), "The Faint Young Sun Problem," in *The Sun in Time*, C.P. Sonnett, M.S. Giampapa, and M.S. Matthews, eds. (Tucson: University of Arizona Press), pp. 447–462.

KASTING, J.F., AND T.P. ACKERMAN (1986), "Climatic Consequences of Very High Carbon Dioxide Levels in the Earth's Early Atmosphere," *Science* **234:** 1383–1385.

KASTING, J.F., D.H. EGGLER, AND S.P. RAEBURN (1993), "Mantle Redox Evolution and the Oxidation State of the Archean Atmosphere," *J. Geol.* **101:** 245–257.

KASTING, J.F., D.P. WHITMIRE, AND R.T. REYNOLDS (1993), "Habitable Zones Around Main Sequence Stars," *Icarus* **101:** 108–128.

KASTING, J.F., K.J. ZAHNLE, AND J.C.G. WALKER (1983), "Photochemistry of Methane in the Earth's Early Atmosphere," *Precambrian Res.* **20:** 121–148.

KASTING, J.F., O.B. TOON, AND J.B. POLLACK (1988), "How Climate Evolved on the Terrestrial Planets," *Sci. Am.* **258 (2):** 90–97.

KAVASMANECK, P.R., AND W.A. BONNER (1977), "Adsorption of Amino Acid Derivatives by d- and l-Quartz," *J. Am. Chem. Soc.* **99:** 44–50.

APPENDIX II—BIOLOGICAL ASPECTS REFERENCES (cont.)

KELLER, H.U., C. ARPIGNY, C. BARBIERI, R.M. BONNET, S. CAZES, M. CORADINI, C.B. COSMOVICI, W.A. DELAMERE, W.F. HUEBNER, D.W. HUGHES, C. JAMAR, D. MALAISE, H.J. REITSEMA, W.K.H. SCHMIDT, P. SEIGE, F.L. WHIPPLE, AND K. WILHELM (1986), "First Halley Multicolour Camera Imaging Results from Giotto," *Nature* **321:** 320–326.

KENNY, R., AND L.P. KNAUTH (1992), "Continental Paleoclimates from δD and δ^{18}O of Secondary Silica in Paleokarst Chert Lags," *Geology* **20:** 219–222.

KERRIDGE, J.F. (1991), "A Note on the Prebiotic Synthesis of Organic Acids in Carbonaceous Meteorites," *Origins of Life* **21:** 19–29.

KERRIDGE, J.F., S. CHANG, AND R. SHIPP (1987), "Isotopic Characteristics of Kerogen-Like Material in the Murchison Carbonaceous Chondrite," *Geochim. Cosmochim. Acta* **51:** 2527–2540.

KHARE, B.N., AND C. SAGAN (1972), "Experimental Interstellar Organic Chemistry: Preliminary Findings," in *Proceedings of the Symposium on Spectroscopic Studies of Astrophysical Interest* (Hyderabad), pp. 13–20.

KHARE, B.N., C. SAGAN, W.R. THOMPSON, E.T. ARAKAWA, F. SUITS, T.A. CALCOTT, M.W. WILLIAMS, S. SHRADER, H. OGINO, T.O. WILLINGHAM, AND B. NAGY (1984), "The Organic Aerosols of Titan," *Adv. Space Res.* **4:** 59–68.

KHARE, B.N., C. SAGAN, W.R. THOMPSON, L. FLYNN, AND M.A. MORRISON (1989), "Amino Acids and Their Polymers in the Lower Clouds of Jupiter?—Preliminary Findings," *Origins of Life* **19:** 495–496.

KHARE, B.N., W.R. THOMPSON, L. CHEUNG, C. CHYBA, C. SAGAN, E.T. ARAKAWA, C. MEISSE, AND P.S. TUMINELLO (1993), "Production and Optical Constants of Ice Tholin from Charged Particle Irradiation of (1 : 6) C_2H_6/H_2O at 77 K," *Icarus* **103:** 290–300.

KIMURA, M. (1968), "Evolutionary Rate at the Molecular Level," *Nature* **217:** 624–626.

KISSEL, J., AND F.R. KRUEGER (1987), "The Organic Component in Dust from Comet Halley as Measured by the PUMA Mass Spectrometer on Board Vega 1," *Nature* **326:** 755–760.

KLISS, R.M., AND C.N. MATTHEWS (1962), "Hydrogen Cyanide Dimer and Chemical Evolution," *Proc. Natl. Acad. Sci. USA* **48:** 1300–1306.

KNAUTH, L.P. (1992), "Origin and Diagenesis of Cherts: An Isotopic Perspective," in *Isotopic Signatures and Sedimentary Records, Lecture Notes in Earth Sciences #43*, N. Clauer and S. Chaudhuri, eds. (Berlin: Springer-Verlag), pp. 123–152.

KNAUTH, L.P., AND S. EPSTEIN (1976), "Hydrogen and Oxygen Isotope Ratios in Nodular and Bedded Cherts," *Geochim. Cosmochim. Acta* **40:** 1095–1108.

KNOLL, A.H. (1985), "The Precambrian Evolution of Terrestrial Life," in *The Search for Extraterrestrial Life: Recent Developments*, M.D. Papagiannis, ed. (Dordrecht: D. Reidel Publishing Company) pp. 201–211.

KOCH, A.L. (1994), "Development and Diversification of the Last Common Ancestor," *J. Theor. Biol.* **168:** 269–280.

KONDEPUDI, D.K. (1987), "Selection of Molecular Chirality by Extremely Weak Chiral Interactions Under Far-from-Equilibrium Conditions," *BioSystems* **20:** 75–83.

APPENDIX II—BIOLOGICAL ASPECTS REFERENCES (cont.)

KONDEPUDI, D.K., AND G.W. NELSON (1983), "Chiral Symmetry Breaking in Nonequilibrium Systems," *Phys. Rev. Lett.* **50:** 1023–1026.

KONDEPUDI, D.K., AND G.W. NELSON (1985), "Weak Neutral Currents and the Origin of Biomolecular Chirality," *Nature* **314:** 438–441.

KREBS, C.J. (1985), *Ecology—The Experimental Analysis of Distribution and Abundance*, 3d ed. (New York: Harper & Row).

KREBS, J.S., AND R.G. BARRY (1970), "The Arctic Front and the Tundra-Taiga Boundary in Eurasia," *Geogr. Rev.* **60:** 548–554.

KRESAK, L. (1978), "The Tunguska Object: A Fragment of Comet Encke?," *Bull. Astron. Inst. Czechosl.* **29:** 129–134.

KRINOV, E.L. (1960), *Principles of Meteoritics* (Oxford: Pergamon Press), pp. 120–137.

KUMADA, Y., D.R. BENSON, D. HILLEMANN, T.J. HOSTED, D.A. ROCHFORD, C.J. THOMPSON, W. WOHLLEBEN, AND Y. TATENO (1993), "Evolution of the Glutamine Synthase Gene, One of the Oldest Existing and Functioning Genes," *Proc. Natl. Acad. Sci. USA* **90:** 3009–3013.

KVENVOLDEN, K.A., J.G. LAWLESS, AND C.E. FOLSOME (1973), "Organic Matter in Meteorites," *Sci. Am.* **227 (6):** 38–54.

LACY, J.H., F. BAAS, L.J. ALLAMANDOLA, S.E. PERSSON, P.J. MCGREGOR, C.J. LONSDALE, T.R. GEBALLE, AND C.P.E.M. VAN DE BULT (1984), "4.6 Micron Absorption Features Due to Solid Phase CO and Cyano Group Bearing Molecules Towards Compact Infrared Sources," *ApJ* **276:** 533–543.

LAKE, J.A. (1988), "Origin of the Eukaryotic Nucleus Determined by Rate Invariant Analysis of rRNA Sequences," *Nature* **331:** 184–186.

LANGEVIN, Y., J. KISSEL, J.-L. BERTAUX, AND E. CHASSEFIÉRE (1987), "First Statistical Analysis of 5000 Mass Spectra of Cometary Grains Obtained by PUMA 1 (Vega 1) and PIA (Giotto) Impact Ionization Mass Spectrometers in the Compressed Modes," *A&A* **187:** 761–766.

LARGE, P.J. (1983), *Methylotrophy and Methanogenesis* (Washington, DC: American Society for Microbiology).

LARSON, H.P., D.S. DAVIS, J.H. BLACK, AND U. FINK (1985), "Interstellar Absorption Features Towards the Compact Infrared Source W33A," *ApJ* **299:** 873.

LASKAR, J., AND P. ROBUTEL (1993), "The Chaotic Obliquity of the Planets," *Nature* **361:** 608–612.

LASKAR, J., F. JOUTEL, AND P. ROBUTEL (1993), "Stabilization of the Earth's Obliquity by the Moon," *Nature* **361:** 615–617.

LAZCANO, A. (1993a), "The RNA World, Its Predecessors and Descendants," in *Early Life on Earth*, Nobel Symposium no. 84, S. Bengston, ed. (New York: Columbia University Press), pp. 69–80.

LAZCANO, A. (1993b), "Biogenesis: Some Like It Hot," *Science* **260:** 1154–1155.

LAZCANO, A. (1994), "The Transition from Non-Living to Living," in *Early Life on Earth*, Nobel Symposium no. 84, S. Bengtson, ed. (New York: Columbia University Press), pp. 60–69.

APPENDIX II—BIOLOGICAL ASPECTS REFERENCES (cont.)

LEACH, W.W., D.W. NOONER, AND J. ORÓ (1978), "Abiotic Synthesis of Fatty Acids," in *Origin of Life*, H. Noda, ed. (Tokyo: Center for Academic Publications), pp. 113–122.

LEE, W., AND T.J. WDOWIAK (1993), "Origins of the Hydrocarbon Component of Carbonaceous Chondrites: The Star-Meteorite Connection," *ApJ* **417:** L49–L51.

LEWIS, R.S., M. TANG, J.F. WACKER, E. ANDERS, AND E. STEEL (1987), "Interstellar Diamonds in Meteorites," *Nature* **326:** 160–162.

LEWIS, R.S., S. AMARI, AND E. ANDERS (1994), "Interstellar Grains in Meteorites: II. SiC and Its Noble Gases," *Geochim. Cosmochim. Acta* **58:** 471–494.

LI, W.-H. (1982), "Evolutionary Change of Duplicate Genes," in *Isozymes: Current Topics in Biological and Medical Research*, M.C. Rattazzi, J.G. Scandalios, and G.S. Whitt, eds. (New York: Alan R. Liss), pp. 55–92.

LIEBMAN, S.A., R.A. PESCE-RODRÍGUEZ, AND C.N. MATTHEWS (1995), "Organic Analysis of Hydrogen Cyanide Polymers: Prebiotic and Extraterrestrial Chemistry," *Adv. Space Res.* **15 (3):** 71–80.

LIMBER, D.N. (1960), "The Universality of the Initial Luminosity Function," *ApJ* **131:** 168–201.

LÖB, W. (1913), "Uber das Verhalten des Formamids unter der Wirkung der stillen Entladung. Ein Beitrag zur Frage der Stickstoff-Assimilation," *Berichte der Deutschen Chemischen Gesellschaft* **46:** 684–697.

LOVE, S.G., AND D.E. BROWNLEE (1991), "Heating and Thermal Transformation of Micrometeoroids Entering the Earth's Atmosphere," *Icarus* **89:** 26–43.

LOVELOCK, J.E. (1988), *The Ages of Gaia* (New York: W.W. Norton Co.).

LOVELOCK, J.E. (1993), "Geophysiology—The Science of Gaia," in *Scientists on Gaia*, S.H. Schneider and P.J. Boston, eds. (Cambridge, MA: MIT Press), pp. 3–10.

LOVELOCK, J.E., AND M. WHITFIELD (1982), "Life Span of the Biosphere," *Nature* **296:** 516–563.

LOWE, C.U., M.W. REES, AND R. MARKHAM (1963), "Synthesis of Complex Organic Compounds from Simple Precursors: Formation of Amino-Acids, Amino-Acid Polymers, Fatty Acids and Purines from Ammonium Cyanide," *Nature* **199:** 219–222.

LOWMAN, P.D.J. (1989), "Comparative Planetology and the Origin of Continental Crust," *Precambrian Res.* **44:** 171–195.

LUNINE, J.I., S. ENGEL, B. RIZK, AND M. HORANYI (1991), "Sublimation and Reformation of Icy Grains in the Primitive Solar Nebula," *Icarus* **94:** 333–344.

LUO, L.F. (1995?), *Acta Scientiarum Naturalium Universitatis Intramongolicae* (Neimenggu) **25**.

MACDERMOTT, A.J. (1994), "Electroweak Enantioselection and the Origin of Life," in *Origins of Life* **25:** 191.

MACDERMOTT, A.J. (1995), "The Weak Force and the Origin of Life and Self-Organization," *Chemical Evolution: Self-Organization of the Macromolecules of Life*, J. Chela-Flores et al., eds., Deepak Publishing, pp. 237–49.

MACDERMOTT, A.J., AND G.E. TRANTER (1989), "Electroweak Bioenantioselection," *Croatica Chemica Acta* **62:** 165–187.

APPENDIX II—BIOLOGICAL ASPECTS REFERENCES (cont.)

MacDermott, A.J., and G.E. Tranter (1990), "Biomolecular Handedness and the Weak Interaction," in *Symmetries in Science IV*, B. Gruber and J.H. Yopp, eds. (New York: Plenum New York), pp. 67–124.

MacDermott, A.J., and G.E. Tranter (1996), "The Electroweak Origin of the Chiral Signature of Life," this volume.

MacDermott, A.J., G.E. Tranter, and S.J. Trainor (1992), "The Search for Large Parity-Violating Energy Differences Finds Fruit in Thiosubstituted DNA Analogues," *Chem. Phys. Lett.* **194:** 152–156.

Macklin, J., D.E. Brownlee, S. Chang, and T.E. Bunch (1987), "Micro-Raman Spectroscopic Measurement of Carbon in Meteorites and Interplanetary Dust Particles," *Microbeam* **87:** 211–212.

Maddox, J. (1984), "New Twist for the Anthropic Principle," *Nature* **307:** 409.

Maher, K.A., and D.J. Stevenson (1988), "Impact Frustration of the Origin of Life," *Nature* **331:** 612–614.

Marden, J.H., and M.G. Kramer (1994), "Surface-Skimming Stoneflies: A Possible Intermediate Stage in Insect Flight Evolution," *Science* **266:** 427–430.

Margulis, L. (1993), *Symbiosis in Cell Evolution* (New York: W.H. Freeman and Company).

Marshall, H.G., J.C.G. Walker, and W.R. Kuhn (1988), "Long-Term Climate Change and the Geochemical Cycle of Carbon," *JGR* **93:** 791–801.

Marshall, L.G. (1994), "The Terror Birds of South America," *Sci. Am.* **270 (2):** 90.

Mason, S.F., and G.E. Tranter (1984), "The Parity-Violating Energy Difference Between Enantiomeric Molecules," *Mol. Phys.* **53:** 1091–1111.

Mason, S.F., and G.E. Tranter (1985), "The Electroweak Origin of Biomolecular Handedness," *Proc. R. Soc. Lond. A* **397:** 45–65.

Mathew, M.K., and P. Balaram (1983), *Mol. Cell. Biochem.* **50:** 47–52.

Matthews, C.N. (1982), "Heteropolypeptides on Titan?," *Origins of Life* **12:** 281–283.

Matthews, C.N. (1984), "Chemical Evolution: Protons to Proteins," *Proc. Roy. Inst. Gt. Britain* **55:** 199–206.

Matthews, C.N. (1985), "Universal Protein Ancestors from Hydrogen Cyanide and Water," in *The Search for Extraterrestrial Life: Recent Developments*, M. Papagiannis, ed. (Dordrecht: D. Reidel Publishing Company), pp. 151–156.

Matthews, C.N. (1986), "Simultaneous Synthesis of Polypeptides and Polynucleotides? Hydrogen Cyanide Polymers as Prebiotic Condensing Agents," *Origins of Life* **16:** 500

Matthews, C.N. (1988), "Cosmic Metabolism: The Origin of Macromolecules," in *Bioastronomy—The Next Steps*, G. Marx, ed. (Dordrecht: Kluwer Academic Publishers), pp. 167–178.

Matthews, C.N. (1990), "Simultaneous Synthesis of Polypeptides and Polynucleotides. Hydrogen Cyanide Polymers as Prebiotic Condensing Agents," in *Prebiological Self Organization of Matter*, C. Ponnamperuma and F.R. Eirich, eds. (Hampton, VA: A. Deepak Publishing), pp. 171–178.

APPENDIX II—BIOLOGICAL ASPECTS REFERENCES (cont.)

MATTHEWS, C.N. (1992a), "Dark Matter in the Solar System: Hydrogen Cyanide Polymers," *Origins of Life* **21:** 421–434.

MATTHEWS, C.N. (1992b), "Origin of Life: Polymers Before Monomers?," in *Environmental Evolution*, L. Margulis and L. Olendzenski, eds. (Cambridge, MA: MIT Press), pp. 29–38.

MATTHEWS, C.N., AND R.A. LUDICKY (1986), "The Dark Nucleus of Comet Halley: Hydrogen Cyanide Polymers," in *20th ESLAB Symposium on the Exploration of Halley's Comet*, ESA Report SP-250, vol. 2, B. Battrick, E.J. Rolfe, and R. Reinhard, eds. (Noordwijk: European Space Agency), pp. 273–277.

MATTHEWS, C.N., AND R.A. LUDICKY (1992), "Hydrogen Cyanide Polymers on Comets," *Adv. Space Res.* **12 (4):** 21–32.

MATTHEWS, C.N., AND R.E. MOSER (1966), "Prebiological Protein Synthesis," *Proc. Natl. Acad. Sci. USA* **56:** 1087–1094.

MATTHEWS, C.N., AND R.E. MOSER (1967), "Peptide Synthesis from Hydrogen Cyanide and Water," *Nature* **215:** 1230–1234.

MATTHEWS, C.N., J.E. NELSON, AND R.D. MINARD (1980), "Deuterolysis of Amino Acid Precursors from the Murchison Meteorite," *Abstracts 6th International Conference on the Origins of Life*, Jerusalem, June 22–27, p. 100.

MATTHEWS, C.N., J.E. NELSON, P. VARMA, AND R.D. MINARD (1977), "Deuterolysis of Amino Acid Precursors: Evidence for Hydrogen Cyanide Polymers as Protein Ancestors," *Science* **198:** 622–624.

MATTHEWS, C.N., R.A. LUDICKY, J. SCHAEFER, E.O. STEJSKAL, AND R.A. MCKAY (1984), "Heteropolypeptides from Hydrogen Cyanide and Water: Solid State ^{15}N NMR Investigations," *Origins of Life* **14:** 243–249.

MAURETTE, M., C. JEHANNO, E. ROBIN, AND C. HAMMER (1987), "Characteristics and Mass Distribution of Extraterrestrial Dust from the Greenland Ice Cap," *Nature* **328:** 699–702.

MAYR, E. (1992), "Is It Worth Looking for Extraterrestrial Intelligence?" (in German), *Naturwissenschaftliche Rundschau* **45:** 264–266.

MCDONALD, G.D., B.N. KHARE, W.R. THOMPSON, AND C. SAGAN (1991), "$CH_4/NH_3/H_2O$ Spark Tholin: Chemical Analysis and Interaction with Jovian Aqueous Clouds," *Icarus* **94:** 354–367.

MCKAY, C.P. (1991), "Urey Prize Lecture: Planetary Evolution and the Origin of Life," *Icarus* **91:** 93–100.

MCKAY, C.P. (1996), "Time for Intelligence on Other Planets," this volume.

MCKAY, C.P., AND C.R. STOKER (1989), "The Early Environment and Its Evolution on Mars: Implications for Life," *Rev. Geophys.* **27:** 189–214.

MCKAY, C.P., AND H. HARTMAN (1991), "Hydrogen Peroxide and the Origin of Oxygenic Photosynthesis," *Origins of Life* **21:** 157–163.

MCKAY, C.P., AND W.L. DAVIS (1991), "Duration of Liquid Water Habitats on Early Mars," *Icarus* **90:** 214–221.

APPENDIX II—BIOLOGICAL ASPECTS REFERENCES (cont.)

McKay, R.A., J. Schaefer, E.O. Stejskal, R.A. Ludicky, and C.N. Matthews (1984), "Double-Cross-Polarization Detection of Labelled Chemical Bonds in HCN Polymerization," *Macromolecules* **17:** 1124–1130.

McLafferty, F.W. (1973), *Interpretation of Mass Spectra* (Reading, MA: W.A. Benjamin).

Melcher, G. (1974), "Stereospecificity of the Genetic Code," *J. Mol. Evol.* **3:** 121–140.

Miller, S.L. (1953), "A Production of Amino Acids Under Possible Primitive Earth Conditions," *Science* **117:** 528–529.

Miller, S.L. (1957), "The Mechanism of Synthesis of Amino Acids by Electric Discharges," *Biochimica et Biophysica Acta* **23:** 480–489.

Miller, S.L. (1974), "The First Laboratory Synthesis of Organic Compounds Under Primitive Earth Conditions," in *The Heritage of Copernicus: Theories "Pleasing to the Mind,"* J. Neyman, ed. (Cambridge, MA: MIT Press), pp. 228–242.

Miller, S.L. (1992), "The Prebiotic Synthesis of Organic Compounds as a Step Toward the Origin of Life," in *Major Events in the History of Life*, J.W. Schopf, ed. (Boston: Jones and Bartlett Publishers), pp. 1–28.

Miller, S.L., and A. Lazcano (1996), "From the Primitive Soup to Cyanobacteria: It May Have Taken Less Than 10 Million Years," this volume.

Miller, S.L., and J.E. Van Trump (1981), "The Strecker Synthesis in the Primitive Ocean," in *Origin of Life*, Y. Wolman, ed. (Dordrecht: D. Reidel Publishing Company), pp. 135–141.

Miller, S.L., and J.L. Bada (1988), "Submarine Hot Springs and the Origin of Life," *Nature* **334:** 609–611.

Miller, S.L., and L.E. Orgel (1974), *The Origins of Life on Earth* (Englewood Cliffs, NJ: Prentice-Hall), p. 229.

Minard, R.D., W. Yang, P. Varma, J. Nelson, and C.N. Matthews (1975), "Heteropolypeptides from poly-α-Cyanoglycine and Hydrogen Cyanide: A Model for the Origin of Proteins," *Science* **190:** 387–389.

Ming, T., and E. Anders (1988), "Isotopic Anomalies of Ne, Xe, and C in Meteorites. III. Local and Exotic Noble Gas Components and Their Interrelations," *Geochim. Cosmochim. Acta* **52:** 1245–1254.

Morris, S.C. (1993), "The Fossil Record and the Early Evolution of the Metazoa," *Nature* **361:** 219–225.

Moser, R.E., and C.N. Matthews (1968), "Hydrolysis of Aminoacetonitrile: Peptide Formation," *Experientia* **24:** 658–659.

Moser, R.E., A.R. Claggett, and C.N. Matthews (1968), "Peptide Formation from Diaminomaleonitrile (HCN Tetramer) and from Aminomalononitrile (HCN Trimer)," *Tetrahedron Lett.*, **9:** 1599–1603.

Mueller, B.E., D.J. Tholin, W.K. Hartman, and D.P. Cruikshank (1992), "Extraordinary Colors of Asteroidal Object (5145) 1992 AD," *Icarus* **97:** 150–154.

APPENDIX II—BIOLOGICAL ASPECTS REFERENCES (cont.)

NAKANISHI, K., AND P.H. SOLOMON (1977), *Infrared Absorption Spectroscopy* (San Francisco: Holden-Day).

NATIONAL RESEARCH COUNCIL (1990), *The Search for Life's Origins* (Washington, DC: National Academy Press).

NOBEL, P.S. (1984), "Extreme Temperatures and Thermal Tolerances for Seedlings of Desert Succulents," *Oecologia* **62:** 310–317.

NORBY, R.J., C.A. GUNDERSON, S.D. WULLSCHLEGER, E.G. O'NEILL, AND M.K. MCCRACKEN (1992), "Productivity and Compensatory Responses of Yellow-Poplar Trees in Elevated CO_2," *Nature* **357:** 322–324.

NORELL, M.A., J.M. CLARK, D. DEMBERELYIN, B. RHINCHEN, L.M. CHIAPPE, A.R. DAVIDSON, M.C. MCKENNA, P. ALTANGEREL, AND M.J. NOVACEK (1994), "A Theropod Dinosaur Embryo and the Affinities of the Flaming Cliffs Dinosaur Eggs," *Science* **266:** 779–782.

OBERBECK, V.R., AND G. FOGLEMAN (1989), "Impacts and the Origin of Life," *Nature* **339:** 434.

OBERBECK, V.R., AND G. FOGLEMAN (1990), "Impact Constraints on the Environment for Chemical Evolution and the Continuity of Life," *Origins of Life* **20:** 181–195.

OBERBECK, V.R., AND H. AGGARWAL (1992), "Comet Impacts and Chemical Evolution on the Bombarded Earth," *Origins of Life* **21:** 317–338.

OBERBECK, V.R., AND J.R. MARSHALL (1992), "Impacts, Flood Basalts, and Continental Breakup," abstract, *Lunar and Planetary Science Proceedings, March 23, 1992*, p. 113.

OBERBECK, V.R., J.R. MARSHALL, AND H. AGGARWAL (1993), "Impacts, Tillites, and the Breakup of Gondwanaland," *J. Geol.* **101:** 1–19.

OGMEN, M., AND W.W. DULEY (1988), "Laboratory Synthesis of Hydrogenated Amorphous Carbon Films," *ApJ* **334:** L117–120.

OHTA, T. (1991), "Multigene Families and the Evolution of Complexity," *J. Mol. Evol.* **33:** 34–41.

OLSEN, G.J., AND C.R. WOESE (1993), "Ribosomal RNA: A Key to Phylogeny," *FASEB Journal* **7:** 113–123.

OPARIN, A.I. (1938), *The Origin of Life* (New York: The MacMillan Company).

ORÓ, J. (1960), "Synthesis of Adenine from Ammonium Cyanide," *Biochem. Biophys. Res. Comm.* **2:** 407–412.

ORÓ, J. (1961), "Comets and the Formation of Biochemical Compounds on the Primitive Earth," *Nature* **190:** 389–390.

ORÓ, J. (1965), "Stages and Mechanisms of Prebiological Organic Synthesis," in *The Origins of Prebiological Systems*, S.W. Fox, ed. (New York: Academic Press Inc.), pp. 137–171.

ORÓ, J. (1988), "Constraints Imposed by Cosmic Evolution Towards the Development of Life," in *Bioastronomy—The Next Steps*, G. Marx, ed. (Dordrecht: Kluwer Academic Publishers), pp. 161–165.

APPENDIX II—BIOLOGICAL ASPECTS REFERENCES (cont.)

ORÓ, J. (1996), "Planetary Formation of Biochemical Compounds in Circumstellar Habitable Zones," this volume.

ORÓ, J., AND A. LAZCANO (1997), "Comets and the Origin and Evolution of Life: A Scientific and Historical Introduction," in *Comets and the Origin and Evolution of Life*, P.J. Thomas, C.F. Chyba, and C.P. McKay, eds. (New York: Springer-Verlag), pp. 3–27.

ORÓ, J., AND A. LAZCANO-ARAUJO (1980), "The Role of HCN and Its Derivatives in Prebiotic Evolution," in *Cyanide in Biology*, B. Vennesland, E.E. Conn, C.J. Knowles, J. Westley, and F. Wissing, eds. (New York: Academic Press), pp. 517–541.

ORÓ, J., AND A.P. KIMBALL (1962), "Synthesis of Purines Under Possible Primitive Earth Conditions. II. Purine Intermediates from Hydrogen Cyanide," *Arch. Biochem. Biophys.* **96:** 293–313.

ORÓ, J., AND S.S. KAMAT (1961), "Amino Acid Synthesis from Hydrogen Cyanide Under Possible Primitive Earth Conditions," *Nature* **190:** 442–443.

ORÓ, J., AND T. MILLS (1989), "Chemical Evolution of Primitive Solar System Bodies," *Adv. Space Res.* **9:** 105–120.

ORÓ, J., G. HOLZER, AND A. LAZCANO-ARAUJO (1980), "The Contribution of Cometary Volatiles to the Primitive Earth," in *Cospar Life Sciences and Space Research*, vol. XVII, R. Holmquist, ed. (Oxford: Pergamon Press), pp. 67–82.

ORÓ, J., K. REWERS, AND D. ODOM (1982), "Criteria for the Emergence and Evolution of Life in the Solar System," *Origins of Life* **12:** 285–305.

ORÓ, J., S.L. MILLER, AND A. LAZCANO (1990), "The Origin and Early Evolution of Life on Earth," *Ann. Rev. Earth Planet. Sci.* **18:** 317–356.

ORÓ, J., T. MILLS, AND A. LAZCANO (1992), "Comets and the Formation of Biochemical Compounds—A Review," *Origins of Life* **21:** 267–277.

OSTROM, J.H. (1992), "A History of Vertebrate Successes," in *Major Events in the History of Life*, J.W. Schopf, ed. (Boston: Jones and Bartlett Publishers), pp. 119–139.

OTAKA, E., AND T. OOI (1989), "Examination of Protein Sequence Homologies: V. New Perspectives on Evolution Between Bacterial and Chloroplast-Type Ferrodoxins Inferred from Sequence Evidence," *J. Mol. Evol.* **29:** 246–254.

OWEN, T. (1982), "Titan," *Sci. Am.* **246 (2):** 98–109.

OWEN, T., AND A. BAR-NUN (1993), "Noble Gases in Atmospheres," *Nature* **361:** 693–694.

OWEN, T., A. BAR NUN, AND I. KLEINFELD (1992), "Possible Cometary Origin of Heavy Noble Gases in the Atmospheres of Venus, Earth and Mars," *Nature* **358:** 43–46.

PACE, N.R. (1991), "Origin of Life—Facing Up to the Physical Setting," *Cell* **65:** 531–533.

PALACHE, C., G. ERMAN, AND C. FRONDEL (1962), *Dana's System of Mineralogy*, 7th ed., vol. 3 (New York: John Wiley & Sons), p. 16.

PASTEUR, L. (1884), "La Dissymetrie Moleculaire," *Rev. Scientifique*, 3eme serie **VII:** 2–6.

APPENDIX II—BIOLOGICAL ASPECTS REFERENCES (cont.)

PAYETTE, S. (1983), "The Forest-Tundra and Present Tree-Lines of the Northern Quebec–Labrador Peninsula," in *Tree-Line Ecology: Proceedings of the Northern Quebec Tree-Line Conference*, Collection Nordicana 47, P. Morisset and S. Payette, eds. (Quebec: Universite Laval), pp. 3–23.

PEEL, D.A. (1993), "Cold Answers to Hot Issues," *Nature* **363:** 403–404.

PELTZER, E.T., J.L. BADA, G. SCHLESINGER, AND S.L. MILLER (1984), "The Chemical Conditions on the Parent Body of the Murchison Meteorite: Some Conclusions Based on Amino-, Hydroxy-, and Dicarboxylic Acids," *Adv. Space Res.* **(4) 12:** 61–74.

PENDLETON, Y.J. (1993), "Organics in the Diffuse Interstellar Medium," in *Astronomical Infrared Spectroscopy: Future Observational Directions*, ASP Conf. Ser., Sun Kwok, ed. (San Francisco: Astronomical Society of the Pacific), **41:** 171–181.

PENDLETON, Y.J. (1994), "Infrared Spectroscopy of Organic Material in Diffuse Interstellar Clouds," in *The First Symposium on the Infrared Cirrus and Diffuse Interstellar Clouds*, ASP Conf. Ser., R. Cutri and W. Latter, eds. (San Francisco: Astronomical Society of the Pacific), **58:** 255–265.

PENDLETON, Y.J. (1995), "Laboratory Comparisons of Organic Materials to Interstellar Dust and the Murchison Meteorite," *Planet. Space Sci.*, **43:** 1359–1364.

PENDLETON, Y.J. (1996), "Interstellar Organic Material: Comparisons with Laboratory and Solar System Spectra," this volume.

PENDLETON, Y.J. (1996), "Organics and Ices in Galactic Dust," in *Proceedings from the Cold Dust Morphology Conference, Johannesburg, South Africa*, D. Block, ed. (Dordrecht: Kluwer Academic Publishers).

PENDLETON, Y.J., AND D.P. CRUIKSHANK (1994), "Life from the Stars?," *Sky & Tel.* **87:** 36–42.

PENDLETON, Y.J., S.A. SANDFORD, L.J. ALLAMANDOLA, A.G.G.M. TIELENS, AND K. SELLGREN (1994), "Near Infrared Absorption Spectroscopy of Interstellar Hydrocarbon Grains," *ApJ* **437:** 683–696.

PETERSON, E., F. HÖRZ, G. HAYNES, AND T. SEE (1991), "Modification of Amino Acids at Shock Pressures of 3 to 30 GPa," *Meteoritics* **26:** 384.

PFENNIG, N. (1989), "Multicellular Filamentous Green Bacteria," in *Bergey's Manual of Systematic Bacteriology*, vol. 3, J.T. Staley, M.P. Bryant, N. Pfennig, and J.G. Holt, eds. (Baltimore: Williams and Wilkins), pp. 1697–1707.

POLLACK, J.B., J.F. KASTING, S.M. RICHARDSON, AND K. POLIAKOFF (1987), "The Case for a Wet, Warm Climate on Early Mars," *Icarus* **71:** 203–224.

POLLEY, H.W., H.B. JOHNSON, B.D. MARINO, AND S.M. HERMAN (1993), "Increase in C3 Plant Water-Use Efficiency and Biomass over Glacial to Present CO_2 Concentrations," *Nature* **361:** 61–64.

PUHLER, G., H. LEFFERS, F. GROPP, P. PALM, H.P. KLENK, F. LOTTSPEICH, R.A. GARRETT, AND W. ZILLIG (1989), "Archaebacterial DNA-Dependent RNA Polymerases Testify to the Evolution of the Eukaryotic Nuclear Genome," *Proc. Natl. Acad. Sci. USA* **86:** 4569–4573.

APPENDIX II—BIOLOGICAL ASPECTS REFERENCES (cont.)

RADICATI DI BROZOLO, F., T.E. BUNCH, R.H. FLEMING, AND J. MACKLIN (1994), "Fullerenes in an Impact Crater on the LDEF Spacecraft," *Nature* **369:** 37–40.

RAISON, J.K., J.A. BERRY, P.A. ARMOND, AND C.S. PIKE (1980), "Membrane Properties in Relation to the Adaptation of Plants to High and Low Temperature Stress," in *Adaptation of Plants to Water and High Temperature Stress*, N.C. Turner and P.J. Kramer, eds. (New York: Wiley Interscience), pp. 261–273.

RAMPINO, M.R. (1992), "Ancient 'Glacial' Deposits Are Ejecta of Large Impacts: The Ice Age Paradox Explained," abstract, *EOS Transactions*, American Geophysical Union, vol. 73, no. 43, October 27, 1992, Supplement A32C-1, p. 99.

RAY, T.S. (1991), "An Approach to the Synthesis of Life," in *Artificial Life II*, Santa Fe Institute Studies in the Sciences of Complexity, vol. XI, J.D. Farmer, C. Langton, S. Rasmussen, and C. Taylor, eds. (Redwood City, CA: Addison-Wesley), pp. 371–408.

RAY, T.S. (1992), "Evolution, Ecology, and Optimization of Digital Organisms," Working Paper 92-08-042 (Santa Fe: Santa Fe Institute).

REINHARD, R. (1986), "The Giotto Encounter with Comet Halley," *Nature* **321:** 313–318.

RETTIG, T.W., S.C. TEGLER, D.J. PASTO, AND M.J. MUMMA (1992), "Comet Outbursts and Polymers of HCN," *ApJ* **398:** 293–298.

REUTER, D.C. (1992), "The Contribution of Methanol to the 3.4 Micron Emission Feature in Comets," *ApJ* **386:** 330–335.

ROBBINS, E.I., K.G. PORTER, AND K.A. HABERYAN (1985), "Pellet Microfossils: Possible Evidence for Metazoan Life in Early Proterozoic Time," *Proc. Natl. Acad. Sci. USA* **82:** 5809–5813.

ROBINSON, J.M. (1991), "Phanerozoic Atmospheric Reconstructions: A Terrestrial Perspective," *Palaeogeogr., Palaeoclimatol., Palaeoecol. (Global Planet. Change Sect.)* **97:** 51–62.

ROTHSCHILD, L.J., AND R.L. MANCINELLI (1990), "Model of Carbon Fixation in Microbial Mats from 3,500 Myr Ago to the Present," *Nature* **345:** 710–712.

RUNNEGAR, B. (1992), "Evolution of the Earliest Animals," in *Major Events in the History of Life*, J.W. Schopf, ed. (Boston: Jones and Bartlett Publishers), pp. 65–93.

RUSSELL, D.A., AND R. SÉGUIN (1982), "Reconstructions of the Small Cretaceous Theropod *Stenonychosaurus inequalis* and a Hypothetical Dinosauroid," in *Syllogeus*, **37:** 1–43 (Ottawa: National Museum of Natural Sciences).

SAGAN, C. (1974), "The Origin of Life in a Cosmic Context," *Origins of Life* **5:** 497–505.

SAGAN, C. (1977), "Reducing Greenhouses and the Temperature History of Earth and Mars," *Nature* **269:** 224–226.

SAGAN, C. (1977), *The Dragons of Eden* (New York: Random House).

SAGAN, C., AND B.N. KHARE (1970), "Production of Organic Molecules in the Interstellar Medium," *Bull. Amer. Astron. Soc.* **2:** 340.

SAGAN, C., AND B.N. KHARE (1979), "Tholins: Organic Chemistry of Interstellar Grains and Gas," *Nature* **277:** 102–107.

APPENDIX II—BIOLOGICAL ASPECTS REFERENCES (cont.)

SAGAN, C., AND C. CHYBA (1996), "Organic Shielding of UV-Labile Greenhouse Gases on Early Earth and Mars," abstract, this volume.

SAGAN, C., AND G. MULLEN (1972), "Earth and Mars: Evolution of Atmospheres and Surface Temperatures," *Science* **177:** 52–56.

SAGAN, C., W.R. THOMPSON, AND B.N. KHARE (1992), "Titan: A Laboratory for Prebiological Organic Chemistry," *Accounts of Chemical Research* **25**: 286–292.

SAGDEEV, R.Z., J. BLAMONT, A.A. GALEEV, V.I. MOROZ, V.D. SHAPIRO, V.I. SHEVCHENKO, AND K. SZEGÖ (1986), "Vega Spacecraft Encounters with Comet Halley," *Nature* **321:** 259–262.

SAKAI, A. (1970), "Freezing Resistance in Willows from Different Climates," *Ecology* **51:** 485–491.

SAKATA, A., AND S. WADA (1989), "Chemical, Optical and Infrared Properties of Quenched Carbonaceous Chondrites," in *Interstellar Dust*, IAU Symp. 135, L.J. Allamandola and A.G.G.M. Tielens, eds. (Dordrecht: Kluwer Academic Publishers), pp. 191–196.

SAKATA, A., S. WADA, T. ONAKA, AND A.T. TOKUNAGA (1987), "Infrared Spectrum of Quenched Carbonaceous Composite (QCC). II. A New Identification of the 7.7 and 8.6 Micron Unidentified Infrared Emission Bands," *ApJ* **320:** L63–L67.

SALAM, A. (1992), "The Role of Chirality in the Origin of Life," *J. Mol. Evol.* **33:** 105–113.

SALVINI-PLAWEN, L.V., AND E. MAYR (1977), "On the Evolution of Photoreceptors and Eyes," in *Evolutionary Biology*, M.K. Hecht, W.C. Steere, and B. Wallace, eds. (New York: Plenum Press).

SÁNCHEZ, R.A., J.P. FERRIS, AND L.E. ORGEL (1966), "Cyanoacetylene in Prebiotic Synthesis," *Science* **154:** 784–786.

SANDFORD, S.A. (1984), "Infrared Transmission Spectra from 2.5 to 25 μm of Various Meteorite Classes," *Icarus* **60:** 115–126.

SANDFORD, S.A., L.J. ALLAMANDOLA, A.G.G.M. TIELENS, K. SELLGREN, M. TAPIA, AND Y. PENDLETON (1991), "The Interstellar C–H Stretching Band Near 3.4 Microns: Constraints on the Composition of Organic Material in the Diffuse Interstellar Medium," *ApJ* **371:** 607–620.

SANDFORD, S.A., Y.J. PENDLETON, AND L.J. ALLAMANDOLA (1995), "The Galactic Distribution of Aliphatic Hydrocarbons in the Diffuse Interstellar Medium," *ApJ* **440:** 697–705.

SCHAEFER, J., E.O. STEJSKAL, G.S. JACOB, AND R.M. MCKAY (1982), "Natural Abundance N-15 NMR of the Solids from the Reaction of HCN and Ammonia," *Appl. Spectr.* **36:** 179–182.

SCHIDLOWSKI, M. (1988), "A 3,800-Million-Year Isotopic Record of Life from Carbon in Sedimentary Rocks," *Nature* **333:** 313–318.

SCHIDLOWSKI, M. (1992), "The Initiation of Biological Processes on the Earth: Summary of Empirical Evidence," *Adv. Space Res.* **12 (4):** 143–156.

SCHNEIDER, S.H., AND P.J. BOSTON, eds. (1993), *Scientists on Gaia* (Cambridge, MA: MIT Press).

APPENDIX II—BIOLOGICAL ASPECTS REFERENCES (cont.)

SCHOPF, J.W., ED. (1983), *The Earth's Earliest Biosphere: Its Origin and Evolution* (Princeton, NJ: Princeton University Press).

SCHOPF, J.W. (1992), "The Oldest Fossils and What They Mean," in *Major Events in the History of Life*, J.W. Schopf, ed. (Boston: Jones and Bartlett Publishers), pp. 54–63.

SCHOPF, J.W. (1993), "Microfossils of the Early Archean Apex Chert: New Evidence of the Antiquity of Life," *Science* **260:** 640–646.

SCHOPF, J.W., AND B.M. PACKER (1987), "Early Archean (3.3-Billion to 3.5-Billion-Year-Old) Microfossils from Warrawoona Group, Australia," *Science* **237:** 70–73.

SCHOPF, J.W., J.M. HAYES, AND M.R. WALTER (1983), "Evolution of Earth's Earliest Ecosystems: Recent Progress and Unsolved Problems," in *Earth's Earliest Biosphere*, J.W. Schopf, ed. (Princeton, NJ: Princeton University Press), pp. 361–384.

SCHUTTE, W. (1988), *The Evolution of Interstellar Organic Grain Mantles*, PhD thesis, Leiden University, Leiden.

SCHUTTE, W.A., AND J.M. GREENBERG (1988), "The Evolution of Organic Refractory Mantles on Interstellar Grains," in *Dust in the Universe*, M.E. Bailey and D.A. Williams, eds. (New York: Cambridge University Press), pp. 403–406.

SCHWARTZ, R.M., AND M.O. DAYHOFF (1978), "Origins of Prokaryotes, Mitochondria and Chloroplasts," *Science* **199:** 395–403.

SCHWARTZMAN, D. (1993), "Microbial Weathering and Gaia," in *Scientists on Gaia*, S.H. Schneider and P.J. Boston, eds. (Cambridge, MA: MIT Press), pp. 320–329.

SCHWARTZMAN, D. (1993), comment on "Weathering, Plants, and the Long-Term Carbon Cycle," by Robert A. Berner, *Geochim. Cosmochim. Acta* **57:** 2145–2146.

SCHWARTZMAN, D. (1995), "Temperature and the Evolution of the Earth's Biosphere," in *Progress in the Search for Extraterrestrial Life*, ASP Conf. Ser., S. Shostak, ed. (San Francisco: Astronomical Society of the Pacific), **74:** 152–161.

SCHWARTZMAN, D., AND M. MCMENAMIN (1993), "A Much Warmer Earth Surface for Most of Geologic Time: Implications to Biotic Weathering," *Chem. Geol.* **107:** 221–223.

SCHWARTZMAN, D., AND T. VOLK (1991a), "Biotic Enhancement of Weathering and Surface Temperatures on Earth Since the Origin of Life," *Palaeogeogr., Palaeoclimatol., Palaeoecol. (Global Planet. Change Sect.)* **90:** 357–371.

SCHWARTZMAN, D., AND T. VOLK (1991b), "Geophysiology and Habitable Zones Around Sun-Like Stars," in *Bioastronomy, Proceedings, Val Cenis, France, 1990, Lecture Notes in Physics 390*, J. Heidmann and M.J. Klein, eds. (Berlin: Springer-Verlag), pp. 155–162.

SCHWARTZMAN, D., AND T. VOLK (1992), "Biotic Enhancement of Earth Habitability," in *Encyclopedia of Earth System Science*, vol. 1, W.A. Nierenberg, ed. (New York: Academic Press), pp. 387–394.

SCHWARTZMAN, D., M. MCMENAMIN, AND T. VOLK (1993), "Did Surface Temperatures Constrain Microbial Evolution?," *BioScience* **43:** 390–393.

APPENDIX II—BIOLOGICAL ASPECTS REFERENCES (cont.)

SCHWARTZMAN, D.W., AND T. VOLK (1989), "Biotic Enhancement of Weathering and the Habitability of Earth," *Nature* **340:** 457–460.

SHAPIRO, R. (1988), "Prebiotic Ribose Synthesis: A Critical Analysis," *Origins of Life* **18:** 71–85.

SHARPTON, V.L., K. BURKE, A. CAMARGO-ZANOGUERA, S.A. HALL, D.S. LEE, L.E. MARIN, G. SUÁREZ-REYNOSO, J.M. QUEZADA-MUÑETON, P.D. SPUDIS, AND J. URRUTIA-FUCUGAUCHI (1993), "Chicxulub Multiring Impact Basin: Size and Other Characteristics Derived from Gravity Analysis," *Science* **261:** 1564–1567.

SHERWOOD, E., A. JOSHI, AND J. ORÓ (1977), "Cyanamide Mediated Synthesis Under Plausible Primitive Earth Conditions. II. The Polymerization of Deoxythymidine 5′-Triphosphate," *J. Mol. Evol.* **10:** 211–220.

SHKLOVSKII, I.S., AND C. SAGAN (1966), *Intelligent Life in the Universe* (San Francisco: Holden-Day).

SHUKLA, J., AND Y. MINTZ (1982), "Influence of Land-Surface Evapotranspiration on the Earth's Climate," *Science* **215:** 1498–1501.

SHUKLA, J., C. NOBRE, AND P. SELLERS (1990), "Amazon Deforestation and Climate Change," *Science* **247:** 1322–1325.

SIMPSON, G.G. (1964), "The Nonprevalence of Humanoids," *Science* **143:** 769–775.

SIMPSON, G.G. (1980), *Splendid Isolation: The Curious History Of South American Mammals* (New Haven, CT: Yale University Press).

SIROIS, L. (1992), "The Transition Between Boreal Forest and Tundra," in *A Systems Analysis of the Global Boreal Forest*, H.H. Shugart, R. Leemans, and G.B. Bonan, eds. (Cambridge: Cambridge University Press).

SLEEP, N.H., K.J. ZAHNLE, J.F. KASTING, AND H.J. MOROWITZ (1989), "Annihilation of Ecosystems by Large Asteroid Impacts on the Early Earth," *Nature* **342:** 139–142.

SMITH, M.W., AND R.F. DOOLITTLE (1992), "Anomalous Phylogeny Involving the Enzyme Glucose-6-Phosphate Isomerase," *J. Mol. Evol.* **34:** 544–545.

SMITH, M.W., D.F. FENG, AND R.F. DOOLITTLE (1992), "Evolution by Acquisition: The Case for Horizontal Gene Transfer," *Trends Biochem. Sci.* **17:** 489–493.

SMITH, W.K. (1978), "Temperatures of Desert Plants: Another Perspective on the Adaptability of Leaf Size," *Science* **201:** 614–616.

SMOLIN, E.M., AND L. RAPOPORT (1959), *S-Triazines and Derivatives* (New York: Interscience Publishers, Inc.), p. 545.

SOGIN, M.L., J.H. GUNDERSON, H.J. ELWOOD, D.A. ALONSO, AND D.A. PEATTIE (1989), "Phylogenetic Meaning of the Kingdom Concept: An Unusual Ribosomal RNA from *Giardia lambia*," *Science* **243:** 75–77.

SONTI, R.V., AND J.R. ROTH (1989), "Role of Gene Duplications in the Adaptation of *Salmonella typhimurium* to Growth on Limiting Carbon Sources," *Genetics* **123:** 19–28.

SPURR, S.H., AND B.V. BARNES (1980), *Forest Ecology*, 3d ed. (Chichester: John Wiley & Sons).

APPENDIX II—BIOLOGICAL ASPECTS REFERENCES (cont.)

STEEL, D. (1992), "Cometary Supply of Terrestrial Organics: Lessons from the K/T and the Present Epoch," *Origins of Life* **21**: 339–357.

STEPHEN-SHERWOOD, E., J. ORÓ, AND A.P. KIMBALL (1971), "Thymine: A Possible Prebiotic Synthesis," *Science* **173:** 446–447.

STRAUSS, H., D.J. DES MARAIS, J.M. HAYES, AND R.E. SUMMONS (1992), "The Carbon-Isotopic Record," in *The Proterozoic Biosphere*, J.W. Schopf and C. Klein, eds. (Cambridge: Cambridge University Press), pp. 117–127.

STRAZZULLA, G., A.C. CASTORINA, M.E. PALUMBO (1995), "Ion Irradiation of Astrophysical Ices," *Planet. Space Sci.* **43:** 1247–1252.

STRIBLING, R., AND S.L. MILLER (1987), "Energy Yields for Hydrogen Cyanide and Formaldehyde Syntheses: The HCN and Amino Acid Concentrations in the Primitive Ocean," *Origins of Life* **17:** 261–273.

SU, Y.-L., Y. HONDA, P.E. HARE, AND C. PONNAMPERUMA (1989), "Search of Peptide-like Materials in Electric Discharge Experiments," *Origins of Life* **19:** 237–238.

SUMMERS, D.P., AND S. CHANG (1993), "Prebiotic Ammonia from Reduction of Nitrite by Iron (II) on the Early Earth," *Nature* **365:** 630–633.

SWANSON, R. (1984), "A Unifying Concept for the Amino Acid Code," *Bull. Math. Biol.* **46:** 187–203.

TAYLOR, R., G.J. LANGLEY, H.W. KROTO, AND R.M. WALTON (1993), "Formation of C_{60} by Pyrolysis of Naphthalene," *Nature* **366:** 728–731.

TEGLER, S.C., D.A. WEINTRAUB, L.J. ALLAMANDOLA, S.A. SANDFORD, T.W. RETTIG, AND H. CAMPINS (1993), "Detection of the 2165 Wavenumber (4.619 micron) XCN Band in the Spectrum of L1551 IRS5," *ApJ* **411:** 260.

TEGLER, S.C., D.A. WEINTRAUB, T.O. RETTIG, Y.J. PENDLETON, C.B. WHITTET, AND C.A. KULESA (1995), "Evidence for Chemical Processing of Precometary Icy Grains in Circumstellar Environments of Pre-Main-Sequence Stars," *ApJ* **439:** 279–287.

TIBONI, O., P. CAMMARANO, AND A.M. SANANGELANTONI (1993), "Cloning and Sequencing of the Gene Encoding Glutamine Synthase I from the Archaeum *Pyrococcus woesei*: Anomalous Phylogenies Inferred from Analysis of Archaeal and Bacterial Glutamine Synthase I Sequences," *J. Bacteriol.* **175:** 2961–2969.

TIELENS, A.G.G.M., AND L.J. ALLAMANDOLA (1987), "Evolution of Interstellar Dust," in *Physical Processes in Interstellar Clouds*, G.E. Morfill and M. Scholer, eds. (Dordrecht: D. Reidel Publishing Company), pp. 333–376.

TLSTY, T.D., A.M. ALBERTINI, AND J.H. MILLER (1984), "Gene Amplification in the *lac* Region of *E. coli*," *Cell* **37:** 217–224.

TOKUNAGA, A.T., AND T.Y. BROOKE (1990), "Did Comets Form from Unaltered Interstellar Dust and Ices? The Evidence from Infrared Spectroscopy," *Icarus* **86:** 208–219.

TRANTER, G.E. (1985), "Parity Violating Energy Differences of Chiral Minerals and the Origin of Biomolecular Homochirality," *Nature* **318:** 172–173.

APPENDIX II—BIOLOGICAL ASPECTS REFERENCES (cont.)

TRANTER, G.E. (1986), "Preferential Stabilization of the D-Sugar Series by the Parity-Violating Weak Interaction," *J. Chem. Soc. Chem. Commun.*, **1:** 60–61.

TRANTER, G.E., A.J. MACDERMOTT, R.E. OVERILL, AND P.J. SPEERS (1992), "Computational Studies of the Electroweak Origin of Biomolecular Handedness in Natural Sugars," *Proc. R. Soc. Lond. A* **436:** 603–615.

TREMAINE, S. (1993), "The Distribution of Comets Around Stars," in *Planets Around Pulsars*, J.A. Phillips, S.E. Thorsett, and S.R. Kulkarni, eds. (San Francisco: Astronomical Society of the Pacific), pp. 335–344.

TURCOTTE, D.L. (1980), "On the Thermal Evolution of the Earth," *Earth Planet. Sci. Lett.* **48:** 53–58.

ULBRICHT, T.L.V., ET AL. (1962), "Attempts to Induce Optical Activity with Polarized β-Radiation," *Tetrahedron* **18:** 629–637.

UMEMOTO, K., M. TAKAHASHI, AND K. YOKOTA (1987), "Studies on the Structure of HCN Oligomers," *Origins of Life* **17:** 283–293.

UREY, H.C. (1952), "On the Early Chemical History of the Earth and the Origin of Life," *Proc. Natl. Acad. Sci. USA* **38:** 351–363.

VESTER, F., ET AL. (1959), "Optical Activity and Parity Violation in β-Decay," *Naturwissenschaften* **46:** 68.

VÖLKER, T. (1960), "Polymere Blausäure," *Angew. Chem.* **72:** 379–384.

VÖLKER, T. (1960), "Polymeric Hydrogen Cyanide," *Angew. Chem.* **72:** 379–384.

WÄCHTERSHÄUSER, G. (1990), "Evolution of the First Metabolic Cycles," *Proc. Natl. Acad. Sci. USA* **87:** 200–204.

WALD, G. (1954), "The Origin of Life," *Sci. Am.* **191 (8):** 44–53.

WALKER, J.C.G. (1985), "Carbon Dioxide Cycle on the Early Earth," *Origins of Life* **16:** 117–127.

WALKER, J.C.G. (1990), "Precambrian Evolution of the Climate System," *Palaeogeogr., Palaeoclimatol., Palaeoecol. (Global Planet. Change Sect.)* **82:** 261–289.

WALKER, J.C.G., P.B. HAYS, AND J.F. KASTING (1981), "A Negative Feedback Mechanism for the Long-Term Stabilization of Earth's Surface Temperature," *JGR* **86:** 9776–9782.

WALKER, J.F. (1964), *Formaldehyde* (New York: Reinhold), p. 511.

WALSH, M.M. (1992), "Microfossils and Possible Microfossils from the Early Archean Onverwacht Group, Barberton Mountain Land, South Africa," *Precambrian Res.* **54:** 271–293.

WALTER, M.R., AND D.J. DES MARAIS (1993), "Preservation of Biological Information in Thermal Spring Deposits: Developing a Strategy for the Search for Fossil Life on Mars," *Icarus* **101:** 129–143.

WANG, J.Y., AND L.F. LUO (1985), "The Origin of Biological Chirality and Weak Interaction," *Scientia Sinica B* **28:** 1265–1277.

APPENDIX II—BIOLOGICAL ASPECTS REFERENCES (cont.)

WARREN, C.B., R.D. MINARD, AND C.N. MATTHEWS (1974), "Synthesis of α-Cyanoglycine N-Carboxyanhydride and α-Cyanoglycine," *J. Org. Chem.* **39:** 3375–3378.

WEISSMAN, P.R. (1985), "Cometary Dynamics," *Space Sci. Rev.* **41:** 299–349.

WESTHEIMER, F.H. (1987), "Why Nature Chose Phosphates," *Science* **235:** 1173–1178.

WETHERILL, G.W. (1990), "Formation of the Earth," *Ann. Rev. Earth Planet. Sci.* **18:** 205–256.

WHIPPLE, F.L. (1974), "The Nature of Comets," *Sci. Am.* **230 (2):** 48–57.

WHITE, C.M. (1986), "Predictions of the Boiling Point, Heat of Vaporization, and Vapor Pressure at Various Temperatures for PAHs," *J. Chem. Eng. Data* **31:** 198–203.

WICKRAMASINGHE, D.T., AND D.A. ALLEN (1980), "The 3.4 μm Interstellar Absorption Feature," *Nature* **287:** 518–519.

WILSON, E.O. (1992), *The Diversity of Life* (Cambridge, MA: Harvard University Press).

WINDLEY, B.F. (1984), *The Evolving Continents*, 2d ed. (Chichester: John Wiley & Sons).

WING, M.R., AND J.L. BADA (1991), "Geochromatography on the Parent Body of the Carbonaceous Chondrite Ivuna," *Geochim. Cosmochim. Acta* **55:** 2937–2942.

WISE, D.U. (1974), "Continental Margins, Freeboard and the Volumes of Continents and Oceans Through Time," in *The Geology of Continental Margins*, C.A. Burke and C.L. Drake, eds. (Berlin: Springer-Verlag), pp. 45–58.

WOELLER, F., AND C. PONNAMPERUMA (1969), "Organic Synthesis in a Simulated Jovian Atmosphere," *Icarus* **10:** 386–392.

WOESE, C.R. (1987), "Bacterial Evolution," *Microbiol. Rev.* **51:** 221–271.

WOESE, C.R., AND G.E. FOX (1977a), "The Concept of Cellular Evolution," *J. Mol. Evol.* **10:** 1–6.

WOESE, C.R., AND G.E. FOX (1977b), "Phylogenetic Structure of the Prokaryotic Domain: The Primary Kingdoms," *Proc. Natl. Acad. Sci. USA* **74:** 5088–5090.

WOESE, C.R., O. KANDLER, AND M.I. WHEELIS (1990), "Towards a Natural System of Organisms: Proposal for the Domains Archaea, Bacteria and Eucarya," *Proc. Natl. Acad. Sci. USA* **87:** 4576–4579.

WOLFROM, J.L., R.U. LUMIEUX, AND S.M. OLIN (1949), "Configurational Correlation of L-(levo)-Glyceraldehyde with Natural (dextro)-Alanine by a Direct Chemical Method," *J. Am. Chem. Soc.* **71:** 2870–2873.

WOLMAN, Y., S.L. MILLER, J. IBAÑEZ, AND J. ORÓ (1971), "Formaldehyde and Ammonia as Precursors to Prebiotic Amino Acids," *Science* **174:** 1039–1040.

WOLSZCZAN, A. (1994), "Confirmation of Earth-Mass Planets Orbiting the Millisecond Pulsar PSR B1257+12," *Science* **264:** 538–542.

WONG, JT-F. (1988), "Evolution of the Genetic Code," *Microbiol. Sci.* **5:** 174–181.

WORSELY, T.R., AND R.D. NANCE (1989), "Carbon Redox and Climate Control Through Earth History, a Speculative Reconstruction," *Palaeogeogr., Palaeoclimatol., Palaeoecol. (Global Planet. Change Sect.)* **75:** 259–282.

APPENDIX II—BIOLOGICAL ASPECTS REFERENCES (cont.)

WRIGHT, G.S., A. BRIDGER, T.A. GEBALLE, AND Y.J. PENDLETON (1996), "Studies of NIR Dust Absorption Features in the Nuclei of Active and IRAS Galaxies," in *Proceedings from the Cold Dust Morphology Conference, Johannesburg, South Africa*, D. Block, ed. (Dordrecht: Kluwer Academic Publishers).

WYCKOFF, S. (1992), "Cometary Constraints on the Planet Forming Environment," *Adv. Space Res.* **12:** 13–20.

YAMAGATA, Y. (1966), "A Hypothesis for the Asymmetric Appearance of Biomolecules on Earth," *J. Theor. Biol.* **11:** 495–498.

YANG, W., R.D. MINARD, AND C.N. MATTHEWS (1976), "Azacyclopropenylidenimine: A Low Energy Dimer of Hydrogen Cyanide," *J. Theor. Biol.* **56:** 111–123.

YOUVAN, D.C., AND B.L. MARRS (1984), "Molecular Genetics and the Light Reactions of Photosynthesis," *Cell* **39:** 1–3.

YOUVAN, D.C., AND S. ISMAIL (1985), "Light-Harvesting II (B800–850-Complex) Structural Genes from *Rhodopseudomonas capsulata*," *Proc. Natl. Acad. Sci. USA* **82:** 58–62.

ZAHNLE, K., AND D. GRINSPOON (1990), "Comet Dust as a Source of Amino Acids at the Cretaceous/Tertiary Boundary," *Nature* **348:** 157–160.

ZAHNLE, K., AND L. DONES (1992), "Impact Origin of Titan's Atmosphere," in *Proceedings Symposium on Titan,* ESA SP-338, pp. 19–25.

ZAHNLE, K.J. (1986), "Photochemistry of Methane and the Formation of Hydrocyanic Acid (HCN) in the Earth's Early Atmosphere," *JGR* **91:** 2819–2834.

ZELDOVITCH, YA.B. AND D.B. SAAKYAN (1980), "Asymmetry in the Separation of Optically Active Molecules by Longitudinally Polarized Relativistic Electrons," *J. Exp. Theor. Phys.* **78:** 2233–2237.

ZHAO, M., AND J. BADA (1995), "Determination of α-Dialkylamino Acids and Their Enantiomers in Geological Samples by High-Performance Liquid Chromatography after Derivatization with a Chiral Adduct of o-Phthaldialdehyde," *J. Chromatog. A* **690:** 55–63.

ZHAO, M., AND J.L. BADA (1989), "Extraterrestrial Amino Acids in Cretaceous/Tertiary Boundary Sediments at Stevns Klint, Denmark," *Nature* **339:** 463–465.

ZILLIG, W., P. PALM, AND H-P. KLENK (1992), "A Model of the Early Evolution of Organisms: The Arisal of the Three Domains of Life from the Common Ancestor," in *The Origin and Evolution of the Cell*, H. Hartman and K. Matsuno, eds. (Singapore, New Jersey, London, Hong Kong: World Scientific), pp. 163–182.

—APPENDIX III—

CONFERENCE AGENDA

FIRST INTERNATIONAL CONFERENCE ON CIRCUMSTELLAR HABITABLE ZONES

JANUARY 19–21, 1994

Sponsored by the Space Science Division, NASA Ames Research Center
and the SETI Office, NASA Ames Research Center

Conference arrangements by
The Space Science Division, NASA Ames Research Center and SETI Institute, Mountain View, CA

AGENDA

Conference Chairman:	**Dr. Laurance R. Doyle** **SETI Institute**
Conference Cochairmen:	**Dr. John Billingham** **Chief, SETI Office**
	Dr. Donald DeVincenzi **Deputy Chief, Space Science Division**

JANUARY 19, 1994:	**PLANETARY ASPECTS OF HABITABLE ZONES**
8:00–9:00	NASA Ames Security Conference Registration
9:00–9:25	Introductory Remarks and Welcome
Morning Session:	***Atmospheric Considerations*** **Cochairpersons: Dr. Carl Sagan** **Cornell University** **Dr. O. Brian Toon** **NASA Ames Research Center**
9:25–9:30	Introduction to Planetary Aspects Dr. Donald DeVincenzi, NASA Ames Research Center
9:30–10:15	Overview Speaker **Dr. James F. Kasting,** Pennsylvania State University *"Habitable Zones Around Main Sequence Stars"*
10:15–10:30	**Carl Sagan,** Cornell University Christopher Chyba, National Security Council *"Organic Shielding of UV-Labile Greenhouse Gases on Early Earth and Mars"*
10:30–10:45	**Robert M. Haberle,** D. Tyler, C.P. McKay, and W.L. Davis, NASA Ames Research Center *"A Model for the Evolution of* CO_2 *on Mars"*
10:45–11:00	**David P. Summers,** SETI Institute Sherwood Chang, NASA Ames Research Center *"Experimental Results on the Prebiotic Nitrogen Fixation by Iron (II) in the Prebiotic Environment"*
11:00–11:15	*Coffee Break*
11:15–11:30	**Jon M. Jenkins,** SETI Institute *"Venus: A Failed Earth?"*
11:30–11:45	**Jan Rosenqvist,** Observatoire de Meudon E. Chassefiere, Service d'Aeronomie A. Léger, Universite de Paris *"Chemistry of Oxygen Compounds in Dense Atmospheres: Implication for the Detection of Life"*
11:45–12:00	**Martin Heath,** Greenwich Community College, England *"The Forest-Habitability of Earthlike Planets"*
12:00–12:30	Discussion and Summation by Session Chairpersons
12:30–1:30	*LUNCH BREAK*

Afternoon Session: ***Geological/Oceanographic Considerations***
Cochairpersons: Dr. David Morrison
NASA Ames Research Center

Dr. Michael Carr
United States Geological Survey

1:30–2:00	Overview Speaker **Dr. Christopher McKay,** NASA Ames Research Center *"The Earth as an Example"*
2:00–2:15	**David W. Schwartzman,** Howard University Steven N. Shore, Indiana University at South Bend *"Biotically Mediated Surface Cooling and Habitability for Complex Life"*
2:15–2:30	**Donald Lowe,** Stanford University *"Planetary Controls of the Habitable Zone: Examples from the Early Earth"*
2:30–2:45	**John Grotzinger,** Massachusetts Institute of Technology *"Earth Analogs of Possible Martian Carbonate Rocks"*
2:45–3:00	*Coffee Break*
3:00–3:15	**J. Peter Gogarten,** University of Connecticut *"The Effects of Heavy Bombardment on the Early Evolution of Cellular Life—A New Look at the Molecular Record"*
3:15–3:30	**Friedemann Freund** and Richard Ho, San José State University and SETI Institute *"Organic Matter Supplied to a Planet by Tectonic and Volcanic Activity"*
3:30–3:45	**Hakobu Nakamura,** Biological Institute, Konan University *"Was the Primitive Seawater Gentle for Life? Its Excess Salt Is Toxic to the Cells"*
3:45–4:00	**Linda L. Jahnke,** NASA Ames Research Center *"The Use of Molecular Fossils for the Interpretation of Paleoenvironments"*
4:00–4:15	**Robert A. Wharton, Jr.,** Desert Research Institute *"Life in Snow and Ice Ecosystems"*
4:15–4:45	Discussion and Summation by Session Chairpersons

JANUARY 20, 1994: **ASTROPHYSICAL ASPECTS OF HABITABLE ZONES**

Morning Session: ***Stellar Considerations***
Cochairpersons: Dr. Frank Drake
University of California at Santa Cruz
and SETI Institute

Dr. Icko Iben
University of Illinois

9:00–9:05	Introduction to Astrophysical Aspects Dr. Laurance R. Doyle, SETI Institute
9:05–10:05	Overview Speaker **Dr. Daniel P. Whitmire,** University of Southwestern Louisiana *"Circumstellar Habitable Zones: Overview of Stellar Considerations"*
10:05-10:20 CANCELLED	**Alan Stern,** Southwest Research Institute *"Main Sequence Stars and Post–Main Sequence Habitability Zones Around Single Stars: Where Is Water Liquid?*
10:20–10:35	**John R. Stauffer,** Harvard-Smithsonian Center for Astrophysics *"Physical Diversity Among Young Low Mass Stars"*
10:35–10:50	**Alan Hale,** Southwest Institute for Space Research *"Determining Habitable Planetary Environments Within Binary and Multiple Systems"*
10:50–11:05	*Coffee Break*
11:05–11:20	**Laurance R. Doyle,** SETI Institute N.C. Heather, Synernet *"Stellar Mass Loss and the Young Solar System"*
11:20–11:35	**C.T. Vanajakshi,** Indian Institute of Astrophysics *"Modeling of Mass Loss in the Young Solar System"*
11:35–11:50	**Yvonne J. Pendleton,** NASA Ames Research Center *"Organic Material in the Interstellar Medium"*
11:50–12:05	**Dale Cruikshank** and Yvonne Pendleton, NASA Ames Research Center *"Organic Solids in Planetary Systems: Remnants of the Interstellar Medium"*
12:05–12:35	Discussion and Summation by Session Chairpersons
12:35–1:35	*LUNCH BREAK*

Afternoon Session: ***Planetary Systems: Stability and Materials Considerations***
Cochairpersons: Dr. Jill Tarter
SETI Institute

Dr. William Borucki
NASA Ames Research Center

1:35–2:35	Overview Speaker **Dr. George Wetherill,** The Carnegie Institution of Washington *"The Formation of Habitable Planetary Systems"*
2:35–2:50	**Jack J. Lissauer,** State University of New York at Stony Brook *"Spacing of Planetary Orbits"*
2:50-3:05 CANCELLED	**Anthony Dobrovolskis,** C. Levit, J. Scargle, J. Cuzzi, L. Dones, M. Showalter, and K. Young, NASA Ames Research Center *"Orbital Stability of Planets of Binary Stars"*
3:05–3:20	**Luke Dones,** University of California at Santa Cruz and NASA Ames Research Center *"Origin and Evolution of Planetary Rotation Rates"*
3:20–3:35	*Coffee Break*
3:35–3:50	**Darren M. Williams,** Pennsylvania State University *"Chaotic Obliquity Variations and Planetary Habitability"*
3:50–4:05	**David Koch** and William Borucki, NASA Ames Research Center *"A Search for Earth-Sized Planets in Habitable Zones Using Photometry"*
4:05–4:20	**Jean Schneider,** Observatoire de Paris—Meudon *"On the Detectability of Habitable Planets"*
4:20–4:35	**Stefan Brandmeier,** Polytechnic University of Konstanz *"Simulation of Planetary Transits in Binary Star Systems"*
4:35–5:05	Discussion and Summation by Session Chairpersons
7:00–9:30	**CONFERENCE BANQUET** **Guest Speaker: Dr. Carl Sagan, Cornell University** **Honorees:** Stephen Dole Ray Reynolds Carl Sagan Charles Seeger

JANUARY 21, 1994:	**BIOLOGICAL ASPECTS OF HABITABLE ZONES**
Morning Session:	***Biogenic Material: Production and Delivery Considerations*** **Cochairpersons: Dr. Sherwood Chang** **NASA Ames Research Center** **Dr. Clifford Matthews** **University of Illinois**
9:00–9:05	Introduction to Biological Aspects Dr. John Billingham, NASA Ames Research Center
9:05–9:50	Overview Speaker **Dr. Christopher F. Chyba,** National Security Council *"Are Comets Necessary for a Habitable Zone?"*
9:50–10:05	**John Oró,** University of Houston *"Planetary Formation of Organics: The Role of Comets"*
10:05–10:20	**Akiva Bar-Nun,** Tel Aviv University Tobias Owen, University of Hawaii *"Delivery of Noble Gases, Water, and Organics to the Inner Planets by Comets"*
10:20–10:35	**Theodore E. Bunch,** NASA Ames Research Center Julie M. Paque, SETI Institute Luan Becker, Scripps Institution of Oceanography *"The Effects of Hypervelocity Impact on Carbonaceous Compounds: Experimental Insights"*
10:35–10:50	*Coffee Break*
10:50–11:05	**David W. Deamer,** University of California at Davis *"Biophysical Properties of Meteoritic Organics"*
11:05–11:20	**Clifford N. Matthews,** University of Illinois at Chicago *"Cyanide Polymers in Circumstellar Habitable Zones"*
11:20–11:35	**Max P. Bernstein,** SETI Institute and NASA Ames Research Center L.J. Allamandola, S.A. Sandford, and S. Chang, NASA Ames Research Center *"The Photo Production of Organic Molecules in Cometary and Interstellar Ice Analogs"*
11:35–11:50	**Gene D. McDonald** (presenting for Jeffrey Bada), University of California at San Diego *"A Search for Extraterrestrial Amino Acids in Polar Ice"*
11:50–12:20	Discussion and Summation by Session Chairpersons
12:20–1:20	*LUNCH BREAK*

Afternoon Session: ***Biological Contributions to Habitability Considerations***
Cochairpersons: Dr. Harold P. Klein
Santa Clara University

Dr. Donald Lowe
Stanford University

1:20–1:50	Overview Speaker **Dr. David J. Des Marais,** NASA Ames Research Center *"How Has Life Shaped Its Environment?"*
1:50–2:05	**Benton C. Clark,** Martin Marietta *"Habitability of the Comet Pond"*
2:05–2:20	**Antonio Lazcano,** Universidad Nacional Autónoma de México Stanley L. Miller, University of California at San Diego *"How Long Did It Take for Life to Appear and to Evolve to Cyanobacteria?"*
2:20–2:35	**Alexandra J. MacDermott,** Oxford Brookes University and Oxford University *"The Electroweak Origin of the Chiral Signature of Life"*
2:35–2:50	*Coffee Break*
2:50-3:05 CANCELLED	**Alexey F. Topunov,** A.N. Bakh Institute of Biochemistry, Russian Academy of Science *"Evolution of Heme-Containing Proteins and Evolution of Atmosphere"*
3:05–3:20	**Alain Léger** and J.L. Puget, Institut d'Astrophysique Spatiale—Orsay J.M. Mariotti, D. Rouan, and J. Schneider, Observatoire de Paris *"How to Evidence Primitive Life on an Exoplanet? The DARWIN Project"*
3:20–3:35	**Rosemarie Swanson** and Jun-Jie Liu, Texas A&M University *"Evolutionary Behavior of a Self-Copying Computer Program"*
3:35–3:50	**Koji Ohnishi,** Niigata University *"Poly-tRNA-Mediated Peptide-Synthesis in Early Terrestrial Organisms Before Genetic Codes: A Perspective on Early Life"*
3:50–4:20	Discussion and Summation by Session Chairpersons

JANUARY 22, 1994: **EXECUTIVE COMMITTEE SUMMARY AND RECOMMENDATIONS**

9:00–12:00	Session Chairpersons and Overview Speakers meet to discuss results of Conference at the SETI Institute

—APPENDIX IV—

SUBMITTED ABSTRACTS OF ATTENDEES

PLANETARY ASPECTS OF HABITABLE ZONES
Atmospheric Considerations

HABITABLE ZONES AROUND MAIN SEQUENCE STARS

James F. Kasting
Department of Geosciences
211 Deike
Pennsylvania State University
University Park PA 16802 USA

A one-dimensional, radiative-convective, climate model was used to estimate the width of the instantaneous and continuously habitable zones around main sequence stars (1). Both zones are defined as the region in space in which liquid water would be present on the surface of a planet orbiting the star. Our calculation differed from previous calculations, e.g. (2), by including a more elaborate radiative transfer model and by considering the stabilizing effects of the carbonate-silicate cycle on atmospheric CO_2 concentrations and planetary climates. According to our model, the inner edge of the current habitable zone (HZ) around our own Sun is determined by the onset of rapid water loss, rather than by an actual runaway greenhouse. If cloud feedback is neglected, rapid water loss would occur at solar fluxes exceeding 1.1 S_0 (where S_0 is the current solar flux at Earth's orbit, 1370 W m^{-2}); this is equivalent to an orbital radius of 0.95 AU. The outer edge of the HZ is determined in our model by the appearance of CO_2 clouds at a solar flux of 0.53 S_0 or an orbital radius of 1.37 AU. These clouds would lower the tropospheric lapse rate, reducing the atmospheric greenhouse effect, and they would increase the planetary albedo. Our estimates are conservative in that the actual effect of clouds should be to widen the HZ on both sides compared to our calculation.

The continuously habitable zone (CHZ) around our Sun is narrower than the HZ because the Sun gets brighter as it evolves. The width of the CHZ depends on the length of time that one requires a planet to remain habitable. A conservative estimate for the width of the 4.6-Gyr CHZ around our Sun is 0.2 AU, or about three times wider than the estimate of reference 2. The main reason that our CHZ is wider is that the atmospheric CO_2 level is assumed to increase as the solar flux decreases, in accordance with current thinking (3) regarding control of atmospheric CO_2 by the carbonate-silicate cycle. Finite CHZs lasting > 2 Gyr are also predicted for other main sequence stars ranging in spectral class from early F to mid K. Stars earlier than F0 have main sequence lifetimes < 2 Gyr, whereas planets orbiting stars later than ~ K5 are subject to tidal locking of their rotation. If planets are spaced logarithmically, as they are in our own Solar because they evolve more slowly than do F or G stars.

Although these one-dimensional climate calculations provide room for optimism in the search for habitable planets around other stars, they do not account for latitudinal climate

APPENDIX IV—ATTENDEE ABSTRACTS: PLANETARY ASPECTS, Kasting (cont.)

System, early K stars should have the best chance of harboring long-lived, habitable planets variations or for seasonal variability. The recent demonstration (4) that Earth's obliquity might undergo large, chaotic, oscillations were the Moon not present may imply that truly Earthlike planets are fairly rare. More complicated, latitudinally resolved, climate models, in combination with theoretical studies of planetary accretion, should be used to calculate the odds of finding habitable planets elsewhere in the Galaxy.

References

1. Kasting, J.F., D.P. Whitmire, and R.T. Reynolds (1993), *Icarus* **101:** 108–128.

2. Hart, M.H. (1978), *Icarus* **33:** 23–39; (1979), *Icarus* **37:** 351–357.

3. Walker, J.C.G., P.B. Hays, and J.F. Kasting (1981), *JGR* **86:** 9776–9782.

4. Laskar, J., F. Joutel, and P. Robutel (1993), *Nature* **361:** 615–617.

Organic Shielding of UV-Labile Greenhouse Gases on Early Earth and Mars

Carl Sagan
Laboratory for Planetary Sciences
Cornell University
Ithaca NY 14853 USA

Christopher Chyba
National Security Council
Old Executive Office Building #365
Washington DC 20506 USA

Mixing ratios $\geq 10^{-5\pm1}$ of NH_3 and other reduced gases in the atmospheres of Earth and Mars 4 Gyr ago, through their 8–14 μm opacity would have raised surface temperatures above 0°C—in agreement with geological evidence—despite the 25 percent lower solar luminosity prevailing then. However, such model atmospheres have been rejected because of their extreme UV-lability (lifetime of unshielded NH_3 against photolysis ~ 10 yr). High-altitude photolysis of CH_4, even as a minor constituent, is the most effective source of organic synthesis yet suggested for early Earth. Expected steady-state amounts of high-altitude organic solids from CH_4 photolysis may have shielded NH_3 sufficiently [τ(2000 Å) ≥ 5] that modest NH_3 resupply rates were able to maintain the required mixing ratio. For Earth a significant fraction of the present N_2 budget may be generated from early NH_3 photolysis.

APPENDIX IV—ATTENDEE ABSTRACTS: PLANETARY ASPECTS (cont.)

A Model for the Evolution of CO_2 on Mars

Robert M. Haberle
D. Tyler, C. P. McKay, and W. L. Davis
Mail Stop 245-3
NASA Ames Research Center
Moffett Field CA 94035-1000 USA

We have constructed a model that predicts the evolution of CO_2 on Mars from the end of heavy bombardment to the present. The model draws on published estimates of the main processes believed to affect the fate of CO_2 during this period: chemical weathering, regolith uptake, polar cap formation, and atmospheric escape. Except for escape, the rate at which these processes act is controlled by surface temperatures which we calculated using a modified version of the Gierasch and Toon (1973 [*Jour. Atmos. Sci.* **30:** 1502] *see* Appendix II, Planetary Aspects) energy balance model. The modifications account for the change in solar luminosity with time, the greenhouse effect, and an equatorial (as well as polar) energy budget.

Using published estimates for the main parameters, we find no evolutionary scenario in which CO_2 is capable of producing a warm (global mean temperatures > 250 K) and wet (surface pressures > 30 mbar) early climate, and then evolves to present conditions with ~ 7 mbars in the atmosphere, < 300 mbars in the regolith, and < 5 mbars in the caps. Such scenarios exist only if the early Sun was brighter than standard solar models suggest, if greenhouse gases other than CO_2 were present in the early atmosphere, or if the polar albedo is significantly lower than 0.75. However, these scenarios generally require the storage of large amounts of CO_2 (> 1 bar) in the carbonate reservoir. If the warm and wet early Mars constraint is relaxed, then we find best overall agreement with present day reservoirs for initial CO_2 inventories of 0.5–1.0 bars.

We also find that the polar caps can have a profound effect on how the system evolves. If the initial amount of CO_2 is less than some critical value, then there is not enough heating of the poles to prevent permanent caps from forming. Once formed, these caps control how the system evolves because they set the surface pressure and, hence, the thermal environment. If the initial amount of CO_2 is greater than this critical value, then caps do not form initially, but can form later on when weathering and escape lower the surface pressure to a point where polar heating is no longer sufficient to prevent cap formation and the collapse of the climate system. Our modeling suggests this critical initial amount of CO_2 is between 1 and 2 bars, but its true value will depend on all factors affecting the polar heat budget.

APPENDIX IV—ATTENDEE ABSTRACTS: PLANETARY ASPECTS (cont.)

Experimental Results on the Prebiotic Nitrogen Fixation by Iron (II) in the Prebiotic Environment

David P. Summers
SETI Institute
Mail Stop 239-4
NASA Ames Research Center
Moffett Field CA 94035-1000 USA

Sherwood Chang
Mail Stop 239-4
NASA Ames Research Center
Moffett Field CA 94035-1000 USA

Theories for the origin of life require the availability of reduced nitrogen. However, under non-reducing atmospheres electric discharges produced NO rather than HCN or NH_3. This raises the questions of: How can ammonia be formed under a neutral atmosphere, and what conditions are needed for such formation to occur? One possibility is the reduction of nitrite, formed from the NO, by aqueous Fe(II) such as was present on the early Earth. We wish to describe experimental results showing that the reduction of nitrite and nitrate to ammonia by Fe(II) (equation 1) is plausible under prebiotic seawater conditions.

$$6\ Fe^{+2} + 7\ H^+ + NO_2^-\ 6\ Fe(III) + 2\ H_2O + NH_3 \qquad (1)$$

Nitrite reduction follows an apparent rate law of the form Rate = $k\ [NO_2^-]\ [Fe]^{1.8}$ and is not significantly inhibited by many of the solutes found in the modern day ocean. The reaction is favored by temperatures $\geq 25°C$, but proceeds only at $pH > 7.3$.

References

1. Schopf, J.W., ed. (1983), *Earth's Earliest Biosphere* (Princeton, NJ: Princeton University Press), and references therein.

2. Walker, J.C.G. (1985), *Origins of Life* **16:** 117–127

3. Mattioli, G.S., and B.J. Wood (1986), *Nature* **322:** 626–628.

4. Chameides, W.L., and J.C.G. Walker (1981), *Origins of Life* **11:** 291–302; Yung, Y.L., and M.B. McElroy (1979), *Science* **203:** 1002–1004; Kasting, J.F. (1990), *Origins of Life* **20:** 199–231.

5. Mancinelli, R.L., and C.P. McKay (1988), *Origins of Life* **18:** 311–325.

6. Walker, J.C.G., and P. Brimblecombe (1985), *Precambrian Res.* **28:** 205–222.

APPENDIX IV—ATTENDEE ABSTRACTS: PLANETARY ASPECTS (cont.)

VENUS: A FAILED EARTH?

Jon M. Jenkins
SETI Institute
Mail Stop 244-11
NASA Ames Research Center
Moffett Field CA 94035-1000 USA

While Earth and its twin, Venus, are roughly the same size and mass, and were formed by the same process of planetary accretion, Earth supports intelligent life, while Venus is not capable of supporting life at all. The obvious difference between life on Earth and no life on Venus is the closer position of Venus to the Sun. This constrains water to appear only in its gas phase in the present atmosphere of Venus. However, the enhancement of deuterium to hydrogen in the Venus atmosphere, as compared to the Earth, and the radically different character of tectonic processes on Venus revealed by Magellan, belie this simple explanation for the absence of life on Venus. How much water was there in the early atmosphere of Venus? Is the Venus water vapor abundance in steady state, or is it a signature of an early ocean? The explanations for these observations and questions, while somewhat elusive, have implications for evolution of atmospheres on terrestrial planets, and their capacity to support life.

CHEMISTRY OF OXYGEN COMPOUNDS IN DENSE ATMOSPHERES: IMPLICATION FOR THE DETECTION OF LIFE

Jan Rosenqvist
DESPA, Observatoire de Paris
92195 Meudon FRANCE

E. Chassefiere
CNRS
Service d'Aeronomie FRANCE

A. Léger
IAS
Universite de Paris XI FRANCE

The relation between the oxygen compounds amounts and the type of atmosphere (i.e., dense-hot-wet . . .) is a crucial point for the search of life on a planet (1, 2). Our purpose is here to give rough estimates of the global O_2 amount that should be present in very different kinds of atmospheres.

We will use a one-dimensional model of inorganic photochemical reactions involving species derived from CO_2 and H_2O. The model was previously described in (3), where the regulation of CO on Mars is studied on a global scale. Therefore, this model, preliminarily based on the current Martian atmosphere, will be adapted to very different conditions such as much more dense (up to 10 bar) and warmer atmospheres.

In particular, we will examine the influence of the thermal profile (with or without tropopause, cold or hot temperatures), the composition of the atmosphere (CO_2–N_2–H_2O), the distance to the "Sun" on the total amount of O_2 in the considered atmosphere. In addition, an arbitrary oxygen flux can be included in order to simulate the oxidation and/or the photosynthesis at the surface.

APPENDIX IV—ATTENDEE ABSTRACTS: PLANETARY ASPECTS, Rosenqvist et al. (cont.)

The results will be discussed in the frame of the detection of O_2 (or O_3) for criterion of life on a planet.

REFERENCES

1. OWEN, T. (1980), *Strategies for the Search for Life in the Universe*, M.D. Papagiannis, ed. (Dordrecht: D. Reidel), pp. 177–185.

2. LÉGER, A., M. PIRRE, AND F.J. MARCEAU (1993), *AA* **227:** 309.

3. CHASSEFIERE, E. (1991), *Geophys. Res. Lett.* **18:** 1055–1058.

THE FOREST-HABITABILITY OF EARTHLIKE PLANETS

Martin Heath
Greenwich Community College
47 Tulsemere Rd.
London SE27 9EH ENGLAND

There has been much discussion about the habitability of possible Earthlike planets. One of the most characteristic features of Earth's biosphere is the existence of forest biomes, which nominally cover about 40% of the planet's land surface, and contain about 90% of Earth's total biomass (Duvigneaud 1979). They create distinctive local environments, and may influence regional climate. It cannot be assumed that all other biospheres would function like ours, but the study of ours offers pointers to modifications necessary if something like Earth's massive land biosphere was to thrive in different global environments.

A recent model (Caldeira and Kasting 1992b) for CO_2 drawdown in response to future increases in solar luminosity suggests that it may be the eventual rise in temperature, not CO_2 starvation, that limits the life span of the biosphere. C4 plants can have high optimum temperatures and function with very low ambient CO_2. I point out that despite the promising tolerances of some species, the massive land biosphere as an entity takes up something on the order of 60×10^9 tons of carbon per year in net primary production (Duvigneaud 1979). No matter how efficient carbon fixation might become, it will not be possible to cycle this quantity through the biomass when there is no longer this much carbon in the atmosphere. The prospects for the forest biomass seem poor. The Caldeira and Kasting (1992b) model has biological productivity declining by an order of magnitude by insolation 1.06 present; 0.97 AU from the Sun today (possible cloud-climate feedback ignored).

Full-sized trees exist on Earth where temperatures exceed 10°C for 30 days/year. The outer limit for forest biomes would depend on the ability of the carbonate-silicate cycle to permit temperatures significantly higher than this over much of a planet's land area. Climatic models differ, but that of Marshall et al. (1988) would permit low global temperature if continents moved into an equatorial configuration, with more severe reductions in temperature associated with lower insolation values. Global freezing events remain possible at present insolation (Caldeira and Kasting 1992a), though CO_2 buildup under conditions of reduced weathering would permit deglaciation above 0.8 this value, 1.12 AU from the Sun. However, global low temperatures and a sharp rise in CO_2 could eliminate forests.

Some trees achieve maximum photosynthesis at relatively low light levels (oak near 0.25 solar constant—equivalent to cloudless skies at 2.00 AU), but pioneer tree species need 0.6 clear sky insolation (1.3 AU). Understory species and saplings surviving near their light compensation points would be disadvantaged under tree canopies as dense as those in Earth's forests, and flecks

APPENDIX IV—ATTENDEE ABSTRACTS: PLANETARY ASPECTS, Heath (cont.)

of sunlight would contribute less energy to the forest floor. There would be an overall reduction in total energy available for photosynthesis during favorable seasons on planets with larger orbits, and longer orbital periods. On the other hand, increases in ambient CO_2 would help the fixation of carbon.

The window of favorable conditions for a massive land biomass in geological history might extend from the apparent low latitude glaciations of the Late Proterozoic until CO_2 starvation, or until some earlier readjustment in the Earth system.

A variety of alternative biologies can be envisaged, both for Earth, planets with a differing tectonic style, and planets orbiting stars of differing mass and metallicity. In principle (Heath 1991, 1993), biology could drive the equivalent of Earth's biogeochemical cycles on geologically inert worlds. Biological membranes, secretions and constructs interposed at the atmosphere-ocean, crust-ocean, and crust-atmosphere interfaces could control all aspects of gas exchange and weathering, and even modify planetary albedo. However, the more extreme the global environment to be overcome, the less likely it is that suitable biological systems could ever evolve.

REFERENCES

1. CALDEIRA, K., AND J.F. KASTING (1992a), *Nature* **357:** 578–581.

2. CALDEIRA, K., AND J.F. KASTING (1992b), *Nature* **360:** 721–723.

3. DUVIGNEAUD, P. (1979), "SCOPE," in *The Global Carbon Cycle*, B. Bolin et al., eds. (Chichester: John Wiley & Sons).

4. HEATH, M.J. (1991), *J. Brit. Interplanet. Soc.* **44:** 1437–1449.

5. HEATH, M.J. (1993), Greenwich Community College report.

6. MARSHALL, H.G., ET AL. (1988), *JGR* **93:** 791–801.

PLANETARY ASPECTS OF HABITABLE ZONES (cont.)
Geological/Oceanographic Considerations

THE EARTH AS AN EXAMPLE

Christopher P. McKay
Mail Stop 245-3
NASA Ames Research Center
Moffett Field CA 94035-1000 USA

The Earth is the only habitable planet currently known. It is not surprising therefore that models of habitable zones and the development and evolution of life on other worlds are based largely on Earth's geological and biological record. The general trends observed are rapid origin of life, extremely slow evolution of an oxygenic atmosphere, rapidly followed by the development of large multicellular organisms, slowly followed by the emergence of intelligence. All of this occurring within the context of a temperate climate. A critical review of Earth's history suggests that the time required for each of these steps can be orders of magnitude different on different worlds.

APPENDIX IV—ATTENDEE ABSTRACTS: PLANETARY ASPECTS (cont.)

BIOTICALLY MEDIATED SURFACE COOLING AND HABITABILITY FOR COMPLEX LIFE

David W. Schwartzman
Department of Biology
Howard University
Washington DC 20059 USA

Steven N. Shore
Department of Physics and Astronomy
Indiana University at South Bend
South Bend IN 46634 USA

The width of the habitable zone for complex life is substantially smaller than that for the appearance of biota. The consequences of this largely ignored fact will be discussed in light of the implications to SETI strategy. Biotic colonization of continents and resultant enhancement of weathering, first by thermophilic bacteria, led to lower steady-state atmospheric carbon dioxide levels and surface cooling on Earth. Evolution of prokaryotes and complex life on terrestrial planets is expected to have similar geochemical and climatic consequences. Biotically mediated cooling increases the width of the habitable zone for the possible occurrence and evolutionary time frame of complex life. We discuss models for this effect in light of astrophysical constraints.

PLANETARY CONTROLS ON THE HABITABLE ZONE: EXAMPLES FROM THE EARLY EARTH

Donald Lowe
Department of Geological and Environmental Sciences
Stanford University
Stanford CA 94305 USA

The effective habitable zone of a planetary system can be enlarged, displaced, or otherwise modified by a number of endogenic planetary processes. Most important among these are crustal evolution, volcanism, atmospheric evolution, and hydrothermal activity. All of these processes were active on the early Earth. The combined effects of volcanism, the lack of large areas of continental crust, and atmospheric evolution may have produced a pre-4.0-Ga Earth that lay effectively near the inner limit of the habitable zone, with a surface temperature only slightly below 100°C. Evolution of the Earth after 4.0-Ga ago and before about 2.5-Ga ago, saw a decline in surface temperatures, mainly as a result of the evolution of the continental crust, and the effective transfer of the planet toward the interior of the habitable zone. Thermal springs represent a means of preserving local biofriendly environments on a planet that otherwise lies outside of the habitable zone.

APPENDIX IV—ATTENDEE ABSTRACTS: PLANETARY ASPECTS (cont.)

EARTH ANALOGS OF POSSIBLE MARTIAN CARBONATE ROCKS

John Grotzinger
Department of Earth, Atmospheric and Planetary Sciences
Massachusetts Institute of Technology
Cambridge MA 02139 USA

Earth and Mars are thought to have experienced similar early histories. The development of carbonate precipitating environments during the Archean on Earth provided a diverse number of rock types not formed during younger times. These are thought to have been deposited during conditions of high atmospheric pCO_2 and high saturation of seawater with respect to calcium carbonate. If carbonate rocks are present on Mars, similar facies may be encountered.

THE EFFECTS OF HEAVY BOMBARDMENT ON THE EARLY EVOLUTION OF CELLULAR LIFE—A NEW LOOK AT THE MOLECULAR RECORD

J. Peter Gogarten
Department of Molecular and Cell Biology
University of Connecticut
Storrs CT 06269-3044 USA

A characteristic of molecular phylogenies (e.g., 16S-like rRNAs, ATPases, RNA polymerases and elongation factors) is that the three domains of life (Bacteria, Archaea, Eucarya) are clearly separated from each other. So far there are no survivors of "missing links" known. Several theories involving biological explanations have been proposed to explain the absence of deep branching lineages. However, the analyses of ancient duplicated genes cast doubts on these scenarios. The last common ancestor of all presently known surviving life forms appears to already have been a sophisticated cellular prokaryote with DNA, RNA, ribosomes, energized membranes, etc. There is no a priori reason to assume that this cellular organism would have had difficulties to undergo speciation and to adapt to different ecological niches present on the early Earth. Analysis of the molecular record suggests an early catastrophic event that wiped out all but the two surviving lineages, i.e. the ancestors to the Bacteria and Archaea, respectively.

(The latter also contributed the major portion to the nucleocytoplasmic component of the Eucarya.) A likely cause for this catastrophe is a late nearly-ocean-boiling impact. This scenario also explains why the deep branches in both the archaeal and the bacterial domain are occupied by extreme thermophiles: only organisms that at the time of the impact were already adapted to extremely hot temperatures were able to survive the post-impact conditions. Calibrating the molecular record with regard to time is still difficult. One possible calibration point is the emergence of Eucarya in the fossil record suggesting that the proposed late impact certainly occurred before 2.3×10^9 years, another is the early fossil stromatolites. If the microfossils in these stromatolites are interpreted as representatives of present day existing groups of bacteria (i.e., as cyanobacteria), then the impact is dated to around 3.7×10^9 years BP.

Conclusion: Life emerged early in Earth's history, even before the time of the heavy bombardment was over. Early life forms already had colonized extreme habitats that allowed at least two prokaryotic species to survive a late nearly ocean boiling impact. The distribution of ecotypes on the rooted universal tree of life cannot necessarily be interpreted to show that life originated in extremely hot environments.

APPENDIX IV—ATTENDEE ABSTRACTS: PLANETARY ASPECTS (cont.)

ORGANIC MATTER SUPPLIED TO A PLANET BY TECTONIC AND VOLCANIC ACTIVITY

Friedemann Freund
Richard Ho
Department of Physics
San José State University
San José CA 95192-0106 USA
and
SETI Institute
2035 Landings Drive
Mountain View CA 94043 USA

In the context of possible existence of Life elsewhere, a planet formed within a habitable zone around a star must have accumulated enough organic matter, either from external or internal sources, to enable Life to emerge. Organic matter that is synthesized in space and carried downward by dust and meteorites is a possible external source. Among the internal sources, pyrolytic, photolytic and catalytic reactions in the atmosphere, hydrosphere, and at gas-liquid-solid interfaces have been intensely studied in the past. However, organic matter also occurs in igneous rocks of upper mantle provenance which are brought to the Earth's surface by volcanic or tectonic activity. The synthesis of this organic matter remains highly controversial. One widely disseminated suggestion had been that volcanic gases (H_2O, CO, CO_2, and N_2) catalyze organic molecules in microcracks that open temporarily due to mechanical or thermal stress. A more fundamental process by which organic matter can be synthesized is based on solid solution formation between $H_2O/CO/CO_2/N_2$ and minerals which invariably form during crystallization in fluid-rich environments. Upon cooling and decompression, such solid solutions become supersaturated, leading to the segregation of reduced species H, C, and N to dislocations, grain, and subgrain boundaries. There C–C, C–H, and C–N bonds form, leading to organic precursors. On a planet with surface water, weathering causes these precursors to hydrolyze and to become full-fledged organic molecules. On a tectonically active planet, such as the Earth, this process can lead to the release of significant amounts of specific organic matter, possibly exceeding the amount that can be provided by meteoric in-fall and atmospheric reactions. We present supporting data from recent studies involving vacuum fracture of olivine single crystals from upper mantle peridotite nodules and their carrier basalts, followed by extraction of the organics and their analysis using gas chromatography and different methods of mass spectroscopy.

WAS THE PRIMITIVE SEAWATER GENTLE FOR LIFE? ITS EXCESS SALT IS TOXIC TO THE CELLS

Hakobu Nakamura
Biological Institute
Faculty of Science, Konan University
8-9-1 Higashinada, Kobe 658 JAPAN

The present seawater contains 3.5% salt (on an average), which had to be very toxic to protocells. Therefore, if the primitive seawater contained such excess salt, then all the life would not be generated on the earth. The cells of all extant organisms possess Na^+-efflux system (ATP-dependent Na^+ pump) in the plasma membranes. We have isolated an *E. coli* mutant that is

APPENDIX IV—ATTENDEE ABSTRACTS: PLANETARY ASPECTS, Nakamura (cont.)

supersensitive to excess salt and that has an optimum concentration at a 0.3% NaCl. This mutant seems to be deficient in the pump. It appears from the evidence that proto-cells did not have such a highly organized pump (partially published in *Viva Origino* **21:** 75–90 [1993]).

THE USE OF MOLECULAR FOSSILS FOR THE INTERPRETATION OF PALEOENVIRONMENTS

Linda L. Jahnke
Mail Stop 239-4
NASA Ames Research Center
Moffett Field CA 94035-1000 USA

Over the course of Earth's history, the most important biological influence has come from the evolution of oxygenic photosynthesis and the development of an oxygen-rich biosphere. Although the availability of free oxygen had profound effects on subsequent biological and planetary evolution, clear paleobiological evidence of the timing of this transition is lacking. Recent technical advances in the microanalysis of organic matter have made it possible to detect residual molecules (biomarkers) in proterozoic (2.5–0.6 billion years) sedimentary rock characteristic of specific groups of microorganisms. When coupled with the carbon isotopic fractionations characteristic of biological systems and the new field of compound specific isotope analysis, biomarkers could prove to be a powerful tool for decoding ancient biochemistry from the geological record. We have been studying the synthesis of organic biomarker molecules in several groups of contemporary bacteria that should have been key to carbon cycling in paleoenvironments after the evolution of oxygenic photosynthesis.

LIFE IN SNOW AND ICE ENVIRONMENTS

Robert A. Wharton, Jr.
Desert Research Institute
P.O. Box 60220
Reno NV 89506 USA

On Earth, microorganisms such as bacteria, algae, and fungi are commonly found thriving in snow, glacial ice, lake ice, and ice-shelves. Snow and ice environments are typical of polar and alpine regions. These environments are suitable habitats for microorganisms only when liquid water is present for at least part of the year, allowing for growth and reproduction of the microbes. Hence, the surface of the vast Antarctic Polar Plateau is virtually lifeless because of the lack of water in the liquid phase. In contrast, the bottom surfaces of the Antarctic glaciers are wet due to basal melting and may serve as an important habitat for microbes. The sea ice of the Arctic Ocean often contains blooms of snow algae during the summer when meltwater is present on the ice surface. Microbes are abundant in snowfields and in depressions and ponds found on glaciers. Microbes found in snow and ice have adapted special growth forms and mechanisms for tolerating the long periods of desiccation during the winter and the high irradiance levels characteristic of the summer. Microorganisms play a fundamental role in the biogeochemistry of snow and ice. The microbes are intimately involved in the primary production, respiration, nutrient, cycling, decomposition, metal accumulation, and food webs associated with snow and ice. The microbes contribute to the melting of snowpacks, formation of cryoconite holes on glaciers, and the mass wastage of glaciers.

APPENDIX IV—ATTENDEE ABSTRACTS: PLANETARY ASPECTS, Wharton (cont.)

Whether life exists in the snow and ice environments of celestial bodies beyond Earth is a subject of much speculation. For example, if there is, or ever was, life on Mars it may be associated with the snow and ice environments observed on the planet. It is conceivable that basal melting of the Martian polar ice cap may provide liquid water necessary to support microbial life. The surface ice of Europa has been proposed as a suitable habitat for microbial life. Snow and ice environments may be a common feature on other appropriately placed circumstellar bodies within the Universe. However, if microbial life (as we know it) is to exist on these bodies then mechanisms to produce liquid water must be invoked.

ASTROPHYSICAL ASPECTS OF HABITABLE ZONES
Stellar Considerations

CIRCUMSTELLAR HABITABLE ZONES: OVERVIEW OF STELLAR CONSIDERATIONS

Daniel P. Whitmire
University of Southwestern Louisiana
Lafayette LA 70504-4210 USA

We review the stellar properties that influence the size and evolution of habitable zones around single main sequence stars. These include luminosity, mass, spectral type, the coverage and distribution of star spots and the possibility of significant mass loss (approximately 0.04 solar masses) during the early main sequence (as has been suggested to explain lithium depletion in the Sun and warm climates on early Earth and Mars). Mid to early K-type stars appear to be as good or better than G-type stars as a priori sites for the evolution of complex life, regardless of whether planetary spacings tend to be logarithmic or random. Beyond the main sequence phase, the shorter evolutionary time scales are countered by correspondingly larger habitable zones. Whether life is likely to exist on planets or satellites around evolved stars depends (in addition to the time scale for the appearance of life) on whether frozen planets and satellites can be cold started as the stellar luminosity increases. Since most solar-type stars (< or = 65%) reside in multiple star systems, we review the evidence suggesting that planets can accrete in these systems and the dynamic constraints on their orbits. We then combine these dynamic constraints with binary star orbital-element statistics and with planetary climate model results to investigate the size and long term evolution of habitable zones in binary star systems.

PHYSICAL DIVERSITY AMONG YOUNG LOW MASS STARS

John Stauffer
Harvard-Smithsonian Center for Astrophysics
60 Garden Street
Cambridge MA 02138 USA

I will review a number of observational programs that have recently been reported that plausibly might illuminate the relationship between the formation and early evolution of stars and the formation of (habitable) planets. Among these programs are (1) estimates of the fraction of pre–main sequence stars that are formed in binary or triple systems, (2) estimates of the fraction of

APPENDIX IV—ATTENDEE ABSTRACTS: ASTROPHYSICAL ASPECTS, Stauffer (cont.)

stars born with relatively massive disks, (3) the influence of disks on the angular momentum evolution of low mass stars and the distribution of rotational velocities for low mass stars when they arrive on the main sequence, and (4) the X-ray luminosity of young stars and evolution of coronal activity with time. Even for a coeval population of stars of the same mass, there is a wide range in rotational velocity and coronal activity until at least an age of a few hundred million years, and therefore one might expect that at least the early evolution of planets might vary depending upon the properties of the star they orbit.

DETERMINING HABITABLE PLANETARY ENVIRONMENTS WITHIN BINARY AND MULTIPLE SYSTEMS

Alan Hale
Southwest Institute for Space Research
P.O. Box 4189
Alamogordo NM 88311-4189 USA

The presence of one or more stellar companions significantly complicates the issues involved in the discussion of suitable planetary environments for life, not the least of which is the question whether or not planets can even exist in such a system. Given that they can, other items that must be addressed include stability of planetary orbits and effects of varying insolation. These and other issues will be discussed in the context of current astronomical knowledge.

STELLAR MASS LOSS AND THE YOUNG SOLAR SYSTEM

Laurance R. Doyle
SETI Institute
Mail Stop 244-11
NASA Ames Research Center
Moffett Field CA 94035-1000 USA

N. C. Heather
Synernet
Mail Stop 245-4
NASA Ames Research Center
Moffett Field CA 94035-1000 USA

The standard solar model which indicates that the young Sun was 70% as bright as it is today leads to a number of problems with our understanding of the Solar System ecoshell. For example, the evidence of liquid water on early Mars indicates that it was within the ecoshell for the first billion years of the origin of the Solar System. However, it appears that atmospheric greenhouse constituents would largely have condensed out of the atmosphere before they could have warmed the planetary surface to this extent (Kasting 1991). Consequently a more massive early Sun model has been suggested based on a number of arguments (Whitmire et al. 1995) with possible observational checks including the direct detection of young solar-type stellar mass loss in the radio region of the spectrum (Doyle et al. 1995). We present here the results of centimeter wavelength observations of several nearby young solar-mass stars using the NRAO Very Large Array telescope. Consequent observational constraints on early mass loss from solar-type stars are then applied to the early history of the solar ecoshell.

APPENDIX IV—ATTENDEE ABSTRACTS: ASTROPHYSICAL ASPECTS, Doyle and Heather (cont.)

REFERENCES

1. DOYLE, L.R., R. VIKRAMSINGH, D.P. WHITMIRE, AND N.C. HEATHER (1995), "Circumstellar Habitable Zones and Mass Loss from Young Solar-Type Stars. II. Observational Considerations," in *Progress in the Search for Extraterrestrial Life*, ASP Conf. Ser., S. Shostak, ed., **74:** 195–204.

2. KASTING, J.F. (1991), "CO_2 Condensation and the Climate of Early Mars," *Icarus* **94:** 1–13.

3. WHITMIRE, D.P., L.R. DOYLE, J.J. MATESE, AND R.T. REYNOLDS (1995), "Circumstellar Habitable Zones and Mass Loss from Young Solar-Type Stars. I. Theory," in *Progress in the Search for Extraterrestrial Life*, ASP Conf. Ser., S. Shostak, ed., **74:** 183–193.

MODELING OF MASS LOSS IN THE YOUNG SOLAR SYSTEM

C. T. Vanajakshi
Indian Institute of Astrophysics
Bangalore INDIA

The mass loss from a young solar system as it loses angular momentum by ejecting high-angular-momentum material through stellar winds is computed through modeling the evolution of the coronal magnetic field and the solar wind plasma. A stationary sun, with a radial magnetic field, is given an initial rotation and the system is allowed to relax to a steady state. The mass loss during the relaxation period as well as after the steady state is reached is calculated. The simulation is performed by solving the time-dependent, magneto-hydrodynamic equations of a rotating, magnetized fluid.

ORGANIC MATERIAL IN THE INTERSTELLAR MEDIUM

Yvonne J. Pendleton
Mail Stop 245-3
NASA Ames Research Center
Moffett Field CA 94035-1000 USA

Near-infrared observations have revealed the presence of short chain aliphatic hydrocarbons in the diffuse interstellar medium. The similarity of the interstellar medium spectra to laboratory residues and to the acid insoluble extract from Murchison meteorite suggests that some of the interstellar material may have survived incorporation into primitive Solar System bodies.

APPENDIX IV—ATTENDEE ABSTRACTS: ASTROPHYSICAL ASPECTS (cont.)

ORGANIC SOLIDS IN PLANETARY SYSTEMS: REMNANTS OF THE INTERSTELLAR MEDIUM

Dale P. Cruikshank
Mail Stop 245-6
NASA Ames Research Center
Moffett Field CA 94035-1000 USA

Yvonne Pendleton
Mail Stop 245-3
NASA Ames Research Center
Moffett Field CA 94035-1000 USA

Icy planetesimals formed in the solar nebula incorporated organic solids and ices from the interstellar medium. This material has changed chemically and physically during the 5 billion years since its accretion into planetesimals, in ways dependent upon the exposure to galactic cosmic rays, heating from direct sunlight, internal heating from radiogenic isotopes, and other events not completely understood. In the inner Solar System, we find organic solids with an isotopic signature from the ISM in carbonaceous meteorites and IDPs, which have been altered by liquid water and heat. Observational evidence is mounting for the occurrence of less altered organic solids in the intermediate and outermost Solar System. Planetesimals 5145 Pholus and 1993 HA2 are key objects in understanding the incorporation of light hydrocarbon–rich organics from the ISM into bodies of a few hundred kilometers in dimension, and the subsequent evolution of these bodies to the parents of the carbonaceous meteorites. This paper presents the evidence from the origin and evolution of these organic-rich planetesimals from the ISM to the carbonaceous meteorites. This sequence of events is expected to occur in other planetary systems that form the way in which the Solar System formed.

ASTROPHYSICAL ASPECTS OF HABITABLE ZONES (cont.)

Planetary Systems: Stability and Materials Considerations

THE FORMATION OF HABITABLE PLANETARY SYSTEMS

George W. Wetherill
Department of Terrestrial Magnetism
Carnegie Institution of Washington
Washington DC 20015 USA

Current work on the formation of the Solar System has led to the realization that this can be done only in the context of a general theory of planetary formation, which in turn is an adjunct of star formation. From this point of view, it is expected that the variety of planetary systems will be very large, both because of the range of stellar masses, the very uncertain, but probably wide variety of preplanetary disks, and the chaotic nature of many of the physical processes involved.

For these reasons, current understanding of this problem is minimal. Nevertheless, for stars of mass similar to our Sun, and with disk surface densities similar to our Solar System, quantitative modeling suggests that "terrestrial" planetary systems similar to ours may be common. Although their exact configuration will be highly variable, the formation of at least one planet within proposed thermal habitability zones is plausible. Formation of giant planets, analogous to Jupiter and Saturn, is much less certain, and this could be important with regard to the habitability of the terrestrial planets of their systems.

Preliminary calculations indicate that the above statements are not likely to be a strong function of stellar masses between 0.5 and 1.5 $M_{\odot}$, nor of disk surface densities within a factor of three of that of our own Solar System. The greatest uncertainties are those resulting from the near-absence of observational and theoretical understanding of the relationship between protostellar disks and the residual planetesimal disks that are hypothesized to be the starting point of planet formation.

SPACING OF PLANETARY ORBITS

Jack J. Lissauer
Department of Earth and Space Sciences
State University of New York
Stony Brook NY 11794-2100 USA

The orbital spacing of planets is an important factor in determining how many planets are likely to exist within habitable zones. Terrestrial planets are believed to grow by pairwise accretion until the spacing of planetary orbits becomes large enough that the configuration is stable for the age of the system. Stability criteria for systems with several planets are not fully understood.

However, scaling laws based on generalizations of the Jacobi-Hill stability and the resonance overlap criterion to many-body systems provide remarkably consistent fits when applied to the planets in our Solar System as well as to planetary satellite systems. These scalings will be described and applied to hypothetical planetary systems that accreted from disks of differing surface mass densities around stars of various masses.

ORIGIN AND EVOLUTION OF PLANETARY ROTATION RATES

Luke Dones
Mail Stop 245-6
NASA Ames Research Center
Moffett Field CA 94035-1000 USA
and
University of California
Santa Cruz CA 95064 USA

The primordial spins of the terrestrial planets likely arose through off-center impacts by large planetesimals (bodies with masses > 1% of the final planetary mass) which occurred during planet formation. A newly formed planet should have, to order of magnitude, a rotation period of 10 hours and a rotation axis that is randomly oriented in space; that is, prograde and retrograde spin are equally likely. In this interpretation, Earth has probably always spun prograde, and Venus has always been retrograde. However, the tides can cause secular changes in spin rate. For instance, the Moon has despun the Earth from its original spin period of 4–5 hours, and Mercury has been despun by solar tides. Mutual planetary perturbations can cause large oscillations in the orientation of spin axes ("obliquities") on time scales as short as 100,000 years. At present, Mars' obliquity has probably varied by tens of degrees over the last 10 Myr, while the Moon helps stabilize the Earth's obliquity, which varies only by a few degrees (Ward 1974, Laskar et al. 1993,

APPENDIX IV—ATTENDEE ABSTRACTS: ASTROPHYSICAL ASPECTS, Dones (cont.)

Touma and Wisdom 1994; *see* Appendix II, Astrophysical Aspects). In some cases, "chaotic" obliquity variations can even convert an initially retrograde spin into prograde, or vice versa. I will discuss the implications of these results for the origin and evolution of life.

Chaotic Obliquity Variations and Planetary Habitability

Darren M. Williams
Department of Astronomy and Astrophysics
525 Davey Lab
Pennsylvania State University
University Park PA 16802 USA

Possible chaotic behavior of terrestrial planetary obliquities may allow global climatic fluctuations that ultimately affect planetary habitability. A simple one-dimensional energy balance climate model has reproduced Earth's present mean annual and seasonal temperatures for current obliquity. Despite current modeling uncertainties, initial experiments using high obliquities have yielded no climate catastrophes. Pending conclusive results, we hope to definitively determine the susceptibility of Earth and Mars to climatic instabilities induced by a time varying obliquity. Assuming the phenomenon is common to many planets in many systems, the question of habitability may be intimately linked to climatic response and biota adaptation to the changes for worlds on which life can otherwise evolve.

A Search for Earth-Sized Planets in Habitable Zones Using Photometry

David Koch and William Borucki
Mail Stop 245-6
NASA Ames Research Center
Moffett Field CA 94035-1000 USA

Several methods have been suggested for the detection of extrasolar planets. In the habitable zone, the radial velocity and astrometric methods are only sensitive to the detection of massive planets (greater than about ten to one-hundred Earth masses). However, the photometric method is sensitive to Earth-sized planets in inner orbits about solarlike stars. The methodology for conducting a photometric space-based search is described. The information that can be directly and indirectly derived from the observations about any planetary systems discovered is discussed.

On the Detectability of Habitable Planets

Jean Schneider
Observatoire de Meudon
92195 Meudon FRANCE

At least five methods are used or planned in the search for extrasolar planets. In the present paper they are reviewed to investigate whether or not each one is suited for the detection of habitable planets. The comparative merits are studied in this respect.

APPENDIX IV—ATTENDEE ABSTRACTS: ASTROPHYSICAL ASPECTS (cont.)

SIMULATION OF PLANETARY TRANSITS IN BINARY STAR SYSTEMS

Stefan Brandmeier
Department of Computer Science
Polytechnic University of Konstanz
Konstanz GERMANY

The photometric method for the detection of Earth-sized planets in binary star systems requires the study of light curves. A simulation of light curves as introduced here helps to reveal the characteristics of planetary transits. We use a numerical Keplerian model to simulate the motion of the binary pair and the orbiting planets. Star type, size and orbital parameters can be varied. The simulation also allows multiple planets. Limb-darkening of the stars is included in the calculations as well as star spots and flickering. The simulation provides animated graphics and simultaneous plotting of the resulting light curve. Comparing the simulated light curve with actual observation data shows the S/N ratio and thus the detectability of a given planet. Learning the characteristics of planetary transits from the simulated light curve will aid the development of signal processing methods to actually detect planetary transits in light curves.

BIOLOGICAL ASPECTS OF HABITABLE ZONES
Biogenic Material: Production and Delivery Considerations

ARE COMETS NECESSARY FOR A HABITABLE ZONE?

Christopher F. Chyba
National Security Council
Old Executive Office Building #365
Washington DC 20506 USA

Liquid water is the *sine qua non* of life as we know it. However, formation temperatures in our Solar System appear to have been such that planets that formed in the zone where liquid water is now possible were strongly depleted in volatiles. Comets served as a mechanism to bring "biogenic elements" from the outer Solar System into this zone. Whether comets played an analogous role in the case of intact prebiotic organics is much more difficult to assess.

PLANETARY FORMATION OF ORGANICS—THE ROLE OF COMETS

John Oró
Department of Biochemical & Biophysical Sciences
University of Houston
Houston TX 77204-5934 USA

Heretofore we have made the assumption that most of the biochemical molecules were prebiotically synthesized from the precursors present in the primitive Earth atmosphere in the already differentiated terrestrial oceans. We are not negating the possibility of such a synthesis, as a continuous steady-state process. However, it now appears that by the time the oceans were

APPENDIX IV—ATTENDEE ABSTRACTS: BIOLOGICAL ASPECTS, Oró (cont.)

differentiated, most of the biochemical molecules (and probably life as well) had already been generated from the chemical precursors contributed in relatively high concentrations to the primitive Earth by comets. This hypothesis will be developed in some detail.

Delivery of Noble Gases, Water, and Organics to the Inner Planets by Comets

Akiva Bar-Nun
Department of Geophysics and Planetary Sciences
Tel Aviv University
Tel Aviv 69978 ISRAEL

Tobias Owen
Institute of Astronomy
University of Hawaii
Honolulu HI 96822 USA

It seems that the abundance of water, C, H, O, N elements, and noble gases on Venus, Earth, and Mars can not be attributed only to chondritic meteorites. The ratios of ^{36}Ar, ^{84}Kr, and ^{132}Xe in the atmospheres of these bodies suggest an additional cometary contribution. The contributions of these materials by the "warm" Uranus-Neptune comets and the "cold" Kuiper Belt comets will be discussed. Atmospheric-oceanic processes, which produced biocompounds, versus intact delivery of these compounds by comets will be discussed as well.

The Effects of Hypervelocity Impact on Carbonaceous Compounds: Experimental Insights

Theodore E. Bunch
Planetary Biology Branch
Mail Stop 245-3
NASA Ames Research Center
Moffett Field CA 94035-1000 USA

Julie Paque
SETI Institute
2035 Landings Drive
Mountain View CA 94043 USA

Luan Becker
Scripps Institution of Oceanography
University of California at San Diego
La Jolla CA 92093-0212 USA

In order to gain knowledge about the potential contributions of comets, meteorites and cosmic dust to the origin of life on Earth, more information is needed about the survivability of carbonaceous compounds and the possibility of forming secondary experiments using a variety of carbonaceous impacts into Al targets at velocities up to 6.4 km/s. Estimated peak shock pressures probably did not exceed 120 GPa and peak shock temperatures were probably < 4000 K for time periods of $\approx$ nano- to microsecs. Noteworthy among the results is the exceptional survivability of aromatic compounds (PAHs) in impacted carbonaceous chondrite particles: some preservation of most species between masses 178 and 236 and a complete loss of PAHs between

APPENDIX IV—ATTENDEE ABSTRACTS: BIOLOGICAL ASPECTS, Bunch et al. (cont.)

128 and 178. Possible formation of additional alkylated PAH species within the mass range of 202 and 236 was also noted. Small amounts of phthalic acid molecules also survived impacts up to the experimental limit. Aliphatic compounds were not investigated. The results of these and other experiments and their implications will be discussed in the context of carbonaceous deliverables to early Earth.

BIOPHYSICAL PROPERTIES OF METEORITIC ORGANICS

David W. Deamer
Section of Molecular and Cellular Biology
University of California
Davis CA 95616 USA

Certain physical properties of organic compounds extracted from carbonaceous meteorites are potentially related to the origin of cellular systems. Organic extracts of the Murchison, Mighei, Murray, and four Antarctic specimens have been examined. Amphiphilic compounds are present in all of these samples, and compounds from the Murchison have the ability to self-assemble into membranous vesicles. One of the amphiphiles—nonanoic acid—has been tentatively identified by gc-ms methods, and also displays membrane formation under defined conditions. We are also investigating photochemical properties of polycyclic aromatic compounds present in such meteorites. For example, pyrene can capture light energy and reduce a second compound by photoreduction, and anthracene derivatives undergo photochemical reactions that are able to produce protons. It follows that relatively common organic hydrocarbons may have functioned as primitive pigment systems, mediating both electron transport and formation of proton gradients.

CYANIDE POLYMERS IN CIRCUMSTELLAR HABITABLE ZONES?

Clifford N. Matthews
Department of Chemistry
MC 111, Rm. 4500
University of Illinois
845 West Taylor St.
Chicago IL 60607-7061 USA

Hydrogen cyanide polymers—heterogeneous solids varying in color from yellow to orange to red to black—may be among the organic macromolecules most readily formed within the Solar System. Current studies point to the presence of polyamidine structures readily converted by water to polypeptides. Implications for prebiotic chemistry are profound. As on the primitive Earth, these powerful condensing agents may be the essential products needed for the emergence of life within circumstellar habitable zones.

APPENDIX IV—ATTENDEE ABSTRACTS: BIOLOGICAL ASPECTS (cont.)

THE PHOTO PRODUCTION OF ORGANIC MOLECULES IN COMETARY AND INTERSTELLAR ICE ANALOGS

M. P. Bernstein
SETI Institute
Mail Stop 245-6
NASA Ames Research Center
Moffett Field CA 94035-1000 USA

L. J. Allamandola
Mail Stop 245-6
NASA Ames Research Center
Moffett Field CA 94035-1000 USA

S. A. Sandford
Mail Stop 245-6
NASA Ames Research Center
Moffett Field CA 94035-1000 USA

S. Chang
Mail Stop 245-6
NASA Ames Research Center
Moffett Field CA 94035-1000 USA

Ice mixtures consistent with current models of comets and interstellar grains (H_2O, CH_3OH, NH_3, CO = 10, 5, 1, 1) were condensed at 10 K and subjected to UV radiation (Lyman Alpha) and annealing, approximating conditions in the interstellar medium. IR, NMR, and GC-Mass Spectral data confirm the production of volatile organic species such as CO_2, H_2CO, and CH_3CH_2OH and a complex nonvolatile residue containing hexamethylenetetramine (below), a nitrogen-rich molecule known to yield amino acids upon hydrolysis.

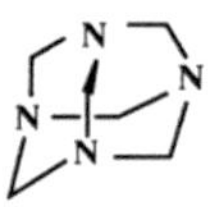

A SEARCH FOR EXTRATERRESTRIAL AMINO ACIDS IN POLAR ICE

Jeffrey Bada
Scripps Institution of Oceanography
University of California at San Diego
La Jolla CA 92093-0212 USA

Extraterrestrial sources have been suggested as an alternative for the Earth-based syntheses of the organic compounds necessary for the origin of life. If extraterrestrial sources were important on the early Earth, they should still be supplying the Earth today with organic matter although at a significantly reduced rate. We have analyzed several polar ice samples for the extraterrestrial amino acids alpha-aminisobutyric acid and isovaline, two of the more abundant amino acids in carbonaceous meteorites, but ones that rarely occur on the Earth. The levels of these two amino acids in polar ice are less than 0.01 ppb, which is an order of magnitude less than the amounts predicted from extraterrestrial organic flux estimates.

BIOLOGICAL ASPECTS OF HABITABLE ZONES (cont.)
Biological Contributions to Habitability Considerations

How Has Life Shaped Its Environment?

David J. Des Marais
Mail Stop 239-4
NASA Ames Research Center
Moffett Field CA 94035-1000 USA

Today, microorganisms, plants and animals alter our environment in countless ways. However, to assess life's major effect upon the habitable zone of our Solar System, we should examine how the most prominent biological effects have operated over much of Earth's history. To give one example, the biosynthesis of organic matter has had a major impact. Before oxygenic photosynthesis arose, life utilized chiefly the reducing power of volcanic emanations to make organic matter. By assimilating volcanic H_2, life substantially lowered atmospheric H_2 levels and thus prevented this highly reduced species from escaping to space. Later, oxygenic photosynthesis created a mechanism whereby reduced carbon could be buried in crustal sediments and, thus, over time, oxidized equivalents of O_2, sulfate, nitrate, etc. could accumulate in the atmosphere and oceans. In addition, as volcanism waned and the subaerial weathering of continents became more important, life helped to lower atmospheric CO_2 levels and thus make global temperatures lower than they would have been otherwise. Therefore, perhaps life's most prominent role has been to alter the long-term trends both in the oxidation state and in the temperature regime of Earth's surface environment.

Habitability of the Comet Pond

Benton C. Clark
Planetary Sciences Lab (B0560)
Martin Marietta
Denver CO 80201 USA

The concept that life could have originated in relatively rare, but inevitable, soft-landings of comets on planetary surfaces to form chemical ponds uniquely suited to abiotic evolution places a different emphasis on the requirements for habitability. No longer are the beginnings of life so narrowly dictated by the chemical environment of the planetary surface. Comets are intrinsically rich in organic compounds, transition metal elements, and colloid-sized ("attodust"), chemically complex particles that can facilitate reaction sequences. Many comets contain abundant methanol (1–5% of H_2O). Processing of CH_3OH ices by UV leads to formaldehyde and more complex organics. Furthermore, in the comet pond, unreacted CH_3OH would be attacked by atmospheric photoproduct •OH to produce H_2CO. In addition to being both a carbon source and energy source in some metabolic systems, methanol itself is a freezing point depressor, is less volatile than H_2CO, and provides an excellent starting point from which to derive higher compounds. Chaos theory predicts not only the singleness of each planetary system, vis-à-vis planet formation and stability of orbits, but also could have important implications for macro- and microscopic metastable states during structured chaotic fluid motion (imperfect mixing) in a comet pond, analogous to chaotic stirring in tidal basins with segregation of components. Stochastic impacts creating transitory and quite exotic environmental conditions are now widely accepted as causing not only

APPENDIX IV—ATTENDEE ABSTRACTS: BIOLOGICAL ASPECTS, Clark (cont.)

the K-T mass extinctions but possibly other extinctions. These events may provide important transitory conditions that also favor the advancement of biotic systems. The conditions for habitability can be broad under the assumption that early life could adapt to new surroundings and invade the macroniche outside the pond.

How Long Did It Take for Life to Appear and to Evolve to Cyanobacteria?

Stanley L. Miller
Department of Chemistry
University of California at San Diego
La Jolla CA 92093-0317 USA

Antonio Lazcano
Departamento de Biología
Facultad de Ciencias, UNAM
Apdo. Postal 70-407
Cd. Universitaria
México 04510 D.F. MÉXICO

There is convincing paleontological evidence showing that stromatolite-building phototactic prokaryotes were already in existence 3.5×10^9 years ago. Late accretion impacts may have killed off life on our planet as late as 3.8×10^9 years ago. This leaves only 300 million years to go from the prebiotic soup to the RNA world and to cyanobacteria.

Three hundred million years should be more than sufficient time. The prebiotic synthesis reactions of amino acids, purines and other organic compounds have relatively short half-lives. Very slow prebiotic reactions are not feasible because the intermediate compounds would have been destroyed due to the passage of the entire ocean through deep sea vents every 10^7 years. Therefore, it is likely that life emerged in a period shorter than the destruction rate of its components (< 5 million years).

The time for evolution from the first DNA/protein organisms to cyanobacteria is usually thought to have been very long because of their complexity. However, the similarities of many enzymatic reactions together with the analysis of the available sequence data suggest that a significant number of the components involved in basic biological processes are the result of ancient duplication gene events. This also appears to be the case for several of the proteins involved in anaerobic and oxygen-releasing photosynthesis, as well as chlorophyll biosynthesis. If the rate of gene duplication and development of new enzymatic specificities were as rapid in the early Archean as they are in present microbial populations, then early biological evolution could have proceeded at a rapid pace. As little as 5 million years may have been sufficient for the development of cyanobacteria from the first DNA/protein organisms.

If life can arise from the prebiotic soup and evolve to cyanobacteria in 10 million years, then O, B, and A stars, which are usually omitted in calculations of the abundance of life because of their short main-sequence lifetimes, may have life on planets surrounding them. Double stars, which are also omitted from SETI calculations because they do not have long-term stable orbits, may also have microorganisms on their planets.

APPENDIX IV—ATTENDEE ABSTRACTS: BIOLOGICAL ASPECTS (cont.)

THE ELECTROWEAK ORIGIN OF THE CHIRAL SIGNATURE OF LIFE

Alexandra J. MacDermott
Physical Chemistry Laboratory
Oxford Brookes University and Oxford University
South Parks Road
Oxford OX1 3QZ ENGLAND

Detection of extraterrestrial optical rotation could indicate life or life in the making. Prebiotic homochirality (all molecules of one hand) is now believed to be a precondition for life since polymerizations don't go in racemic solution. But *which* hand? Why are we made of L amino acids and not D? We believe this is due to the weak force, and we present calculations showing that L amino acids and other natural biomolecules are slightly more stable than their "unnatural" mirror images. This effect could be amplified over long periods in suitable atmospheres, such as Titan, and detection of the same hand in different extraterrestrial systems would confirm the universal symmetry breaking effect of the weak force.

HOW TO EVIDENCE PRIMITIVE LIFE ON AN EXOPLANET? THE DARWIN PROJECT

A. Léger
Institut d'Astrophysique Spatiale
91405 Orsay FRANCE

J. L. Puget
Institut d'Astrophysique Spatiale
91405 Orsay FRANCE

J. M. Mariotti
Observatoire de Paris
92195 Meudon FRANCE

D. Rouan
Observatoire de Paris
92195 Meudon FRANCE

J. Schneider
Observatoire de Paris
92195 Meudon FRANCE

DARWIN is a space mission concept that revisits the well-known idea of detection of exoplanets with a space thermal-infrared interferometer. But, contrary to previous propositions (e.g., Bracewell, Angel, Bely, et al.) the emphasis is not put on detection by imagery, but on spectroscopic capabilities. The main objective of the mission is the search for primitive life on extrasolar, Earthlike planets, i.e. with temperature ~ 300 K.

The presence of abundant O_2, or O_3, in the atmosphere of an exoplanet has been shown to be a criterion for photosynthetic activity. The main detectable signatures of these species are in the visible (O_2: 760, 680 nm), the IR (O_3: 9.6 μm) and the radio (O_2: 5 mm).

APPENDIX IV—ATTENDEE ABSTRACTS: BIOLOGICAL ASPECTS, Léger et al. (cont.)

The main challenge when searching for these signatures in a planet atmosphere is the presence of the nearby, much brighter, star. In the Sun/Earth case, the luminosity ratio is 7×10^9 in the visible and 10^7 at 10 μm, pointing out the advantage of observing in the IR. After the starlight is rejected as much as possible, the signatures, if present, appear as dips in the spectrum. Considering the quantum noise alone, the feasibility of the detection can be evaluated by calculating the integration time required for a 5 σ detection of a 50% deep spectral band. It is shown that the IR detection of O_3 is the *only reasonable approach* considering today's technology.

The proposed concept is a null interferometer (rejection of the on-axis starlight is achieved by phase opposition in an interferometric set-up) associated with an IR spectrometer. It has two versions: (1) the "cheap" DARWIN mission made of 2 × 1.3 m telescopes with a 10 m baseline (~ 35 G and K "accessible" stars), and (2) the "full" DARWIN mission, with 4 × 0.9 m telescopes in a cross pattern achieving a better rejection of the starlight, and 30 m baselines (~ 110 G and K "accessible" stars).

The spectral range is 5–25 μm or broader. At 9.6 μm, the number of photons received per hour in one spectral element from the planet is

$$N_{pl} = 9.6\ 10^2 \left(\frac{D}{6.3pc}\right)^{-2} \left(\frac{\phi tel}{1.3m}\right) \left(\frac{\lambda/\Delta\lambda}{20}\right)^{-1} \left(\frac{R_{pl}}{R_\oplus}\right) \text{photons·hours}^{-1}$$

The problem is to extract such a tiny flux out of strong ones. Several sources of noise and spurious signals are discussed, in particular: starlight leaks, zodiacal light likely to be associated to the planetary system, IR cirrus background, Solar System zodiacal emission, telescope emissivity, and detector noise.

DARWIN has been proposed to ESA as a space mission concept for the post–Horizon 2000 era.

EVOLUTIONARY BEHAVIOR OF A SELF-COPYING COMPUTER PROGRAM

Rosemarie Swanson
Biochemistry-Biophysics
Texas A&M University
College Station TX 77843-2128 USA

Jun-Jie Liu
Biochemistry-Biophysics
Texas A&M University
College Station TX 77843-2128 USA

A nonbiological alternative "world" of self-reproducing computer programs offers an opportunity to consider questions about evolution from a broadened perspective. Must particular features exist in any system capable of evolution, regardless of the medium in which it is expressed? Is it necessary or beneficial to have a genetic code? The observed natural genetic code has an error-resistant form (it is almost a Gray code). What would be the effects on evolutionary diversity and progress of a more error-sensitive or more error-tolerant code? Some discussion of the benefits and liabilities of a genetic code and some intermediate results on both the diversity and the features in common of "creatures" produced in independent evolutionary runs will be described and evaluated. The original human-written ancestral program will be compared with some of the programs that evolved to illustrate the nature and limits of creativity generated by mutation and selection.

APPENDIX IV—ATTENDEE ABSTRACTS: BIOLOGICAL ASPECTS (cont.)

On Production of the Chirality of Biogenetic Material in the Early Earth

Liaofu Luo
Department of Physics
Inner Mongolia University
Huhehote CHINA

Crick said that the chirality of biological macromolecules is the first principle of biochemistry. In fact, the universal chirality of amino acid and ribose molecules is the most mysterious and striking event at the molecular level in the evolution of life. One may say with no exaggeration that chirality is one of the most important marks of life. The origin of biological chirality is closely related to the evolution of life on Earth which provides a clue in exploring the origin of life. How to explain the origin of the chirality of biogenetic material? There are two possible answers. The first is, it is delivered by comets, meteorites, or other loads of interstellar organics. The second is, it is produced in the Earth. The analysis of the chirality of amino acids on 1969 Murchison meteorite seems to support the latter since all amino acids with asymmetrical carbon on it are racemic. If there is no evidence on the L-R asymmetry of amino acids in meteorites in further observations, we must accept the point that life is not nonnative. In this note, we shall investigate the possibility of the production of chirality of biogenetic material in the early Earth. We will point out that the weak interaction as an asymmetrical force can drive the amino acids into one of the two probable configurations opposite to each other. The left-right asymmetry of chemical reaction rate induced by inelastic scattering of beta-electrons with macromolecules is the dominant factor for producing biogenetic chirality. We will demonstrate that the rate asymmetry is on the order of 10^{-6}. The asymmetry introduces a term of asymmetrical force to the action-diffusion equation and gains application in the solution of the equation. So the concentration difference between L- and D-type molecules may reach a considerable figure. At last, the process of the chemical evolution of life (as a kind of evolution of chiral system) pronounces the victory of one kind of chirality.

Poly-tRNA-Mediated Peptide-Synthesis in Early Terrestrial Organisms Before Genetic Codes: A Perspective on Early Life

Koji Ohnishi
Department of Biology
Faculty of Science
Niigata University
Ikarashi-2
Niigata 950-21 JAPAN

An RNA transcript from the *Bacillus subtilis trrnD* operon has a structure of 5′(16S rRNA-23S rRNA-5S rRNA-$[\text{tRNA}]_{16}$)3′ where $(\text{tRNA})_{16}$ is tandem repeat of tRNA ("poly-tRNA") in which the order of amino acid specificity of tRNA is 5′SEVMDFTYWHQGCLL3′. A peptide having this order of amino acids in sequence is termed "*trrnD*-peptide." This poly-tRNA structure has been found to be not only a relic of an early RNA making *trrnD* peptide but also a relic of most primitive mRNA ancestral to those encoding aminoacyl- (especially glycyl-) tRNA synthetases, F_0-ATPase a subunit, and other housekeeping proteins possessing *trrnD* peptidelike sequence motif. Recently proposed model of early protein-synthesis by *trrnD* poly-tRNA (poly-tRNA model) can well explain the origin and early evolution of protein-encoding mRNA and genetic codes. This result suggests important information about early terrestrial organisms before genetic codes. F_0-like proton-channel might have been an earliest protein made by poly-tRNA, which suggests that early metabolism might have evolved in protocells having primitive proton-channel.

—APPENDIX V—

SUBMITTED ABSTRACTS OF NONATTENDEES

CONDITIONS NECESSARY FOR THE EMERGENCE OF LIFE AND LIVING OF SYSTEMS IN CIRCUMSTELLAR HABITABLE ZONES

Krishna Bahadur and S. Ranganayaki
Professors of Chemistry, retired
University of Allahabad
68 Dilkusha NewKatra
Allahabad 21102 INDIA

Taking the hint from life on our Earth, we can safely conclude that for life to exist in some circumstellar zone, it is essential that the zone must have a liquid in it which can exist in all the three states in the prevailing environment of the zone and is a fairly good solvent, like water, considering it on the basis of its physical properties and not the chemical ones. This liquid in the zone should mostly be stable in the liquid state because in this state it can take part in various chemical transformations and is not diffused out in gaseous state or does not become rigid and confined to a fixed locale as in solid. There should be a possibility of formation of a semipermeable membrane with selective permeability and an energy release system that takes the energy from some perpetual source, stores it in some chemical form, and releases it slowly by some chemical transformation possible in that zone.

Under these conditions the synergesis will form discrete structures with boundary wall and internal structures, and utilizing the catalytic functions of the chemicals present therein may show nonlinear properties and if it has an energy release system trapped in it, some of the inherent properties of matter, i.e., (1) matter duplicates under suitable conditions and (2) a system of matter in equilibrium has the inherent property of adaptability, will be manifested and it is probable that some of these discrete structures will show the properties of growth from within, multiplication by budding and will have metabolic activity. These will be living in the exobiological sense and will evolve in the mildly changing environment of the zone.

JEEWANU are one of such systems which were synthesized by a variety of materials by us in 1963 and the work was independently confirmed by M. H. Briggs in 1964 and again in 1965.

APPENDIX V—NONATTENDEE ABSTRACTS (cont.)

Can One Detect Primitive Forms of Life on an Extrasolar Planet from the Ground?

Bruno Bezard
Observatoire de Paris—Meudon
92195 Meudon FRANCE

It has been suggested that O_2 of biological origin can be searched for during transit of an extrasolar planet in front of its parent star. We investigated the conditions under which such a detection could be performed in observations from the ground.

The Double Role of Extreme Environmental Factors in Life Origin and Evolution

Oleg V. Davydov
Belorussian Research Institute for
Epidemiology and Microbiology
4 Clara Tsetkin Street
Minsk 220050 REPUBLIC OF BELARUS

Some environmental conditions (cosmic cold and radiation, high level of temperature and concentration of some chemicals, the lack of oxygen in the atmosphere, etc.) are incompatible with normal functioning of the majority of modern organisms. However, during primary biogenesis, these factors probably promoted advantages and stimulated the origin of life. This is why extreme ecosystems with their peculiar habitants (R-stable viroids, heat- and acid-stable microorganisms and proteins) are to be classified not as exotic but as principal and inevitable steps of life's origin.

Orbital Stability of Planets of Binary Stars

A. Dobrovolskis
C. Levit, J. Scargle, J. Cuzzi, L. Dones,
M. Showalter, K. Young
Mail Stop 245-3
NASA Ames Research Center
Moffett Field CA 94035-1000 USA

Since so many stars are binary or multiple, the frequency of habitable planets involves the stability of planetary orbits in binary systems. Previous studies of this problem have concentrated on varying the stellar and planetary masses, distances, and sometimes orbital eccentricities. However, very few have examined the effects of inclination; these are generally dismissed as unimportant unless the planet's orbit is nearly perpendicular to the plane of the binary. In some cases, the combination of inclination with eccentricity may be important to the stability of the orbit; a familiar example would be the complicated resonance between Neptune and Pluto. We plan to perform extensive numerical integrations of the Three-Body Problem on the CM-5 Connection Machine in order to explore the mutual effects of eccentricity and inclination on the stability of planets of binary stars.

APPENDIX V—NONATTENDEE ABSTRACTS (cont.)

AN ESTIMATE OF THE PREVALENCE OF BIOCOMPATIBLE AND HABITABLE PLANETS

Martyn J. Fogg
Probability Research Group
c/o 44 Hogarth Court, Fountain Drive
London SE19 1UY UNITED KINGDOM

A Monte Carlo computer model of extrasolar planetary formation and evolution, which includes the planetary geochemical carbon cycle, is presented. The results of a run of one million galactic disk stars are shown where the aim was to assess the possible abundance of both biocompatible and habitable planets. (Biocompatible planets are defined as worlds where the long-term presence of surface liquid water provides environmental conditions suitable for the origin and evolution of life. Habitable planets are those worlds with more specifically Earthlike conditions.) The model gives an estimate of one biocompatible planet per 39 stars, with the subset of habitable planets being much rarer at one such planet per 413 stars. The nearest biocompatible planets may thus lie approximately 14 light years distant and the nearest habitable planet approximately 31 light years away. If planets form in multiple star systems then the above planet/star ratios may be more than doubled. By applying the results to stars in the solar neighborhood, it is possible to identify 28 stars at distances of less than 22 light years with a nonzero probability of possessing a biocompatible planet.

(See *Journal of The British Interplanetary Society* **45:** 3–12, 1992.)

OXYGEN AND CARBON DIOXIDE IN THE PRE-1.1 GA TERRESTRIAL ATMOSPHERE

Heinrich D. Holland
Department of Earth and Planetary Sciences
Harvard University
Cambridge MA 02138 USA

The composition of paleosols and of sedimentary rocks indicates that the O_2 content of the atmosphere was $\leq 10^{-3}$ atm prior to 2.2 Ga, and that it rose to ≥ 0.03 atm by 1.9 Ga. The mineralogy of paleosols suggests that atmospheric PCO_2 was $\geq 10^{-2}$ between 2.75 Ga and 2.2 Ga. Such low CO_2 pressure implies that greenhouse gases other than CO_2 were important during this time interval, and/or that solar luminosity was considerably greater than predicted by most solar evolution models.

APPENDIX V—NONATTENDEE ABSTRACTS (cont.)

THE INTERSTELLAR CHEMISTRY OF NITROGEN AND SULFUR AND LINKS TO THE PROTOSOLAR NEBULA

William Irvine
Five College Radio Astronomy Observatory
University of Massachusetts
Amherst MA 01003 USA

Nitrogen and sulfur are key biogenic elements. In order to understand the chemical processing which relates interstellar molecular material to that surviving in the protosolar nebula, it is important to determine the chemical reservoirs for nitrogen and sulfur in interstellar molecular clouds. Recent studies of the abundance and distribution of several nitrogen-containing and sulfur-containing molecules will be described. Among these are NO, NS, H_2CN, H_2S, and more exotic nitriles that may have played a role in prebiological evolution. We also describe the results of a search for a potentially significant phosphorus-containing molecule, PO^+.

HABITABLE ZONES: PHOTOCHEMICAL, GEOCHEMICAL, AND BIOCHEMICAL CONSIDERATIONS

Joel S. Levine
Atmospheric Sciences Division
NASA Langley Research Center
Hampton VA 23681-0001 USA

To a large extent, "circumstellar habitable zones" are controlled by the composition of the gaseous envelopes or atmospheres surrounding planets. For example, a "habitable zone" must be shielded from biologically lethal solar ultraviolet radiation and must possess a hospitable temperature range. On Earth, ultraviolet radiation shielding and planetary temperature are controlled by the presence of trace gases in the atmosphere: ozone for ultraviolet shielding and the "greenhouse" gases, carbon dioxide, water vapor, methane, and nitrous oxide which control global temperature. Geochemical, biochemical, and photochemical processes are responsible for the presence of these gases in the atmosphere. Ozone is formed via atmospheric photochemical processes involving atmospheric oxygen. Oxygen itself results from photosynthesis, a biochemical process. Atmospheric water vapor and carbon dioxide are products of volcanic emissions, a geochemical process. In addition, carbon dioxide results from respiration and decay, biochemical processes. Methane and nitrous oxide also result from biochemical processes: methanogenesis for methane and denitrification and nitrification for nitrous oxide.

The modification of the atmosphere by geochemical, biochemical, and photochemical processes over geological time will be investigated. Chemical disequilibrium in planetary atmospheres due to biochemical processes will also be examined with particular emphasis on the role and impact of evolving atmospheric oxygen and its effects on photochemical and biochemical processes on the evolving Earth. The evolution of atmospheric oxygen changed the planet from an anaerobic microbial world to an aerobic world.

APPENDIX V—NONATTENDEE ABSTRACTS (cont.)

THE WINDOW FOR LIFE: HOW BROAD OR NARROW WOULD IT BE?

Koichiro Matsuno
Department of Bioengineering
Nagaoka University of Technology
Nagaoka 940-21 JAPAN

Suppose there are certain amounts of water, carbon dioxide, and luminosity with a diurnal cycle supplemented by a small amount of trace elements. Then the emergence of photosynthesis would be expected to occur in a limited window region in the three-dimensional space spanned by the three quantities of water, carbon dioxide, and luminosity. Once photosynthesis happened to emerge somewhere in there, the process would be self-focusing in the sense that only a tiny region in the window would eventually be occupied in the subsequent evolution.

MURCHISON METEORITE HYDROPHOBIC MATERIAL: PHOTOPROTECTIVE FUNCTION OF POLYANNULAR HYDROCARBONS IN RELATION TO LINEAR ALIPHATICS

Sergi Seleznev
Department of Medbiophysics
Institute of Medicine
97300 Akmola KAZAKHSTAN

Recently, in collaboration with D. W. Deamer (University of California at Davis), we studied the influence of the cyclic hydrocarbons pyrene and fluoranthene on photo-oxidation of hexadecane. Such hydrocarbons were discovered by many investigators in Murchison carbonaceous chondrite, and photo-oxidative reactions, as it was argued by us earlier, served as a source of amphiphilic material delivered on prebiotic Earth and generated on it in secondary processes.

In this study, we shall discuss the mechanism of the photoprotective action of cyclic hydrocarbons which could play a role in the preservation of linear aliphatics from destruction on celestial bodies and prebiotic Earth.

MAIN SEQUENCE AND POST–MAIN SEQUENCE HABITABILITY ZONES AROUND SINGLE STARS: WHERE IS WATER LIQUID?

S. Alan Stern
D. C. Boyce and H. F. Levison
Space Science Department
Southwest Research Institute
6620 Culebra Rd.
San Antonio TX 78238 USA

The condition for liquid water is probably as good a prerequisite for the habitability of planetary environments as is known. We investigate the range of astrocentric distance for liquid H_2O stability as a function of stellar type around single stars. For each type of star we calculate the liquid H_2O stability range as a function of time as the star evolves from H ignition to stellar

APPENDIX V—NONATTENDEE ABSTRACTS, Stern et al. (cont.)

remnant on the Hertzsprung-Russell diagram. Liquid H_2O stability calculations are made for various assumptions about planetary albedo and atmospheric greenhouse magnitude. We show that in addition to the well-known, inner planets region, stellar luminosity alone can generate "thermally habitable zones" in the icy, distant outer Solar System for several hundred million years during post–main sequence evolution.

ROLE OF UV RADIATION IN THE PRODUCTION AND EVOLUTION OF BIOGENIC MATERIAL

Taisya A. Telegina
A. N. Bakh Institute of Biochemistry
Russian Academy of Sciences
33 Leninsky Prospect
117071 Moscow RUSSIA

The evidence for the important role of short-wave UV in abiosynthesis of organic low-molecular compounds and the role of long-wave UV in abiosynthesis and biosynthesis of biologically important substances with a considerable number of carbon atoms will be discussed. We accept the long-wave UV as an essential factor for evolution of biogenic material in relation to the evolution of living systems and the preservation of ecosystems.

HABITABLE EARTH'S EARLY ATMOSPHERIC-OCEANIC SYSTEM AND ITS IMPLICATION ON THE ORIGIN AND EVOLUTION OF LIFE: SOME EVIDENCE FROM INDIAN LITHOSPHERE (ARCHEAN-PROTEROZOIC SEDIMENTS OF THE PENINSULAR AND LESSER HIMALAYAN ZONES)

Vinod Chandra Tewardi
Wadia Institute of Himalayan Geology
203 Vanasthali South Mandir Lane
Ballupur Dehradun 248001 INDIA

Early Earth's abiotic atmosphere was reducing and dominated by hydrogen since it condensed from a dust cloud about 5 billion years ago. The atmospheric and oceanic conditions for early habitability (biotic) on Earth suggest that around 3.8–3 billion years ago oxygenic atmospheric conditions had developed and the earliest unicellular microorganisms (molecular chemical fossils) appeared on Earth, which were also responsible for development of microbialites (stromatolites), the first megascopic evidence of life on Earth. Recent biogeochemical and carbon isotopic evidence from the peninsular shield and the Lesser Himalayan Zone of the Indian Lithosphere indicates a transition from a primitive reduced atmosphere to a highly oxidized one like the modern atmosphere, development of the ozone layer and the photosynthesis of the cyanobacteria (hence development of stromatolites) in the early toxic ocean. Accordingly the multicellularity of life reflects varying rates of photosynthesis, diversification and evolution in the Deoban-Blaini-Krol-Tal succession. An oxic oceanic event is recorded at the Terminal Proterozoic in the Lesser Himalaya.

APPENDIX V—NONATTENDEE ABSTRACTS (cont.)

Evolution of the Heme-Containing Proteins and Evolution of Atmosphere

Alexey F. Topunov
A. N. Bakh Institute of Biochemistry
Russian Academy of Sciences
33 Leninsky Prospect
117071 Moscow RUSSIA

Evolution of heme-containing proteins is in correlation with the evolution of the early Earth atmosphere. The movement from "simple" heme-containing proteins through "hemoglobin-like" proteins and "almost" hemoglobins to "real" hemoglobins occurred simultaneously with the changing of the atmosphere from anaerobic through microaerobic to a contemporary aerobic one. Two of these processes are also connected with the evolution of living organisms including complication from prokaryotic to eukaryotic ones.

From HCN to Polynucleotides

Geoffrey Zubay
Department of Biological Science
Columbia University
New York NY 10027 USA

Our laboratory is studying the reactions leading from HCN to polynucleotides. We visualize these reactions as starting at clay-liquid interfaces leading to purines. This is followed by reactions leading to activated nucleotides which may occur without association with solid surfaces. The final reactions leading to polymerization of activated nucleotides might proceed best at mineral-liquid interfaces. Wet-dry cycles are crucial to the reactions occurring in these three different phases. We have made significant improvements in the conditions for some of these reactions. Once optimal prebiotic conditions have been found for all of the steps in this pathway, we will attempt to link the various steps together.

Hubble Observes the Planet Uranus

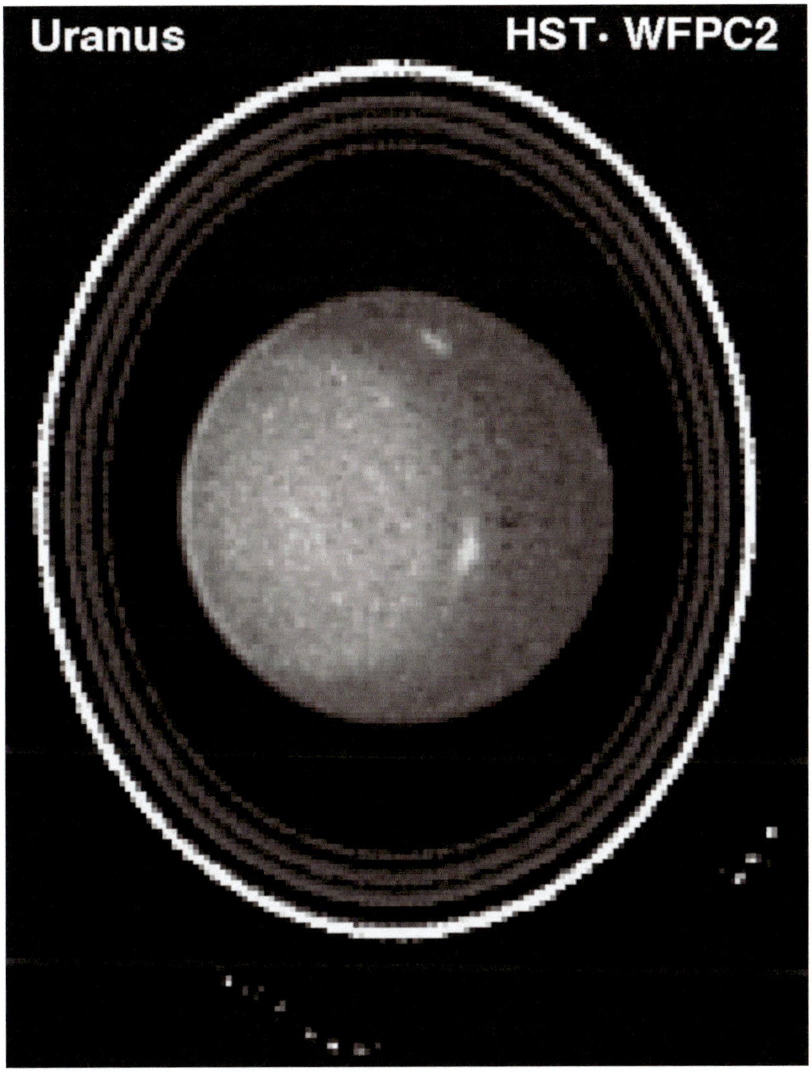

Photo Credit: Kenneth Seidelmann, U.S. Naval Observatory, and NASA
https://science.nasa.gov/asset/hubble/hubble-observes-the-planet-uranus/

[retrieved 12-12-25]

This NASA HST image of the planet Uranus was taken in 1994, and shows the rings, moons, bright clouds and a high-altitude haze, appearing as a bright cap, above the planet's south pole. Uranus was 1.7 billion miles from Earth at the time. These details were previously seen only by Voyager 2, which flew by Uranus in 1986. Uranus' rings were discovered indirectly in 1977 through stellar occultation observations, but have never been seen in visible light through a ground-based telescope. Uranus has 11 concentric rings of dark dust. It is tipped such that its rotation axis lies in the plane of its orbit, so the rings appear nearly face-on. Three of the inner moons appear as a string of dots at the bottom of the picture; the picture is a composite of three images. From left to right, they are Cressida, Juliet, and Portia. They move much more rapidly than our Moon does.

See the news release at https://science.nasa.gov/missions/hubble/hubble-observes-the-moons-and-rings-of-the-planet-uranus/

Earth's largest habitable zone: the oceans, home to the largest and smallest of all Earth's creatures, birds, fish, plant species, crustaceans, zooplankton, and many other species, and the largest mammal ever known to have lived on Earth, the blue whale.

Blue whale blows

A blue whale blow

Sea otter feeding on a sea urchin

Photo Credit: © Jodi Frediani (all three photos) From *Wild Monterey Bay—Up Close and Personal: Stories of Memorable Wildlife Encounters*

Trees and Mountains

Forests are Earth's vital organs, covering roughly one-third of our land. They act as the planet's primary lungs and water filters, regulating global temperatures, and sustaining the vast majority of land-based biodiversity. Forests store vast amounts of CO_2. Their preservation is critical to curbing climate change; deforestation releases vast amounts of greenhouse gases. Trees absorb CO_2 and emit fresh oxygen. Forests regulate local and global climates through shade, water-energy cycling, and release of water vapor. —Panda.org

Mountains are vital to Earth's survival. As the planet's 'water towers,' they store precipitation as snow and ice, supplying freshwater to over half of humanity. Covering about 25% of Earth's land, mountains host more than 85% of global bird, mammal, and amphibian species. They create unique microclimates that shelter diverse life.

—One Earth, World Meteorological Organization.

Photo Credit: © Ellen Blue (both photos)

Colusa Sunset, Sacramento National Wildlife Refuge

Photo Credit: © Ellen Blue
https://www.fws.gov/refuge/sacramento/about-us

The Colusa National Wildlife Refuge is part of the Sacramento National Wildlife Refuge Complex northwest of Sacramento, California. The complex encompasses 70,000 acres of wetland, grassland, and riparian habitats for a wide array of waterfowl, shorebirds, raptors, waterbirds, songbirds, reptiles, and mammals. This rich habitable zone currently supports nearly 300 species of birds.

As many as 150,000 ducks and 60,000 geese spend their winters in the Colusa refuge, which lies in the Colusa Basin. These refuges are a spectacular example of the magnificent habitable zones to be found on Earth, but there are millions or billions or more of magnificent habitable zones on our little planet.

This particular area is part of the 4,000- to 10,000-mile Pacific Flyway, which stretches from the Arctic tundra in Alaska to Patagonia. Annually, millions of waterfowl, shorebirds, and songbirds, including snow geese, ducks, and swans, use this natural "highway in the sky" to spend the spring and fall/winter in these protected wetlands and the surrounding flooded fields to breed, rest, and feed. The Sacramento National Wildlife Refuge Complex is one of the most important wintering areas along the entire Pacific Flyway, supporting roughly 44% of the Flyway's wintering duck and goose population.

By the way, modern birds are the only surviving members of the dinosaur family!

—U.S. Fish & Wildlife Service

thp TRAVIS HOUSE PUBLICATIONS

Travis House Publications was established in 1994 and in 1996 published its first offering, *Circumstellar Habitable Zones—Proceedings of the First International Conference*, edited by Dr. Laurance R. Doyle.

The name Travis House Publications is used in memory of the Travis family and the Travis House, which was located on the Travis Ranch, in Forestville, California. It was built in 1873 by Annie Travis (ne'e O'Calahan of County Cork, Ireland), who came West in the 1840s in an oxcart, and built the three-gabled, Victorian-style house out of redwood, using square nails and wooden pegs. The property was part of a Spanish land grant and when acquired in 1865 contained 320 acres. The house stood as a Sonoma County landmark for 109 years before being designated a Sonoma County Historical site in 1982. It was occupied by the Travis family continuously until 1986.

Burton M. and E. Pearl Travis lived in the home from 1927 to 1972 and 1963, respectively. By then it had become a 180-acre sheep ranch. Bert, a gentle, intelligent man without a formal education, had a deep love of learning and served for over two decades on the Analy Union High School Board of Trustees. He was highly respected by his colleagues and in the community, and after a number of years, friends and associates convinced him to run for the State Senate. Pearl, a woman of great wit and loving nature, was very active in the Christian Science church in Sebastopol, and in the Garden Club in Graton. They had three children, Russell Burton Travis, Jewel Travis, and Jacqueline Travis. Russell's oldest daughter, Ellen Travis Blue, founded and operates Travis House Publications.

OTHER WORKS BY TRAVIS HOUSE PUBLICATIONS

The Relevancy of Christian Healing Today
[paperback and ebook]
By Anne Hindmarsh Cooling

La relevancia de la curación cristiana hoy
By Anne Hindmarsh Cooling
Translated by Elaine Douglas-Harrison

Wild Monterey Bay—Up Close and Personal: Stories of Memorable Wildlife Encounters
Jodi Frediani and Katlyn Taylor, Editors

Cambria Cookin': I Don' Wanna Cook, I Jus' Wanna Eat
By h. numossi

IN THE WORKS BY TRAVIS HOUSE PUBLICATIONS

The History of El Molino Rancho and Forestville Town
By Burton M. Travis [author]
Russell B. Travis [coauthor/editor]
Ellen Travis Blue [coauthor/editor/publisher]

Observatories of California
By Ellen Blue and Laurance R. Doyle

Along the Great San Andreas Fault
By Ellen Blue

Along California Highways
By Ellen Blue

Three Exciting New Developments in Our Circumstellar Habitable Zones Publications

- In collaboration with Dr. Laurance R. Doyle, we will be uploading the professionally videotaped proceedings of the 1994 three-day First International Conference on Circumstellar Habitable Zones, including all the talks, along with slide presentations.
- We will also be publishing a Teacher's Guide written by Dr. Doyle for Frank Drake's UC Santa Cruz Life in the Universe class, for both the 1996 First Edition and the 2026 Thirtieth Anniversary Edition. See the 2026 Foreword in this volume for details.
- We have initiated a collaboration with Dr. Doyle and the authors of the papers given at the 1994 conference to look into the feasibility of offering updated information on how their fields have progressed over the last thirty years and to determine how best to present that information.

Please contact Ellen Blue at travishousepublications@gmail.com for more information.

Made in the USA
Monee, IL
15 September 2026

63312028R00347